Frontiers
in
Bioprocessing II

Frontiers in Bioprocessing II

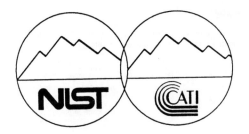

EDITED BY

Paul Todd
National Institute of Standards and Technology

Subhas K. Sikdar
U.S. Environmental Protection Agency

Milan Bier
University of Arizona

Proceedings of Frontiers in Biochemistry II,
Boulder, Colorado,
June 17–21, 1990

American Chemical Society, Washington, DC 1992

Library of Congress Cataloging-in-Publication Data

Frontiers in bioprocessing II: proceedings of Frontiers in bioprocessing II, Boulder, Colorado, June 17–21, 1990 /   edited by Paul Todd, Subhas K. Sikdar, Milan Bier.

p.   cm.—(Conference proceedings series)

Includes bibliographical references and index.

ISBN 0–8412–2181–2

1. Biotechnology—Congresses.

I. Todd, Paul, 1936–   . II. Sikdar, Subhas K. III. Bier, Milan IV. Series: Conference proceedings series (American Chemical Society)

TP248.14.F76   1992
660'.06—dc20   91–34769
                                                    CIP

The paper used in this publication meets the minimum requirements of American National Standard for Information Sciences—Permanence of Paper for Printed Library Materials, ANSI Z39.48–1984.   ∞

Conference Proceedings Series

M. Joan Comstock, *Series Editor*

1992 ACS Books Advisory Board

Conference Sponsors

- ➢ National Institute of Standards and Technology –Boulder, Colorado

- ➢ Colorado Advanced Technology Institute

- ➢ University of Colorado at Boulder–Boulder, Colorado

- ➢ Center for Separation Science–University of Arizona

Organization of Conference

Conference Chairman
Milan Bier
Center for Separation Science
University of Arizona

Keynote Speaker
Ronald Cape
Cetus Corporation
Emeryville, California

Conference Committee

Paul Todd
Center for Chemical Technology
National Institute of Standards
  and Technology

Virginia Orndorff
Colorado Advanced Technology
  Institute

Subhas K. Sikdar
U.S. Environmental Protection Agency
Cincinnati, Ohio

Eric Dunlop
Colorado State University

Milan Bier
Center for Separation Science
University of Arizona

Dhinakar Kompala
University of Colorado

Contents

PART II: BIOSENSORS AND IN SITU
MEASUREMENT SYSTEMS

PART III: CELL-CULTURE SYSTEMS

PART IV: BIOREACTOR ENGINEERING AND CONTROL

PART V: BIOSEPARATIONS SCALEUP

PART VI: EMERGING TECHNOLOGIES IN BIOSEPARATIONS

Preface

FRONTIERS IN BIOPROCESSING II presents the proceedings of the second in a series of conferences. The conference concept consists of considering, once every three years, those recent advances that are expanding the envelope of bioprocess technology, whether by technology pull or by market push. For this conference, five or six speakers were chosen in each of six major areas in which recent notable advances had been made. Each speaker was asked to prepare a chapter for the conference book, which is not only a proceedings volume, but also a coherent interdisciplinary update. All participants were invited to present posters and to submit short chapters based on their posters, so that most sections of the book also contain one or two poster-based chapters.

The product of the conference, as Chapter 1 clearly indicates, is not just a book; the assembly of the most significant investigators from an unusual combination of disciplines inevitably results in a flow of ideas that affects the future of a field. The principal author of each chapter heard the stories of the principal authors of all of the other chapters. Unfortunately, the interdisciplinary dialogue of the conference cannot be captured in the specialized chapters that follow; however, the spirit of their interdependency can be captured by reading the first chapter by Alan S. Michaels, who, through lively discussions, also contributed substantially to the success of the conference.

We wish to thank all those who made the conference a success and this book possible. Special recognition is deserved by the technical peers of the chapter authors who gave freely of their time to review and evaluate all of the chapters. The Office of Conference Services of the University of Colorado, the student assistants, and the corporate and institutional sponsors—Schering-Plough, Genentech, Syntex, Synergen, Colorado State University, and the University of Colorado—were all essential to the smooth operation of the conference. The primary sponsors were the Colorado Advanced Technology Institute and the National Institute of Standards and

Technology. Their combined sponsorship is symbolized in the logo of the conference, which appears on the cover of this book.

PAUL TODD
Chemical Science and Technology Laboratory
National Institute of Standards and Technology
325 Broadway, 831.02
Boulder, CO 80303

SUBHAS K. SIKDAR
Director, Water and Hazardous Waste Treatment Research Division
U.S. Environmental Protection Agency
Cincinnati, OH 45268

MILAN BIER
Center for Separation Science
University of Arizona
Building 20, Room 157
Tucson, AZ 85721

September 2, 1991

Frontiers in Bioprocessing: An Assessment

Alan S. Michaels, President, Alan Sherman Michaels, Sc. D., Inc., Chestnut Hill, Massachusetts; Distinguished University Professor, Emeritus, North Carolina State University, Raleigh, North Carolina

The past five years have witnessed, in my opinion, major advances in the process technology of synthesis, isolation, and purification of biological products, resulting in large measure from fundamental changes in the cross-disciplinary interactions between life scientists (particularly biologists, biochemists, microbiologists, and geneticists) on the one hand, with physical scientists and chemical/biochemical engineers on the other, involved in the solution of industrial problems relating to the manufacture of products (principally bioactive proteins and peptides) evolving from the "new biology".

I recall quite vividly the proceedings of the First Engineering Foundation Conference on Recovery of Biological Products, held in Banff, Alberta in the Spring of 1981 (of which I had the privilege of serving as co-organizer), where it had been possible to interest only very few life scientists working at the cutting edge of recombinant DNA technology to participate in the program; the prevailing notion of the life science community at that time was that they had precious little to say that would either aid or be of interest to engineers, nor did they have anything of value themselves to learn from engineers. That state of mind had already been reflected at a 1980 international conference on "Genetic Engineering", which was distinguished by a program studded with luminaries from the life science community, but without a single engineering participant. The dominant message taken away from the Banff Conference, however, was that only by early and continuous involvement of engineers in all stages of development of a recombinant organism for synthesis of a biological would the course to economic industrial production of such a compound be navigable.

How times have changed: Today, at conferences such as Frontiers in Bioprocessing, were it not for the disciplinary identification of participants in the printed program, one would be hard-pressed to deduce whether a speaker were a molecular biologist, protein chemist, geneticist, or chemical engineer! This is a tribute, I believe, not only to the willingness of engineers to learn both the language and *modus operandi* of the life scientist and to integrate the perspectives of the life scientist with the basic principles of physical and engineering science, but also to the recognition by life scientists that engineers do, indeed, have something to contribute to the understanding and solution of problems they have long regarded as their sacred domain and may hold the key to "getting the product out of the cell and into the bottle". These changes have not occurred overnight, of course -- it has been a slow, tortuous, and often painful process, but today there are noteworthy examples of the fruits of this constructive collaboration. One can cite Eli Lilly's development of Humulin, and Genentech's development of tissue-type plasminogen activator, as unequivocal evidence of the success of such efforts. Suffice it to say that logjam of effective communication between life scientist and engineer seems to have been broken, and with it, the major obstacle to rapid and efficient bioprocess development. Since this assessment is intended to address "frontiers of bioprocessing", let me suggest that **diffusion of the life science/engineering interface** has been a signal accomplishment in advancing those frontiers.

Now for my thoughts on other frontiers; I have attempted to organize my comments according to four areas of bioprocess technology: These are (1) synthesis of biological products; (2) characterization and control of biopolymer (protein) structure; (3) biosensors and process control strategies; and (4) separation and purification of bioproducts. My thesis is that developments in each of these areas will pace the development of commercially successful bioproducts, and that recent advances in each of these areas, as well as their unsolved problems, will define the frontiers for the next decade.

THE FIRST FRONTIER: SYNTHESIS OF BIOLOGICAL PRODUCTS

While considerable progress has been made over the past 10-15 years in our ability to marshal microorganisms and mammalian

2181–2/92/0001$06.00/0 © 1992 American Chemical Society

cells for the industrial-scale biosynthesis of biological products (including proteins, oligopeptides, and secondary metabolites such as antibiotics), we still have much to learn about the genetic, intracellular, and environmental factors influencing the kinetics of product biosynthesis, the propensity (or lack thereof) to form inclusion bodies within the cell, the mechanism and dynamics of product excretion, and (for proteins and peptides) post-translational changes in composition or structure of the synthesized products.

The recognition that protein tertiary or quaternary structure is essential for biological activity has focused attention on mammalian cells for production of regulatory proteins and peptides; this, despite the characteristically low biosynthesis kinetics of such cells, and concerns about possible product contamination by oncogenic or antigenic nucleotides or proteins. However, growing insight into the intracellular processes leading to protein folding, and the environmental conditions and genetic information required to initiate and modulate those processes, has opened the possibility of programming microbes such as yeast or streptomyces to perform these transformations and of employing conventional fermentation systems to produce such products in bioactive form.

In many situations involving cell-mediated biosyntheses, both the rate of generation of product and product quality are governed by transport-limited supply of essential nutrients or removal of metabolites. Novel strategies for mitigating these mass-transport barriers, and thus for improving both yields and product purity in bioreactors, are being pursued; these include techniques for improved micromixing in suspension cultures, hollow-fiber-membrane devices for immobilized cell culture, and continuous extractive bioreactor systems for efficient removal of products and/or inhibitory metabolites.

As an alternative to the use of viable cells for synthesis of certain classes of biological products, there has been a resurgence of interest in the use of **immobilized enzyme bioreactors** as routes to such products. This interest stems in large part from a growing recognition by both the pharmaceutical industry and drug regulatory agencies of the importance of **chirality** in determining the efficacy and safety of many new (and old) pharmaceutical products. Of particular importance, in my opinion, is the use of enzymes with stereospecific catalytic activity to perform **chiral transformations** on enantiomers of bioactive products, or to

permit highly selective homochiral synthesis of the desired stereoisomer from an achiral precursor. This in many cases allows the synthesis of the fully active stereoisomer of a pharmaceutical, and/or the elimination of an at best inactive, or at worst, toxic byproduct. Often forgotten is the fact that it was the therapeutically inactive enantiomer of thalidomide which was the tragic teratogen.

The use of microporous-wall hollow-fiber, immobilized enzyme bioreactors for continuous, high-capacity, continuous biotransformations of this type (Sepracor, Inc.) is receiving particular attention today. Other microporous, surface-activated matrices for enzyme immobilization, including thin film and particulate structures, are also being evaluated for this purpose. When these developments are combined with the techniques of **site-directed mutagenesis** to effect changes in enzyme structure and composition, in order to **controllably alter enzyme activity, selectivity, and specificity**, it seems evident that enzyme chemistry and enzyme engineering will for some time remain an important frontier of biotechnology.

Lastly, one of the most intriguing developments in protein synthesis that has come to my attention of late is that being pursued by Alexander Spirin and his colleagues in the USSR, where the *ab initio*, **cell-free synthesis of pure proteins** is being carried out by continuous translation of mRNA *in vitro* in a membrane-modulated bioreactor from which the desired protein is continuously harvested, while all the translational machinery for the synthesis is retained. The translational machinery, comprising the necessary ribosomal nucleotides, tRNA's, and other enzymatic components, is readily extracted (devoid of nuclear material) from animal, microbial, or plant cell cytoplasm; the mRNA coding for the protein of interest is added to this mixture. The reactor need then only be supplied continuously with the appropriate amino acids and the essential energy sources (ATP and GTP) to initiate continous production of the desired protein. The attractiveness of this synthesis route is the **total commitment** of the bioreactor system to the production of a single protein, if only one mRNA clone is present, thereby making isolation and purification of the protein a trivial exercise.

The major limitation of the process is, of course, the consumption of ATP and GTP, without provision for its regeneration; if a practical biological or synthetic chemical route for reconversion of the nucleoside di- and mono- nucleophosphates to the triphosphates were to be developed, this could become a preferred route for production of (at least small to medium sized) peptides and proteins. Success has evidently also been realized in carrying out **concurrent transcription and translation** in such a bioreactor, using a specific DNA.

When one thinks about combining the capabilities of the Polymerase Chain Reaction for multiplicative DNA synthesis with those of this protein synthesis procedure, one becomes mind-boggled at the prospect of being able to synthesize, devoid of any living cell, in immense quantities if desired, a specific protein beginning with a single molecule of DNA coding for that protein. While clearly such cell-free protein synthesis technique may be of questionable utility for production of bioactive proteins requiring glycosylation or other post-translational transformations, it surely could become the preferred route for producing enzymes, simple regulatory oligopeptides, and selected peptide hormones such as insulin. This unquestionably must be regarded as a true frontier of biosynthesis and bioprocessing.

THE SECOND FRONTIER: CHARACTERIZATION AND CONTROL OF BIOPOLYMER STRUCTURE

Today, it is quite generally accepted that the majority of biological macromolecules, including bioactive proteins and oligopeptides, enzymes, and probably polysaccharides, owe their specific activity to highly distinctive features present on or near their external surfaces. The surface characteristics of such molecules are governed not only by their primary structure, but also by the conformation imposed upon the molecule subsequent to or in the course of its biosynthesis. Usually, the bioactive form of a biopolymer is but one of an enormous number of possible conformations that can be assumed by that molecule as it undergoes intramolecular ionic, hydrogen bonding, and hydrophobic interactions. When such molecules are produced in their natural intracellular environment, these interactions are controlled and modulated by yet-poorly-understood processes occurring within the living cell to assure that only the bioactive conformation is assumed. When genetic manipulative techniques are employed to promote biosynthesis in a foreign host-cell, these subtle conformational controls are usually absent, and the resulting macromolecule, while constitutionally identical to the natural product, is frequently improperly folded and biologically inactive.

Differentiating between a "native", bioactive form of a protein, and an "unnatural", inactive form of the same molecule requires rather precise knowledge of its size and shape, of the spatial arrangement of molecular segments within the coil, and of the disposition of functional groups over its external surface. Until recently, gaining this knowledge would have been a impossible task except for the simplest molecules. Fortunately, however, recent progress in our ability to understand and control the processes of nucleation and growth of protein crystals has made possible the preparation of essentially defect-free crystals (of even very large protein molecules) amenable to precise X-ray diffraction analysis which, when combined with nuclear magnetic resonance spectrometry and sophisticated computer analysis of such data, permit a very detailed and reliable analysis of molecular conformation . Thus, it may soon be possible to identify unequivocally a protein as "native" and thus bioactive without having to resort to tedious and often ambiguous bioassay methods, and to rely upon such analysis as "proof of identity" of a biological product for regulatory purposes.

Surely of no less importance than our improved ability to characterize biomacromolecules is our growing awareness of the intracellular processes which induce proper folding of proteins and cause their secretion from the cell in proper "native" conformation. Today it is possible by physicochemical control of the solution environment to promote renaturation of denatured or improperly folded proteins produced in and recovered from recombinant organisms, and -- in some cases -- to introduce DNA sequences into recombinant genes coding for a specific protein which dictate proper post-translational folding of that protein. Hence, these important advances in molecular characterization and conformation control are very likely to pace our progress in the commercial production of ever more complex biologicals.

Yet, no sooner than our understanding of structure/property relationships in biomacromolecules advances, do we discover complexities of interactions of these substances with living systems which confront us with even greater problems and challenges: During the past few years, it has become clear that, for many regulatory peptides and proteins, intracellular post-translational glycosylation processes yield a bewildering array of **glycoforms** of these proteins which, in addition to complicating their characterization and analysis, seem to alter significantly (and in some instances dominate) their bioactivity, specificity, duration of action, and stability. It would appear that these individual glycoforms may target for specific cell receptors present only on certain cell types, or otherwise modulate the interaction of these molecules with cell surfaces or enzymes, thereby altering either their spatial or temporal patterns of activity. At present, our knowledge about how these oligomeric sugar appendages to proteins are formed, and what their intended function is, is still quite rudimentary. As we gain more insight into the nature of these phenomena, we may perhaps be able ultimately to "tailor" bioactive proteins by controlled, site-directed glycosylation to achieve unique therapeutic or prophylactic activity. Recent progress in the development of conjugate vaccines, comprising covalently linked recombinant protein and polysaccharide components, which display much elevated immunogenic activity and potency, is but a hint of what the development of controlled glycoprotein synthesis may hold in store for us.

THE THIRD FRONTIER: BIOSENSORS AND BIOPROCESS CONTROL STRATEGIES

The ultimate objective of bioreactor design and engineering is to be able simultaneously to monitor the key parameters revealing the cellular metabolic state, rate of production of products and other metabolites, and rate of consumption of nutrients, and to employ that information for appropriate adjustment of operating conditions to optimize reactor performance. At present, our bioprocess control capabilities are paced primarily by our existing monitoring capabilities, which are still limited to relatively few bioreactor variables, such as oxygen/carbon dioxide tension, pH, certain microion concentrations, and temperature. While assay techniques for measuring more sophisticated parameters controlling metabolic activity and bioproduct generation (e.g., cytokine concentration, ATP turnover rate, product concentration) are in many cases available, they involve invasive sampling procedures and time consuming analytical protocols that are hardly adaptable to real-time monitoring for process control. Hence, the current thrust of most biosensor research and development work is toward non-invasive, real-time, high-sensitivity devices which can accurately measure continuously trace quantities of complex biologically active substances in the reactor medium.

Many of these devices embody conventional optical or electrochemical sensing systems, in combination with biospecific elements which respond to changes in concentration of bioactive components by causing corresponding changes in conductivity, microion or microsolute concentration, pH or optical density. These include immobilized enzyme sensors, such as those used for measurement of glucose or urea concentrations, and immobilized living-cell sensors which respond to specific antibiotics or cytotoxins by changes in local pH or oxygen tension consequent to changes in cell metabolic rate. Of more recent development are sensors based on thin-film semiconductor devices which contain immobilized monoclonal antibodies tailored specifically to bind selected antigens, and which undergo changes in surface potential, conductance, capacitance, or photoelectric response in the process of formation of the immunocomplex. Sensors of this type often have impressively high sensitivity and specificity but lack reversibility, reproducibility, and/or the durability required for dependable in-field service. Progress toward development of such biospecific sensors, while encouraging, remains slow, and along with it, progress toward reliable, industrial-scale bioprocess control. Suffice it to say that a breakthrough in biosensor development would have a major impact on our capacity to control and automate many important industrial biosynthetic processes.

THE FOURTH FRONTIER: SEPARATION AND PURIFICATION OF BIOPRODUCTS

Recent developments in separation/purification processes for biological products have drawn rather heavily upon the cooperative enterprise between life scientists and engineers. Most have evolved from analytical and micropreparative techniques pioneered in the life science research laboratory, which have been submitted to engineering analysis and scaled up and adapted to industrial manufacturing practices. These processes are principally intended to recover and isolate proteins and peptides from multicomponent aqueous solutions or solid/water dispersions where they are present in quite low concentrations, and where isolation in high purity without denaturation is an important objective. They fall into four categories:

1. Selective Precipitation Processes

2. Two-Liquid-Phase Extractive Processes

3. Membrane Separation Processes

4. Selective Liquid/Solid Adsorptive Separations

5. Electrokinetic Separations

Selective precipitation as a technique for protein separation has been a long-standing practice for recovery of individual blood plasma proteins from whole plasma by the Cohn plasma-fractionation process. Capitalizing on differences in isoelectric point, molecular weight, and hydrophobicity, controlled changes in pH, ionic strength, organic solvent (e.g., alcohol) concentration and temperature of multicomponent protein solutions often permit differential precipitation of selected proteins in variable degrees of purity. Most such precipitates are amorphous hydrogels which inescapably entrain significant amounts of impurities; this has led bioprocess specialists to regard precipitation as at best a crude and inefficient separation process, and thus ill-suited for recovery of bioactive proteins whose **ultra-high purity** is an essential requirement for safe use. However, recent progress which has been made in facilitating **crystallization** of proteins from solution (as mentioned earlier) suggests that these same techniques may very well be extended to a preparative process for the selective precipitation of proteins in a state of purity comparable to that of other crystalline products. This is clearly a process worthy of serious evaluation.

Separation/purification by liquid/liquid extraction has long been a standard practice in the pharmaceutical industrial for recovery and purification of low-molecular weight biological products such as antibiotics, vitamins, and steroid hormones, and has recently been extended to protein recovery through the process of **two-phase aqueous extraction.** This process, which employs either aqueous mixtures of simple electrolytes and water-soluble polymers (e.g., polyethylene glycol), or of incompatible water-soluble polymers (e.g., dextran/PEG) to form two immiscible water-rich phases, has been shown to be quite effective in facilitating the separation of cells or cellular debris from soluble proteins in fermentation broths, and permitting recovery of enzymes and other useful biomacromolecular products from such mixtures. Unfortunately, however, efforts to develop a predictive rationale for selection of appropriate multiphase-forming polymer systems useful for specific protein recovery requirements have yet to be developed, so that the process remains a trial-and-error prospect for any specific separation. Efforts aimed at increasing the selectivity of such extractions by attaching affinity ligands to the phase-splitting polymers have yielded promising results and may ultimately lead to broadened utility of this procedure for protein isolation.

Membrane separation processes such as microfiltration, ultrafiltration, and reverse osmosis have over the past decade become workhorse operations in bioprocessing, for dewatering of cellular suspensions, for solid/liquid separations involving separation of cells and cellular debris from dissolved bioproducts, for concentration of aqueous protein solutions by UF, and for removal or replacement of electrolytes and other microsolutes in biomacromolecule-containing solutions by diafiltration. Recent developments in membrane fabrication and surface-treatment technology have yielded microporous and ultramicroporous membrane structures (prepared from polymeric, ceramic, metallic, and composite materials) with very low sorptivity for proteins and other biopolymers, which are proving to be nonfouling membranes with superior flux performance and lifetime, and enhanced clearance of dissolved solutes, in crossflow filtration applications. Even more intriguing are novel non-sorptive, narrow-pore-size-distribution ultrafiltration membranes (Millipore) which are capable of virtually complete removal of virus particles from solutions of high molecular weight proteins without significant protein retention -- a development of particular importance for the production of safe therapeutic or prophylactic biologicals from mammalian tissue sources. These developments raise the prospect of broadening the utility of membrane separations to permit rather precise fractionation of protein mixtures based on **moderate differences in molecular size** -- a capability generally perceived to be limited to gel permeation chromatography.

Steady progress continues to be made in the **engineering of membrane process systems** toward improved fluid-management aimed at reducing solute polarization and stress-induced damage to sensitive biomolecules. Rotary membrane devices (Membrex, Pall/Sulzer) are now available which utilize shear-induced vortexing at low pressures to depolarize membranes and materially enhance both flux and solute clearance; and a recently-developed ceramic/metallic composite

5

membrane (Alcan Separations) with unusual surface configuration has displayed astonishingly high flux and solute clearance with high-solids-content feed streams at low pressures and tangential fluid velocities. These developments, when combined with the advances in membrane fabrication technology for control of membrane microstructure mentioned above, promise to expand significantly the range of applications of membrane separations in industrial bioseparation/purification operations.

The use of adsorptive solids to extract specific components selectively from multicomponent liquid mixtures is a technique long practiced in industry for recovery of low-molecular-weight products, and in the life science research laboratory for the analysis and micropreparation of macromolecular biological products. Column chromatography is a mainstay technique in molecular biology for protein identification and analysis, and its adaptation to large scale isolation and purification of polymeric bioproducts has been under intensive development for several years; unfortunately, however, the scaleup problems of column chromatographic separations are formidable, and success in adapting the process to industrial practice has been limited.

What has, however, been until recently overlooked regarding adsorptive processes for macromolecular separations, however, is the fact the **binding affinities of large molecules to solid substrates are usually very large**, and often **exceedingly sensitive to minor changes in such solution composition variables as pH, ionic strength, or cosolvent concentration**. Thus, it is frequently possible to choose a solid adsorbent which will selectively and nearly quantitatively adsorb a specific protein (or class of proteins) from a multicomponent mixture in which it is present in very low concentration, to the exclusion of all other components present. If the adsorbent is then washed free of the unbound species, and treated with a solution favoring the complete desorption of the bound product(s), the released components can be recovered in high concentration and purity in the displacing medium, while the adsorbent (after washing to remove residual eluting solvent) is rendered suitable for reuse. The "bind/release" approach to sorptive separations, while an intermittent, cyclical process, is nonetheless readily adaptable to scaleup, and while obviously lacking the resolving power of classical column chromatographic separation, is nonetheless a very attractive and potentially economic recovery/purification strategy for proteins and other macromolecular biologicals.

A paradox encountered in bind/release adsorptive separation of biopolymers, however, is the fact that most high-surface-area adsorbents are ultramicroporous solids whose pore sizes are so small that polymeric sorbates diffuse exceedingly slowly into and out of the solid phase, and thus the adsorption and desorption steps become unacceptably slow. While small adsorbent particle size can ameliorate this problem, the high hydraulic resistance to flow through packed beds of such particles makes the cure almost worse than the disease.

What may in large measure have proved to solve this problem is the recent development of **macroreticular** microporous solid adsorbents, in either particulate or monolithic (sheet, hollow fiber) form, which contain micron- or submicron-sized channels through which liquids can flow convectively at relatively low driving pressures, and also possess ultramicroporosity and high sorptive surface area. The surfaces of these macroreticular adsorbents can be functionalized to render them anionic or cationic, or hydrophilic or hydrophobic, and thus capable of adsorbing solutes by ionic, hydrogen bonding, or van der Waals interaction mechanisms. Moreover, surface chemical coupling techniques can also be employed to attach **affinity ligands** to these sorbent surfaces, thereby rendering them highly selectively sorptive for particular proteins or protein classes.

Employing these sorbents (in shallow, packed-particle beds, as proposed by PerSeptive Biosystems; in hollow fiber membrane modules, as taught by Sepracor and Kinetek Systems; or as single-sheet or sheet-stack forms as recommended by Millipore [MemSep], FMC Bioproducts [Actidisk/Actimod], and Nygene [MASS]), it has been possible to pass liquids through the solids at rates sufficient to saturate the adsorbent, to flush the solid, and then to release the product from the solid, in a matter of minutes. Because of these short cycletimes, the productive capacity of these adsorbents for isolation of specific proteins is often measured in hundreds of grams per day per liter of adsorbent volume. For recovery of monoclonal IgG from tissue culture media, the use of a Protein A affinity-ligand adsorbent of this type appears to permit concentration and purification of the protein far more efficiently and economically that any other process. Ion exchange macroreticular adsorbents of these various geometric forms are now available, and their potential utility for recovery and purification of specific proteins seems great indeed.

Another unique characteristic of monolithic (i.e., sheet- and hollow-fiber) forms of these adsorbents is that they can, by tangential flow fluid-contacting techniques, be employed to extract selectively proteins and other biopolymers from **particle-containing fluids** such as fermentation broths, cell suspensions, or lysed cell dispersions, since the particles are unable to enter the fluid-permeable micropores. Hence, selective adsorption may become the preferred first step in recovery of biological products from fermentation or culture media. Moreover, these adsorbent devices can be used to carry out continuous, extractive fermentations and cell culture operations in which the cell-containing medium depleted of the adsorbent component is recycled to the bioreactor -- a bioprocessing technique particularly appealing for products derived from slow-growing, fastidious, mammalian-, plant-, or insect-cell cultures.

Another important advantage attributable to the rapidity of the bind/release cycle for these adsorbents is the concomitant short transit time of the product from its source to its final isolated state. Thus, if the product is **highly labile**, or if the medium into which the product is elaborated contains **proteolytic enzymes which degrade it**, the rapidity of isolation will result in significant improvement in **overall product yield.**

It is my view that these high-speed, short-cycle-time, bind/release sorptive separation processes are likely to become the methods of choice for the industrial-scale recovery and purification of proteins and peptides, for reasons of efficiency, simplicity, and overall economy. I also see important opportunities for this technique for the removal of trace amounts of objectionable impurities (e.g., pyrogens, nucleic acids, etc.) from therapeutic or prophylactic biologicals as a terminal step in product purification.

In many respects, **electrokinetic separation** of biomacromolecules is regarded as the "holy grail" of bioseparations, because it has proved to be such a powerful laboratory technique for identifying and isolating biopolymers of widely disparate complexities, and for differentiating between macromolecular species so closely related as to defy discrimination by any other means.

Electrophoretic separation is, in principle at least, capable of resolving charged macromolecules differing but slightly in size, shape, or charge density, and may be the method of choice for separating different glycoforms of the same protein molecule. Isoelectric focusing is believed to be potentially capable of resolving mixtures of proteins differing in constitution by as little as one amino acid in one thousand. Pulsed-field gel electrophoresis of nucleic acids has proved to be extraordinarily successful in separating DNA fragments of enormous length, and has revolutionized procedures for genetic mapping of nuclear material.

Yet, efforts to adapt these sophisticated techniques to macro-scale isolation and purification of biopolymers have been frustrated by such randomizing phenomena as molecular diffusion, thermally-induced convection, and fluid turbulence. While many of these disturbances can be minimized by, for example, operating in microgravity environments, or in stabilizing laminar shear-fields, these solutions hardly seem practical for industrial-scale production of biological products. Considerable progress has been made with **recycling free flow isoelectric focusing** (mainly by Milan Bier and his colleagues at University of Arizona), and the scalability of this rather intriguing process for large-scale protein purification seems achievable.

With this exception, however, efforts to engineer electrokinetic processes into useful moderate- to large-scale systems for biopolymer isolation have been fragmentary, and relatively unsuccessful. What is needed here is some revolutionary new thinking about electric-field-induced molecular transport, and how these and other force-fields can be collectively harnessed to achieve the desired separations. This is clearly a fruitful research frontier for life scientist, colloid chemist, and engineer together.

A LOOK INTO THE FUTURE: WHITHER BIOPROCESSING?

Certainly today, and probably also for the next decade at least, the primary focus of industrial biotechnology will be upon highly potent, bioactive proteins, peptides and related biopolymers intended for use as human therapeutic, prophylactic, and diagnostic products. Such products will in large measure be administered in miniscule amounts, and will command exceedingly high prices on a dollars-per-kilogram basis; moreover, "industrial production" levels for such products will probably be measured in grams, or at most, a few kilograms per year. For such products, the actual costs of synthesis and isolation/purification are likely to represent a small fraction of the market price, meaning that there will be little economic incentive either to introduce new and less costly processing operations, or otherwise to attempt to improve the recovery efficiency or cost of production. The major criteria for downstream process selection and improvement during this period will continue to be, as it is now, **product quality and purity.**

In due course, however, the focus of industrial biotechnology is certain to shift to (or broaden to cover) bioproducts intended for use as veterinary therapeutics and growth regulators, pesticides and plant growth regulators, and food-components and -additives. For such products, both the scale of production, and the market structure, will demand the development of high-tonnage, cost-effective manufacturing procedures which will not only meet the required quality and purity standards, but can **meet price competition in the market**. Few of the bioprocessing techniques now in use, or contemplated for use, in the production of human therapeutic biologicals are likely to satisfy these requirements. How will we respond to this challenge? It is my hope and expectation that, now that the log-jam of interdisciplinary communication between life scientist and engineer has been broken, together we can eagerly and effectively attack these new bioprocess problems, and speedily solve them.

Part I
Protein and Enzyme Engineering

THE PROMISE OF AN ABILITY TO SYNTHESIZE any compound using enzymes, thus liberating biotechnologists from the necessity to resort to catalysts requiring production and maintenance of adventitious biomass, lies just beyond the visible frontier of bioprocessing. After the opening chapter indicates the importance and utility of determining protein structure, we discuss the determination of several protein structures and their applications. Three exciting new developments are then described in detail, progressing from the practical equivalent of an artificial cell—an ultimate bioreactor that can use isolated genes as information for sustained synthesis of specific peptides—through an organic-phase technique for synthesizing enantiomeric hydrophobic materials to a totally synthetic enzyme built on principles of catalysis and peptide structure.

The Role of Protein Crystals in Biotechnology and Industry

Alexander McPherson, Department of Biochemistry, University of California—Riverside, Riverside, CA 92521

The technique of X-ray crystallography as applied to biological macromolecules (for discussions see Blundell and Johnson, 1976; McPherson, 1982) has in recent years made enormous strides in terms of both speed and precision. It has become, for molecular biology, a practical, reliable, and relatively rapid means of obtaining straightforward answers to perplexing questions. In particular, it has made possible the directed use of recombinant DNA techniques to produce proteins of modified structure and function. Similarly, our ability to describe and utilize protein structure and to visualize interactions with ligands has made possible the systematic design of new drugs and pharmacological agents. For the power of the method to be realized, however, the macromolecule of interest must first be crystallized. Not only must crystals be grown, but they must be of exceptional quality and size in order to be adequate for a high resolution X-ray diffraction analysis. Crystallization has become the rate limiting step. The explanation as to why protein crystallization has proven such an obstacle lies to a great extent in the inherent complexity of the systems under study which necessitates an essentially empirical approach. The cause lies as well in the poor state of our current understanding of macromolecular crystallization phenomena and the forces that promote and maintain protein and nucleic acid crystals. As a substitute for the precise and reasoned approaches we commonly apply to scientific problems, we are forced to employ trial and error methodology. Macromolecular crystallization is, therefore, a matter of searching, as systematically as possible, the ranges of the individual parameters that impact upon crystal formation, finding a set of these factors that yield some kinds of crystals, and then optimizing the variable sets to obtain the best possible crystals for X-ray analysis. Because the number of variables is so large, and their ranges so broad, intelligence and intuition in designing and evaluating the individual and collective trials becomes essential.

The New Polymer Chemistry

A variety of industrial processes utilize enzymes to catalyze bulk chemical reactions. Brewing, baking and food processing are chief among the users of this relatively old technology. Other commercial and agricultural processes could be more expeditiously carried out if the fragile properties of specific proteins could be strengthened by structural alterations at the molecular level that otherwise preserve activity. Improvements in heat stability and pH tolerance are examples, increase in solubility is another. Such enhancements can, in general, now be readily achieved by altering the genes that specify the proteins through the normal cellular protein synthetic process. This is illustrated schematically in Figure 1. Structural alteration could, for example, involve the introduction of additional disulfide cross links to instill rigidity and render the enzyme more robust. Solubility might be increased by addition of hydrophilic peptide sequences.

Another objective might be to produce enzyme molecules with altered substrate specificities, that is, enzymes that catalyze reactions not normally carried out in living systems. Thus complicated, low yield, intermediate reactions in the synthesis of certain medicinal drugs or pesticides might be replaced by straightforward, high yield, stereospecific, enzyme catalyzed reactions. From enzymes having the correct catalytic activity and whose structures are known from X-ray crystallography, analogous enzymes might be created. This could be done by examining the disposition of amino acid residues of the known enzyme and

2181–2/92/0010$06.00/0 © 1992 American Chemical Society

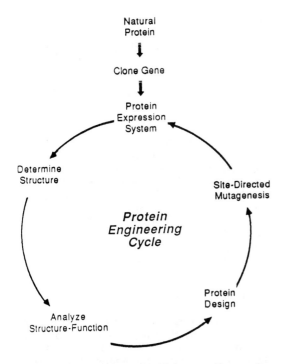

Natural
Protein

↓

Clone Gene

↓

Protein
Expression
System

Determine
Structure

Site-Directed
Mutagenesis

*Protein
Engineering
Cycle*

Protein
Design

Analyze
Structure-Function

Fig. 1. The process by which a protein may be rationally redesigned by combining genetic engineering and X-ray crystallography.

genetically altering the steric and electrostatic features at the active site by intentional substitution of other amino acids. A novel enzyme could be constructed from one with the required activity but having a spatial array of atoms at the active center compatible with the binding and orientation of the desired substrate.

Many protein molecules are not enzymes, but serve other roles in living systems such as signal transducers, protective agents, transport molecules, or as structural materials. The protein from common barnacles responsible for their amazing adhesive properties, the anticoagulant protein from leeches, the antiinsecticidal proteins from many species of bacillus are all examples of proteins highly specialized to carry out specific functions that might be of potential value to man. Many of these functions could well be transposed from cells to *in vitro* processes or products if the molecules were modified to maintain their structure and function and to accomplish an intended chemical or physical task.

Monoclonal antibodies could be designed and conjugated with reporter groups to act as molecular sensors for detecting and signaling the presence of specific compounds at otherwise immeasurably low concentrations. These would have uses ranging from medical diagnosis to pollutant monitoring and the detection of carcinogens in the environment. Transport proteins could be designed to carry specified drugs in the blood and constructed to deliver them only to designated target tissues. Proteins with unique physical properties such as collagen and elastin that make up the connective tissues of the body might be modified in useful ways to create new materials for a broad spectrum of products ranging from prostheses to electronics.

A final example of this "new polymer chemistry" is that of protein engineering in agriculture and nutrition. The storage proteins found in cereal grains, beans and other plant seeds comprise major protein food sources worldwide. Unfortunately, many of these proteins are lacking in certain essential amino acids. By redesigning these storage proteins to contain a better balance of nutritionally important components without otherwise impairing their physiological roles, enhanced crops could be created. By similar means, the abilities of plants to resist disease and withstand environmental extremes might be fortified and the palatability and taste of plant proteins altered, thereby generating new classes of food materials. The use of the enzyme cellulase to convert wood products to sugar and invertase to make sucrose from corn syrup are currently under intense investigation by biotechnology and industrial concerns.

The New Pharmacology

The discovery of new drugs, pesticides, herbicides and other molecules efficacious in the prevention of disease or the elimination of pests currently depends on the trial and error synthesis and physiological assay of potentially useful compounds. This is a lengthy, extremely expensive, and painstaking process because correlation between precise chemical structure of a drug and its physiological effect is frequently difficult to discern and often entirely absent. When a new compound of promise is found the usual approach is to synthesize all possible variants and to test each for improved effectiveness. This often requires thousands of syntheses and even more assays.

The targets of most drugs are proteins, nucleic acids, or complexes of the two. These may be in the aqueous cytosol of cells, embedded in cellular membranes, or found

11

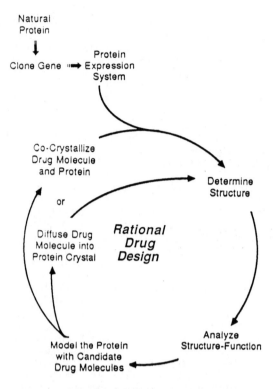

Natural
Protein

Clone Gene ⇒ Protein
Expression
System

Co-Crystallize
Drug Molecule
and Protein

Determine
Structure

or

*Rational
Drug
Design*

Diffuse Drug
Molecule into
Protein Crystal

Analyze
Structure-Function

Model the Protein
with Candidate
Drug Molecules

Fig. 2. A diagram showing the path by which X-ray crystallography in combination with computer graphics analysis and synthetic chemistry may be rationally applied to create new chemical compounds such as pesticides.

in extracellular fluids such as blood serum. To exert its physiological effect, the drug must bind to its target macromolecule and express its activity by altering the structure or the function of the macromolecule. Effectors of macromolecules, which in this case are drugs, herbicides or pesticides, are referred to as specific ligands. They are bound with high affinity, usually at the active site of the target macromolecule, by non covalent chemical forces. A salient determinant in the association is a close chemical and geometric complementarity between the ligand and the binding site on the enzyme; the classical lock and key analogy applied at the atomic level.

It requires little imagination to see that detailed knowledge of the target macromolecule could be of great service in directing a chemist more rapidly toward synthesis of the most effective compound. As shown schematically in Figure 2, if the precise atomic structure of the target macromolecule is known, which is possible by applying the X-ray diffraction technique to single crystals of proteins and nucleic acids, then rational and

systematic design of the chemical compound could follow. New drugs to fight antibiotic-resistant bacterial infection could be designed and synthesized, efficient and highly species-specific pesticides created, or anti-tumor agents could be optimized and even individualized by targeting enzymes of pronounced importance to neoplastic growth. In agriculture, the impact would again be profound. From the structures of target enzymes, new classes of fungicides, insecticides and herbicides could be devised that are safe and nondestructive to the environment but otherwise more effective. Using X-ray diffraction analysis of crystals of the appropriate macromolecules, the rapid identification, evaluation and classification of potential or known mutagens and carcinogens could be carried out by atomic level visualization of their effects.

The Role of Protein Crystals

The objectives of the new polymer chemistry and the new pharmacology are attainable only for those cases where the synergy of the synthetic-analytic loop can be completed and effectively exploited, that is, where the gene for the target macromolecule can be isolated for subsequent manipulation, and the target protein can be crystallized and subjected to X-ray diffraction analysis to reveal its structure at atomic resolution. To genetically engineer a protein of medical, agricultural or industrial significance, as illustrated in Figure 1, the protein's gene must be isolated from the genome of the organism, altered by site-directed mutagenesis, and expressed in living organisms as the new, altered protein.The modified protein must then be crystallized, subjected to diffraction analysis, the structure of the altered protein compared with that of the native structure, and further cycles of biosynthesis-X-ray analysis initiated.

To design pharmacologically important ligands, that is drugs, as shown in Figure 2, the target protein or enzyme must again be crystallized and analyzed by X-ray diffraction analysis and its structure determined. Examination of that portion of the macromolecule of interest, the binding site for the drug, is then undertaken using three dimensional computer graphics. From the arrangement of protein atoms that make up the binding site improved ligands can be designed, synthesized, and assayed. The new drug can then be diffused into crystals of the target protein and this complex then subjected to diffraction analysis as well. By detailed examination of the complex between protein and drug, using what are known as difference Fourier techniques, the degree of chemical and steric complementarity can be defined, an improved drug designed, and a new round of chemical synthesis-X-ray analysis conducted (McPherson, 1987)

An important question is why the new polymer chemistry and new pharmacology has not developed at a more rapid pace. What, currently, is the main obstacle to closing the biosynthetic-bioanalytic feedback loop. The methodologies for isolating the genes, altering their structures, and expressing protein products have now reached an advanced state of development. Success in this regard is seldom a question, being only dependent on the amount of time and degree of effort expended.

X-ray crystallography, the keystone of the analytical effort, has also progressed dramatically in recent years. At one time, the analysis of the structure of a single crystalline protein might have occupied the efforts of several crystallographers for many years. Collection of X-ray diffraction data was extremely time consuming and computing procedures were similarly tedious. Few techniques existed for detailed examination of structure once it had been determined. This has changed in striking fashion. X-ray sources, both rotating anode generators and synchrotrons, have been developed that yield X-ray flux densities several orders of magnitude more intense than those only ten years ago. New area detector systems for the measurement of diffracted X-rays have come into use and these have had an extraordinary impact on the rate and accuracy with which X-ray data can be accumulated. What was formally a process requiring several years of intense effort has now been reduced to no more than several weeks. Computing speed and precision has advanced proportionately, accompanied by the invention of mathematical procedures that greatly enhance our ability to utilize X-ray diffraction to determine and study macromolecular structure.

It is now true that once protein crystals have been obtained that are suitable for X-ray data collection, then structure solution, refinement and display in three dimensions is possible in no more than a half a year. Even this period is rapidly shrinking. When suitable protein crystals are available, and when the structure of the protein that makes up the crystals is known, then the detailed structures of its complexes with drugs and ligands can, even now, be precisely determined by difference Fourier techniques in only a few

days. The methodologies of X-ray crystallography match those of the genetic engineer in power, the synthetic-analytic loop should now be closed.

The orphan child of X-ray crystallography, the growing of protein crystals, remains the problem. Because its value to the process was never fully appreciated, its practice considered as much art as science, the process of protein crystal growth was ignored and relegated to obscure corners of laboratories. Not only did the phenomenon of protein crystal growth remain a mystery but its practitioners languished. There are, therefore, few people trained in the techniques and methods for growing protein crystals, there are only a limited number of empirical approaches, and the inherent difficulties of growing protein crystals are many and varied. It now appears that the key to realizing the promise of the new polymer chemistry and the new pharmacology lies in the development of new and reliable techniques, more systematic and scientific, for obtaining suitable protein crystals.

The reason that the crystallization step has become the primary obstacle to expanded structural knowledge is the necessarily empirical nature of the methods employed to overcome it (see McPherson, 1976, 1982, 1985, 1989, 1990a, 1990b). Macromolecules are extremely complex physical-chemical systems (see Creighton, 1984; Richardson, 1981; Schulz and Schirmer, 1979) whose properties vary as a function of many environmental influences such as temperature, pH, ionic strength, contaminants, and solvent composition, to name only a few. They are structurally dynamic, microheterogeneous, aggregating systems, and they change conformation in the presence of ligands. Superimposed on this is the poor state of our current understanding of macromolecular crystallization phenomena and the forces that promote and maintain protein and nucleic acid crystals.

As a substitute for the precise and reasoned approaches that we commonly apply to scientific problems, we are forced, for the time being at least, to employ a strictly empirical methodology. Macromolecular crystallization is, thus, a matter of searching, as systematically as possible, the ranges of the individual parameters that impact upon crystal formation, finding a set or multiple sets of these factors that yield some kind of crystals, and then optimizing the variable sets to obtain the best possible crystals for X-ray

13

analysis. This is done, most simply, by conducting a long series, or establishing a vast array, of crystallization trials, evaluating the results, and using information obtained to improve matters in successive rounds of trials. Because the number of variables is so large, and their ranges so broad, intelligence and intuition in designing and evaluating the Individual and collective trials becomes essential. Every factor that might aid us in this quest must be considered and evaluated.

Recently a number of new approaches have been applied to the protein crystal growth problem with some measure of success. Although each contributes incrementally to our under-standing and mastery of the process, no one method or component has emerged as a broad solution. Each simply supplies another tact, another opportunity, when others fail.

The chemical components of a crystallization trial may, for example, be augmented with biochemically gentle neutral detergents such as β-octylglucoside. These detergents reduce the effects of hydrophobic interactions that frequently lead to non specific aggregation, and hence to precipitate formation. While these surfactants have shown particular value in the case of membrane protein crystallization (Michael, 1990), they have improved the probability of success for proteins normally soluble in aqueous solvents as well (McPherson et al., 1986).

Undoubtedly, further investigation of the role played by detergents in protein crystallization is warranted, and a wider definition of useful compounds may substantially increase the number of proteins successfully crystallized.

Because nucleation is a critical component in the crystallization phenomenon, some appreciable effort has also gone into devising means to promote nuclei formation and to monitor its development. Among the techniques recently utilized are the inducement of protein crystal growth at lower levels of protein supersaturation by the use of heterogeneous and epitaxial nucleants. In particular, mineral surfaces of various sorts have been able to serve as promoters of protein crystal nuclei, apparently because their ordered, periodic surfaces provide two dimensional initiation arrays of protein molecules that then lead to the formation of three dimensional matrices (McPherson and Shlichta, 1988).

Monitoring of protein crystal nucleation even prior to direct microscopic visualization of crystals is also being attempted using the technique of inelastic light scattering (Kam et al., 1978; Carter et al., 1988; Kadima et al., 1990). This method, which detects aggregate formation and quantifies the distribution of hydrodynamic radii, it is hoped, will allow an investigator to identify much earlier the optimal conditions for the crystallization of any specific protein.

Robotic systems are also being devised and built by a number of investigator teams, principally in industrial laboratories, to alleviate much of the tedium and manual inaccuracy associated with protein crystallization screens (Cox and Weber, 1988; Morris et al., 1989). It is anticipated that robotics systems may not only speed the definition of optimal conditions, but will provide a greater degree of reproducibility as well.

Finally, one of the most exotic methods currently being investigated for its potentially positive effect on protein crystal growth is the use of microgravity as an environment (DeLucas et al., 1986; Littke and John, 1984). The justification for attempting to crystallize protein crystals in space is threefold. First, density driven convective flow at the surfaces of growing crystals, known to produce defects and disorder in conventional crystals, is effectively eliminated. Second, sedimentation is eliminated, thus, preventing crystal contact during growth and consequent defect formation. Third, it is possible in microgravity to crystallize proteins in a completely containerless way, thus, doing away with the unfavorable influences of vessel surfaces.

While definitive conclusions have not yet been reached regarding the efficacy of microgravity environment, many experiments have now been performed and data on a number of different protein crystals is accumulating. Reasonably convincing evidence has now, in fact, appeared (DeLucas et al., 1989) and it seems likely that protein crystals grown in space do, indeed, enjoy a superiority in terms of order and perfection over those grown in laboratories on earth.

If it can be conclusively demonstrated that elimination of convection and sedimentation effects in microgravity have a significant positive effect on the phenomenon of protein crystal growth, then it will have a major impact on the commercial prospects for materials processing in space, and it will contribute substantially to the ground based biotechnology revolution now taking place. Because protein crystals are the key to revealing macromolecular structure, they are the essential component in allowing the application of logic and rational design to the engineering of protein structure and the design of new pharmaceutical agents directed against target proteins of known structure.

REFERENCES

Blundell, T. L.; Johnson, L. N. *Protein crystallography*; Academic Press: New York, NY, **1976**.

Carter, C. W.; Baldwin, E. T.; Frick, L. *J. Cryst. Growth* **1988**, *90*, 60-73.

Cox, M. J.; Weber, P. C. *J. Cryst. Growth* **1988**, *90* , 318-324.

Creighton, T. E. *Proteins: Structures and Molecular Properties*; Freeman Co.: New York, NY, **1984**.

DeLucas, L.J.; Smith, C. D.; Smith, H. W.; Vijay-Kumar, S.; Senadhi, S. E.; Ealick, S. E.; Carter, D. C.; Snyder, R. S.; Weber, P. C.; Salemme, F. R.; Ohlendorf, D. H.; Einspahr, H. M.; Clancy, L. L.; Navia, M. A.; McKeever, B. M.; Nagabhushan, T. L.; Nelson, G.; McPherson, A.; Koszelak, S.; Taylor, G.; Stammers, D.; Powell, K.; Darby, G; Bugg, C. E. *Science* **1989**, *246*, 651-654.

DeLucas, L. J.; Suddath, F. L.; Snyder, R.; Naumann, R.; Broom, M. B.; Pusey, M.; Yost, V.; Herren, B.; Carter, D.; Nelson, B.; Meehan, E. J.; McPherson, A.; Bugg, C. E. *J. Cryst. Growth* **1986**, *76*, 681-693.

Kadima, W.; McPherson, A.; Dunn, M. F.; Jurnak, F. A. *Biophys. J.* **1990**, *57*, 125-132.

Kam, Z.; Shore, H. B.; Feher, G. *J. Mol. Biol.* **1978**, *123*, 539.

Littke, W.; John, Chr. *Science* **1984**, *225*, 209.

Membrane Protein Crystallization, CRC Reviews; Michael, H., Ed.; CRC Press: Orlando, FL, **1990**; (in press).

McPherson, A. In *Crystallography Reviews;* More, M., Ed.; Gordon and Breach: United Kingdom, **1987**, Vol. 1; 191-250.

McPherson, A. *Eur. J. Biochem.* **1990a**, *189*, 1-23.

McPherson, A. *Sci. Am.* **1989**, *260*, 62-69.

McPherson, A. In *Methods of Biochemical Analysis;* Glick, D., Ed.; Academic Press: New York, NY, **1976**, Vol. 23; pp. 249-345.

McPherson, A. *The Preparation and Analysis of Protein Crystals*; John Wiley and Sons: New York, NY, **1982**.

McPherson, A. In *Methods in Enzymology: Diffraction Methods;* Hirs; Timasheff; Wyckoff, Eds.; Academic Press: New York, NY, 1985, Vol. 114; 112-120.

McPherson, A. In *CRC Reviews - The Crystallization of Membrane Proteins;* H. Michael, Eds.; CRC Press: Orlando, FL, **1990b** (in press).

McPherson, A.; Shlichta, P. *Science* **1988**, *2398*, 385-387.

McPherson, A.; Koszelak, S.; Axelrod, H.; Day, J.; Williams, R.; Robinson, L.; McGrath, M.; Cascio, D. *J. Biol. Chem.* **1986**, *261*, 1969-1975.

Morris, D.; Kim, C. Y.; McPherson, A. *Biotechniques*, **1989**, May, 1989.

Richardson, J. In *Advances in Protein Chemistry*; Academic Press: New York, NY, **1981**, Vol. 34, 167-339.

Schulz, G. E; Schirmer, R.H. *Principles of Protein Structure;*, Springer-Verlag: New York, NY, **1979**.

Center for Advanced Research in Biotechnology: A Multidisciplinary Approach to Protein Engineering

G. L. Gilliland,[1] B. Veerapandian, J. Dill, O. Herzberg, C. Chiu, G. Kapadia, J. Moult, D. Bacon, M. Toner, K. Fidelis, F. Schwarz, P. Reddy, K. McKenney, M. Kantorow, J. Moore, J. Hoskins, W. J. Stevens, D. Garmer, E. Eisenstein, K. Fisher, P. Bryan, P. Alexander, S. L. Edwards, M. Mauro, K. Darwish, K. Choudhury, H. Li, R. Raag and T. L. Poulos, Center for Advanced Research in Biotechnology of the Maryland Biotechnology Institute, University of Maryland, Shady Grove and the National Institute of Standards and Technology, 9600 Gudelsky Drive, Rockville, MD, 20850, U.S.A.

Protein engineering at the Center for Advanced Research in Biotechnology (CARB) is based upon the ability to determine the detailed atomic structures of proteins and to understand the relationship between that structure and the properties of the particular proteins before and after engineering. Toward this aim CARB has established programs in four critical research areas: molecular biology, macromolecular crystallography, physical biochemistry, and computational chemistry and modeling. Currently a variety of protein engineering projects utilizing a multidisciplinary approach are underway at CARB. Several examples of such projects are presented along with the strategy implemented for attacking such problems.

Protein engineering, the specific modification of the atomic structure of a protein, is based upon developments resulting from the molecular biology revolution of the past two decades. These developments include the ability to isolate any particular gene coding sequence, modify the DNA, incorporate the gene into a high expression system, and produce large quantities of the protein for purification. Protein engineering, although in its infancy, has produced important products that are being used in a variety of agricultural, medical and commercial applications.

Protein engineering requires that we understand how the three-dimensional structure of a protein is determined by the amino acid sequence, and how the three-dimensional structure of a particular macromolecule relates to its function. Currently a large body of information is being accumulated on the structure of proteins from X-ray crystallographic and NMR studies, nevertheless structural biologists are a long way from accurately predicting the structures of proteins. However, enormous insight can be gained for protein engineering studies from knowledge of the structures of a proteins. Moreover, a multidisciplinary approach to the study of structure-function relationships of proteins provides the means to assist in understanding the fundamental principles that govern protein function. Thus, CARB was founded on the premise that a multidisciplinary approach to the study of structure-function relationships would accelerate our understanding of protein function.

---

[1]Address correspondence to this author.

CARB:Organization and Activities

The University of Maryland, the National Institute of Standards and Technology (NIST) and Montgomery County, Maryland have jointly established the Center for Advanced Research in Biotechnology, a unique forum for collaborative research among academic, government and industry scientists. CARB is also one of six biotechnology centers under the umbrella of the Maryland Biotechnology Institute (MBI).

The CARB facility is located on the University of Maryland's Shady Grove Campus at the Shady Grove Life Sciences Center in Rockville, Maryland. The facility was designed to meet the needs of the biotechnology community in the field of protein structure, function, and design. In addition to the laboratory environment at CARB, facilities for seminars, workshops and scientific meetings are present on-site.

One unique aspect of CARB's mission is its relationship with private industry. Companies can participate in CARB research by becoming industrial affiliates, jointly sponsoring specific projects or establishing collaborative research projects under agreements which permit the company to obtain a license to any proprietary technology resulting directly from the research.

Certain commercial equipment, instruments, and materials are identified in this chapter in order to specify experimental procedures. Such identification does not imply recommendation or endorsement by the National Institute of Standards and Technology, nor does it imply that the materials or equipment identified are necessarily the best available for the purpose.

Molecular Biology-An essential component of protein structure/function research at CARB is the ability to clone, express and modify genes for the production of large quantities of natural and engineered proteins for detailed genetic, biochemical and biophysical studies. Molecular biology research at CARB focuses upon the efficient use of prokaryotic cells to yield specific proteins as well as the study of protein function in prokaryotic DNA regulation, cellular chemistry and metabolism.

CARB has well equipped molecular biology laboratories with specialty equipment required for DNA and protein synthesis, purification and characterization. This includes the EG&G Molecular Biology Corporation DNA Sequencer and an Applied Biosystems DNA synthesizer for oligonucleotide production. CARB also has small scale fermentation facilities, 1.5 and 12 liter New Brunswick Scientific fermentors with microprocessor control, and access to the large bioprocessing facility at the University of Maryland, College Park, Maryland.

Physical Biochemistry-At CARB a variety of physical biochemistry techniques are used to study the properties of proteins in solutions in order to relate the structural aspects of macromolecules determined by X-ray crystallography or NMR and the functions evident from molecular biology and biochemical studies. These techniques include a number of spectroscopic probes such as fluorescence and absorption spectroscopy, ultracentrifugation studies and stopped-flow kinetics. The thermal stability of proteins is analyzed with biothermodynamic studies utilizing differential scanning calorimetry and titration calorimetry. The thermodynamics of a biological macromolecule in equilibrium with its "functional" state is described in terms of changes in the state functions, the Gibbs Free Energy (ΔG), the Enthalpy (ΔH), and the Entropy (ΔS). The simplest method of measuring the transition between the functional state and the nonfunctional state of an enzyme is by heating the protein in a differential scanning calorimeter (DSC). The Hart 7707 DSC heat conduction scanning microcalorimeter in the Biothermodynamics laboratory at CARB utilizes cell volumes of 1 ml and is capable of scanning from 5 to 60 K/hr.

Titration calorimetry measures the heat absorbed or released upon binding to a protein of a ligand which can act as a co-factor or inhibitor of the protein. Aliquots of the ligand solution are injected into the protein solution isothermally and the amount of heat absorbed or released is measured by a thermocouple device. The equilibrium constant and, thus, ΔG can be calculated from the rate of the change of the heat measurements upon successive injections of the ligand solution. In one experiment, ΔG, ΔH, and, thus, ΔS can be determined for the binding reaction at a given temperature. Such information can be used to determine whether a binding reaction on the protein is enthalpically or entropically driven. The titration calorimeter in the Biothermodynamics laboratory at CARB is a Microcal Omega titration calorimeter which can be operated isothermally from 10 to 80°C.

Protein Crystallography-Protein engineering requires detailed information about the atomic structure of the macromolecule. At CARB the three-dimensional structures of proteins are determined once crystals and diffraction data are obtained by standard computational procedures such as multiple or single isomorphous replacement or molecular replacement techniques. Protein crystallogra-phy also provides a powerful tool with which to measure the structural changes of the protein resulting from amino acid substitutions. The X-ray structure determination of genetically altered molecules is quite straightforward, providing the crystals are isomorphous (virtually identical) to those of the native protein and the conformation of the macromolecule has not been dramatically changed. When the wild-type structure is available, difference Fourier analysis followed by model building or molecular replacement methods are used once X-ray data have been collected.

The CARB crystallography laboratory includes complete wet lab facilities for protein purification and crystal growth. For X-ray data collection there are two X-ray sources, a Rigaku RU200 rotating anode X-ray generator and a Phillips XRG100 X-ray generator. Data acquisition is done with three Siemens electronic area detectors, two mounted on Supper 3-axis cameras, one mounted on an Huber 4 circle goniostat. Each instrument is driven by a PCS computer with a modified UNIX operating system. One of the data collection stations is equipped with a modified Enraf-Nonius cryostat for low temperature studies. Interactive graphics and computations are performed on various graphics systems, workstations, and computers which are mentioned below.

Computational Chemistry and Modeling-Computer-assisted molecular design and computational chemistry play a prominent role in the research at CARB. Experimentalists as well as modelers and theoreticians use the available computational facilities for protein structure refinement, molecular dynamics, molecular modeling and quantum chemistry to guide experimental development of altered protein structures and therapeutic agents. The computational facilities are centered about a fully configured Alliant FX/80. Visual analysis of the computational results is performed with a number of color graphics workstations, including three Evans & Sutherland PS390 systems and 10 Silicon Graphics Iris systems. All computer systems are networked via Ethernet, including 11 MicroVAX II minicomputers, a variety of PC's and other special purpose minicomputers used for experimental data acquisition. The CARB network is connected to the main NIST broadband network by a 56 kB telephone link. This provides access to the NIST computing facility as well as a gateway to national and international networks. The molecular modeling programs implemented on these computer systems range from commercially available packages such as Polygen's QUANTA to in-house programs being developed as part of research efforts in quantum chemistry, molecular dynamics, and protein modeling and analysis. Several molecular dynamics packages such as AMBER, GROMOS and CHARMm and standard quantum chemistry programs such as HONDO, GAMESS and GAUSSIAN-90 are also being used.

The long range goal of designing proteins for specific functions requires a complete understanding of all of the processes protein molecules can undergo, from the incorporation of single amino acid changes to the folding of the complete polypeptide chain into a biologically active conformation. Although much development work remains to be done, techniques in the field of molecular modeling have now reached the point where useful results can be obtained in many areas of protein biochemistry, guiding the choice of the most appropriate experiments, or helping interpret their outcome. The emphasis at CARB is on further development of modeling algorithms, together with applications to practical problems posed by ongoing experimental work.

Protein Engineering Research Studies

Hyperexpression of Gene Products in _Escherichia coli_-The ability to express any gene at high levels to produce large quantities of protein in bacteria using recombinant DNA methods is one of the biotechnology industry's core technologies. However, not all genes can be cloned in expression systems for reasons that until now have been unclear. We have recently demonstrated that current high expression systems fail to work with genes that produce protein products that are lethal to bacterial cells because of "leaky" vector transcription.

Working with the gene that codes for the enzyme adenylate cyclase, it was demonstrated that increasing the level of enzyme in the cell from 0.004% to 0.20% of total protein was lethal to the cell (Reddy et al., 1985). The lethality problem could be overcome in the specific case of adenylate cyclase by working in a modified bacterial cell that had a defective receptor protein for cyclic AMP, the product of adenylate cyclase catalysis. To solve the problem presented by other lethal gene products we used DNA technology to engineer a new generation of expression vectors that block "leaky" transcription (Reddy et al., 1989). Using this system it was demonstrated that adenylate cylclase could be expressed in normal bacterial cells at very high levels (up to 30% of total cell protein). This means that the amount of this protein once obtained from a 1,000 liter fermentor full of cells can now be produced by one liter of cells. This technology has been used to express a variety of procaryotic and eucaryotic gene products to very high levels, typically 10% to 50% of total cell protein.

Protein Engineering of Threonine Deaminase, an Allosteric Enzyme-A current and important goal of protein engineering is to incorporate new attributes into enzymes to enable them to switch into active or inactive forms on specific signals. One approach to this complex problem is to elucidate in detail the interactions between polypeptides and ligands that control the activity of allosteric enzymes that regulate metabolism. Biosynthetic threonine deaminase from Escherichia coli provides a discrete model for the study of the molecular mechanisms that regulate catalysis. Cloning of ilvA, the structural gene for threonine deaminase, high-level expression, and the development of a rapid and efficient protocol for purification of gram quantities of enzyme have made possible the growth of large single crystals suitable for the X-ray diffraction studies required to elucidate the unique structural features of threonine deaminase responsible for cooperativity and feedback inhibition. The sigmoidal dependence of wild-type enzyme activity on L-threonine concentration can be attributed to an interconversion between low-activity and high-activity conformations of the enzyme (see Fig. 1), and is described by a Hill coefficient, n_H, of 2.3 and a $K_{0.5}$ of 8.0 mM. The negative allosteric effector L-isoleucine strongly inhibits the enzyme, yielding a value for n_H of 3.9 and $K_{0.5}$ of 74 mM, whereas enzyme activity is maximally increased by L-valine, which yields hyperbolic kinetics characterized by a value for n_H of 1.0 and a $K_{0.5}$ of 5.7 Mm. Thus, these effectors shift almost completely the equilibrium between the more active and less active forms of the tetrameric enzyme. Classical

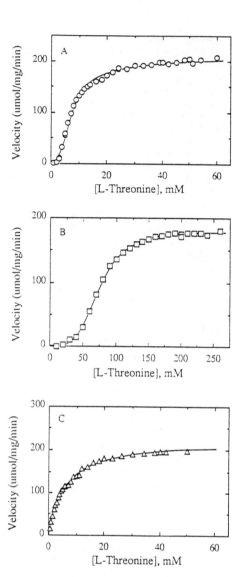

Fig. 1. Effect of feedback modifies on the steady-state kinetics of wild-type threonine deaminase. Assays were performed in 0.05 M potassium phosphate, pH 7.5 at 20°C. Formation of the product, 2-ketobutyrate, was followed continuously in a Shimadzu UV265 recording spectrophotometer using an extinction coefficient of 540 $M^{-1}cm^{-1}$. Reactions were initiated by the addition of threonine deaminase to a final concentration of 0.6ug/ml. (A) No effectors present; (B) 0.05 mM isoleucine; (C) 0.5 mM valine.

genetic selections have been developed to identify amino acid residues involved in catalysis and feedback inhibition. Thus far, lysine at position 62 has been identified as the active site residue that forms a Schiff base with the essential cofactor pyridoxal phosphate. A reduction of enzyme activity by at least a factor of 10^5 occurs when lysine 62 is replaced by alanine. Many other single site replacements have been found that reduce activity only slightly relative to the wild-type enzyme. This information contributes to a structural map that will help define the domains responsible for the functional properties of the enzyme and will be valuable in the interpretation of the three-dimensional structure once it is in hand.

β-Lactamase-The production of β-lactamases in bacteria is the major mechanism by which bacteria resist β-lactam antibiotics. As a consequence, the effectiveness of this powerful class of therapeutic compounds has been diminishing rapidly since their introduction some sixty years ago. Understanding the principles of the catalytic mechanism and the factors that determine substrate specificity of β-lactamases is essential for finding practical solutions to the clinical problems imposed by these enzymes. The high resolution crystal structure of a β-lactamase from S. aureus PC1 has been determined (Herzberg and Moult, 1987) and refined (Herzberg, submitted). Analysis of the architecture of the active site, along with available biochemical information, indicates that the catalytic mechanism resembles that of the serine proteases, with some important differences. Mutants have been designed, based on the crystal structure, to test the proposed catalytic mechanism and to affect substrate specificity by altering either the binding affinities of various substrates or the catalytic rates along the hydrolysis pathway. The mutants are being prepared by our collaborator Dr. Andrew Coulson from the University of Edinburgh. They will be analyzed biochemically, and the interesting ones will be crystalized and their structure determined.

Troponin C-The contraction of skeletal muscle is regulated by calcium binding to troponin C (TnC). This dumbbell shaped protein binds two metal ions in each of its domains. Binding of calcium to the N-terminal domain is associated with a series of conformational changes which triggers muscle contraction. The high resolution crystal structure of turkey skeletal muscle TnC has been determined (Herzberg and James, 1985) and refined (Herzberg and James, 1988). It represents the protein in the relaxed state of muscle, when the regulatory domain is not occupied by calcium. Based on this structure we have proposed a model for TnC in the contracted state of muscle, consisting of conformational changes in the low-affinity calcium binding loops and relative shifts of some helices in the regulatory domain (Herzberg et al., 1987). We have identified two residues remote from the calcium-binding site whose replacement should affect the affinity to calcium and the development of muscle tension. In collaboration with Dr. Fernando Reinach (Univ. of Sao Paulo), a potential salt bridge was introduced across two of the helices, some 20 Å away from the calcium positions, reducing calcium affinity and necessitating an increase in calcium concentration for tension recovery in reconstructed muscle (Fujimori et al., 1990). Future mutants will probe the properties of TnC further and will be aimed at increasing calcium affinity and identifying the sites of interaction with other members of the muscle system.

The Phosphoenolpyruvate:Sugar Phosphotransferase System (PTS)-Bacterial PTS is responsible for the transport of sugar substrates across the cytoplasmic membrane. This multicomponent system, consisting of four enzymes (Enzyme I, Enzyme II, Enzyme III, and HPr), uses a protein phosphoryl transfer chain to mediate sugar transport. Two of the proteins, HPr and Enzyme III from _B. subtilis_ have been cloned and expressed (Reizer _et al_., 1989; Sutrina, Saier, and Reizer, in preparation), and crystals have been obtained in our laboratory. Mutants of HPr have been designed to mimic a phosphorylated state of the protein which plays a regulatory role in the PTS, and to replace the catalytic histidine by a residue that cannot be phosphorylated. These mutants have been crystallized as well, and determination of their structure is in progress(Kapadia _et al_., 1990). To aid with the crystallographic work, cysteine mutants have been prepared to facilitate heavy atom derivative work.

In addition to their function in sugar transport, the PTS proteins in _E. coli_, also modulate the activity of adenylate cyclase which synthesizes an important molecule cAMP. In order to understand the regulation of adenylate cyclase activity, it is necessary to reconstitute the system _in vitro_ with purified sugar transport proteins (Reddy, _et al_., 1985). The corresponding genes were cloned into pRE, an expression vector, and each of the proteins were overproduced to represent 30-40% of the total protein (Reddy, Fredd, and Peterkofsky, unpublished data). From such overproducing strains gram quantities of proteins were purified. Crystals of Enzyme III have been obtained (Kapadia, Reddy, and Herzberg,

unpublished data). We have also created site-directed mutants at the catalytic histidine of Enzyme III with a view to understanding the structure-function relationship with a particular reference to adenylate cyclase. All the site-directed mutants of Enzyme III created to date have been expressed at high levels.

Catabolite Activator Protein-Attempts are being made at CARB to elucidate the molecular mechanism involved in the activation of the cyclic AMP receptor protein (**crp**) by the small molecule ligand, CAMP. We have cloned, sequenced, engineered, hyperexpressed, and developed both _in vivo_ and _in vitro_ functional test systems for the crp gene product. The progress made at CARB during this past year has three components. First we developed a new generation of M13 vectors that have potential broad application in biotechnology. These vectors have drug resistance markers that permit one to work in a variety of cell lines and genetic backgrounds while maintaining the essential biochemical components of M13 based vectors for efficient gene modification and characterization (see Fig. 2). Second, we have overcome (in a specific case) one of the major problems in protein engineering, that of proper protein folding in the case of a nonfunctional amino acid substitution. Third, we have made and tested a number (54) of amino acid substitutions that demonstrate the important amino acid side chain contacts required for ligand recognition and provide evidence for a model to explain the CAMP activation mechanism.

Activation of a protein not only depends on its structural conformation but on the presence of other species in the solution which may induce its activity, a co-factor, or inhibit its activity, an inhib-

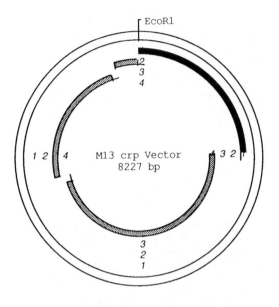

EcoR1

M13 crp Vector
8227 bp

1. Cyclic AMP receptor protein coding sequence
2. Pi promoter
3. Kanamycin resistance gene
4. M13 Vector

Fig. 2. An M13 drug resistant vector expressing the crp gene from the Pi promoter is shown.

itor. The binding of an inhibitor or co-factor can conformationally stabilize a protein, for example, the transition temperature and enthalpy increase upon binding of the inhibitor 3'-cytidine monophosphate to ribonuclease A (Schwarz, 1988). Binding of the co-factor cyclic adenosine monophosphate to **crp** induces transitions of two or more domains unfolding at higher temperatures as observed in this laboratory and by Sturtevant's laboratory (Ghosaini et al., 1988). The **crp** co-factor apparently activates the **crp** by

inducing changes in the interaction between domains. Additional insight into the effect of co-factors and inhibitors on protein function can be obtained from titration calorimetry studies.

A recent study (S. DeLauder, F.P. Schwarz and D. Atha, unpublished results) of the binding of polysaccaride inhibitor heparin to anti-thrombin, a protein in the blood responsible for clotting, shows that the heparin consists of tight binding and weak binding species, with the former being essentially entropically driven. Similar studies are underway with **crp** and should provide insight into the activation mechanism for **crp**.

Microcalorimeter studies of **crp** have also been initiated to study protein folding stability changes as a result of amino acid substitution. The substitution of threonine for serine at position 128 decreases the transition temperature from 66°C for the natural protein to about 54°C for the threonine substituted protein. Thus, the conformational stability is reduced by this one substitution.

Heme Enzyme Structures-The overall goal of this project is to determine the structure of new and interesting heme-containing enzymes. Currently our efforts are directed toward determining the crystal structure of ligninase. This is an enzyme capable of degrading lignin in a peroxide dependent reaction (for a review see Tien and Kirk, 1988). Lignin is a complex aromatic polymer that provides the coating around the cellulose of woody plants. Lignin is the second most abundant polymer in the biosphere and the most abundant renewable aromatic material. Therefore, the recovery of lignin is a major energy and environmental issue, and there is considerable interest in understanding

23

the molecular details of this process. The white rot fungi, Phanerochate chrysosporium, produces ligninase, and this enzyme is capable of degrading lignin to a state where microorganisms can utilize the breakdown products as energy sources. Ligninase has been crystallized at CARB in a form suitable for high resolution structural studies. A comparison of the ligninase structure with other heme enzyme structures which are under investigation at CARB will provide detailed information on the elements of structure control that function in these enzymes.

Heme Enzyme Protein Engineering-Recombinant DNA techniques are used to address questions in heme enzyme structure and function. Cytochrome c peroxidase or CCP is a well known enzyme that catalyzes the peroxide dependent oxidation of cytochrome c. The X-ray structure is known (Finzel et al., 1984) and at CARB the gene has been cloned and expressed in Escherichia coli. Site-directed mutants are characterized using a variety of biophysical techniques, for example, spectroscopic probes such as fluorescence and absorption spectroscopy, stopped-flow kinetics, and X-ray crystallography. One goal of the project is to utilize the information derived from newly determined X-ray structures as a guide to the redesign of substrate specificity in CCP and/or other heme enzymes that can be cloned and expressed. A long-term goal is to be able to produce heme enzymes with novel properties.

Cytochrome P-450-Cytochrome P-450's (P-450) are a group of enzymes that hydroxylate a wide range of aromatic and aliphatic compounds (Hayaishi, 1974). These enzymes are important in the degradation and elimination of toxic compounds and the pro-

duction of important metabolic intermediates like steroids. P-450's exhibit a wide range of specificities. On one extreme are the liver microsomal P-450s which catalyze the hydroxylation of a large number of aromatic and aliphatic compounds. These P-450s represent one of the primary means of removing toxic substances from the body. On the other extreme are mitochondrial and bacterial P-450s which are highly specific since these enzymes participate in the production of specific and important metabolic intermediates. Despite this range in specificities, P-450s share a common mechanism. Reducing equivalents ultimately derived from NADPH and NADH are transferred to P-450 via a reductase. The reduced P-450 heme activates molecular oxygen resulting in the cleavage of the oxygen O-O bond. The "activated" oxygen atom linked to the P-450 heme iron atom then is inserted into the substrate C-H bond.

At CARB X-ray crystallography is being used to understand the structural basis of P-450 substrate diversity and to provide a structural database for the rational design of P-450 inhibitors. Thus far these efforts have led to the solution of the first P-450 X-ray structure, P-450cam, from the bacterium Pseudomonas putida (Poulos et al., 1987). A series of inhibitor/substrate-enzyme complexes has revealed the structural basis for specificity and some aspects of the enzyme reaction (Raag and Poulos, 1989a; Raag and Poulos, 1989b; Raag et al., 1990; Raag and Poulos, 1990). Work currently is underway on new P-450 structures which exhibit different specificities.

Interleukin 1-Interleukin 1 (IL1) belongs to the family of cellular mediators known as cytokines. They are hormone--like proteins that mediate immunologic and inflammatory responses to infection or tissue damage. Two distinct species of IL1 have been characterized, (IL1α and IL1β), and both of them are expressed as 31 kDa precursors which are processed to give 17.5 kDa mature proteins. The current understanding of the biological and biochemical functions of IL1 has been recently reviewed (Dinarello et al., 1989; Oppenheim et al., 1986; di Giovine et al., 1990).

In order to understand the interactions of the IL1 molecule with its receptor, the mode of its operation, and also to develop an antagonist in order to control IL1-induced inflammation, the three-dimensional structure determination of IL1β and IL1α were initiated. Crystals of recombinant human IL1β (Gilliland, et al., 1987) were used to collect X-ray diffraction data and the Multiple Isomorphous Replacement method was employed to solve the three-dimensional structure at 2 Å resolution and refined to a crystallographic R factor of 19.0% (Veerapandian, Raag, Poulos, Svensson, and Gilliland, unpublished data). Four heavy atom derivatives were used and among them one was a mutant derivative where one cysteine was mutated to alanine. The single polypeptide chain with 153 residues is folded into a six stranded β-barrel, exhibiting approximately a 3-fold symmetry about the axis of the barrel. It looks like a tetrahedron with hydrophobic residues filling up its core. Similar results was also reported by Finzel et al.(1989).

In order to make this interleukin molecule an "immunosuppressant" or an "immunostimulant" the functional part of the molecule needs to be identified. Comparison of the amino acid sequences of IL1α and IL1β of various species led to the idea that these two molecules might have similar structures but bind to the receptor

in different ways (Priestle et al., 1989). Site-directed mutagenesis of IL1 has been used to probe the molecule for residues that may be important for activity(McDonald et al., 1986; Gronenborg et al., 1986; Kamogashira et al., 1988). The structural information thus obtained along with biochemical and biophysical data will play a major role in understanding the IL1 function.

At CARB, in addition to the crystal structure of the mature IL1β, crystal structures of three mutants of IL1β (supplied by Otsuka Pharmaceutical Company Ltd.,Japan) have been determined and refined to reasonable R factors (19.0% for wild-type IL1β and 20.0% for the mutant IL1β structures; Veerapandian, Raag, Poulos, and Gilliland, unpublished data). Knowledge of these structures and their comparisons with the wild-type protein in conjunction with the analysis of spectroscopic and other biochemical data, will assist in understanding the structure/function relationships of this protein. Such studies will play a crucial role in the drug design for immune therapy.

Chymosin-Bovine chymosin is the primary enzyme used in cheese production. The structure of recombinant bovine chymosin, which was cloned and expressed in Escherichia coli by Genex Corporation scientists, has been determined at CARB using X-ray data extending to 2.3 Å resolution (Gilliland et al., 1990). A comparison of recombinant chymosin with other acid proteases reveals the high degree of structural similarity with other members of this family of proteins as well as the subtle differences which make chymosin unique. In particular, Tyr77 of the flap region of chymosin does not hydrogen bond to Trp42 but rather protrudes out into the P1 pocket forming hydrophobic interactions with Phe119 and Leu32.

This may have important implications concerning the mechanism of substrate binding and
substrate specificity.

The three-dimensional
structure of chymosin (shown in
Fig. 3) has provided new insight into the molecular properties which make chymosin a
good enzyme for cheese production. For example, near the
N-terminus of the polypeptide
chain, six positive charges are
located within residues 48 to
62 forming a positive patch on
the surface of the molecule.
This sequence of positive residues is not found in other acid
protease molecules. This patch
of positive residues effectively polarizes the chymosin molecule (J. Moult, M. Toner, D.
Bacon, and G. Gilliland, unpublished data). This polarization may play a role in determining the efficiency of chymosin cleavage of the Phe105-
Met106 peptide bond of κ-casein. The polarization of the
molecule may predispose the
correct orientation of chymosin
as it nears the substrate; this
may be particularly important
if the micelle surface to which

κ-casein is associated is negatively charged. Alternatively
the patch of positive residues
may interact favorably with
other components of the casein
micelles to facilitate attacking the preferred peptide bond
of κ-casein.

The results of the struc-
ture determination and comparison studies in conjunction with
the electrostatic calculations
provide a basis for making in-
telligent changes in homologous
proteins which may convert microbial acid proteases into
more chymosin-like enzymes.

Subtilisin BPN'-The ability to
engineer more stable proteins
should broaden their utility
for many industrial and therapeutic purposes (Bryan,1987;
Chen, et al., 1990). Because
most proteins are not optimized
for stability over the course
of evolution, the engineering
of substantial increases in
stability is often achievable
with relatively minor modifications in a starting structure.

Over 20 stabilizing amino
acid substitutions in subtilisin BPN' have been identified

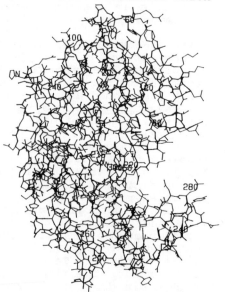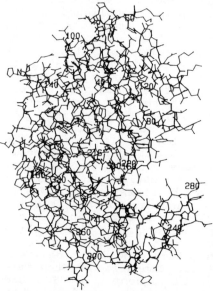

Fig. 3. A stereosopic view of
all 2511 non-hydrogen atoms of
the recombinant chymosin molecule. The first (1), last

(323), the active site
aspartates (34 and 216) and
every twentieth residue are
labeled.

using six different approaches: 1) random mutagenesis (Bryan, et al., 1988; Rollence, et al., 1988); 2) introduction of disulfide cross-links (Pantoliano, et al., 1987); 3) design of improved electrostatic interactions at calcium ion binding sites (Pantoliano, et al., 1988); 4) design of buried hydrophobic side chains; 5) sequence homology; 6) serendipity (Pantoliano, et al., 1989). A combination variant enzyme, containing six of these amino acid substitutions, was created using in vitro mutagenesis and characterized by differential scanning calorimetry and X-ray crystallography. Individually the six amino acid substitutions increase the free energy of unfolding between 0.3 and 1.3 Kcal/mol. In the combination variant, changes in free energy appear to accrue in an independent and additive manner resulting in a net increase of 4.3 Kcal/mol in the free energy of unfolding. Therefore, an extremely stable version of subtilisin has been created in a step by step manner. Thermodynamic stability of subtilisin was shown to be related to resistance to irreversible inactivation at high temperature (65°C) and high alkalinity (pH 12.0). Under these conditions, the combination variant subtilisin is inactivated ~300 times slower than the wild-type subtilisin BPN'.

X-ray crystallographic analysis of several combination mutants reveals that conformational changes associated with each mutation tend to be highly localized with minimal distortion of the backbone structure. Thus, very large increases in stability can be achieved with no radical changes in the tertiary protein structure but rather minor independent alterations. Contributions to the free energy of stabilization appear to be gained in many different ways including improved hydrogen bonding, van der Waal and hydrophobic inter-

actions in the folded form and decreased chain entropy of the unfolded enzyme (Pantoliano, et al., 1988).

Molecular Modeling Research

Protein Folding: Systematic Conformational Search-Understanding protein folding is the greatest challenge facing modelers of protein structure. Established simulation techniques, most notably molecular dynamics, have taught us a great deal about protein flexibility, but they are largely ineffective for determining structure, because of the long time scales involved in structural changes. About ten orders of magnitude more computer time than is now available on supercomputers would be needed to fold a protein via molecular dynamics simulation. Consequently, we are concentrating on other methods for predicting structure such as systematic conformational search (SCS). The SCS approach generates and evaluates a large number of possible protein conformations, using discriminatory functions to choose the most accurate ones. The number of conformations is restricted by applying rules that are observed to hold for known structures, such as allowed dihedral angle values and favorable electrostatic environments. So far, the method has proved effective in homologous modeling situations, and work is now concentrated on extending its scope (Moult and James, 1986).

Protein Folding: Homologous Modeling-It is possible to predict realistic structures of proteins with sequences related to molecules of known structure. The rapidly growing number of protein structures determined by protein crystallography has greatly increased the incidence of new sequences found to be homologous to those of one or more proteins of known structure. When the degree of sequence identity is high (at present better than about 30%), the techniques of computer modeling may be used to construct a usefully realistic picture of the new protein (Moult, 1989). Homologous modeling has proved particularly useful in the field of drug design, where any structural information about the intended receptor can be critical in guiding the search for active compounds. Current work in this area is focused on improving algorithms for the production of more accurate models (reliable sequence alignment in particular), identification of conserved structural features and the building of new regions of structure using the SCS technique.

Design of Inhibitors and Site-Directed Mutants of Proteins-Practical applications of molecular modeling techniques include the design of novel enzyme inhibitors either as tests of theories of enzyme mechanisms or as potential drugs. Semi-quantitative predictions of the behavior of such inhibitors is already possible, and methods are under development that will make possible the calculation of relative binding constants of enzyme-inhibitor complexes. These methods are currently being used to develop inhibitors for the enzyme β-lactamase, which is the primary bacterial defense mechanism against β-lactam antibiotics (Herzberg and Moult, 1987).

It is also possible to predict the effect of single amino acid replacements on protein structure, thus guiding the application of site-directed mutagenesis to resolve questions of function, stability, folding, and intermolecular interactions. Such predictions are made with the aid of computer graphics, energy minimization, molecular dynamics, and systematic conformational search. Systems currently un-

der study at CARB include β-lactamase and troponin C (Fujimori et al.,1990) described above.

Applications of Quantum Chemistry-In addition to modeling of molecular structures, methods are under development at CARB which will allow the application of computational chemistry and quantum mechanical methods to the study protein structure, electronic properties, and enzyme mechanisms. Such calculations are extremely computationally intensive and have been restricted historically to the study of small molecules and simple reactions. Applications to biochemical problems are possible, however, through the selection of proper prototypes. This approach was used recently to deduce the mechanism of calcium ion selectivity in Ca-binding proteins by analyzing the energetics of small clusters of ligands surrounding Ca, Mg, and Na ions (Krauss and Stevens, 1990). It was found that the selectivity is not a result of the protein binding calcium more tightly, but rather that the bulky protein groups are not as effective as water in binding the smaller Mg and Na ions, and therefore cannot pull those ions out of solution.

While prototypes can be useful, the application of computational chemistry to more realistic models is imperative in order to understand how a protein environment modifies the energetics and pathways of chemical reactions at the active site. A new method is under development at CARB which will allow structures and reactions in the active site to be studied quantum mechanically while taking into account the influence of the surrounding protein molecule. The quantum mechanical Hamiltonian of the molecular fragment under investigation is modified to include electrostatic, polarization, and electron exchange interactions with nearby "spectator" fragments which influence the chemistry but are not chemically active themselves. The structure of the surrounding protein is taken from crystallographic analysis or modeling results, and the interaction with the chemically active fragments and substrates is based on transferrable functional group properties (Garmer and Stevens, 1989). Current applications of this new method include a study of the mechanism of action of the enzyme carbonic anhydrase and an analysis of the mechanism and specificity of the aspartyl proteinase family of enzymes.

Concluding Remarks

As an interagency, inter-institutional, multidisciplinary research resource in biotechnology, CARB has demonstrated success in blending molecular biology, protein chemistry, thermodynamics, X-ray crystallography, macromolecular modeling, and quantum chemistry to accelerate the science of protein engineering.

Acknowledgments

Partial support was provided by NIH R01 GM41034 and R01 LM05102 to J. Moult, by NIH R01 AI27175 to O. Herzberg, and by NIH R01 GM33688, NIH R01 GM42614, and NSF DMB8716316 to T. Poulos.

References

Bryan, P.N., Biotechnology Advances, **1987**, _5_, 221.

Bryan, P.N., Rollence, M.L., Pantoliano, M.L., Wood, J.F., Finzel, B.C., Gilliland, G.L., Howard, A.J., Poulos, T.L., _Proteins: Struc. Func. Gen._, **1986**, _1_, 326-334.

Chen, S.-T., Hennen, W.J., Bibbs, J.A., Wang, Y.-F., Liu, J.L.-C., Wong, C.-H., Pantoliano, M.W., Whitlow,

M., Bryan, P.N., _J. Am. Chem. Soc._, **1990**, _112_, 945-953.

di Giovine, F.S. and Duff, G.W., _Immunology Today_, **Jan.**, **1990**, 13-20.

Dinarello, C.A., Savage, N., _CRC Crit. Rev. Immunol._, **1989**, _9_, 1-20.

Finzel, B.C., Poulos, T.L., Kraut, J., _J. Biol. Chem._, **1984**, _259_, 13027-13036.

Finzel, B.C., Clancy L.L., Holland, D.R., Muchmore S.W., Watenpaugh K.D., Einspahr, H.M., _J. Mol. Biol._, **1989**, _209_, 779-791.

Fujimori, K., Sorenson, M., Herzberg, O., Moult, J., Reinach, F.C., _Nature_, **1990**, _345_, 182-184.

Garmer, D.R., Stevens, W.J., _J. Phys. Chem._, **1989** _93_, 8263-8-270.

Ghosaini, L.R., Brown, A.M., and Sturtevant, J.M., _Biochemistry_, **1988**, _27_, 5257-5-261.

Gilliland, G.L., Winborne, E.L., Masui, Y., Hirai, Y., _J. Biol. Chem._, **1987**, _262_, 12323-12324.

Gilliland, G.L., Winborne, E.L., Nachman, J., Wlodawer, A., _Proteins: Struc. Func. Gen._, **1990**, _8_, 82-101.

Gronenborn, A. M., Clore, G. M., Schmeissner, U., Wingfield, P., _Eur. J. Biochem._, **1986**, _161_, 37-43.

Hayaishi, O., _Molecular Mechanisms of Oxygen Activation_, Academic Press, New York, NY, 1974.

Herzberg, O., James, M.N.G., _Nature_, **1985**, _313_, 653.

Herzberg, O., James, M.N.G., _J. Mol. Biol._, **1988**, _203_, 761.

Herzberg, O., Moult, J., _Science_, **1987**, _236_, 694-701.

Kamogashira, T., Masui, Y., Ohmoto, Y., Hirato, T., Nagamura, K., Mizuno, K., Hong, Y.-M., Kikumoto, Y., Nakai, S., Hirai, Y., _Biochem. Biophys. Res. Commun._, **1988**, _150_, 1106-1114.

Kapadia, G., Reizer, J., Sutrina, S., Saier, M., Reddy, P., Herzberg, O., _J. Mol. Biol._, **1990**, _212_, 1-2.

Krauss, M., Stevens, W.J., _J._

Am. Chem. Soc., **1990**, <u>112</u>, 1460-1466.

McDonald, H.R., Wingfield, P., Schmeissner, U., Shaw, A., Clore, G.M., Gronenborn, A.M., <u>FEBS Lett.</u>, **1986**, <u>209</u>, 295-298.

Moult, J., <u>J. Res. NIS</u>, **1989**, <u>94</u>, 79-84.

Moult, J., James, M.N.G., <u>Proteins: Struc. Func. Gen.</u>, **1986**, <u>1</u>, 146-163.

Oppenheim, J.J., Kovacs, E.J., Matsushima, K., Durum, S.K., <u>Immunol. Today</u>, **1986**, <u>7</u>, 45-56.

Pantoliano, M.W., Ladner, R.C., Bryan, P.N., Rollence, M.L., Wood, J.F., Poulos, T.L., <u>Biochemistry</u>, **1987**, <u>26</u>, 2077-2082.

Pantoliano, M.W., Whitlow, M., Wood, J.F., Finzel, B.C., Gilliland, G.L., Poulos, T.L., Rollence, M.L., Bryan, P.N., <u>Biochemistry</u>, **1988**, <u>27</u>, 8311-8317.

Pantoliano, M.W., Whitlow, M., Wood, J.F., Dodd, S.W., Hardman, K.D., Rollence, M.L., Bryan, P.N., <u>Biochemistry</u>, **1989**, <u>28</u>, 7205-7213.

Poulos, T.L., Finzel, B.C., Howard, A.J., <u>J. Mol. Biol.</u>, **1987**, <u>195</u>, 687-700.

Priestle, J.P., Schar, H.P., Grutter, M.G., <u>Proc. Natl. Acad. Sci. USA</u>, **1989**, <u>86</u>, 9667-9671.

Raag, R., Poulos, T.L., <u>Biochemistry</u>, **1989a**, <u>28</u>, 917-922.

Raag, R., Poulos, T.L., <u>Biochemistry</u>, **1989b**, <u>28</u>, 7586-7592.

Raag, R., Poulos, T.L., In <u>Frontiers in Biotransformations</u>; Ruckpaul, K. Ed.; AKademie-Verlag, Berlin, 1990; Vol. 8.

Raag, R., Swanson, B., Poulos, T.L., Ortiz de Montellano, P.R., <u>Biochemistry</u>, **1990**, in press.

Reddy, P., Peterkofsky, A., McKenney, K., <u>Proc. Natl. Acad. Sci. USA</u>, **1985a**, <u>82</u>, 5656-5660.

Reddy, P., Roseman, S., Peterkofsky, A., <u>Proc. Natl. Acad. Sci. USA</u>, **1985b**, <u>82</u>, 8300-8304.

Reddy, P., Peterkofsky, A., McKenney, K., <u>Nucleic Acid Research</u>, **1989**, <u>17</u>, 10473-10488.

Rollence, M.L., Filpula, D., Pantoliano, M.W., Bryan, P.N., <u>Critical Reviews in Biotechnology</u>, **1988**, <u>8</u>, 217-224.

Schwarz, F.P., <u>Biochemistry</u>, **1988**, <u>27</u>, 8429-8436.

Tien, M., Kirk, T.K., <u>Methods in Enzymology</u>, **1988**, <u>161</u>, 238-249.

Cell-Free Protein Synthesis Bioreactor

Alexander S. Spirin, Institute of Protein Research, Academy of Sciences of the USSR, Pushchino, Moscow Region, USSR

In vivo expression of foreign or synthetic genes can be subject to certain restrictions such as protein aggregation, degradation and toxicity. Conventional in vitro systems can overcome these problems, but in turn suffer from other limitations, in particular short life-time and low protein yield. In this review, two types of gene expression system are described. Both are based on the novel concept of enhanced expression from cell-free extracts where incubation is performed in the continuous flow of a feeding solution, rather than in a fixed volume of a test-tube. The first makes use of cell-free translation of mRNA templates in either prokaryotic or eukaryotic cell lysates. The second utilizes coupled transcription-translation of DNA templates, with genes transcribed by either endogenous or bacteriophage RNA polymerases. In both systems translation can be carried out over tens or hundreds of hours resulting in high protein yields.

Protein synthesis is a basic feature of all living beings. It is a highly complex process comprising interrelated events of nucleic acid transcription, its regulation, post-transcriptional alterations, nucleoprotein assembly, nucleic acid and nucleoprotein transport, mRNA translation, translational regulation, co- and post- translational polypeptide modifications, co- and post-translational protein folding, assembly, transport and secretion.

Translation is at the heart of this process and is often considered protein synthesis proper. It is a finely regulated multi-step sequence of events, performed by ribosomes moving along a mRNA chain. Many ribosomes moving along the same mRNA form a polyribosome. Ribosomes are specialized ribonucleoprotein particles with binding, catalytic and mechanical functions. Each translation cycle of the ribosome results in the synthesis of one polypeptide and involves three successive stages: initiation, elongation and termination.

It is remarkable that such a complex process can be entirely reproduced after disruption of a living cell, i.e. in a cell homogenate or extract. The first cell-free systems of protein synthesis were developed in the 1950's. Experiments in the 1960's demonstrated that exogenous template polynucleotides, added either directly as polyribonucleotide messengers (Nirenberg and Matthaei, 1961) or as polydeoxyribonucleotide genes (Lederman and Zubay, 1967; De Vries and Zubay, 1967) could be expressed into polypeptide and protein products in cell-free translation or coupled transcription-translation systems, respectively. Since then cell-free systems have played the main role in unravelling molecular mechanisms of protein synthesis on ribosomes.

The 1970's and 80's witnessed the development of recombinant DNA technology and methods for in vivo expression of foreign and synthetic genes leading to the so-called Biotechnology revolution. Today, biologically active polypeptides and proteins of different origin, composition and properties can be produced from cell cultures harboring the appropriate cloned genes.

However, in vivo expression of foreign and synthetic genes inevitably encounters certain limitations. Heterologous genes may be unstable or poorly expressed because of host regulatory mechanisms. Many gene products are insoluble and aggregate as inclusion bodies. Other proteins are unstable and are degraded by intracellular proteases. Finally, some gene products are toxic to the host cell and therefore are not expressed.

In principle these and other problems of gene expression could be solved by the use of cell-free systems. Absence of cellular control mechanisms, ability to manipulate incubation mixture composition, i.e. introduction of protease and nuclease inhibitors, selective removal of undesired proteins and substances, as well as purification of protein-synthesizing components make cell-free systems look attractive. However, short life-time and low protein yield have been the major limitations preventing use of classical cell-free systems for preparative gene expression.

Recently it has been demonstrated that continuous translation over long periods resulting in high protein yields can be attained in cell-free systems when the incubation is performed not under static conditions in a fixed volume but in the flow of a feeding solution through the cell-free extract (Spirin et al., 1988; Ryabova et al., 1989; Baranov et al., 1989).

Flow-through Bioreactors for Cell-Free Protein Synthesis

In our first experiments with translation of viral RNAs the standard Amicon 8 MC micro-ultrafiltration chamber was used as a bioreactor for establishing a continuous flow cell-free system (Spirin et al., 1988). A 1 ml incubation mixture in the working chamber of the thermostated instrument was supplied with a constant flow (1 ml/hour) of feeding solution for translation (amino acids, ATP, GTP and buffer), while the reaction products were continuously removed at the same rate through an ultrafiltration membrane (Fig. 1). Both feeding and incubation solutions were sterile. When MS2 bacteriophage RNA was translated in DNA- and RNA-free extracts of *Escherichia coli* (0.06 nmoles of MS2 RNA per 0.6 nmoles of ribosomes) the phage coat protein was synthesized linearly over 20 hours at 37°C and passed through a PM-30 membrane as the only detectable protein of the eluate (Fig. 2). The yield was 100 copies of the protein per MS2 RNA molecule, i.e.

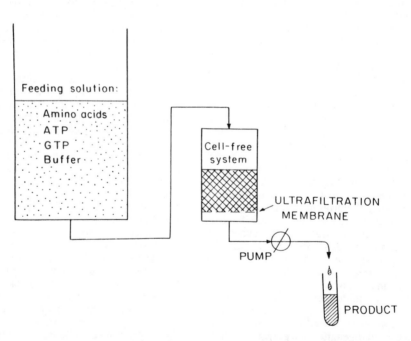

Fig. 1. Diagram of a bioreactor for translation based on direct flow-through ultrafiltration membrane.

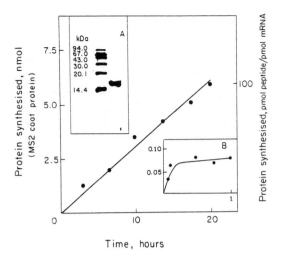

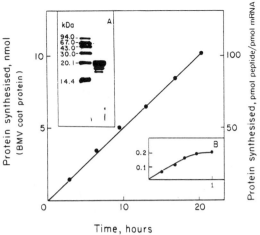

Fig. 2. Kinetics of MS2 coat protein synthesis in the *E.coli* continuous cell-free translation system. 0.06 nmol of MS2 RNA was added to a 1-ml reaction mixture containing 70S ribosomes, tRNA, S100 protein fraction and protease and ribonuclease inhibitors. The feeding solution containing ATP, GTP and amino acids was passed through the reactor at a constant rate of 1 ml/hour, 37°C. *Inset A:* Electrophoretic pattern of translation products. Here and in Figs. 3-5 the left lane shows reference polypeptide bands. *Inset B:* Kinetics of protein synthesis in standard test-tube cell-free system of the same composition and volume. Here and in subsequent figures the axes are in the same units as on the main graph.

Fig. 3. Kinetics of BMV coat protein synthesis in the wheat germ continuous cell-free translation system. 0.1 nmol of BMV RNA was added to a 1-ml reaction mixture containing S30 extract from wheat embryos and protease and ribonuclease inhibitors. The feeding solution containing ATP, GTP and amino acids was passed through the reactor at a constant rate of 1 ml/hour, 27°C. *Inset A:* Electrophoretic pattern of translation products. *Inset B:* Kinetics of protein synthesis in standard test-tube cell-free system of the same composition and volume.

ca 6 nmoles or 0.1 mg from 1 ml incubation mixture. In the case of brome mosaic virus (BMV) RNA translation in wheat embryo extract at 27°C synthesis again continued for up to 20 hours; it was linear and yielded *ca* 10 nmoles or 0.2 mg of the BMV coat protein. Product was visualized as a predominant 20 kDa polypeptide in the filtrate which had passed through an XM-50 membrane; two additional bands, probably corresponding to abortive peptides, were also present (Fig. 3).

The same ultrafiltration-based bioreactor was used for the synthesis of unmodified calcitonin polypeptide, both in the prokaryotic (*E. coli*) and eukaryotic (wheat) cell-free systems (Spirin et al., 1988). mRNAs were pre-

synthesized by transcription from synthetic Val 8-calcitonin genes, with and without the Shine-Dalgarno sequence, using bacteriophage SP6 RNA polymerase. Polypeptide synthesis was recorded for up to 40 hours of incubation at either 37°C (*E. coli* lysate) (Fig. 4) or 27°C (wheat embryo extract) (Fig. 5). Yields were about 300 copies of the polypeptide per mRNA molecule (18 nmoles or 60 μg of the product) and about 150 copies of the polypeptide per mRNA molecules (9 nmoles or 30 μg of the product), respectively. The product was the only detectable polypeptide passed through the PM-10 membrane used in the experiment, although some dimerisation was observed in the wheat germ system.

An important consideration concerns the functional activity of proteins synthesized in such systems. To test functional activity, a 0.5 ml S30 wheat germ extract containing 0.1 nmoles of

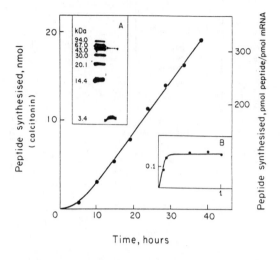

Fig. 4. Kinetics of calcitonin synthesis in the *E.coli* continuous cell-free translation system. 0.06 nmol of calcitonin mRNA was added to a 1-ml reaction mixture containing 70S ribosomes, tRNA, S100 protein fraction and protease and ribonuclease inhibitors. The feeding solution containing ATP, GTP and amino acids was passed through the reactor at a constant rate of 1 ml/hour, 37°C. *Inset A:* Electrophoretic pattern of translation products. *Inset B:* Kinetics of protein synthesis in standard cellfree systems of the same composition and volume.

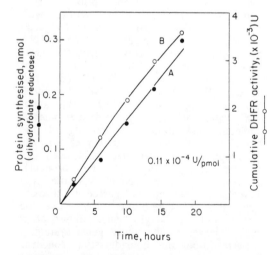

Fig. 5. Kinetics of calcitonin synthesis in the wheat germ continuous cell-free translation system. 0.06 nmol

of calcitonin mRNA was added to a 1-ml reaction mixture containing S30 extract from wheat embryos and protease and ribonuclease inhibitors. The feeding solution containing ATP, GTP and amino acids was passed through the reactor at a constant rate of 1 ml/hour, 27°C. *Inset A:* Electrophoretic pattern of the translation products. *Inset B:* Kinetics of protein synthesis in standard cellfree system of the same composition and volume.

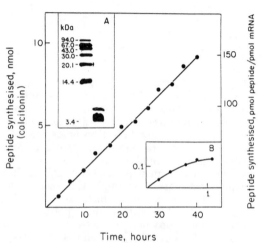

Fig. 6. Kinetics of dihydrofolate reductase synthesis in the wheat germ continuous cell-free translation system. 0.1 nmol of *E.coli* dihydrofolate reductase mRNA was added to a 0.5-ml reaction mixture containing S30 extract from wheat embryos and protease and ribonuclease inhibitors. The feeding solution containing ATP, GTP and amino acids was passed through the reactor at a constant rate of 1 ml/hour, 24°C. A: Protein synthesis. B: DHFR activity of the product.

dihydrofolate reductase (DHFR) mRNA was incubated at 24°C for 18h in the flow (1ml/hour) of feeding solution. In parallel with monitoring protein synthesis, its activity was continuously measured (Fig.6). Throughout the whole period of synthesis, the protein was found to possess functional activity.

More recently, an automated large-scale bioreactor (100 ml of incubation mixture) for cell-free translation based on the principle of ultrafiltration through a membrane has been constructed and tested with the synthesis of interleukin 4 in the Institute of Protein Research, Pushchino (Yu.B.Alakhov, S.Yu.Ovodov and L.M.Vinokurov). However, instead of direct flow, a reverse pulse principle was utilized: feeding solution was first pumped under pressure into the reaction chamber through the membrane; the ultrafiltrate then was forced from the chamber through the same membrane by a reverse flow (Fig. 7). Such pulses prevented obstruction of membrane pores by protein aggregates. After 20 hours of translation the interleukin 4 of 85% purity was obtained giving a yield of 50 mg per liter.

A further development involves a hollow fiber bioreactor invented jointly by H.Bauer, Tübingen University, and Yu.B.Alakhov, S.Yu.Ovodov and V.I.Baranov, Institute of Protein Research, Pushchino. Here the feeding solution flows along an ultrafiltration hollow fiber coil within the chamber containing the incubation mixture. Exchange of the translation substrates and the products between incubation mixture and feeding solution takes place across the fiber membrane. A bioreactor of this type can also operate in a reverse pulse mode where the feeding solution in-flow under pressure alternates with the product ultrafiltrate out-flow.

One of the most promising types of bioreactors for cell-free protein synthesis utilizes a column technique. Here the incubation mixture is embedded in gel granules (e.g. alginate). Alternatively drops of the liquid incubation mixture can be encapsulated in polysaccharide vesicles (e.g. zosterin). In both cases the resulting beads are packed into a column, and feeding solution is passed through the beads (Fig. 8). Translation of synthetic calcitonin mRNA in wheat embryo extract was performed in a column reactor for 100 hours at $25\,^\circ$C (Fig. 9). The kinetics of the synthesis was shown to be strictly linear producing 260 copies of the polypeptide per mRNA molecule, i.e. 80

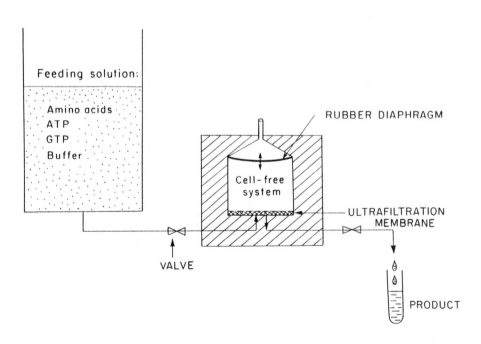

Fig. 7. Scheme of bioreactor for translation based on reverse pulse flow through ultrafiltration membrane.

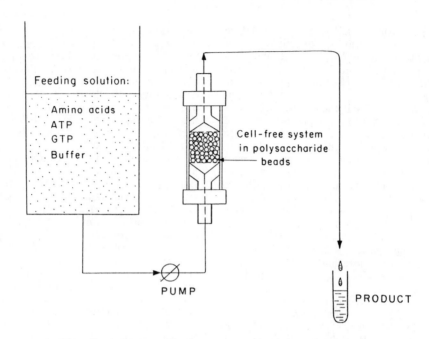

Fig. 8. Scheme of column bioreactor for translation.

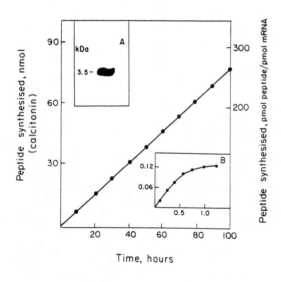

Fig. 9. Kinetics of calcitonin synthesis in the wheat germ continuous cell-free translation system with the use of polysaccharide encapsulated incubation mixture. 0.3 nmol of calcitonin mRNA was added to a 5-ml reaction mixture containing S30 extract from wheat embryos and protease and ribonuclease inhibitors. The mixture was encapsulated in zosterin beads of 0.1-0.2 mm in diameter. The beads were packed in a column. The feeding solution containing ATP, GTP and amino acids was passed through the column at a constant rate of 2.5 ml/hour, 25 °C. *Inset A:* Electrophoretic pattern of the translation products. *Inset B:* Kinetics of protein synthesis in standard cell-free system of the same composition and volume.

nmoles or 250 μg from 5 ml incubation mixture. Only one polypeptide band, corresponding to the molecular mass of calcitonin (3.5 kDa) was detected in the eluate.

Development and construction of an ATP-GTP-regeneration block and an immuno-adsorption concentration block are in progress. For all types of bioreactor special attention must be paid to the maintenance of sterility during incubation.

Characteristics and Peculiarities of the Flow-through Cell-Free Systems

Flow-through cell-free translation systems based on bacterial or plant cytoplasmic extracts have been described above. Fig. 10 demonstrates the application of the same flow principle for the translation of isolated globin mRNA in RNA-free rabbit reticulocyte extract (Ryabova et al., 1989). A 1 ml incubation mixture in Amicon 8 MC micro-ultrafiltration chamber with

XM-100 membrane was used at 30 $^\circ$C. By chance the rate of the feeding solution flow changed during the above experiment from 1 ml/hour to 2 ml/hour and the rate of globin synthesis unexpectedly more than doubled. Therefore, in the next experiment the flow rate was increased to 3 ml/hour from the beginning (Fig. 11). This resulted in maximum protein yield of 100 nmoles or 2 mg of product from 0.5 ml incubation mixture after 100 hours (Ryabova et al., 1989). Thus *flow rate is critical for maximizing yield of* cell-free protein synthesis.

Prolonged incubation of mRNA during enhanced cell-free translation was expected to lead to problems with mRNA

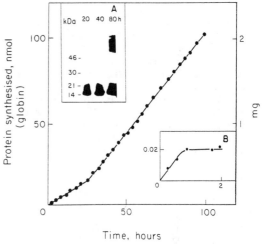

Fig. 11. Kinetics of globin synthesis in the rabbit reticulocyte continuous cell-free translation system. 3 μg of globin mRNA was added to a 0.5-ml reaction mixture containing micrococcal nuclease-treated rabbit reticulocyte lysate and protease and ribonuclease inhibitors. The feeding solution containing ATP, GTP and amino acids was passed through the membrane reactor at a constant rate of 3 ml/hour, 30 $^\circ$C. *Inset A:* Electrophoretic pattern of translation products at 20, 40 and 80 hours of incubation. *Inset B:* Kinetics of protein synthesis in standard cell-free system of the same composition and volume. Reproduced from Ryabova, L. A. et al. *Nucl. Acids Res.* **1989**, *17*, 4412 with permission.

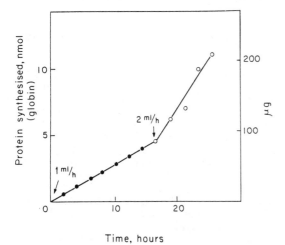

Fig. 10. Kinetics of globin synthesis in the rabbit reticulocyte continuous cell-free translation system. 15 μg of globin mRNA was added to a 1-ml reaction mixture containing micrococcal nuclease-treated rabbit reticulocyte lysate. The feeding solution containing ATP, GTP and amino acids was passed through the membrane reactor at a rate of 1 ml/hour for the first 16 hours and then at a rate of 2 ml/hour, 30 $^\circ$C.

degradation since RNAase inhibitors such as human placental ribonuclease inhibitor cannot completely protect mRNA. This is because of non-complete inhibition of RNAases, presence of different types of non-specific phosphodiesterase activities in the extracts, and the removal of the inhibitor by continuous flow during incubation. This apprehension seems to be the main psychological barrier preventing use of cell-free translation systems in reactions with long incubation times. Unexpectedly, *no mRNA degradation* was observed in any type of cell-free translation system during tens and even hundreds of hours at elevated temperatures. In all experiments limiting amounts of mRNA were present compared to other components of the protein-synthesizing system. This means that any degradation of mRNA would be immediately reflected in the decline of a kinetic curve. One plausible explanation is that the actively working protein-synthesizing machinery, or ribosomes themselves, protect the mRNA against degradation.

More surprising was the *absence of significant leakage* of translation factors and tRNAs through the ultrafiltration membrane during incubation. Initially, ultrafiltration membranes with pore sizes as small as possible were chosen to retard the leakage of the components of the protein-synthesizing machinery: PM-10 in the case of calcitonin polypeptide synthesis, or PM-30 in the case of MS2 coat protein synthesis. Eventually it was realized, however, that there was no need for this precaution. For example, when globin synthesis was performed using a high flow rate for 100 hours with an XM-100 membrane (Ryabova et al., 1989) no serious leakage of proteins involved in translation was recorded. Now YM-100 and XM-300 membranes are routinely used by our group in all types of system (bacterial, plant and animal origin) for the synthesis of various proteins with different molecular masses. Polysaccharide capsules may have even larger pores, but tRNAs and proteins required for translation are still retained. This leads us to speculate that the actively working translational machinery is organized in dynamic multi-component complexes, so that individual elements are not free for most of the time.

During the first few minutes of incubation many ballast proteins and other substances were eluted from the protein-synthesizing mixture. Subsequently, however, the outflow contained only or predominantly the product polypeptide. This results in a unique advantage of the flow-through cell-free system: the protein or polypeptide synthesized is found in an *almost pure state*, rather than in a complex cell homogenate. Hence there is no need for special purification procedures. This product homogeneity in the eluate is especially important when the properties of the expressed protein or polypeptide are unknown or modified, e.g., in the case of the expression of an unidentified open reading frame or at some steps of protein engineering.

Finally, the *long life-time* of flow-through cell-free protein-synthesizing systems is worth emphasizing. As a consequence, a *high yield* of proteins or polypeptides is achieved. This makes such cell-free systems highly promising for preparative syntheses of polypeptides and proteins.

Direct Gene Expression in Preparative Cell-Free Systems

Availability of specific mRNA is the limiting factor in preparative synthesis of a protein or a polypeptide in cell-free translation systems. Extrapolation from Figs. 2-5 shows that synthesis of 1 mg of protein over 100 hours requires approximately 20 μg of specific mRNA, assuming about 500 polypeptide copies are synthesized per mRNA molecule. Correspondingly, 20 mg of mRNA will be required to synthesize 1 g of a protein. Such amounts of mRNA can be synthesized in vitro by bacteriophage SP6 or T7 RNA polymerases. However, this approach is expensive.

An alternative approach makes use of coupled transcription-translation in the cell-free mixture. Bacterial extracts are known to contain DNA-dependent RNA polymerase necessary for transcription. Several versions of the coupled transcription-translation cell-free system based on crude bacterial extract (Lederman and Zubay, 1967; De Vries and Zubay, 1967) or on more or

38

less purified components of it (Gold and Schweiger, 1969; Schweiger and Gold, 1969; Kung et al., 1977; 1979) have been described. The system proved useful but was short-lived, and protein yield was low. However, application of the flow-through principle to coupled transcription-translation using endogenous RNA polymerase and exogenous DNA (genes) has increased the feasibility of using such systems (6). Fig. 12 demonstrates the DNA-directed synthesis of both β-lactamase (Bla) and dihydrofolate reductase (DHFR) in nucleic acid-free *E. coli* extract supplemented with plasmid pDF34 (pUD18) carrying the two corresponding genes (N.V.Murzina, A.T.Gudkov). The feeding solution containing all four nucleoside triphosphates and twenty amino acids was passed through the incubation chamber (1 ml volume, XM-100 membrane, in the Amicon 8 MC micro-ultrafiltrator). Protein synthesis at 37°C was monitored for 50 hours. Protein production was continuous and synthesis rate was directly dependent on the flow rate: the switch from 3 ml/hour to 2 ml/hour decreased the synthesis rate about two fold, and the subsequent switch from 2 ml/hour to 3 ml/hour restored the previous synthesis rate. Total protein yield was more than 0.2 mg. Only two bands, corresponding to β-lactamase and dihydrofolate reductase were detected in approximately equimolar amounts by electrophoretic analysis.

Hence, the coupled transcription-translation system described above retains the advantages of the flow-through cell-free translation systems, resulting in the production of preparative amounts of almost pure proteins. It uses DNA molecules directly as templates for mRNA transcription by endogenous RNA polymerase, thereby reducing the number of steps involved in in vitro gene expression. In addition, this system creates new possibilities for the study of transcriptional and translational regulation mechanisms, as well as of the coupling of transcription and translation.

However, there are several limitations in the use of the endogenous RNA polymerase of bacterial extracts. Most of them are caused by problems of transcriptional regulation, proper promoters, termination of transcription, etc. As a rule, a complete circular plasmid with the gene of interest (e.g., DHFR), a selection gene (e.g., Bla) and all necessary regulatory elements is used in coupled transcription-translation experiments. This means the selection gene is also transcribed and translated.

A simpler version of the coupled transcription-translation system can be produced utilizing bacteriophage SP6 or T7 RNA polymerases instead of endogenous cellular RNA polymerase. Fig. 13 shows expression in an *E. coli* extract with added SP6 polymerase of plasmid DNA specifying the DHFR gene under the control of the SP6 polymerase promoter and of the Bla gene under the control of the *E. coli* RNA polymerase promoter. Rifampicin was added to

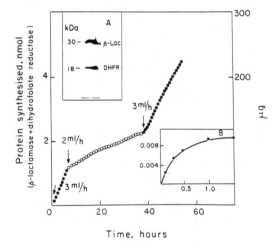

Fig. 12. Kinetics of expression of ß-lactamase and dihydrofolate reductase-encoding genes in the *E.coli* continuous cell-free system with endogeneous RNA polymerase. 150 μg of plasmid (pDF34) was added to a 1-ml reaction mixture containing *E.coli* S30 extract and additional tRNA. The feeding solution containing ATP, GTP CTP, UTP and amino acids was passed through the reactor at rates of 3 ml/hour and 2 ml/hour, 37°C. *Inset A:* Electrophoretic pattern of translation products. *Inset B:* Kinetics of protein synthesis in standard cell-free system of the same composition and volume. Reproduced from Baranov, V. I. et al. *Gene*, **1989**, *84*, 463 with permission.

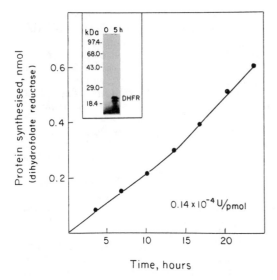

Fig. 13. Kinetics of expression of dihydrofolate reductase-encoding gene in the *E.coli* continuous cell-free system with bacteriophage SP6 RNA polymerase. 50 µg of plasmid (pSP65) with the *E.coli* DHFR gene under SP6 promotor control and Bla gene under *E.coli* RNA polymerase promotor control was added to a 0.5-ml reaction mixture containing *E.coli* S30 extract, SP6 polymerase, additional tRNA and protease and ribonuclease inhibitors. The feeding solution containing ATP, GTP CTP, UTP and amino acids was passed through the membrane reactor at a constant rate of 1.5 ml/hour, 37°C. *Inset:* Electrophoretic pattern of translation products at 0 and 5 hours of incubation.

suppress the endogenous RNA polymerase activity. The feeding solution containing all four nucleoside triphosphates and twenty amino acids was passed through the working chamber (0.5 ml volume, YM-100 membrane). Protein synthesis at 37°C was monitored for 24 hours. Proteins were synthesized continuously. One main band corresponding to DHFR was detected by electrophoretic analysis indicating that only dihydrofolate reductase was synthesized (L.A. Ryabova, O.B. Yarchuck, V.I. Baranov, A.S. Spirin). Enzyme activity was measured (Baccanari and Joyner, 1981) in parallel with protein yield from the eluate. Total

protein yield was about 0.6 nmol and the specific activity of DHFR was about 0.14×10^{-4} U/pmol.

A further major limitation of coupled transcription-translation systems seemed to be their sole application to bacterial (prokaryotic) extracts. In eukaryotes, transcription and translation are known to be uncoupled and even spatially separated. Following the experiments described above, coupling of bacteriophage enzymatic activity with eukaryotic translational machinery was proposed (V.I. Baranov, L.A. Ryabova, O.B. Yarchuck, A.S. Spirin). Fig. 14

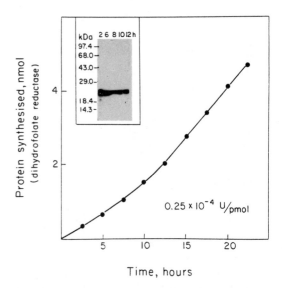

Fig. 14. Kinetics of expression of dihydrofolate reductase-encoding gene in the wheat germ continuous cell-free system with bacteriophage SP6 RNA polymerase. 50 µg of plasmid (the same as in Fig. 13) was added to a 0.5-ml reaction mixture containing S30 extract from wheat embryos, SP6 polymerase and protease and ribonuclease inhibitors. The feeding solution containing ATP, GTP CTP, UTP and amino acids was passed through the membrane reactor at a constant rate of 1.5 ml/hour, 24°C. *Inset:* Electrophoretic pattern of translation products at 2, 6, 8, 10 and 12 hours of incubation.

illustrates expression of the DHFR gene in wheat embryo extract supplemented with SP6 RNA polymerase. The feeding solution containing all four nucleoside triphosphates and twenty amino acids was passed through the working chamber (0.5 ml volume, YM-100 membrane). Protein synthesis at 24° C was monitored over 24 hours and again the system continuously synthesized the protein giving a final yield of about 5 nmol. Enzyme activity of DHFR was measured in parallel with protein yield from the eluate; the specific activity was of about 0.25×10^{-4} U/pmol.

In a second experiment a chloramphenicol acetyl transferase (CAT) gene was expressed in a rabbit reticulocyte lysate supplemented with SP6 RNA polymerase (Fig. 15). Feeding solution containing all four nucleoside triphosphates and twenty amino acids was passed through the working chamber (0.5 ml volume, XM-300 membrane) and synthesis rate in the continuous action system was constant for at least 35 hours at 34° C. Total protein yield was about 2.5 nmol. CAT activity was measured (Hodges and Hruby, 1987) in parallel with protein yield from the eluate and again the enzyme was functionally active.

The advantages of combining bacteriophage RNA polymerase with cellular extracts (prokaryotic or eukaryotic) are obvious. The simplest genetic constructions, e.g. those containing an SP6 or T7 promoter upstream from any gene on a linear DNA fragment, can be used for expression. Products of chemical DNA synthesis or DNA polymerase chain reaction can be directly added to the system for expression. Expression products appear as single pure proteins or polypeptides in the eluate throughout prolongated incubation. The system is stable and effective enough to provide preparative amounts of the expression product. The cell-free character of the expression system allows manipulation of all conditions of incubation, regardless of cell control.

Conclusions

Preparative cell-free protein synthesis opens up new possibilities and areas in science and biotechnology. Firstly, this technology will be indispensable for biosynthesis of polypeptides and proteins which cannot be produced in cells because of instability, e.g., in vivo protease degradation or induced aggregation. The same applies to cytotoxic proteins and to a number of key proteins where overproduction is lethal. Proteins poorly expressed in cells because of genetic control also can be produced in cell-free systems. Difficulties with

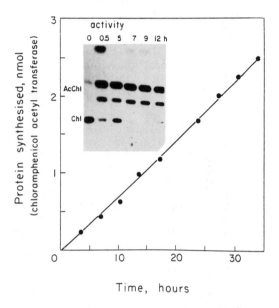

Fig. 15. Kinetics of expression of chloramphenicol acetyl transferase-encoding gene in the rabbit reticulocyte continuous cell-free system with bacteriophage SP6 RNA polymerase. 50 μg of plasmid (pSP65) with the TN5 CAT gene under SP6 promotor control and Bla gene under E.coli RNA polymerase promotor control added to a 0.5-ml reaction mixture containing micrococcal nuclease-treated rabbit reticulocyte lysate and protease and ribonuclease inhibitors. The feeding solution containing ATP, GTP CTP, UTP and amino acids was passed through the membrane reactor at a constant rate of 1.5 ml/hour, 34° C. *Inset:* Thin-layer chromatography of chloramphenicol acetyl transferase assay products. Functional activity of protein synthesized in the continuous system was measured after 0, 0.5, 5, 7, 9, and 12 hours of incubation.

41

isolation or in vivo synthesis of some antigenic proteins and polypeptides can also be easily overcome by the use of the preparative cell-free systems.

A second potential use concerns easy identification and investigation of expression products from open reading frames, since the outflow of the reactor contains the product as the sole polypeptide. Similarly, polypeptides or proteins with invented amino acid sequences and therefore unknown properties can be collected in a pure state and studied, thus permitting wide screening of newly designed proteins.

Thirdly, a specifically scientific application of continuous flow cell-free systems involves the in vitro study of transcriptional and translational regulation mechanisms. The long life-time of such systems and their dynamic (flux) character provide special advantages for such studies. For example, different effectors can be given transiently, imitating the in vivo condition, and after-effects can be traced for a long time.

One more important scientific application could be preparative syntheses of protein folding inter-mediates. Protein folding is a physical problem whose solution requires large amounts of the material for physical studies. It is likely that preparative cell-free translation systems can serve as a source of polypeptides terminated at definite stages of their natural folding pathways.

Preparative biosynthesis of proteins with massive or specified substitutions of unnatural or modified amino acids is a very promising application for the use of flow-through cell-free systems. This can be achieved by introducing amino acid derivatives recognizable by aminoacyl-tRNA synthetases, aminoacyl-residue-modified aminoacyl-tRNA, or artificially formed aminoacyl-tRNA with unnatural amino acid and special anticodons, etc. In any case, such substitutions are important for physical studies of proteins, for functional investigations of molecular mechanisms of protein activities, and for practical protein engineering.

Finally, preparative cell-free protein synthesis systems may initiate a new era in protein engineering: (1) a gene specifying an engineered protein can be directly expressed in vitro, without the need for in vivo expression; multiple, engineered versions of a chemically synthesized gene can be expressed directly; (2) all variants of a gene under investigation can be transcribed and translated without limitations imposed by cellular control mechanisms, proteolytic degra-dation, etc; (3) all engineered versions of a protein can be easily visualized and tested directly in the outflow of the reactor without requiring isolation and purification. From this, rapid screening of altered proteins can be achieved; in addition protein variants with altered physico-chemical properties will not be excluded and protein versions toxic to cells can be obtained.

The invention of flow-through systems allows total automation of cell-free protein synthesis. It is hoped that automated protein-synthesizing bioreactors will be soon constructed and introduced into practice. A future development would link a DNA synthesizer, a gene multiplication instrument based on the DNA polymerase chain reaction, and an automated cell-free protein synthesis bioreactor.

Of course, the most difficult and critical problem in cell-free protein biosynthesis concerns the correct co-translational and post-translational modification of a synthesized poly-peptide. Solutions to this problem are under investigation. One approach might involve cell-free protein synthesis in vesicles made of natural endoplasmic reticulum membranes, with the system of polypeptide transport and modification intact. In principle, a flow-through protein synthesis bio-reactor could be developed based on such vesicles placed in a column. Correct in vitro glycorylation may require yet another type of bioreactor.

In any case, I believe that future biotechnology will be mainly cell-free biotechnology, including cell-free gene cloning, cell-free gene multiplication, cell-free gene expression, and cell-free protein modification.

Acknowledgements

I wish to express my deepest gratitude to my co-authors and

42

colleagues who contributed to this work and helped me in writing the manuscript - Dr. Vladimir Baranov, Dr. Lubov Ryabova, Prof. Yuly Alakhov, Sergey Ovodov, Igor Morozov, and Oleg Yarchuk. I am also indebted to Dr. Stephen Ortlepp for his scientific contribution and for correcting my English.

References

Baccanari, D. P.; Joyner, S. S. *Biochemistry*, **1981**, *20*, 1710.

Baranov, V. I.; Morozov, I. Yu.; Ortlepp, S. A.; Spirin, A. S. *Gene*, **1989**, *84*, 463.

De Vries, J. K.; Zubay, G. *Proc. Nat. Acad. Sci. USA*, **1967**, *57*, 1010.

Gold, L. M.; Schweiger, M. *Proc. Nat. Acad. Sci. USA*, **1969**, *62*, 892.

Hodges, W. M.; Hruby, D. E. *Anal. Biochem.* **1987**, *160*, 65.

Kung, H.-F.; Redfield, B.; Treadwell, B. V.; Eskin, B.; Spears, C.; Weissbach, H. *J. Biol. Chem.* **1977**, *252*, 6889.

Kung, H.-F.; Redfield, B.; Weissbach, H. *J. Biol. Chem.* **1979**, *254*, 8404.

Lederman, M.; Zubay, G., *Biochim. Biophys. Acta*, **1967**, *149*, 253.

Nirenberg, M. W.; Matthaei, J. H. *Proc. Nat. Acad. Sci. USA*, **1961**, *47*, 1588.

Ryabova, L. A.; Ortlepp, S. A.; Baranov, V. I. *Nucleic Acids Res.* **1989**, *17*, 4412.

Schweiger, M.; Gold, L. M. *Proc. Nat. Acad. Sci. USA*, **1969**, *63*, 1351.

Spirin, A. S.; Baranov, V. I.; Ryabova, L. A.; Ovodov, S. Yu.; Alakhov, Yu. B. *Science*, **1988**, *242*, 1162.

Recombination in Replicating RNAs

Alexander B. Chetverin, Leonid A. Voronin, Alexander V. Munishkin, Larisa A. Bondareva, Helena V. Chetverina and Victor I. Ugarov
Institute of Protein Research, Academy of Sciences of the U.S.S.R., Pushchino, Moscow Region, U.S.S.R.

Small RNAs replicated by Qβ phage replicase appear to be natural recombinant molecules arising as a result of multiple recombinations of different RNA sequences. This means that RNA recombination occurs in bacterial cells infected with Qβ phage, and this finding can be promising for elaboration of an RNA recombination cell-free system compatible with Qβ replicase-effected RNA amplification.

Qβ replicase is a unique viral RNA-dependent RNA polymerase in the sense that it can be easily purified and is long living in the *in vitro* RNA amplification reaction. Like other replicases of single strand RNA-containing viruses it synthesizes RNA exponentially. Since both plus and minus strands of a cognate RNA are almost equally effective as templates, the template number doubles in each round of replication, and RNA grows autocatalytically until its molar amount equals that of the enzyme. For RNA of, say, 1000 nucleotides (nt) in length the doubling time in the exponential phase will be some 3 min long (instead of 30 min as is the case for bacterial growth). Starting with a single RNA molecule replication will result in 1 μg ($\approx 10^{12}$ molecules) of RNA in 2 hours, 1 mg in 2.5 hours, etc.

Because of the extremely high performance of the *in vitro* Qβ replicase reaction it can be used to produce desired sequences or whole genes in large amounts. Moreover, it can also be used for fast cloning of individual RNA molecules as an alternative to gene cloning in the cell. The advantages of the Qβ RNA amplification reaction over the popular polymerase chain reaction (PCR) are that it is much faster, and it does not require expensive equipment and synthetic primers. Futhermore, in contrast to PCR, the product of Qβ replicase reaction is RNA which can be directly used for translation in, for example, a protein synthesis bioreactor (see chapter by A. S. Spirin in this volume).

The greatest problem in employing Qβ replicase for the above purposes is the strict template specificity of this enzyme. It is highly adapted for amplification of the Qβ bacteriophage genomic RNA and of a class of non-genomic RNAs which occur in Qβ phage-infected *Escherichia coli* cells and grow spontaneously (i.e. without template addition) in an *in vitro* Qβ replicase reaction (Banerjee et al., 1969; Kacian et al., 1972). Qβ replicase ignores all other RNAs (Haruna & Spiegelman, 1965). To overcome this difficulty, the insertion of desired sequences into an effective natural Qβ replicase template was proposed. This idea was realized by means of site-directed sequence recombination at the RNA (Miele et al., 1983) and DNA (Lizardi et al., 1988; Mills, 1988) level. However, the approaches used so far are time-consuming, restricted to certain sequences and do not utilize the potential of the Qβ replicase reaction for *in vitro* RNA cloning. Besides, it is impossible to decide *a priori* which site of the carrier Qβ replicase template will be most suitable to accomodate a particular foreign sequence in order to ensure maximally effective replication of the resulting recombinant RNA.

In principle the problem could be solved if one would allow RNA molecules (say those capable of replication by Qβ replicase and those targeted for insertion) to recombine randomly so as to obtain a bank of replicating RNA molecules not existing before. Those recombinants could be selected from this bank which both carry the desired insertion and manifest the highest replication capacity. Theoretically, even a single replicating molecule could be selected and amplified, or cloned, with the help of Qβ replicase.

However, this approach has not yet been realized because effective systems of direct RNA recombination compatible with the Qβ replicase reaction are not known. Moreover, for a long time it was thought that no RNA recombination can occur in an *E. coli* cell, which is the host for Qβ phage (Horiuchi, 1975).

In this paper, we report our findings of natural recombinant RNA molecules among the products of spontaneous Qβ replicase-effected RNA synthesis, as well as possible practical implications of this fact.

Synopsis of Methods

The replicating RNAs were isolated from the products of *in vitro* RNA synthesis which occur spontaneously (without template addition) in Qβ replicase reactions containing all four ribonucleoside triphosphates. The Qβ replicase preparation used in these experiments was purified from Qβ phage-infected *E. coli* Q13 cells and contained no detectable RNA impurities. Gel-purified RNA was either sequenced directly after strand separation under non-denaturing conditions (for details see Munishkin et al., 1988) or first cloned in the form of cDNA and then sequenced by the dideoxy chain-terminator

method. RNA secondary structure was probed by limited digestion with single-strand-specific and helix-specific ribonucleases (Munishkin et al., 1988). Sequence comparison was done using the Genbank® database (release #59) and MicroGenie® software.

Recombinant RNA Sequences

The view that RNA recombination does not occur in bacterial cells was unexpectedly disproved by our finding (Munishkin et al., 1988) of an *in vivo*-formed Qβ replicase template, RQ120 RNA, that appeared to be a recombinant made of two pieces, one derived from Qβ genomic RNA and the other from tRNA$_1$^{Asp}, a host sequence (Figures 1 & 2). Since Qβ RNA only exists in the RNA form, recombination must have occurred at the RNA level.

This finding was recently corroborated by the discovery of a highly efficient Qβ replicase template, RQ135 RNA, that seems to be even more interesting. This RNA molecule consists entirely of segments highly homologous to non-cognate RNAs. Furthermore, its sequence

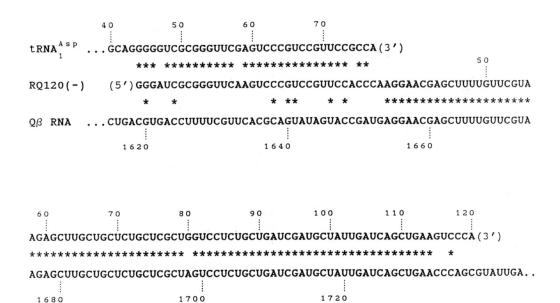

Fig. 1. Homology of the (-) strand of RQ120 RNA to Qβ RNA and to *E. coli* tRNA$_1$^{Asp}.

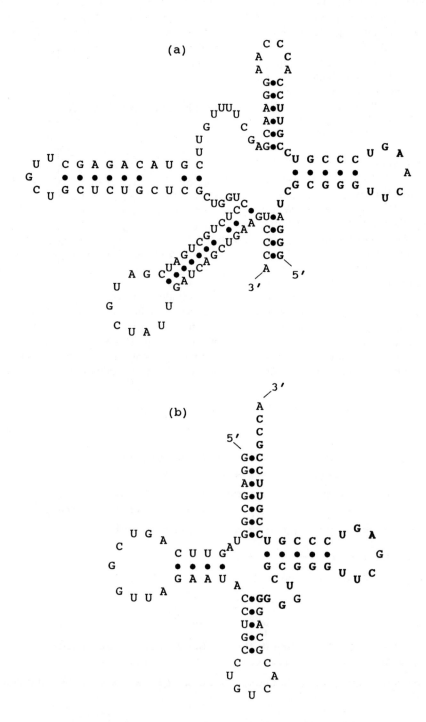

Fig. 2. Comparison of secondary structures of (a) RQ120(-) RNA and (b) *E. coli* tRNA₁^{Asp}.

contains a long direct repeat (Fig. 3). It follows that RQ135 RNA arose as the result of multiple recombinations between foreign RNA sequences. The putative primary recombinant consisted of segments of *E. coli* 23S RNA and λ phage *O*-protein mRNA. This ancestor RNA would possess structural features similar to other known Qβ replicase templates (Chetverin et al., in preparation) and could have evolved into the final RQ135 structure by elongation of one of its hairpins (Fig. 4) via additional recombinations. A possible pathway for RQ135 generation would include insertion of one more foreign sequence and two subsequent duplications, resulting in the sequence shown in Figure 5.

It can be seen from this figure that there are only a few differences between the real RQ135 RNA sequence and a sequence composed only of pieces of non-cognate RNAs. This means that there is no mystique behind the ability of Qβ

replicase templates to replicate. They are made of the same structural elements as other RNAs, and the only difference resides in the spatial arrangement of those elements, which can be changed via RNA recombination.

It is likely that extensive promiscuous recombination among various RNAs occurs in bacterial cells (at least those infected with Qβ phage), and some of the recombination products acquire the ability to be amplified by Qβ replicase. It is not yet known whether it is Qβ replicase that is responsible for these recombinations or whether they are performed by another enzyme (or ribozyme?). This is a subject for future investigations. Nonetheless, it is already obvious that there exists an RNA recombination system compatible with Qβ replicase-effected RNA amplification, which, we hope can provide a basis for a future cell-free RNA engineering technique.

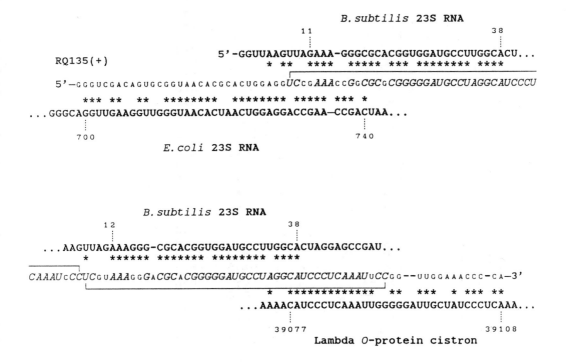

Fig. 3. Homology of the (+) strand of RQ135 RNA to *E. coli* 23S RNA, *B. subtilis* 23S RNA and *O*-protein mRNA of λ phage. Brackets indicate the repeating segments.

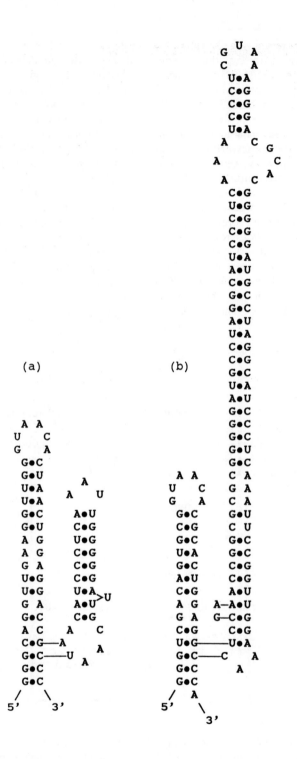

Fig. 4. Secondary structure models of (a) putative primary recombinant made from segments of *E. coli* 23S RNA and λ mRNA and (b) RQ135(+) RNA.

```
GGGCAGGUUGAAGGUUGGGUAACACUAACUGGAGGACCGAAA  GGGCGCACGGUGGAUGCCUUGGCAUCC
***       *  **   **  ********  ******** ******  ***** *** ******** ******
·GGG----UCGACAGUGCGGUAACACGCACUGGAGGUCCGAAACCGGCGCGCGGGGGAUGCCUAGGCAUCC
```

RQ135(+) RNA

```
CUCAAAUCCCUC  AAAGGG  CGCACGGUGGAUGCCUUGGCAUCCCUCAAAUUGGGGGAUUGCUAUCCC
************  ******  *******  ******* ***************  **  * *  * ***
CUCAAAUCCCUCGUAAAGGGACGCACGGGGGAUGCCUAGGCAUCCCUCAAAUUCCGGU-UGGAAACCCCA
```

Fig. 5. Comparison of the RQ135 sequence with a sequence obtained upon insertion into the primary recombinant of a relevant segment of *B. subtilis* 23S RNA and two consecutive duplications of a 33 nt-long and a 9 nt-long segment of the resulting sequence. Black, blank, and shaded bars indicate segments derived from *E. coli* 23S RNA, *B. subtilis* 23S RNA, and λ phage *O*-protein mRNA, respectively.

References

Banerjee, A.K.; Rensing, U.; August, J. T. *J. Mol. Biol.* **1969,** *45,* 181-193.

Haruna, I.; Spiegelman, S. *Proc. Nat. Acad. Sci. U.S.A.* **1965,** *54,* 1189-1193.

Horiuchi, K. In *RNA Phages*; Zinder, N. D., ed.; Cold Spring Harbor Laboratories: Cold Spring Harbor, New York, 1975, 29-50.

Kacian, D. L.; Mills, D. R.; Kramer, F. R.; Spiegelman, S. *Proc. Nat. Acad. Sci. U.S.A.* **1972,** *69,* 3038-3042.

Lizardi, P. M.; Guerra, C. E.; Lomeli, H.; Tussie-Luna, I.; Kramer, F. R. *BioTechnology* **1988,** *6,* 1197-1202.

Miele, E. A.; Mills, D. R.; Kramer, F. R. *J. Mol. Biol.* **1983,** *171,* 281-295.

Mills, D. R. *J. Mol. Biol.* **1988,** *200,* 489-500.

Munishkin, A. V.; Voronin, L. A.; Chetverin, A. B. *Nature* **1988,** *333,* 473-475.

Enantioselective Enzymatic Synthesis of Prostaglandin–Synthons in Multiphase Reaction Media

S. Schapöhler, T. Scheper, K. Schügerl, Institut für Technische Chemie, Universität Hannover, Callinstr. 3, 3000 Hanover 1, Germany, and **E.-R. Barenschee**, Degussa AG, Rodenbacher Chaussee 4, 6450 Hanau 1, Germany

Up to now aqueous reaction media seemed to be necessary for bioconversions. During the last few years more and more reports on successful enzymatic reactions in multiphase media were published. Two different examples of multiphase reaction media are introduced and characterized. The liquid membrane emulsion technique and an organic solvent system demonstrate their applicability for the production of two enantiomerically pure key prostaglandin-precursors, the 4-hydroxy-2-cyclopenten-1-one and its 2',2'-dimethyl-1',3'-diol-ketal derivative. Pig liver esterase and some lipases from *Mucor*, *Pseudomonas*, porcine pancreas etc. exhibit high enantioselectivities for these substrates. Different influences (the kind of organic solvent, water content, etc.) had been under investigation.

The enantioselective synthesis of chiral chemical substances is often necessary to avoid side effects, which can sometimes be very hazardous. A very negative example was the use of the racemate of the sedative contergan, where one of the enantiomers caused frightful deformations of newborns. The main applications for enantioseparations as a way of enantioselective synthesis lie in the area of pharmaceuticals, insecticides and flavours. Biotransformations using isolated enzymes exhibit a unique substrate specificity which makes them ideal for these purposes.

Multiphase reaction media

It is an actual insight that the optimal medium for an enzymatic biotransformation needs not to be an aqueous buffer system. Most *in vivo* reactions take place in multiphase systems. There are different possibilities of technically usable reaction media and ways to characterize them (Laane, 1987; Schapöhler, 1990). Here three cases shall be distinguished:

- aqueous media (mainly aqueous solution, if necessary addition of water soluble organic solvent)

- stabilized mixed phases (reverse micelles and liquid membrane emulsions)

- organic media (pure organic solvent, normally not water-miscible)

In the following the principles of the latter two: liquid membrane emulsion

2181–2/92/0050$06.00/0 © 1992 American Chemical Society

(as one possible example) and organic solvent systems will be explained and their applicability for the preparation of prostaglandin-synthons demonstrated.

Prostaglandin-synthons

Prostaglandins are optically active C_{20} fatty acid derivatives. They play a key role in all organisms as ubiquous hormones. Their pharmacological effects are numerous. They are influencing the blood pressure, the blood plateletts, etc. . All methods for the production of prostaglandins (e.g. PGE_2 and PGD_2) use stable precursors or synthons (Harre, 1982; Noyori, 1984, 1989; Okamoto, 1989). Here the enantioselective synthesis of two key-synthons: 4-Hydroxy-2-cyclopenten-1-one **1** and its 2',2'-dimethyl-1',3'-diol-ketal **2** (Fig. 1) will be described.

Fig. 1. The two prostaglandin-synthons used

Materials and methods

Lipases from the following organisms were used: *Alcaligenes species* (Meito Sangyo Co.; P-1-1), *Candida cylindracea* (Amano AY-30; LAY Mo3517), *Chromobacterium viscosum* (Toyo Jozo Co.; LP-251-S), *Mucor*

miehei (Gist-Brocades; E 30,000; Ref 0282), pig liver esterase (Sigma E-3128), porcine pancreas (Sigma L-3126; 74F-0470) and *Pseudomonas fluorescens/cepacia* (Amano P/PS; LPL 05518; Röhm EI 220-88). The reaction and enantiomeric excess analysis was performed by HPLC (Daicel Chemical Industries Ltd. Chiralcel OA[R]; eluent 90:10 n-hexane : isopropanol; 0.5 ml/min and Macherey & Nagel RP-18 Nucleosil 5 C 18; eluent 75:25 water : methanol; 1 ml/min) and GC (Lipodex[R] A; Macherey & Nagel; quartz-capillary 60 m).

The enzyme activities were determined in a Metrohm autotitration system which kept the pH constant. Pig liver esterase activity was measured by hydrolysis of 200 μl ethyl butyrate, mixed through vigorously stirring with 10 ml of a potassium phosphate buffer solution, pH = 8 (thermostated to 25° C). Lipase activities were determined by hydrolyzing different glycerol triesters in the same apparatus (3-5 g triester, 10 μl Emulan P (emulsifier from BASF) 50 ml 0.05 M potassium phosphate buffer pH = 8, thermostated to 25° C).

Principles of liquid membrane emulsion systems

Liquid membrane emulsions are being prepared through emulsification of an enzyme containing buffer solution with an organic membrane phase (composed of kerosene or paraffin as main constituents, emulsifiers

51

and additives) and dispersion of this emulsion in an outer buffer solution as demonstrated in Fig. 2 (Meyer 1987; Scheper 1987, 1989 and 1990). In this way a three phase system is established in which the two aqueous phases are separated by the organic phase (liquid membrane). The liquid membrane immobilizes the enzyme and keeps it in its native environment. Substrate and product have to permeate through the liquid membrane (Fig. 3). This transport is due to the physical solubility of these substances in the membrane phase, what is demonstrated for ACP (substrate) and HCP (product) in Fig. 4 . The comparison of the permeation courses through a solid supported 100 % paraffin membrane shows a great difference between both substances. ACP permeates nearly 50 times faster than HCP (Meyer, 1987). Through the addition of water-insolu-

ble anion exchangers like quarternary ammonium salts to the organic phase the individual permeation rates can be selectively accelerated. In this way it is possible to generate and extend a separation of the substrate and product. One of the advantages of liquid membrane emulsions is therefore the possibility to integrate reaction and down stream processing in one step. As long as the enzyme is not inhibited, the product for example can be accumulated in the inner phase.

Experimental set-up and results

A process using the liquid membrane emulsion can be performed batchwise or in a continuous way as illustrated in Fig. 5. A continuously stirred tank reactor in which the enzyme containing emulsion has been

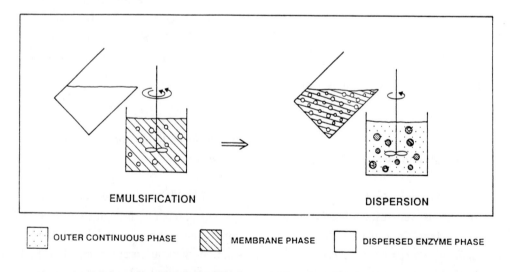

EMULSIFICATION   DISPERSION

OUTER CONTINUOUS PHASE   MEMBRANE PHASE   DISPERSED ENZYME PHASE

Fig. 2. Preparation of a liquid membrane emulsion

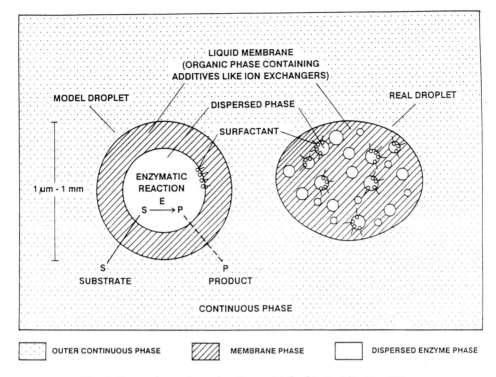

Fig. 3. Comparison of a real and a model liquid emulsion droplet

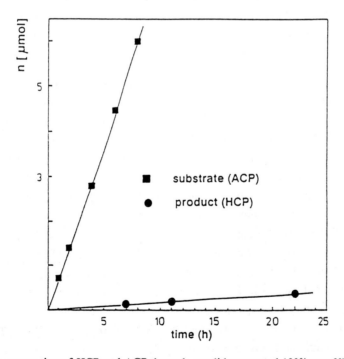

Fig. 4. The permeation of HCP and ACP through a solid supported 100% paraffin membrane

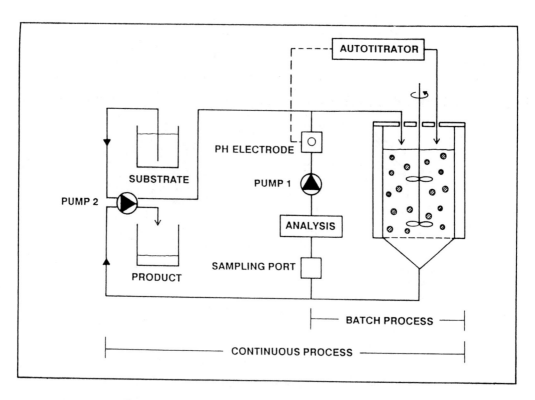

Fig. 5. Process scheme for the liquid membrane emulsion

dispersed is fed with substrate solution. The product solution is pumped off through a hydrophilic membrane filter (Sartorius SM 11303). In the batch-process the outer solution is re-circulated through the analysis loop, which is necessary to keep the pH constant (Metrohm autotitration system).

In this system the enantioselec-tive hydrolysis of 4-acetoxy-2-cyclopen-ten-1-one (ACP) with pig liver esterase was carried out (Fig. 6). The corresponding alcohol (-)-4-hydroxy-2-cy-clopenten-1-one (HCP, **1**) is produced with high enantiomeric excesses, strongly increasing with decreasing tempe-rature, especially below 20° C (Fig. 7).

The maximum of 80 %ee for the S-(-)-enantiomer was measured at 10° C. An emulsion composition of 80 % paraf-fin, 10 % cyclohexane, 9 % Span 80 (Sorbitanmonooleate) and 1 % Tomac (trioctyl-methyl-chloride) proved to be optimal for long term membrane sta-bility, membrane transport properties and minimal enzyme desactivation. HCP could be enriched in the inner phase of the enzyme emulsion. The in-fluences of different possible solvents for the membrane phase on pig liver esterase are shown in Fig. 8. They were determined after a one minute vor-texing of a mixture of 100 μl PLE sus-pension, 3 ml organic solvent and 3 ml

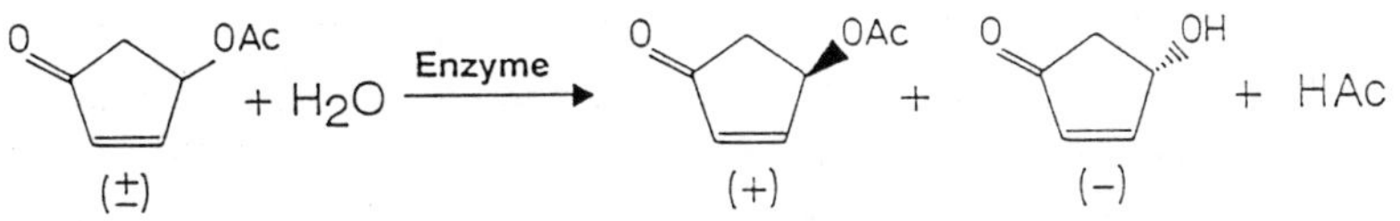

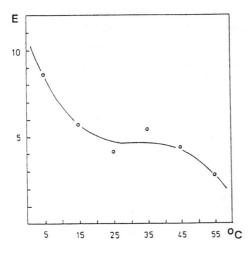

Fig. 6. Reaction scheme for the hydrolysis of 4-acetoxy-2-cyclopenten-1-one (1) with pig liver esterase

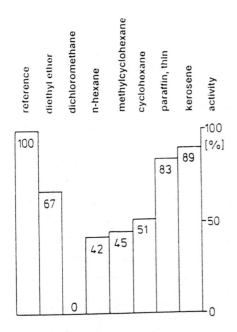

Fig. 7. The enantioselectivity of the HCP-hydrolysis as a function of temperature

Fig. 8. The influence of different organic solvents on the pig liver esterase activity

buffer solution (50 mM potassium phosphate buffer, pH=8) and a waiting period of another 15 minutes. Significantly is the strong loss of activity of the enzyme when it came into contact with dichloromethane. The main constituents of the organic phase, paraffin or kerosene, gave only a small decrease of activity. It is very important to notice that these activity losses took place in a two phase system with high water content (50 %).

In contrast to this the transition to pure organic solvents, described next, leads to a stabilization of the enzymes (higher retaining activity but lower reaction rates). A phenomenon which is still not fully understood and closely related to the discussion of the optimal water content (see below).

Principles of the organic solvent systems

A totally different concept is the use of enzymes, especially lipases, in pure organic phases (Chen, 1989; Klibanov, 1989). Here we get a reversed reaction direction. Alcohols will be esterified. Principally three mechanisms of a lipase catalyzed reaction are imaginable in these media: esterification, transesterification and interesterification. Their possible application on the substrates of this work is exemplified in Fig. 9.

The amount of water in the reaction mixture strongly affects the reaction performance. It is still not clear, how much water is optimal for every of these mechanisms. For transesterifications for example the water content should be in the range from 0 to 1 % by weight of the reaction mixture. Like the reviews of Dordick (1989) and Chen (1989) most of the literature is rooted in only one article (Zaks, 1984), which claims that the wa-

ter tries to form a monolayer around the enzyme. The use of crude enzyme preparations and its heterogeneous, suspended particles makes it at least very difficult to confirm this concept by experiments.

Following Yamane's (1989) detailed examination and modelling the water content for the esterification should be in the same magnitude as mentionened before for the transesterification, while many other authors take much more water (Baratti, 1986; Lazar, 1986). It seems very probable

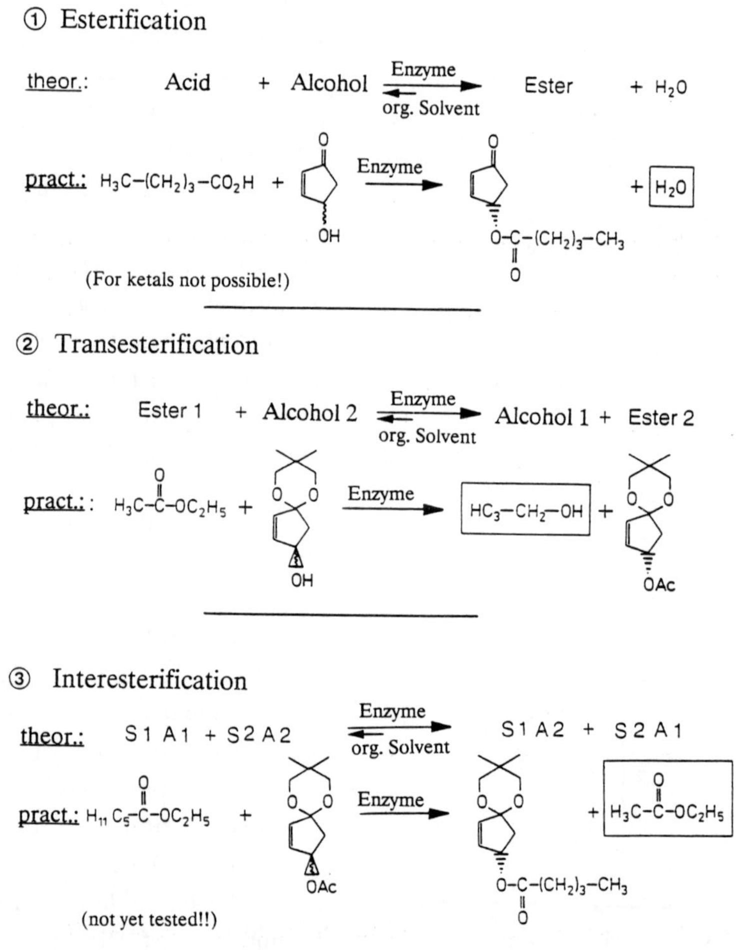

① Esterification

② Transesterification

③ Interesterification

Fig. 9. Possible lipase reactions in organic media

56

that every system has its own optimal water content.

Experimental set-up and results

Instead of the keto-synthon **1** the ketalized derivative **2** was chosen as substrate. It is far more stable, easier to synthesize and gave higher enantiomeric excesses. With a ketal-compound a direct esterification is not senseful and beyond this, transesterifications are often faster. So the transesterification was used (Fig. 10) to separate the two enantiomers (Schapöhler, 1990).

The reaction conditions were: 50-500 mg substrate (**2**), 250-500 mg enzyme powder ("straight from the bottle"), 2-50 ml acyl donor (serves simultaneously as solvent) and eventually additional n-heptane as cosolvent. The reaction mixture was stirred with a magnetic stirrer and thermostated in the range from 25-60° C.

Four of the six tested lipases exhibited high enantioselectivities with a maximum of 100 %ee for the remaining alcohol (measured after deketalization by addition of a catalytic amount of formic acid as HCP) and the enzymes from *Pseudomonas* (Fig. 11). All enzymes preferred the S-(-)-enantiomer of the racemic substrate. The combination of high enantioselectivity and preferation of the (-)-enantiomer suggests high similarities of the active sites of these lipases. This is strongly supported by very recent x-ray structure determinations of porcine pancreas (Winkler, 1990) and *Mucor miehei* (Brady, 1990) lipase.

The time courses of the conversion and the enantiomeric excess for a special experiment are demonstrated in Fig. 12 (50 mg substrate, 50 ml c-hexyl acetate, 250 mg *Pseudomonas* lipase, 25° C). As expected for a reaction with a high enantioselectivity the conversion nearly stops after reaching about 50 % and an enantiomeric excess of 100 %ee for the substrate.

In transesterification reactions there are two substrates, the acyl donor (ester) and the attacking nucleophil (alcohol). Here effects of the va-

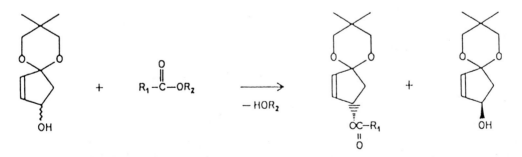

Fig. 10. Reaction scheme for the transesterification of the ketal-alcohol (**2**)

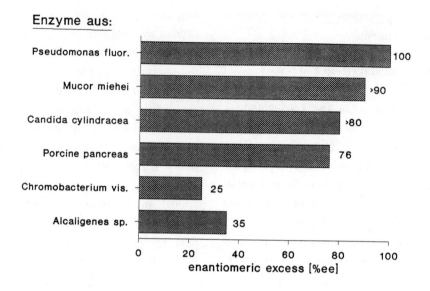

Fig. 11. Enantioselectivity of the enzymes used in the transesterification of the ketal-alcohol (2)

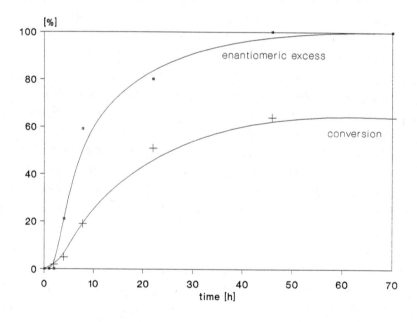

Fig. 12. Comparison of the time courses of the conversion and the enantiomeric excess

riation of the alcohol rest of the acyl donor on the reaction were investigated. The reaction course was strongly influenced. From the straight chain n-butyl acetate to the bulky t-bu-tyl acetate the reaction rate fell to nearly zero (50 mg substrate, 50 ml acyl donor, 250 mg *Pseudomonas* lipase, 30° C, Fig. 13).

The water content of the reac-

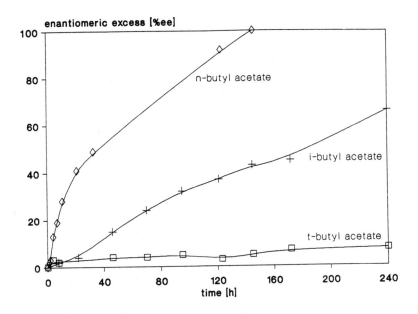

Fig. 13. The time course of the enantiomeric excess as a function of the acyl donor

tion solution and enzyme exhibits a very strong influence, too. This was demonstrated by the addition of molecular sieve to the reaction mixture (50 mg substrate, 2 ml c-hexyl acetate, 250 mg *Pseudomonas* lipase, 40° C), which accelerates the reaction and gives a substantially higher final conversion. The conversion rate reaches nearly zero after only 3 hours. This means that the more reactive enantiomer is consumpted and only the less reactive is remaining (Fig. 14). The remaining enzyme activity is very high. After a stronger loss of 20 % of the initial activity after 70 to 90 hours of use, the further decrease was very low (Fig. 15).

All these results lead to a proposed procedure for the chiral economic (Fischli, 1976) production of the two possible PG-synthons (Fig. 16). It consists mainly of the transesterification step and is followed by the chromatographic separation of the remaining substrate ((+)-ketal alcohol) and the product ((-)-ketal ester) - after removing the enzyme by filtration or centrifugation. If only one enantiomer is needed, it is possible to racemise and recirculate the other one (Hirohara, 1985). Another applicable method would be the inversion of the configuration of the not usable enantiomer (Hirohara, 1985). To get the enantiomerically pure ACP a preparative deketalization without any decrease in the enantiomeric excess is simple to carry out by acid catalysis (for example by addition of a small amount of formic acid).

Summary and outlook

The application of two multiphase reaction media for the enantioselective synthesis of prostaglandine

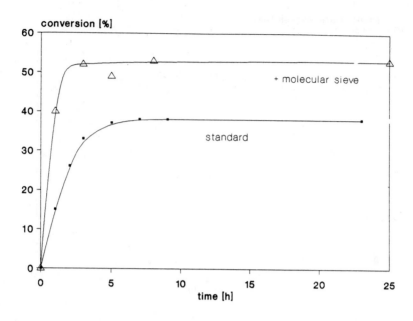

Fig. 14. Comparison of a reaction under standard conditions and one with the addition of molecular sieve

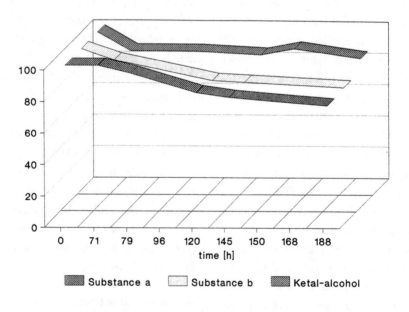

Fig. 15. The remaining enzyme activity as a function of reaction time

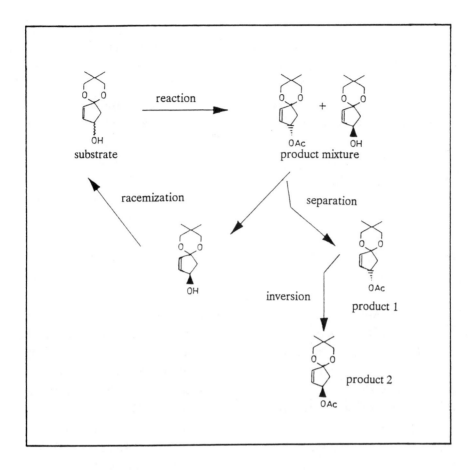

Fig. 16. Procedure for the chiral economic production of ACP and its Ketal-derivative

synthons was demonstrated. Liquid membrane emulsions and organic solvent systems proved to be novel ways for the non-toxic, effective synthesis of these compounds. Multiphase reaction media therefore represent a great potential for the performance of biotransformations that has not yet been fully acknowledged.

Acknowledgement

The authors thank Martina Weiß and Jens Bode for their great help, Prof. Winterfeldt and his coworkers in the Institute for Organic Chemistry of the University of Hannover for providing us with the substrates, all the companies which supplied us with enzymes and the Deutsche Forschungsgemeinschaft for their financial support.

References

Baratti, J.; Buono, G.; Deleuze, H.; Langrand, G.; Secchi, M.; Triantaphylides, C. *Ann. Oils Chem. Soc.* **1986**, 355-358

Brady, L.; Brzozowski, A.R.; Derewenda, Z.S.; Dodson, E.; Dodson, G.; Tolley, S.; Turkenburg, J.P.; Christiansen; L.; Huge-Jensen, B.; Norskov, L.; Thim, L.; Menge, U. *Nature* **1990**, 343, 767-770

Chen, C. S.; Sih, C. J. *Angew. Chem.* **1989**, 101, 711-723

Dordick, J.S. *Enzyme Microb. Technol.* **1989**, 11, 194-211

Fischli, A. *Chimia* **1976**, 30, 4-9

Harre, M; Raddatz, P.; Walenta, R.; Winterfeldt, E. *Angew. Chem.* **1982**, 94, 496-508

Hirohara, H.;Mitsuda, S.; Ando, E.; Komaki, R. *in: Tramper, J.; Plas, H.C.v.d.; Linko, P. "Biocatalysts in organic synthesis"; Elsevier, Amsterdam* **1985**

Klibanov, A. M. *Trends Biochem. Sci.* **1989**, 14, 141-144

Laane, C. *Biocatalysis* **1987**, 1, 17-22

Lazar, G.; Weiss, A.; Schmid, R.D. *Ann. Oils Chem. Soc.* **1986**, 346-354

Meyer, E.-R; *PhD - thesis* **1987**, University of Hannover

Noyori, R.; Suzuki, M. *Angew. Chem.* **1984**, 96, 894-882

Noyori, R. *Chem. in Britain* **1989**, 9, 883-888

Okamoto, S.; Kobayashi, Y.; Sato, F. *Tetrahedron Lett.* **1989**, 30, 4379-4382

Schapöhler, S.; Scheper, T.; Schügerl, K. *Patent Application* No. 87593

Schapöhler, S.; *PhD - thesis* **1990**, University of Hannover

Scheper, T.; Makryaleas, K.; Nowottny, K.; Likidis, Z.; Tsikas, D; Schügerl, K. *Enz. Eng.* **1987**, 8, 165-170

Scheper, T.; Barenschee, E.-R.; Hasler, A.; Makryaleas, K.; Schügerl, K. *Ber. Bunsenges. Phys. Chem.* **1989**, 93, 1034-1038

Scheper, T. *Adv. Drug Delivery Rev.* **1990**, 4, 209-231

Winkler, K.F.; D'Arcy, A.; Hunziker, W. *Nature* **1990**, 343, 771-774

Yamane, T.; Kojima, Y.; Ichiryu, T.; Nagata, M.; Shimizu, S. *Biotechnol. Bioeng.* **1989**, 34, 838-843

Zaks, A.; Klibanov, A.M. *Science* **1984**, 224, 1249-1251

Design and Synthesis of a Peptide Having Chymotrypsin-like Catalytic Activity

John M. Stewart, Karl W. Hahn, Wieslaw A. Klis, John R. Cann, Michael Corey,
Department of Biochemistry, University of Colorado School of Medicine,
Denver, Colorado 80262

A bundle of four short, amphipathic helical peptides has been designed to hold the serine protease catalytic triad of amino acids (serine, histidine, aspartic acid) and a hydrophobic substrate-binding pocket in the same spatial arrangement as they are in chymotrypsin. The designed molecule was synthesized by solid phase peptide synthesis. The product, called "Chymohelizyme-1," contains 73 amino acid residues and shows chymotrypsin-like esterase activity. The overall structure of the molecule bears no resemblance to the protein structure of chymotrypsin. Chymohelizyme is the first example of a synthetic enzyme designed totally from basic principles.

Design and production of selective, effective and affordable enzyme catalysts are major goals of modern chemistry. To reach those goals, several approaches are currently being investigated: 1) native enzymes have been modified by genetic engineering procedures (Kuroki et al., 1989; Pantoliano et al., 1989), 2) molecules such as catalytic antibodies have been developed (Lerner and Iverson, 1989; Schultz, 1988), and 3) totally new types of molecules have been designed to catalyze chemical reactions (Dugas, 1989; Sasaki and Kaiser, 1989). We have made significant progress in category 3: the design and synthesis of a peptide having catalytic properties resembling those of chymotrypsin but having a completely new kind of molecular structure (Hahn et al., 1990). Organic molecules such as cyclodextrins and paracyclophanes have been previously synthesized and shown to have some enzyme-like properties (Dugas, 1989). In one case peptide chains were attached to heme to construct a molecule having some hydroxylase activity (Sasaki and Kaiser, 1989). Our synthetic enzyme differs from molecules such as those in that it is composed entirely of amino acids.

Several investigators have assembled amino acid chains in sequences designed to adopt specific types of three-dimensional structures. Erickson and his collaborators (Unson et al., 1984) have designed and synthesized "beta bellin," a two-chain peptide designed to adopt a conformation analogous to that of the "beta barrel" structure found in several proteins (Richardson, 1985). DeGrado (1988) has synthesized a long peptide chain having four repeating alpha-helical zones separated by turns, and has presented evidence that this chain does fold into the desired antiparallel four-helix bundle. Mutter and coworkers (Mutter and Vuilleumier, 1989) have synthesized "template-assisted synthetic proteins" in which four identical helical chains were grown on lysine sidechains of a "template" peptide. A second model had four beta-predicted sequences appended to the template backbone sequence between the helical chains. None of these investigators has reported incorporation of enzyme active site residues into these structures. In contrast to these projects which have used repeated amino acid sequences predicted to adopt the desired conformation in solution, the amino acids in the chains of chymohelizyme were all designed by molecular graphics in unique sequences predicted to fit most precisely the desired structure.

Enzyme design requires accurate knowledge of the mechanism of the catalytic process, the topography of the active site and the mechanism by which recognition and binding of a specific substrate are accomplished. For the practical realization of a new enzyme, tools must be available for design of the supporting structure and synthesis of the designed molecule. The pancreatic protease chymotrypsin possesses many characteristics that make it a suitable candidate for modeling. The essential features of the structure and mechanism of action of chymotrypsin are well known, and modern computer molecular graphics and solid phase peptide synthesis provide the necessary tools for design and synthesis of a chymotrypsin mimic.

Chymotrypsin (ChTr) is a well-studied serine protease which hydrolyzes protein chains at the carboxyl groups of the aromatic amino acids phenylalanine, tyrosine and tryptophan. It also hydrolyzes simple amides and esters of acylated derivatives of these same amino acids. Chymotrypsinogen is synthesized in the pancreas as a single chain of 245 amino acids; four of these are removed in the process of activation, leaving 241 amino acid residues in the active enzyme. The full three-dimensional structure of ChTr is known from x-ray crystallography studies (Blow, 1971). Its chain assumes a considerable amount of beta and random structure, with but 5% alpha-helix in the entire molecule. The "catalytic triad" of amino acids which constitutes the "charge relay" complex of the active site is widely distributed along the chain (histidine-57, aspartic acid-102 and serine-195), but the tertiary structure of the protein is such that they are brought into close proximity. The His-57 and Asp-102 side chains serve to activate the hydroxyl group of Ser-195 so that it can readily attack the carbonyl carbon of the substrate. In addition, a glycine residue at position 193 performs a critical role in catalysis; it provides hydrogen bonding from the backbone of the protein chain at this point to stabilize the oxyanion tetrahedral intermediate derived from the carbonyl carbon during hydrolysis. This so-called "oxyanion hole" which binds and stabilizes the reaction intermediate is thought to be the most important feature of the active site of the serine proteases (Kraut, 1988; Warshel et al., 1989). It functions critically to lower the energy barrier in the reaction pathway between substrate and products.

The substrate is held in the appropriate position in the active site by virtue of a "hydrophobic pocket" made up of hydrophobic side chains of several amino acids. This hydrophobic pocket can adapt to the sizes of the various aromatic rings in the side chains of substrate residues.

When the substrate is properly positioned in the active site of the enzyme, the unshared electron pair on the hydroxyl group of Ser-195, made especially nucleophilic by the nearby imidazole of His-57 and the carboxylate of Asp-102, attacks the carbonyl carbon of the substrate and forms a new -O-C- bond, the carbonyl oxygen becoming an anion in the process. Stabilization of this anion by the "oxyanion hole" promotes departure of the amino or alcohol component of the substrate, leaving the hydroxyl of Ser-195 acylated by the aromatic amino acid carboxyl of the substrate. This highly reactive ester is rapidly hydrolyzed spontaneously by water, and the carboxyl component of the product then dissociates readily from the enzyme. In order that a model ChTr function satisfactorily, all these components must be positioned in a manner analogous to that in ChTr, and must have similar electronic characteristics in addition to the proper steric milieu.

Design

A structure of four short, parallel amphipathic alpha helices was designed to hold amino acids analogous to the active site residues of ChTr (His-57, Asp-102 and Ser-195 side chains to provide the catalytic triad and backbone NH of Gly-193 to provide the "oxyanion hole" for the tetrahedral intermediate in the reaction sequence) and to provide a hydrophobic "pocket" for recognition and binding of the aromatic side chain of the substrate.

The amino acids necessary to provide catalysis were placed at the amino ends of the four helices, and the substrate binding pocket was provided by the hydrophobic core of the 4-helix bundle (see Fig. 1). Computer molecular graphics (Tripos MENDYL software on a Silicon Graphics IRIS computer) was used to display the full three-dimensional structure of ChTr from the Brookhaven National Laboratory data base of x-ray structures (Bernstein et al., 1977; Blevins and Tulinsky, 1985). The protein was discarded, leaving the catalytic triad in correct 3-dimensional space. The substrate, acetyltyrosine ethyl ester (ATEE), was "docked" in the active site in the correct conformation. Four amphipathic helical peptide chains were then designed to hold all these structural features in the correct alignment. These chains have unique, specifically-designed amino acid

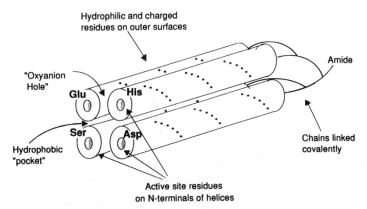

Fig. 1.   Schematic general structure of a "chymohelizyme." The fourth chain, bearing the N-terminal glutamic acid residue, provides the "oxyanion hole" as a hydrogen bond to the peptide backbone.

sequences and vary in length from 15 to 22 amino acid residues. The four chains were linked covalently at their carboxy-terminal ends via sidechain amino groups of ornithine and lysine residues to provide stabilization. In amphipathic helices the amino acid sequence is such that one side of the helix is hydrophobic, while the other is hydrophilic. The hydrophobic sides of the four helices in the helizyme design were given unique amino acid sequences to provide maximum interaction for stabilization of the framework of the molecule. Additional stabilization was provided by interdigitating leucine side chains in the core of the bundle near the carboxyl end of the chains; such "leucine zippers" have been observed in native proteins (Landschutz et al., 1988) and are considered to be important for stabilization of the coiled-coil structure in tropomyosin and other proteins (O'Shea et al., 1988). At the amino ends of the chains the hydrophobic pocket was provided in the core of the four-helix bundle by using amino acids having smaller side chains, such as alanine and glycine. Limited use of glycine appears to be satisfactory in designed helices (Hodges et al., 1988), although it is not common in helical segments of most native proteins.

A schematic representation of the helix backbone structure of "Chymohelizyme-1" (CHZ-1) is shown in Fig. 2, and the entire amino acid sequence is given in Fig. 3. The single ornithine residue which constitutes the carboxyl end of the CHZ-1 molecule was used because its chain length provides the exact separation

needed for optimal interaction of the four helices. This ornithine terminates in an amide; the amide increases the helix stability of the bundle by blocking the helix-destabilizing property of a free carboxyl group in peptides (Fairman et al., 1989). The amino end of each of the four peptide chains is terminated by acetylation, which blocks the helix-inhibiting positive charge of the free amino groups at the amino terminus. Additional stability is given the overall structure by the numerous ionic bonds between side chains of glutamic acid and lysine residues on the hydrophilic outside of the structure. Such ionic "salt bridges" have been found to be particularly effective for stabilization of helical structure in synthetic peptides (Marqusee and Baldwin, 1987). Incorporation of all these features into the design promotes a strong helix-forming tendency in the structure and promotes association of the four helices in a parallel manner.

Synthesis and Purification

The entire 73-residue branched peptide was assembled in one operation by stepwise solid phase peptide synthesis (Stewart and Young, 1984) on methylbenzhydrylamine polystyrene resin in a Beckman 990B synthesizer. Four levels of selective protection were utilized: Boc, Fmoc, Npys and benzyl-related. Benzyl-related groups were used for the link to the resin and "permanent" sidechain protection. The relatively

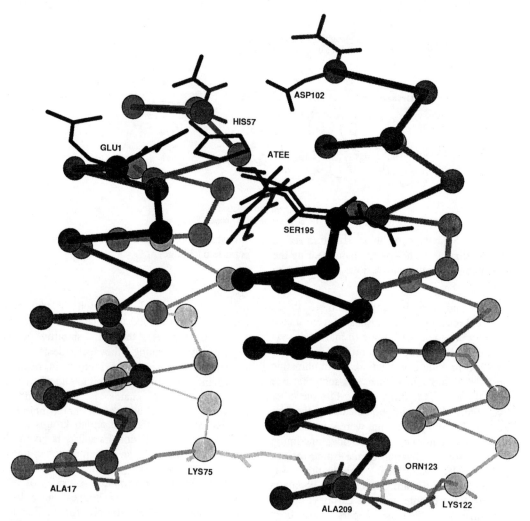

Fig. 2. Computer drawing of a side view of "Chymohelizyme-1," showing the helix backbones and amino acid alpha carbons; side chains of the "catalytic triad" residues are also shown. The darkest chain is in the foreground; the lightest is in the background. N-Terminal and C-terminal residues of each chain are labeled. The substrate, acetyltyrosine ethyl ester, is shown "docked" in the active site.

new Npys group (Matsueda and Walter, 1980) is stable to acid and base, but is removed selectively by treatment with triphenylphosphine. Used in conjunction with the standard acid-labile Boc and base-labile Fmoc goups (Stewart and Young, 1984), it allowed selective protection of the ornithine and lysine sidechain amino groups to permit synthesis of the peptide chains individually. The "chaotropic salt" modification of standard solid phase coupling procedures was used to promote effective coupling reactions of these highly structured peptide chains (Klis and Stewart, 1990). Qualitative and quantitative ninhydrin monitoring was used to insure complete coupling at every step. The peptide was cleaved from the resin with HF, which also simultaneously removed all side-chain blocking groups.

The product was prepurified by ultrafiltration in $6M$ guanidine hydrochloride solution over Amicon YM5 membrane. It was chromatographed successively on Sephadex LH-60 in 70% ethanol and on Sephadex G-50 in 20% acetic acid. Various other types of chrom-

```
Ac–E–E–A–E–E–K–A–K–R–L–L–E–E–L–K–K–A ┐
     (1)                              (17) │
                                           │
Ac–H–E–E–A–K–K–K–A–E–K–L–L–E–E–L–K–K–L–K ┐ │
     (57)                            (75)  │ │
                                           │ │
Ac–D–E–A–G–K–K–A–E–E–E–L–K–K–L–L–E–E–L–K–K–K–Orn–amide
     (102)                           (123) │
                                           │
     Ac–S–E–K–A–K–K–L–L–E–E–L–K–K–L–A ─────┘
          (195)                  (209)
```

Fig. 3. Amino acid sequence of Chymo-helizyme-1, using standard one-letter codes for the amino acids; Ac = acetyl. The catalytic triad His, Ser and Asp residues are given the same sequence numbers as in chymotrypsin, although the sequences of the chains bear no relationship to sequences in chymotrypsin. The fourth chain bearing the N-terminal Glu residue, which adds helix stability, is arbitrarily numbered from residue 1.

atography have been explored. The purified "chymohelizyme-1" showed a single peak on analytical reversed-phase HPLC and the expected amino acid analysis following hydrolysis.

Properties of Chymohelizyme-1

Chymohelizyme-1 is freely soluble in water and in 95% ethanol, but shows a marked tendency to precipitate in intermediate concentrations of ethanol. The helix content (by CD spectroscopy) at pH 8.3 is 76% in water and 85% in 95% ethanol. This helix content observed in ethanol is near the theoretical maximum for the design of the molecule. Helix content is increased about 30% in 0.05M NaCl solution over that seen in water. Molecular sizing chromatography revealed a tendency of CHZ-1 to dimerize, but the exact nature of the intermolecular interaction in this dimer is not known at this time. This dimerization is, however, promoted by acetate; following this discovery, all work on CHZ-1 avoided use of acetic acid, trifluoroacetic acid or their salts.

CHZ-1 migrates as a single band in urea-SDS gel electrophoresis and in velocity ultracentrifugation. Gel electrophoresis and sedimentation equilibrium ultracentrifugation indicated a molecular weight around 9kDa, as expected.

CHZ-1 hydrolyzes the chymotrypsin substrates acetyltyrosine ethyl ester (ATEE), benzoyl-tyrosine ethyl ester (BTEE) and benzyloxy-carbonyltyrosine p-nitrophenyl ester (ZTONP). These are all standard much-studied ester substrates for CHTr (Walsh and Wilcox, 1970). The pH optimum for CHZ-1-mediated hydrolysis of these esters is about 8.5, similar to that of ChTr. The observed decrease in the rate at higher pH demonstrates that the mechanism of hydrolysis is not simply attack on the substrate by hydroxide ion. Hydrolysis of ATEE has been followed for more than 100 turnovers; the product is acetyltyrosine, as demonstrated by analytical HPLC. Hydrolysis of ATEE at 24°, pH 8.2, shows $k_M=1.0M$, $k_{cat}=0.042/sec$. For hydrolysis of ATEE by ChTr, $k_M= 0.7M$, $k_{cat}= 190/sec$ (Berezin et al., 1971; Caplow and Jencks (1964) give $k_{cat}= 150/sec$. The rate of hydrolysis is accelerated about 30% in 0.05M NaCl solution, consistent with the higher helix content of CHZ-1 in this medium. Since ATEE is quite stable in solution at pH 8.5, this rate of CHZ-1 catalyzed hydrolysis represents an acceleration of at least 1000-fold over spontaneous hydrolysis. The rate of hydrolysis of ZTONP by CHZ-1, corrected for spontaneous hydrolysis of this unstable ester, is approximately 3% that of hydrolysis of this substrate by ChTr.

Hydrolysis of all three substrates is

67

proportional to CHZ-1 concentration and shows saturation kinetics, although hydrolysis of BTEE and ZTONP is complicated by strong product inhibition. Although studies on this aspect are not yet complete, the dimer of CHZ-1 also appears to be catalytically active. CHZ-1 shows very slow hydrolysis of acetyltyrosine *p*-nitroanilide, an activated amide substrate.

CHZ-1 does not hydrolyze the standard trypsin substrate benzoylarginine ethyl ester. Hydrolysis is thus specific for ChTr substrates, consistent with the design of the substrate recognition site in CHZ-1. The hydrolytic activity of CHZ-1 is blocked by phenylmethylsulfonyl fluoride, an irreversible inhibitor of ChTr, and is inhibited reversibly by the chymotrypsin inhibitors indole and *p*-cresol.

Hydrolysis catalyzed by CHZ-1 shows enzyme-like temperature characteristics, but with an interesting variation. The rate of hydrolysis increases with temperature, roughly doubling for each 10-degree rise up to 65°; this rate increase is consistent with an energy of activation of about 12kcal/mole. Above 65°, a progressive loss of activity is seen. Thermally-inactivated samples of CHZ-1 remain essentially inactive upon cooling, although when such inactive solutions are lyophilized from weakly acidic solution, full catalytic activity is regained. This is a marked difference from nearly all native protein enzymes, which are permanently inactivated by heating. Whereas the complex three-dimensional structures of most native proteins cannot be restored upon cooling, the much simpler structure of CHZ-1 allows for spontaneous recovery of the designed catalytically-active structure when the molecule is placed in the appropriate solution. Interestingly, the heat-stable catalytic activity does not seem to be enzyme-like.

The catalytic activity of CHZ-1 is stable in solution at room temperature for many days. The structure of CHZ-1 contains no aromatic amino acids which would allow it to digest itself, again in contrast to most natural proteolytic enzymes.

Conclusions

Successful synthesis of "Chymohelizyme-1" demonstrates that the present state of knowledge of enzyme structure and mechanism, of protein and peptide folding, of computer molecular graphics and of solid phase peptide synthesis is adequate for the design and synthesis of peptides having enzyme-like characteristics. While the first helizyme designed and synthesized shows only modest activity, its properties can doubtless be much improved. Indeed, the first modifications already incorporated into the molecule do yield a marked rate enhancement. Additional modifications have been designed in the computer graphics system and are already in synthesis.

Acknowledgments

The authors thank Robert Coombs for the circular dichroism spectroscopy and Robert Binard for amino acid analyses. The research was made possible by contract N00014-86-K-0476 from the Office of Naval Research. Computer purchase was assisted by UCHSC award BRS-888.

References

Berezin, N.F.; Kazanskaya, A.A.; Klyosov, A.A. *FEBS Lett.* **1971**, *15*, 121-124.

Bernstein, F.C.; Koetzle, T.F.; Williams, G.J.B.; Meyer, E.F.; Brice, M.D.; Rodgers, J.R.; Kennard, O.; Shimanouchi, T.; Tasumi, M. *J. Mol. Biol.* **1977**, *112*, 535-542.

Blevins, R.A.; Tulinsky, A. *J. Biol. Chem.* **1985**, *260*, 4264-4275.

Blow, D.M. in *The Enzymes* ; Boyer, P.D., Ed.; Third edition. Academic Press: New York, NY, **1971**, Vol. 3, pp 185-212.

Caplow, M.; Jencks, W.P. *J. Biol. Chem.* **1964**, *239*, 1640-1652.

DeGrado, W.F. *Adv. Protein Chem.* **1988**, *39*, 51-124.

Dugas, H. *Bioorganic Chemistry* ; Springer-Verlag, New York, NY, **1989.**

Fairman, R.; Shoemaker, K.R.; York, E.J.; Stewart, J.M.; Baldwin, R.L. *Proteins Struct. Funct. Genet.* **1989**, *5*, 1-7.

Hahn, K.W.; Klis, W.A.; Stewart, J.M. *Science* **1990**, *248*, 1544-1547.

Hodges, R.S.; Semchuk, P.D.; Taneja, A.K.; Kay, C.M.; Parker, J.M.R.; Mant, C.T. *Peptide Res.* **1988**,*1* ,19-30.

Klis, W.A., Stewart, J.M. in *Peptides, Chemistry, Structure, Biology* , Rivier, J.E.,

Marshall, G.R., Eds.; ESCOM, Leiden, **1990**, pp. 904-906.

Kraut, J. *Science* **1988**, *242*, 533-540.

Kuroki, R.; Taniyama, Y.; Seko, C.; Nakamura, H.; Kikuchi, M.; Ikehara, M. *Proc. Nat. Acad. Sci. USA* **1989**, *86*, 6903-6907.

Landschutz, W.H.; Johnson, P.F.; McKnight, S.L. *Science* **1988**, *240*, 1759-1764.

Lerner, R.A., Iverson, B.L., *Science* **1989**, 243, 1184-1188.

Marqusee, S.; Baldwin, R.L. *Proc. Nat. Acad. Sci. USA* **1987**, *84*, 8898-8902.

Matsueda, R.; Walter, R. *Int. J. Peptide Protein Res.* **1980**, *16*, 392-401.

Mutter, M.; Vuilleumier, S. *Angew. Chem. Int. Ed. Eng.* **1989**, *28,* 535-554.

O'Shea, E.K.; Rutkowski, R.; Kim, P.S. *Science* **1989**, *243*, 538-542.

Pantoliano, M.W.; Whitlow, M.; Wood, J.F.; Dodd, S.W.; Hardman, K.D.; Rollence, M.L.; Bryan, P.N. *Biochemistry* **1989**, *28*, 7205-7213.

Richardson, J.S. *Methods in Enzymol.* **1985**, *115*, 341-380.

Sasaki, T.; Kaiser, E.T. *J. Am. Chem. Soc.* **1989**, 111, 381-383.

Schultz, P.G. Science **1988**, 240, 426-433.

Stewart, J.M.; Young, J.D. *Solid Phase Peptide Synthesis ;* Pierce Chemical Corp., Rockford, IL, **1984.**

Unson, C.B.; Erickson, B.W.; Richardson, D.C.; Richardson, J.S. *Federation Proc.* **1984**, *43*, 1837.

Walsh, K.A.; Wilcox, P.E. *Methods in Enzymology* **1970**, *19*, 31-41.

Warshel, A.; Naray-Szabo, G.; Sussman, F.; Hwang, J.-K. *Biochemistry* **1989**, *28*, 3629-3637.

Part II
Biosensors and In Situ
Measurement Systems

SPECTROSCOPIC TECHNIQUES that cover a considerable range of the electromagnetic spectrum are emphasized in this section. For the first time, the power and sensitivity of surface-enhanced Raman scattering have been applied to bioprocess problems. Fluorescence detectors and fiber optics are shown to be ascendent, and a genuinely biological sensor, the invertebrate neuron, is shown to function in isolation in response to specific analytes. Nuclear magnetic resonance can reveal metabolic events in living cells if they are properly maintained in a waveguide cavity.

Fiber-optic Sensors Using Raman and Surface-enhanced Raman Spectroscopy

S. M. Angel, M. L. Myrick, and F. P. Milanovich, Environmental Sciences Division, Lawrence Livermore National Laboratory, Livermore, CA 94550

We are investigating applications of remote Raman and surface-enhanced Raman spectroscopy (SERS) using fiber optics. Emphasis is placed on developing a fiber-optic Raman sensor that can be used with very long optical fibers. We are also investigating portable near-infrared excitation sources including diode and diode-laser-pumped Nd:YAG lasers with hopes of developing a field-portable Raman spectrometer. Some specific environmental, earth science, and biomedical applications are also considered in this paper.

This paper deals with a very general technique, Raman spectroscopy, and presents solutions to measurement problems encountered when using this technique with long optical fibers. Also, an enhanced Raman technique, surface-enhanced Raman spectroscopy, SERS, is presented and new excitation sources and fiber-optic sensor designs are discussed. Although these techniques are being developed for environmental and biomedical applications, they are also directly applicable to bioprocess monitoring.

Optical fiber chemical sensors (optrodes) are being developed by a number of different groups for groundwater monitoring. Optrodes are designed for specific target analytes, and are designed to provide limited information beyond the concentration of the target molecule. Because of this, they are most useful when the sample matrix is well understood. Remote fiber spectroscopic sensors, on the other hand, are more generally useful and can provide considerably more information than optrodes. Vibrational spectroscopies (e.g., Raman and infrared spectroscopy) are particularly useful for the detection of analytes in complex matrices because they provide a molecular "fingerprint" that, in many cases, can distinguish a particular molecule in a mixture. These very general techniques complement the more specific optrode sensors.

Raman Spectroscopy and SERS

Raman spectroscopy (as well as variants such as resonance Raman and surface-enhanced Raman spectroscopy) is attractive for remote sensing applications because the Raman spectrum contains molecular structural information. Thus, the vibrational information inherent in a Raman spectrum (like an infrared–IR–absorption spectrum) can potentially be used to discriminate among a large number of analytes, allowing molecules to be identified and concentrations to be determined. However, unlike an IR spectrum where the energy of the excitation light must be tunable, a Raman spectrum can be obtained using convenient single-wavelength visible and NIR laser sources. This is an advantage for fiber-optic sensor applications.

An IR spectrum is obtained by tuning an IR excitation source through energies that correspond to vibrational energies in the molecule (see Figure 1). When the excitation energy corresponds to the energy of a particular molecular vibration, light is absorbed by the molecule and a resonance peak is observed in the spectrum. Because vibrational energies are in the IR region (450 to 3500 cm^{-1}), light with a wavelength between 3 and 20 μm must be used. Unfortunately, for applications where measurements are to be made remotely, fiber optics have extremely high losses in this wavelength range. Thus, fiber-optic sensors based on IR spectroscopy are usually not desirable.

A Raman spectrum is obtained by measuring the inelastic scattering of photons from a molecule. Although this is a very inefficient process any wavelength of light can be used as long as the energy is higher than the molecular vibrational

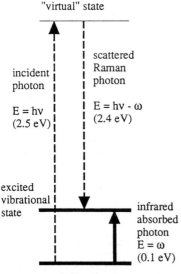

"virtual" state

incident photon

$E = h\nu$
(2.5 eV)

scattered
Raman
photon

$E = h\nu - \omega$
(2.4 eV)

excited
vibrational
state

infrared
absorbed
photon
$E = \omega$
(0.1 eV)

ground vibrational state

Fig.1   Energy level diagram showing the mechanisms for Raman scattering and IR absorption. Raman and IR both give vibrational information. IR requires a tunable excitation source whereas Raman requires a single-frequency source.

energy. In this case the molecule is excited into a short-lived transient or "virtual" state (Figure 1). The molecule loses energy or relaxes within a few vibrational time periods and can end up in the ground state or in an excited vibrational state. In the former case, the photon is scattered elastically with no loss of energy (Rayleigh scattering). In the latter case, the photon loses energy to the molecule (inelastic or Raman scattering) and the wavelength of the scattered photon is longer than the wavelength of the incident photon. The difference between the incident and scattered photon energies is the same as the energy of the excited vibrational mode in the molecule. Thus, a Raman spectrum, like IR, is a vibrational spectrum.

In general, Raman spectroscopy suffers from low sensitivity, so Raman analysis is typically performed on fairly concentrated samples. However, there are methods to obtain enhanced Raman signals, including resonance Raman spectroscopy (RRS) and surface-enhanced Raman (SER) spectroscopy. Both of these techniques provide greatly enhanced Raman signals (enhancements of 10^4 to 10^6) and, therefore, high sensitivity.

In the RRS technique, Raman enhancement requires absorption of the incident photon by the molecule. Many environmental contaminants are colorless and only absorb light in the UV. Thus, a UV laser would be required to obtain enhanced Raman spectra using the RRS technique. However, for highly colored molecules, RRS gives typical enhancements of 10^4 or larger and can be a useful technique for measuring low concentrations of these analytes.

In the SER technique, greatly enhanced Raman signals are observed for certain types of molecules adsorbed on rough metal substrates (Jeanmarie and Van Duyne, 1977). In this case, direct absorption of light by the molecule is not required for large enhancements; instead, there is an indirect interaction of the molecule with the electromagnetic field through the metal surface. Details of the SER phenomenon and general analytical applications will not be elaborated upon in this paper and have appeared in recent review articles (Chang and Furtak, 1982; Garrel, 1989). From an experimental standpoint, it is necessary to have molecules directly adsorbed to a rough metal surface and in many cases, a potential is applied to the metal to attract the molecule. Also, the excitation light must be of an appropriate wavelength to launch a surface electromagnetic wave (surface plasmon) on the metal surface. Experimentally, for commonly used metals such as Ag, Cu, or Au, this usually means using green or red wavelength excitation . These are the minimum requirements for observing large SER enhancements.

SERS has been a focus of much study since it was first reported in 1974 (Fleischmann et al., 1974,) but only recently has attention been given to its analytical applications (Vo-Dinh, et al., 1984; Meier et al., 1985; Enlow et al., 1986; Shen   et al., 1986; Tran, 1986; Torres and Winefordner, 1987; Berthod et al. 1987). Most of our work has involved the use of metal electrodes and colloids. The colloids are particularly attractive because of the ease of sample preparation that they allow. Figure 2 shows an example of the kind of sensitivity that can be obtained using SERS. This figure shows a SER spectrum of 10-ppm 3-chloropyridine on a Cu colloid compared to an unenhanced Raman spectrum of neat 3-chloropyridine. Both spectra were taken

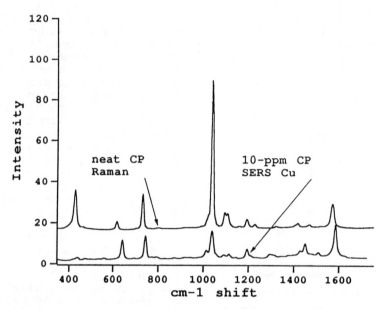

Fig.2 SERS of 10-ppm 3-chloropyridine on a Cu colloid using 1.064 mm excitation (bottom) compared to normal Raman of the neat liquid (top). All spectroscopic conditions are the same for the two spectra.

under identical experimental conditions using 1.064-μm excitation. However, only molecules that are adsorbed on the Cu colloid contribute to the SER spectrum. The signal-to-noise ratio in this case was such that ppb levels could be measured. The 1.064-μm excitation used here is not typically used for SERS (excitation at visible wavelengths is more common). However, as will be discussed in a later section, we are interested in NIR-excited SER and Raman spectroscopies because this spectral region offers many advantages for remote measurements using optical fibers.

Fiber-Optic Sensor Designs

Fiber-Optic Spectroscopic Sensors

Many types of fiber-optic systems have been described for making spectroscopic measurements (Trott and Furtak, 1980), and a central issue in the design of such a system is the geometry of the excitation and collection optics at the distal end of the fiber. The simplest optical configuration is a single bare polished fiber. This type of probe can be useful for measuring reflectance or luminescence , and, because of its

simplicity, it is the most common geometry for optrode measurements. A single-fiber probe also has the advantage of providing perfect overlap between the excitation and collection fields at the end of the fiber. This results in easy alignment and high signals.

Major interferences in fiber-optic luminescence measurements are the background spectral signals that are generated in the fiber. These result primarily from Raman scattering and fluorescence of the fiber core and cladding materials. Thus, one major disadvantage of the single-fiber geometry for luminescence measurements is that the background generated in the optical fiber is difficult to remove and, in many cases, limits the length of fiber that can be used. For this reason, it is advantageous to use separate fibers for excitation and collection.

Figure 3 shows two dual-sampling geometries that have been used by many researchers. Figure 3A shows how one fiber can be used for excitation and another fiber used for collection of luminescence or Raman scattered light. This geometry solves the background problem for non-scattering samples, although it has a significantly lower optical collection efficiency than a single fiber. Furthermore, while it may be useful for

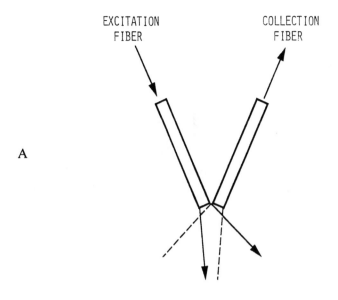

A

B

EXCITATION
FIBER

COLLECTION
FIBER

Fig.3 Two dual-fiber sensor configurations that are useful for reflectance and fluorescence (top) and transmission measurements (bottom).

reflectance and fluorescence measurements, it is not useful for transmission measurements. Figure 3B shows another useful dual-fiber geometry. In this case, the two fibers are pointed at each other, and lenses are attached to each fiber to collimate the light and form a well-defined optical path. The light that exits one fiber is collected by the other. This configuration is ideal for measuring transmission, and it also works well for luminescence spectroscopy. The lenses are necessary for luminescence or Raman measurements if optical path lengths greater than a few millimeters are used. Some advantages of this geometry include large overlap between the excitation and collection volumes and a well-defined optical path. Also, this very efficient design allows high signal levels to be obtained.

Fiber-Optic Raman Measurements

Raman measurements over optical fibers are more difficult than fluorescence measurements because Raman signals are generally much weaker than fluorescence

signals. Also, the wavelengths of the Raman bands are usually closer to the laser excitation wavelength than fluorescence bands, thus requiring high spectral rejection of the Rayleigh-scattered light. It has been shown that a major obstacle in the successful exploitation of Raman spectroscopy with long optical fibers is interference from the large Raman background emission of the fiber itself (Dakin and King, 1983). This background emission makes the detection of Raman signals with single-fiber probes impossible with all but the shortest fibers. Although a lot of work has appeared describing transmission and luminescence measurements using optical fibers, relatively little has been published describing remote Raman spectroscopy (Trott and Furtak, 1980; Dakin and King, 1983; McCreery et al., 1983; Schwab and McCreery, 1984, 1986, and 1987; Newby et al., 1984; Dao et al 1986; Lewis et al., 1988; Archibald and Honigs, 1988; Myrick et al., 1990). In fact, there have only recently been reports of Raman measurements in the "signature" region using very long optical fibers (>100 m),

75

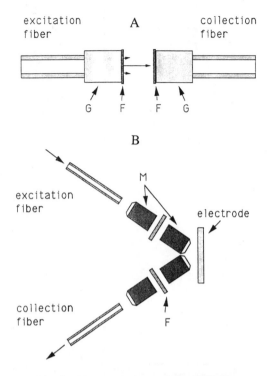

excitation fiber   A   collection fiber

G   F   F   G

B

excitation fiber

M

electrode

collection fiber

F

Fig. **4**. Two variations of the OFF fiber-sensor configuration used for Raman or fluorescence measurements.

and the technique has not been widely applied to date (Myrick and Angel, 1990).

Figure 4 shows two dual-fiber configurations that we have found especially useful for fiber-optic Raman measurements. Figure 4A shows a 180° dual-fiber configuration (optical fiber with forward scattering — OFF) that works well using long fibers. This design is very similar to the one shown in Figure 3B. Miniature graded-index (GRIN) rod lenses (G) are placed at the end of each fiber to collimate the light, and optical filters (F) are used at the end of each GRIN lens to remove background emission that originates at the laser source (plasma emission, etc.) or in the optical fibers. This optrode geometry has very good collection efficiency compared to that of a single fiber because the excitation and collection volumes almost completely overlap (Myrick et al., 1990; Myrick and Angel, 1990a).

In the OFF design, the excitation light is directed toward the collection fiber, so that the advantages of dual-fiber measurements can be lost unless optical filters are used that reject Raman or

fluorescence signals originating in the laser, fibers, or lenses. Plasma emission from the laser source, fiber background emission (Raman and fluorescence), and lens fluorescence are all eliminated by a narrow band-pass filter placed immediately after the excitation lens (see Figures 4A and B). Laser light is prevented from entering the collection optics by a long-pass filter immediately before the collection lens. This prevents generation of Raman or fluorescence in the collection optics and fiber. The filters also serve another minor role; because they reflect light at wavelengths that they reject, the laser light and the Raman signal both make two passes through the cell, resulting in increased sensitivity. However, due to the close proximity of the filters to the collection fiber in the OFF configuration, filters must be selected that generate the least possible luminescence so they do not interfere with the measurements. The proper choice of filters permits Raman spectra to be measured over a wide spectral range with little interfering background. This probe is ideal for highly scattering conditions such as those found in natural water samples and colloidal solutions, and it is also easily miniaturized. A variation of the OFF configuration is shown in Figure 4B. This configuration is especially useful for opaque samples or for front-surface illumination. We are currently using this configuration for making SER measurements on metal electrodes.

Figure 5 shows Raman spectra of neat toluene (A) and chloroform (B) using the OFF sensor and 5-m optical fibers. These spectra demonstrate the wide spectral range available using this type of sensor. Figure 6 shows the normal Raman spectrum of neat benzene using both 5-m fibers (A) and 100-m fibers (B) in the OFF configuration. There is essentially no difference between the two spectra, even though there is considerable background emission generated in the 100-m fiber. The larger background in the top spectrum is the result of a fluorescent impurity in the sample. The filters effectively remove this background with little attenuation of the Raman signal. The very broad background seen in these spectra results from a luminescent impurity in the benzene and does not originate in the optical fibers. This is indicated by the fact that its intensity is independent of the length of the fiber.

The size of the OFF sensor is

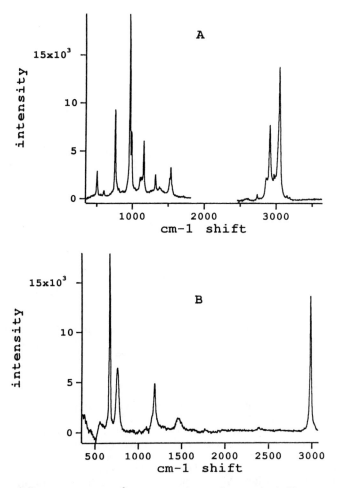

Fig.5. Normal Raman spectra of neat toluene (A) and chloroform (B) measured using a 5-m optical fibers sensor in the OFF configuration.

limited by the diameter of the GRIN lenses and filters. This diameter is 1.8 mm for the lenses used for the majority of our work. Although this is considerable larger than many fiber-optic sensors, the OFF configuration has many advantages. Among these are high sensitivity and the ability to easily control the sampling volume. The latter is very important for applications where the sample is some distance from the probe tip such as measuring fluid composition in a pipe.

Fiber-Optic SERS Measurements

As stated previously, the OFF geometry is well suited to measuring SER spectra with highly scattering colloid solutions. This is illustrated in Figure 7.

Figure 7A shows the SER spectrum of 10^{-5} M pyridine while Figure 7B shows the SER spectrum of 10^{-5} M [Ru(2,2'-bipyridine)$_3$]$^{2+}$,RB3, on a Ag colloid measured over optical fibers. In these experiments, a 4-m-long 200-μm-diameter fiber was used to deliver the excitation light and a 4-m-long 400-μm-diameter fiber was used to collect the Raman scattered light.

Most of our SER studies have been on metal electrodes and the OFF configuration shown in Figure 4B has been used for these measurements. Figure 8A shows SER spectra of 0.01 M pyridine obtained on a Ag electrode using 250-m fibers and an excitation wavelength of 496 nm. This was a worst-case test because

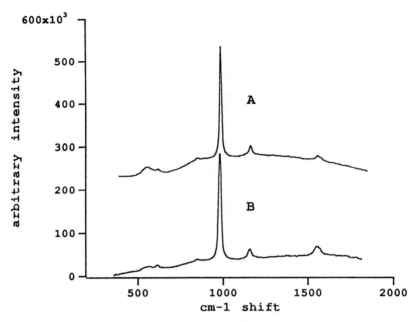

Fig.6. Normal Raman spectra of neat benzene measured using 5-m (A) and 100-m (B) optical fibers in the OFF configuration. All spectroscopic conditions are the same for the two spectra.

the 496-nm excitation caused relatively intense filter and fiber luminescence. Even when such long fibers are used there is no indication of an increase in the background signal, indicating that the filters are completely eliminating the background that results from fiber emission. Most of the broad background shown in Figure 8A is due to the SER substrate and is also seen without the optical fiber (Curve B).

Applications of Remote Raman Spectroscopy

On-Line Raman Measurements

A long-path design, similar to the one described above, allows sampling to be done through cell walls or transparent pipes containing the fluid to be tested with little loss of sensitivity. This allows the optrode to be isolated from the measurement area. To investigate the possibility of making measurements of fluid composition during a flowing intermix of miscible liquids, we performed Raman studies on a flowing mixture of chloroform and toluene. In these experiments, a 1-cm-diameter flow cell was used and the optrode was completely isolated from the fluids by the cell walls (see Figure 9). Spectra were measured every 0.3 s for approximately 1.5 min for a total of 320 spectra. The results of this experiment are shown in Figure 10 (every 10th spectrum is shown).

In this experiment, chloroform in the cell is displaced by flowing toluene, which is then itself displaced by fresh chloroform. Initially, only the two main chloroform Raman bands are seen around 720 and 830 cm^{-1}. These disappear upon introduction of toluene and are replaced by toluene Raman bands around 590, 855, and 1100 cm^{-1}. The introduction of fresh chloroform is seen as the reappearance of the 720 and 830-cm^{-1} bands. This simple experiment shows that this technique can be used to characterize the extent of mixing in a mixing chamber or, alternatively, to determine the concentrations of and identify chemicals in real time with a totally nonintrusive probe.

Oceanographic Measurements

One field application of remote fiber spectroscopy (RFS) that we have been exploring involves the real-time measurement of algae fluorescence in sea

78

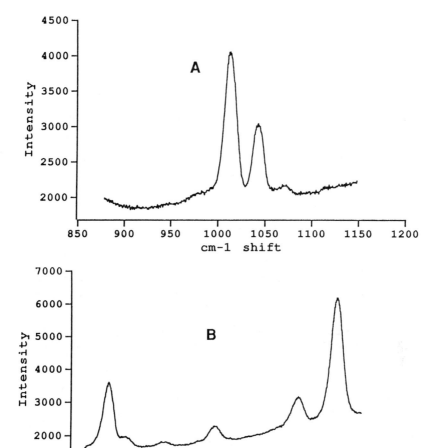

Fig. 7 SER spectra of 5×10^{-5} M pyridine (A) and 10^{-5} M RB3 (B) on Ag colloid using 4-m-long optical fibers and the OFF configuration.

water using long optical fibers (manuscript in preparation). In this application, the probe is attached to a freely falling microstructure profiler called the Rapid Sampling Vertical Profiler or RSVP (Cowles et al., 1989). This oceanographic probe is deployed behind a ship and it is allowed to free fall to a depth of 110 to 130 m at a speed of about 0.5 m per second. Spectra are acquired every 33 msec using a 200-m optical fiber cable for a total of 6600 to 7800 complete spectra per drop. Thus, it is possible to measure microstructure at a resolution of less than 2 cm.

Figure 11 shows the design of the probe. It is based on the OFF configuration shown in Figure 4A. In order to minimize its size, a 200-μm excitation fiber is bent in a 180° loop at the end of the ~1-cm diameter probe and a 400-μm collection fiber is used. Water flows freely through the probe as it falls. The entire probe is pressure sealed using epoxy, and the fibers are protected against breakage by sealing them into teflon sleeves. The fibers extend about 2 m beyond the probe, where they are terminated in stainless steel connectors (SMA 905 Environmental). These are then attached to a 200-m cabled optical fiber that is coupled to a laser-based spectrometer on the ship.

Optical filters in the probe were selected to allow the simultaneous

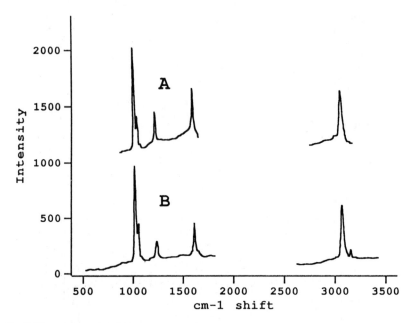

Fig. 8 SER spectra of 0.01 M pyridine on a Ag electrode using 250-m optical fibers (A) and without fibers (B). All spectroscopic conditions are the same for the two spectra.

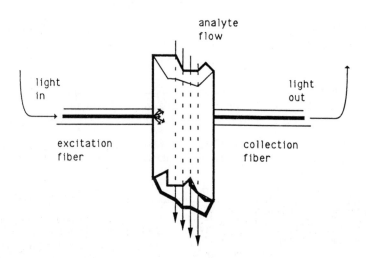

Fig. 9 Fiber-optic cell configuration used to measure real-time Raman spectra of flowing liquids.

measurement of the algae chlorophyll emission band at 680 nm and the 3500-cm^{-1} water Raman band at ~580-nm, using 488-nm excitation. For this application, the water Raman band is used as an internal reference to normalize spectra for laser intensity and fiber transmission fluctuations. Figure 12 shows an example of one such spectrum. The chlorophyll fluorescence and water Raman bands are clearly visible. This spectrum is a single 33 msec exposure obtained over a 200-m fiber-optic cable with the probe at a depth of 38.2 m using

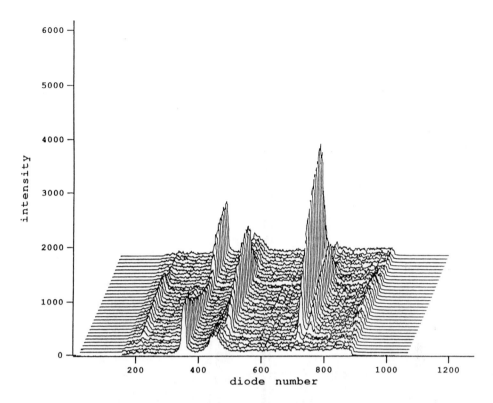

Fig.10 Real time Raman measurements of chloroform and toluene using 5-m optical fibers and the OFF configuration. Diode numbers correspond to wavelength of transmitted light.

about 50-mW of the 488-nm laser line at the input end of the fiber. The sensitivity of the probe is such that chlorophyll concentrations below 0.2 µg/L were routinely measured.

Figure 13 shows a depth profile of chlorophyll fluorescence and water temperature. Approximately 6600 separate spectra were used to generate this plot. These data are being used along with RSVP measurements of salinity, water shear forces (i.e., mixing), density, and fiber-optic measurements of available solar light intensity as a function of depth to determine how microstructures of ocean algae form. For example, it is immediately apparent from this figure that the highest concentrations of algae were present at a depth of about 45 m in this particular area. This corresponded to a sharp temperature gradient, and shear measurements indicated a large amount of mixing in this region. This is evidence that conditions are optimal (in terms of nutrients, sunlight, etc.) for algae growth. Also, the intense fluorescence spikes correspond to small clumps of algae. Depth profiles were found to be dependent upon many parameters including time of day, season, and water temperature.

Groundwater Contaminant Measurements

As mentioned above, analytical applications of SERS have only begun to be explored. One use that has already been mentioned is the measurement of environmental contaminants. Of particular interest is the characterization of water contaminated with fuels. In this case, relatively high levels of compounds such as benzene, toluene, and xylenes would be present as well as low levels of polyaromatic compounds. Many of these are ideal candidates for SER analysis.

Chemical effects are very important and should be considered in any investigation of analytical applications of SERS. For example, quinoline shows a large SER

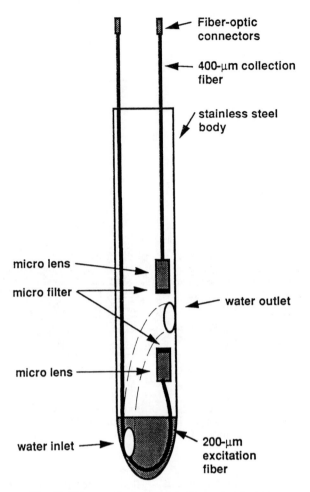

Fiber-optic connectors

400-μm collection fiber

stainless steel body

micro lens

micro filter

water outlet

micro lens

water inlet

200-μm excitation fiber

Fig. 11 Fiber-optic oceanographic probe design.

enhancement on Ag (see bottom of Figure 14) but relatively little enhancement on Cu under the same conditions. This suggests that quinoline does not stick to the Cu surface. The SER spectrum of pyridine is shown in the top of this figure for comparison. Pyridine shows large SER enhancements on Ag or Cu electrodes under these same conditions. Also, 3-Cl-pyridine and 3-methyl-pyridine show large SER enhancements on Cu but little enhancement on Ag (see Figure 15).

SERS of Mixtures

Another chemical effect that must be addressed in developing analytical applications of SERS is the analysis of samples containing mixtures of different compounds. SERS is a surface phenomenon and only molecules close to the SER substrate surface show enhanced spectra. In a mixture, there will be competition for surface sites. Obviously, if one species is more highly attracted to the surface than another, the latter will not show much SER activity. Therefore, it is important to understand the experimental parameters that affect surface adsorption and competitive binding. This area of research has not yet received much attention by spectroscopists interested in developing analytical applications of the SER technique.

As an example of how experimental parameters can effect the SER spectra of mixtures, we have investigated simple two-component mixtures. Figure 15 shows the SER spectrum of a mixture of 3-chloropyridine (CP) and 3-

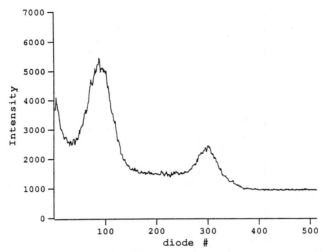

Fig.12 Ocean algae chlorophyll fluorescence spectrum (680 nm) and water Raman spectrum (~580 nm) using the fiber-optic oceanographic probe attached to 200-m-long optical fiber cable. This is a single 33-msec scan using 50-mW of the 488-nm laser line into the input end of the optical fiber. Diode numbers correspond to transmitted light.

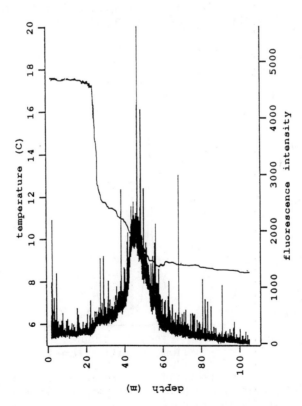

Fig.13 Depth profile of ocean algae fluorescence intensity obtained with a single drop of the fiber-optic oceanographic probe.

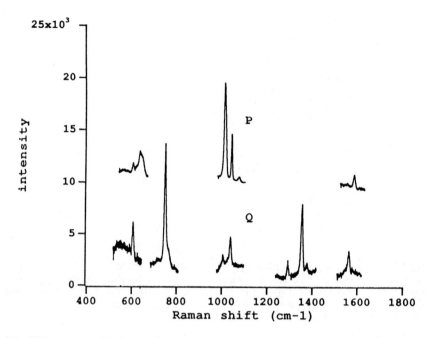

Fig **14** SER spectra of 0.01 M pyridine (top, P) and 0.01 M quinoline (bottom, Q) on Ag electrodes using 807 nm excitation from a diode laser.

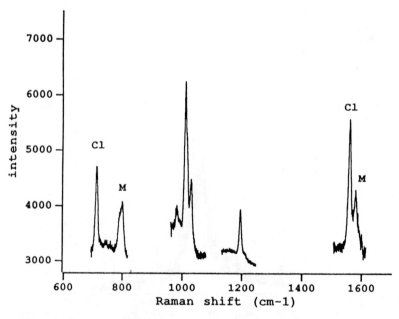

Fig.**15** SER spectrum of a mixture of 0.01 M 3-chloropyridine and 0.01 M 3-methylpyridine on a Cu electrode using 807 nm excitation from a diode laser. The electrode potential was 1.0 V versus SCE.

84

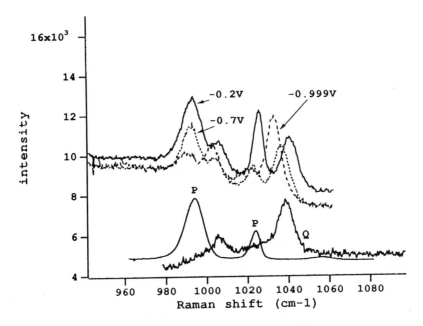

16×10^3

Fig.16 Potential dependent SER spectra of a mixture of 0.01 M pyridine (P) and quinoline (Q) on a Ag electrode (top). The

SER spectra of the individual components are also shown (bottom) at a potential of - 0.6 V versus SCE.

methylpyridine (MP) on a Cu electrode. This spectrum was obtained at an optimal potential of -1.0 V versus saturated calomel electrode (SCE) even though the optimal potential for either component alone is -0.6 V versus SCE. The optimal electrode potential for obtaining a SER spectrum is an important parameter that must be controlled and understood for analytical applications of SERS.

Figure 16 shows a more interesting example of the complex behavior of mixtures. In this case, a mixture of quinoline and pyridine was measured on a Ag electrode at different potentials. For either component alone the optimal potential is about -0.6 V versus SCE. However, for the mixture different bands were observed to show different potential effects. In the 1000 to 1100-cm^{-1} region, shown in Figure 16, the pyridine band at 1000 cm^{-1} decreased while the quinoline band at 1045 cm^{-1} increased in intensity as the potential was made more negative. But, in other regions of the spectrum this was not observed. This obviously is related to competitive adsorption and potential-dependent surface sites; however, it is not completely understood even for simple mixtures.

Drug Measurements

Another application we are exploring is the development of a drug assay based on Raman and SERS. This would involve the identification of certain types of drugs using portable Raman instrumentation. Normal Raman would be used for high concentrations of the compound (e.g., powders) while SERS would be used for low concentration solutions. The goal is to use a diode laser for excitation, Ag or Au colloids as the SER substrate, and an optical fiber probe. We have demonstrated the potential for this type of application by measuring 0.5-mM cocaine and methamphetamine solutions with 40-nm commercial Au colloids (Angel et al., 1990).

Figure 17A shows the normal Raman spectrum of a saturated cocaine solution. This spectrum contains many bands that are unique to cocaine and could be used to identify it. Figure 17B shows the SER spectrum of 0.5 mM cocaine using a commercially available Au colloid solution in a 1-cm disposable sample cuvette. The use of the commercial Au colloid simplifies the assay, but sensitivity is reduced compared to those made in the

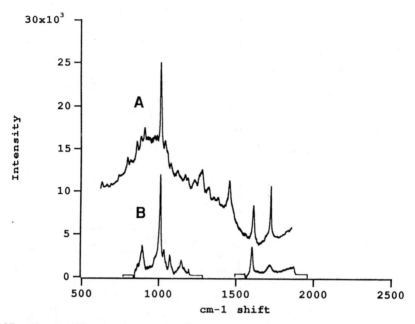

Fig.17 Normal Raman spectrum of cocaine saturated solution (A) compared to the SER spectrum of 0.5 mM solution (B) on a 40-nm Au colloid using 647-nm excitation.

laboratory. The sample was prepared by adding the analyte to the Au colloid solution, 40-nm 0.01%, and adjusting the pH to 8 by addition of carbonate. The SER spectrum appeared almost immediately. Although these spectra were measured using 647-nm excitation, the optimal excitation wavelength for this particular Au colloid was about 632 nm.

The optical fiber setup described above (Figure 4A) is being used to measure SER spectra of the drugs with metal colloids as the SER substrate. Figure 18 shows the spectrum of 0.5 mM cocaine using 4-m optical fibers and a Ag colloid (B) compared to the SER spectrum measured on Au colloid without optical fibers (A). In this figure only the most intense band at 1000-cm^{-1} is shown. Sample preparation for the Ag-colloid solution was the same as that described for the Au-colloid solution except the Ag colloid was first made by reduction of $AgNO_3$ with $NaBH_4$ using standard literature methods.

SERS Using Diode Lasers

Recent developments in diode laser technology have made them attractive NIR excitation sources for Raman spectroscopy. Powerful 830-nm single-mode GaAlAs lasers, up to 100-mW, are now commercially available and complete systems can be obtained for less than the cost of an air-cooled ion laser of similar power. The advantage of the diode laser is stability and long life. Also, the wavelength range of GaAlAs lasers is such that photomultiplier tubes and charge-coupled-device arrays can be used. These detectors are more sensitive and have a higher S/N ratio than those typically used for Fourier Transform (FT) Raman measurements. Thus, a diode-laser-based NIR Raman spectrometer may be an attractive alternative for FT Raman in the near future for certain applications.

We recently published a quick, approximate approach to determine the optimal excitation wavelength for Raman spectroscopy over optical fibers and concluded that NIR excitation can actually result in higher Raman signal levels when long optical fibers are used (Myrick and Angel, 1990b). This is because for long optical fibers, fiber losses may dominate the normal λ^{-4} dependency and result in a preference for longer-wavelength excitation. NIR excitation is especially

86

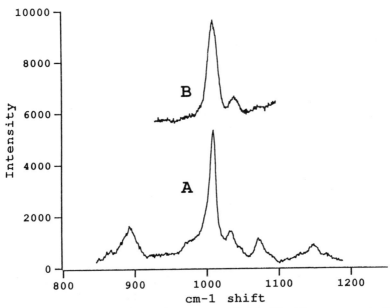

Fig.18 SER spectra of 0.5 mM cocaine solutions using 40-nm Au colloids (A) and Ag colloids (B). The spectrum in B was measured over 5-m optical fibers.

advantageous for fiber-optic SER measurements. We have investigated the excitation-wavelength dependence of the SER enhancement in the NIR and extended the wavelength range for SER-enhancement measurements for Ag, and Cu electrodes to 1.064 μm. Our preliminary data indicates that the SER enhancement is much larger for Ag and Cu electrodes in the 780- to 840-nm wavelength range available from diode lasers than it is at 1.064 μm or in the visible region available from ion lasers. Excitation in the 780- to 840-nm range is attractive for a number of reasons: (1) it gives large SER enhancements, (2) luminescence is reduced for most samples, (3) good detectors are available, (4) diode lasers provide high output in this range, and (5) optical fibers transmit well in this range.

We recently reported the use of a 785-nm GaAlAs diode laser for SER measurements of pyridine and RB3, on Ag and Cu electrodes using a GaAs-type photomultiplier tube detector (Angel and Myrick, 1989), Although only ~4 mW of laser power was used, high quality spectra were measured for these two compounds because the SER enhancement for these two metals is greatest in the NIR wavelength region. Elimination of sample luminescence using the 785-nm diode laser is shown by the SER spectrum of RB3 in Figure 19. Figure 19A shows the SER spectrum of 6 mM RB3 on a Cu electrode. The measurement was made with the electrode placed in the bulk RB3 solution. For comparison, curve B was measured under identical conditions using a 19-mW 632-nm HeNe laser for excitation, rather than the diode. Note the vast reduction in luminescence for the diode-excited spectrum.

The diode laser used for the work described above was of moderately low power compared to others that are now commercially available. We recently obtained a 100-mW 811-nm GaAlAs diode laser. In fact, the SER spectra shown in Figures 14 to 16 were obtained using this laser. Even at this power level the diode laser is already an important research tool. It seems likely that as more powerful diode lasers become available they will become very important in Raman spectroscopy. A diode-laser based spectrometer with a CCD detection system would make an attractive and relatively low-cost alternative to an FT-Raman system for measuring SER (or normal Raman) spectra of highly

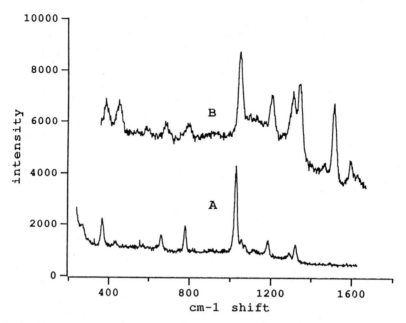

Fig. **19** SER spectrum of 6 mM RB3 on a Cu electrode using a 4-mW 785-nm diode laser (A) and a 19-mW 632-nm HeNe laser (B).

fluorescent compounds. Such a system could also be made small enough to be useful for field applications of RFS.

CONCLUSIONS

Raman spectroscopy provides structural information about molecules and can be used to identify molecules even in mixtures, and surface-enhanced spectroscopy provides the same type of information for certain types of compounds with high sensitivity. The fiber-optic Raman sensor described here can be used for sensitive remote *in situ* measurements even for fibers that are very long. It is sensitive and versatile enough to be useful for a number of different applications. Diode lasers are already being used for measuring Raman spectra and will probably become important excitation sources in the future.

ACKNOWLEDGMENTS

The work at Lawrence Livermore National Laboratory was conducted under the auspices of the U.S. Department of Energy under contract W-7405-Eng-48. The authors would like to express thanks to Paul Duhamel of the Office of Health and Environmental Research for supporting their research under RPIS No. 003906.

References

Angel, S.M.; Roe, J.N.; Andresen, B.D.; Myrick, M.L.; Milanovich, F.P. SPIE, Optical Fibers and Sensors in Medical Diagnostics Raman Applica- tions in Fiber Diagnostics, Los Angeles, CA, Vol. 1201, paper 1201-66, 1990.
Angel S.M.; Myrick, M.L. Anal. Chem., 1989, 61, 1648.
Archibald, D.; Honigs, D. Appl. Spectrosc., 1988, 42, 1558.
Berthod,A.; Laserna, J.J.; Winefordner, J.D. 1987 A. Appl. Spec., 1987, 41, 1137.
Jeanmarie,D.; van Duyne, R.J.; J. Electroanal. Chem. Interfacial Electrochem., 1977, 84, 1.
Chang, R.K.; Furtak, T.E., Eds; Surface Enhanced Raman Scattering, Plenum Press, NY, 1982.

Cowles, T. Moum, J. Desiderio, R.; Angel, S.M. Appl.Optics, 1989, 28, 595.

Dakin, J.; King, A. Proc. Optical Fibre Sensors Conf. IEEE p. 195, 1983.

Dao,N.Q.; Prod'homme, M.; Plaza, M. P.; Joyeux, M. C.R. Acad. Sci. Paris, 1986, 302, 313.

Enlow, P.D.; Buncick, M.; Warmack, R.J.; Vo-Dinh, T. Anal. Chem., 1986, 58, 1119.

Fleischmann, M.P. Hendra, P.; McQuillan, A. Chem. Phys. Lett., 1974, 26, 163.

Garrell, R.L.; Anal. Chem., 1989, 61, 401A.

Lewis, E.N.; Kalasinsky, V.F.; Levin, I.W. Anal. Chem., 1988, 60, 2658.

McCreery, R.L.; Fleischmann, M.; Hendra, P. Anal. Chem., 1983, 55, 146.

Meier, M.; Wokaun, A.; Vo-Dinh, T. J. Chem. Phys., 1985, 89, 1843.

Myrick, M.L.; Angel, S.M.; Desiderio, R. Appl. Optics, 1990, 29, 1333.

Myrick, M.L.; Angel, S. M. Appl. Spectrosc., 1990, 44, 565.

Myrick, M.L.; Angel, S.M. Appl. Opt., 1990, 29, 1350.

Newby, K.W.; Reichert, M.; Andrade, J.D.; Benner, R.E. Appl. Optics, 1984, 23, 1812. Schwab, S.D.; McCreery, R.L. Anal. Chem., 1984, 56, 2199.

Schwab, S.D.; McCreery, R.L. Anal. Chem., 1986, 58, 2486.

Schwab, S.D.; McCreery, R.L. Appl. Spectrosc., 1987, 41, 126.

Sheng, R.S.; Zhu, L.; Morris, M.D. Anal. Chem., 1986, 58, 1116.

Tran, C.D. Anal. Chem., 1984, 56, 824.

Torres, E.L.; Winefordner, J. D. Anal. Chem., 1987, 59, 1626.

Trott, G.R.; Furtak, T.E. Rev. Sci. Instrum., 1980, 51, 1493.

Vo-Dinh, T.; Hiromoto, M.Y.K.; Begun, G.M; Moody, R.L. Anal. Chem., 1984, 56, 1667.

Non-Invasive Spectroscopic Monitoring of a Bioprocess

Anna G. Cavinato, David M. Mayes, Zhihong Ge, and **James B. Callis**, Center for Process Analytical Chemistry, Department of Chemistry, BG-10, University of Washington, Seattle, WA 98195.

A sensor based on Visible (400–700 nm) and Short-Wavelength Near-Infrared (SW-NIR: 700–1100 nm) spectroscopies is used for on-line monitoring of ethanol concentration and the aerobic/anaerobic status of a batch *Saccharomyces cerevisiae* fermentation. Measurements are acquired non-invasively by means of a fiber-optic equipped photo-diode array spectrophotometer. Analysis of ethanol is performed in the SW-NIR region of the spectrum. The aerobic/anaerobic status of the fermentation is diagnosed by monitoring absorbances of cytochromes c, b, and cytochrome oxidase (a, a_3) in the visible region. These *in-situ* indicators of the concentration of oxygen appear to be more sensitive than conventional oxygen sensors applied in fermentation technology.

The ability to control a fermentation process is completely dependent on the degree to which physical and chemical parameters of the fermentation can be measured. Direct information on the physiological status of the microbial population as well as biomass, substrates, secondary metabolites and product concentration are therefore essential for accurate process control. Despite the wide recognition of the lack of suitable sensors for this purpose (Humphrey, 1974; Clark 1985), to date only a few efforts have been made to develop on-line analyzers for fermentation broths (Dinwoodie, 1985; Luli, 1987; Schugerl, 1988; Garn, 1989; Scheper, 1986). Unfortunately, most of these sensors still require invasive sampling techniques such as membrane filtration. A non-invasive method would be desirable, because sterility problems can be avoided and sampling is simplified (Clark, 1985). Spectroscopy has received little attention as a tool for on-line monitoring of fermentations. Both visible and Short Wavelength Near Infrared (SW-NIR) have the potential for real-time, non-invasive, multiparameter monitoring. In the visible region of the spectrum (400–700 nm) the redox level of the cytochromes of intact monocellular organisms can be determined by diffuse transmittance spectrophotometry as shown by Chance and colleagues (Chance,

1956). They showed that there is a gradient of oxidation-reduction level along the respiratory chain which is a precise indicator of the rate and nature of metabolism of the microbial population. The application of this spectrophotometric technique to continuous monitoring of a bioreactor has been explored by Nagel (Nagel, 1986). In the short-wavelength near-infrared (SW-NIR) region of the spectrum (700–1100 nm), one can observe the low energy electronic states of hemes and cytochromes (Eaton, 1981), upon which are superimposed the second and third vibrational overtones of CH, OH, and NH stretches, together with combination bands from other types of vibrations (Weyer, 1985). The exact position of the bands depends on the chemical environment giving rise to a high degree of uniqueness of the spectra for different organic molecules.

Although the low extinction coefficients of short-wavelength near-infrared transitions may seem a disadvantage, they can actually be of great utility for analysis of major constituents in the 0.1–100% concentration range. Long pathlengths can be used, ensuring that a spectrum is more representative of the bulk and that a thin layer of adsorbed materials on the optical window will not fatally degrade the results. Fiber-optic components, conventional monochromators, tungsten lamps and silicon detectors can provide signal-to-noise ratios on the order of 10,000:1, thus allowing very subtle changes in the spectra to be reliably used for analysis. Because this sensing technology is readily implemented in multichannel form (Mayes, 1989), multiple characteristics of a bioprocess can be potentially monitored simultaneously.

One apparent disadvantage of SW-NIR spectroscopy is that the spectral resolution is not high enough to ensure that absorbance bands arising from different components in a mixture will be free of interference. This necessitates the use of multivariate statistical calibration such as Multilinear Regression (MLR), Principal Component Regression (PCR), and Partial Least Squares (PLS) (Sharaf, 1986). Results obtained by Alberti, et al. (Alberti, 1985) using FT-IR data show the advantages of this quantitative technique. These authors measured the infrared spectra of fermentation broths and extracted quantitative information about glucose, ethanol and glycerol, using multivariate analysis. At the Center for Process Analytical Chemistry, SW-NIR spectroscopy and multivariate calibration have been used for gasoline quality evaluation (Kelly, 1989); for measurement of caustic and caustic brine (Phelan, 1989) and for the determination of ethanol in a fermentation broth (Cavinato, 1990). NIR technology is routinely used in agricultural industry to measure the protein, moisture and starch content of grain (Norris, 1983) and has already been used to follow solid phase fermentations (Norris, 1983).

In this paper, the feasibility of continuously analyzing ethanol during production and of monitoring the aerobic/anaerobic status directly through the walls of a glass fermentation vessel is demonstrated. Monitoring and control of ethanol production is of obvious interest in the fermentation industry. Currently, alcohol is measured off-line by specific gravity (pycnometry and hydrometry) or refractometry (Hormitz, 1980). Although these methods are reproducible to 0.1 to 0.2%, they suffer from several drawbacks such as a long and complicated anal-

ysis and large sample size (Kovar, 1985). Other analytical methods for the measurement of ethanol in beverages and fermentation broths include gas chromatography (Cutaia, 1984), Fourier Transform Infrared spectroscopy (Kuehl, 1984), nuclear magnetic resonance (Guillou, 1988), near-infrared spectroscopy (Buchanan, 1988; Halsey, 1985), laser Raman spectrometry (Gomy, 1988), immobilized enzymes (Walters, 1988), and flow injection analysis (Worsfold, 1981). Though most of these techniques are amenable to on-line implementation, some of them do not have enough precision, while others are lengthy and/or expensive; most have no potential for non-invasive analysis. Monitoring of the metabolic state of cell populations has mostly been limited to an indirect measurement of dissolved oxygen (pO_2) in the fermentation broth. More recently, the on-line fluorescent measurement of intracellular NADH levels to monitor cell metabolism has been reported (Armiger, 1986; Zabriskie, 1978; Harrison, 1970). However, the interpretation of total fluorescence from whole culture broth is quite involved since changes in the fluorometric signal of NADH may be affected by changes of various parameters other

Table 1: Defined Medium for *Saccharomyces Cerevisae* **Fermentation**

Component	Amount (g/L)
Glucose	150–200
Yeast Extract	10.4
Ammonium Sulfate	9.2
Potassium Hydrogen Phosphate	2.9
Magnesium Sulfate	1.25

than the extent of dissolved oxygen, such as cell growth and availability of C-source (Zabriskie, 1978; Armiger, 1986).

Materials and Methods

Fermentation. The yeast *Saccharomyces cerevisiae* (Novo-Nordisk A1339-3) was grown on the defined medium listed in Table 1 with glucose as the carbon source. Fermentations were run in batch mode in a Bioflow III reactor from New Brunswick Scientific, Inc., of 5L capacity. Ethanol production was carried out under anaerobic conditions at 30°C with an agitation speed of 300 rpm. The aerobic/anaerobic status of the culture was modified by varying agitation speed and O_2 flow rate. The dissolved oxygen concentration was measured with a polarographic probe (Phoenix Electrode Company). Samples of 2 mL volume were withdrawn periodically over the course of the fermentation for off-line analysis of ethanol by gas chromatography (Martin, 1981).

Visible and NIR Spectroscopy. Visible and SW-NIR spectra were collected on a Hewlett Packard 8452A photodiode array based spectrophotometer with near-infrared option (470–1100 nm). The instrument was modified for use with a bifurcated fiber-optic probe (Sterngold Corporation; 1 m in length, 6 mm outside bundle diameter, 2 mm inside bundle diameter) (Fig. 1). With this configuration, the instrument attains a spectral resolution of 4 nm and a signal to noise ratio of 10^4:1 at zero absorbance with a twenty-five second averaging time and a

0.1 second acquisition time. The ethanol-water mixtures were analyzed in transmission mode using a 1 cm quartz absorption cell. The ethanol/yeast/water mixtures as well as the fermentation were monitored by placing the bifurcated fiber-optic probe (Weyer, 1985; Mayes, 1989) up to the side of a $1 \times 1 \times 4$ cm quartz cuvette or the fermentor glass vessel wall. Some of the light scattered by the particles in the medium is collected by the inner fiber bundle. This light is guided to the HP8452A and collimated by a 42 mm focal length asymmetric lens. The collimated light is reflected onto the existing monochromator focusing lens by a first surface mirror. For the visible range a 150-watt Osram Xenophot HLX lamp and for the NIR a tungsten/halogen lamp (Osram # 64635) with gold reflector, both operated from a highly stable DC power supply, were employed.

Data Analysis. Data analysis was carried out on an IBM PC-AT. Spectra were first smoothed and a second derivative transformation was calculated using a 26 nm window with software developed in-house (Burns, 1989). In analyzing and comparing spectral data from different experiments, the same smoothing and second derivative parameters were used. For calibration and prediction of

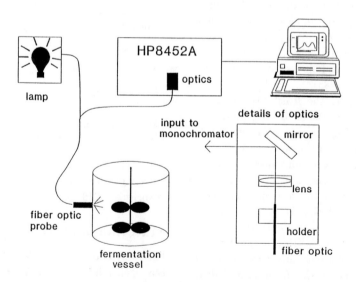

Fig. 1. Diagram of instrument.

ethanol content in a sample, stage-wise multiple linear regression (MLR) provided by Near Infrared Systems (NIRS, 1989) was used.

The standard error of prediction was from a cross validation estimate (Sharaf, 1986) that uses all but one sample as a calibration set to form a prediction equation, and then a prediction is made of the remaining sample. This "leave-one-out" exercise was repeated for each sample in the training set and the standard error of prediction was determined from the predicted and actual values for samples omitted.

Results and Discussion

On-line monitoring of ethanol. The development of a method of analysis based on our spectroscopic sensor consists of the following steps:

1. Qualitative spectral analysis to assess the major spectroscopic features of the analytes of interest;

2. Development of a calibration model: this is performed by submitting to spectral and reference chemical analysis several samples of the analyte of interest in amounts that span the concentration range to be predicted. Correlation at significant wavelengths is achieved by multivariate statistical analysis.

3. Determination of unknown sample.

A series of preliminary experiments was run to characterize the spectroscopic behavior of ethanol in aqueous solutions and assess the nature of the problems encountered in quantifying ethanol in presence of a changing background, particularly with regard to cell density variations during the course of the fermentation process.

Spectroscopic characterization of ethanol-water mixtures. The spectra of pure water and pure ethanol in the spectral region 800–1050 nm in a 1 cm cuvette are shown in Fig. 2. The most prominent band at 960 nm arises from the overtone combination motion $(2\nu_1 + \nu_3)$, where ν_1 is the symmetric OH stretch and ν_3 the bending mode. The

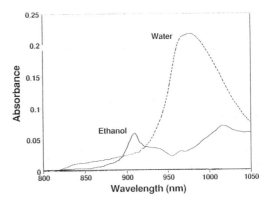

Fig. 2. Short-Wavelength Near-Infrared absorbance spectra of pure ethanol and water.

band is particularly broad due to the presence of two or more types of hydrogen bonded molecular complexes (Scherer, 1979). In addition, a weaker absorption at 829 can be distinguished. The ethanol spectrum is more complex due to the presence of three NIR active functional groups (methyl, methylene, and hydroxide). The band at 905 nm is particularly important to this study and is assigned to the third overtone of the CH stretch on the methyl group. The shoulder at 935 nm is assigned to the third overtone CH stretch on the methylene group. The band at 960 nm is assigned to the OH stretch. The first two assignments are consistent with those previously made for hydrocarbons (Weyer, 1985) and the ratio of intensities reflects the ratio of protons on the methyl and methylene functional groups, respectively.

To exemplify the basic concepts of the use of SW-NIR spectroscopy and multivariate calibration, a series of ethanol-water mixtures in the 0–15% (w/w) concentration range were analyzed. A stage-wise Multilinear Regression (MLR) analysis of the second derivative spectra show a very high correlation (Correlation Coefficient R = 0.999) at 905 nm corresponding to the methyl stretch of ethanol. The use of second derivative spectra eliminates baseline offsets due to cuvette placement in the single beam spectrophotometer and provides better spectral resolution. Since additional wavelengths do not improve the correlation, a single wavelength model using $\lambda = 905$ nm was selected. The standard error of calibration (SEC) with this model was 0.17% (w/w). The standard error of prediction (SEP) obtained from a cross validation estimate was 0.19% (w/w) (R = 0.998) (Fig. 3).

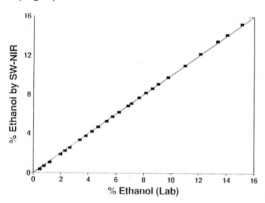

Fig. 3. Measurement of ethanol concentration by Short-Wavelength Near-Infrared in ethanol-water mixtures.

noise to the measurements as demonstrated by a SEP of 0.27% (w/w) (R = 0.980) obtained from a cross-validation analysis which is significantly greater than that of the ethanol-water mixtures (Fig. 4).

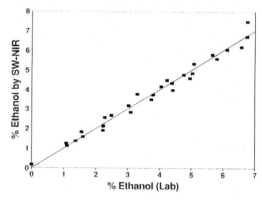

Fig. 4. Measurement of ethanol concentration by Short-Wavelength Near-Infrared in water-ethanol-yeast mixtures

Effect of Light Scatter. During the time course of a fermentation, the number of yeast cells may increase by several orders of magnitude. This will result in greater light scattering, which may either increase or decrease the apparent absorbance through changes in the effective pathlength (Reynolds, 1976). This pathlength change could also affect ethanol determination. In order to investigate the influence of cell mass, a series of water-ethanol-yeast mixtures were analyzed in which the ethanol was varied between 0–7% (w/w) and the yeast concentrations varied between 0–17 g/l. Results from an MLR analysis on the second derivative spectra indicate that a one wavelength ($\lambda = 905$ nm) model similar to that constructed for ethanol-water solutions, but with different slope and offset, successfully predicts ethanol concentration despite the presence of yeast. This remarkable result suggests that the absorbance at 905 nm is not influenced by pathlength changes at least in the cell mass concentration range investigated.

However, the presence of these highly scattering organisms contributes considerable

Quantitative Determination of Ethanol in Fermentation. The production of ethanol in the medium defined in Table 1 was followed in real time. Spectra were taken at half-hour intervals for a 30-hour period. A representative set of spectra is shown in Fig. 5. The most obvious change in the spec-

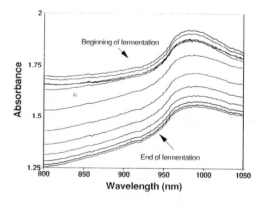

Fig. 5. Short-Wavelength Near-Infrared absorbance spectra of fermentation broth.

tra is the decrease in relative baseline absorbance with increased time of the reaction. With increased scattering material, relative reflectance increases, causing more light to return to the detector (Reynolds, 1976), thus decreasing the relative absorbance. At the beginning of the fermentation, when yeast production is very slow, the baseline offset remains constant. However, as the yeast cells begin to rapidly reproduce, the baseline offset decreases, and when yeast reproduction stops, the offset again becomes constant. While there is no ethanol information evident in this set of spectra because of the baseline-offset variation and the large size of the water peak, calculation of the second derivative enhances the third overtone methyl CH stretch (Fig. 6). The spectral information is further

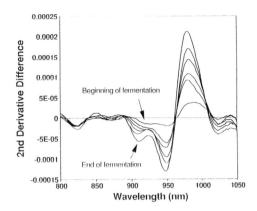

Fig. 7. Short-Wavelength Near-Infrared second derivative difference spectra of fermentation broth.

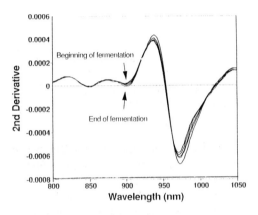

Fig. 6. Short-Wavelength Near-Infrared second derivative absorbance spectra of fermentation broth.

enhanced if a spectrum, from early in the fermentation process, is subtracted from the data set (Fig. 7). A calibration model was constructed using a single wavelength (905 nm) of the second derivative spectra by correlating SW-NIR concentration values with independent reference values obtained by off-line GC analysis of the fermentation broth. This correlation resulted in a SEC of 0.19% (w/w) (R = 0.993). Prediction of ethanol production over the time course of the fermentation is shown in Fig. 8. Also shown is the concentration of ethanol measured by GC on samples withdrawn from the reaction vessel immediately after the recording of a spectrum. The shape of this curve follows the ethanol production pattern of a typical batch fermentation (Wang, 1979).

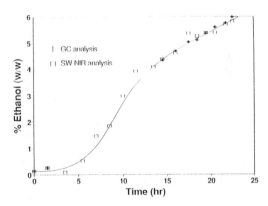

Fig. 8. Measurement of ethanol production during the time course of a fermentation: + Gas Chromatography analysis; □ Short-Wavelength Near-Infrared analysis.

Applicability of Calibration Constants to Succeeding Fermentation. The final series of studies was designed to investigate whether a model developed for one fermentation could be applied to succeeding runs. Accordingly, two further experiments were undertaken, which were duplications of the first. Fig. 9 shows that the model developed on the first fermentation could be successfully applied to the second and third fermentation. The linear regression equation for the production of ethanol was:

$$\%\text{ethanol } (w/w) = K(0) + K(1) * A_{2nd\lambda=905}$$

where $K(0) = 2.82$ and $K(1) = -70888$.

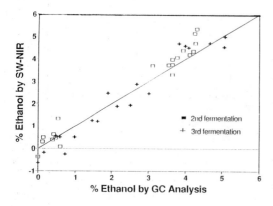

Fig. 9. Prediction of ethanol concentration during other fermentations: □ 2nd fermentation; + 3rd fermentation.

The standard error of prediction was 0.42% (w/w) and R = 0.95. Conversely, models constructed with either the second or third fermentations could be used to predict the other two with similar values of SEP and R. These results are all the more remarkable when one considers that the probe was removed and replaced between each fermentation.

On-line monitoring of the aerobic/ anaerobic status. The physiological status of the fermentation was diagnosed in real-time by monitoring the cytochromes bands in the 500–650 nm region. The spectral properties of cytochromes *c*, *b* and *aa₃* in intact yeast suspension under anaerobic (reduced) and aerobic (oxidized) conditions are shown in Fig. 10. The absorptions are as-signed respectively to the α bands of cytochrome *c* (552 nm), cytochrome *b* (564 nm) and cytochrome *aa₃* (604 nm) (Estabrook, 1956). Changes in the extracellular environment from aerobic to anaerobic conditions cause large intensity changes in the cytochromes bands. Fig. 11 shows a set of spec-

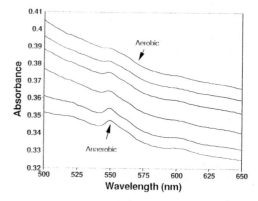

Fig. 11. Absorbance spectra of yeast cytochromes curing a fermentation.

tra recorded on a fermentation broth with a cell density of approximately 20 g/L. The increase in cytochromes absorption bands due to the transition from the aerobic to anaerobic state is visible along with baseline off-sets due to changes in the scattering properties of the medium. A second derivative transformation of the same set of spectra is shown in Fig. 12. A cross section of the sec-

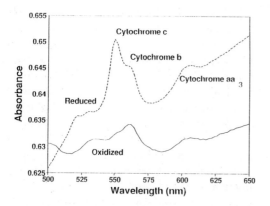

Fig. 10. Absorbance spectra of yeast cytochromes: (- - -) reduced, (-) oxidized.

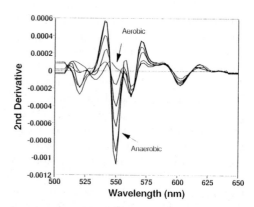

Fig. 12. Second derivative spectra of yeast cytochromes during a fermentation.

ond derivative absorption at 552 nm relevant to cytochrome *c* shows high correlation with the degree of cytochrome oxidation (Fig. 13).

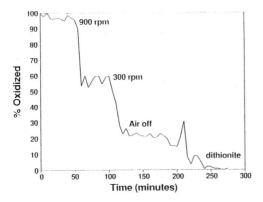

Fig. 13. Redox profile of cytochrome c (552 nm) during a fermentation.

At $t = 0$ the medium was inoculated with 20 g/L of yeast and aerobic conditions were insured by high stirring (900 rpm) and air flow (3 L/min) rates. After 1 hour the stirring rate was decreased to 300 rpm, causing a sharp reduction in the system and then a stabilization at approximately 55% oxidized; at 2 hours the air to the fermentor was turned off. This caused an additional sharp reduction to 20% oxidized, followed by a slow decline due to complete depletion of oxygen. Complete anaerobicity was attained by addition to the broth of sodium dithionate, an inhibitor of the respiratory chain. It should be noted that the step-increases in absorbance actually occur faster than changes recorded by the DO probe and that a certain degree of aerobic metabolism is still present at DO levels next to zero. These findings suggest that the spectroscopic sensor has more rapid response relative to the DO probe and that its response to oxygen concentration still occurs at oxygen levels which are below the sensitivity range of membrane oxygen electrodes (Harrison, 1970).

Conclusion

These experiments demonstrate that visible and SW-NIR spectroscopies provide a valuable tool for on-line monitoring of a bioprocess. The major advantage of this sensor is that it is non-invasive and requires only that the reactor vessel have a quartz or glass window. Thus, the need for elaborate sterile sampling systems and anti-fouling probes is eliminated. Another advantage of this approach is the apparent robust characteristic of the calibration, as shown by the ability to apply a model developed on one fermentation to the prediction of succeeding fermentations. The method in itself represents an excellent tool for monitoring and control of ethanol production fermentations. However, the present technology does not yet allow monitoring of ethanol trace level, as in the case of aerobic fermentations.

Preliminary results on monitoring aerobic/anaerobic yeast physiology based on cytochromes spectroscopy suggest that these *in-situ* sensors are better indicators of the microscopic intracellular environment than commercially available sensors.

Acknowledgment

We are grateful to Michael deBang (Novo-Nordisk, Inc.) for useful discussions.

Funding for this research was provided by the Center for Process Analytical Chemistry. We are indebted to the Instrumentation Donation Committee of Hewlett-Packard for the gift of the NIR spectrophotometer and to Novo-Nordisk, Inc., for the gift of the Bioflow III fermentor.

References

Alberti, J.C.; Phillips, J.A.; Fink, D.J.; Wacasz, F.F. *Biotechnol. Bioeng. Symp.* **1985**, *15*, 689–722.

Armiger, W.B.; Forro, J.F.; Montalvo, L.M.; Lee, J.F. *Chem. Eng. Commun.* **1986**, *45*, 197–206.

Buchanan, B.R.; Honigs, D.E.; Lee, C.J.; Roth, W. *Appl. Spectrosc.* **1988**, *42*, 1106–1111.

Burns, D.H., personal communication, Department of Bioengineering, FL-10, University of Washington, Seattle.

Cavinato, A.G.; Mayes, D.M.; Ge, Z.; Callis. J.B. *Anal. Chem.* **1990**, *62*, 1977–1982.

Chance, B.; Hess, B. *Ann. N.Y. Acad. Sci.* **1956**, *63*, 1008–1016.

Clark, D.J.; Calder, M.R.; Carr, R.J.G.; Blake-Coleman, B.C.; Moody, S.C.M.; Collings, T.A. *Biosensors* **1985**, *1*, 213–320.

Cutaia, A.J. *J. Assoc. Anal. Chem.* **1984**, *67*, 192–193.

Dinwoodie, R.C.; Mehuert, D.W. *Biotechnol. Bioeng.* **1985**, *26*, 1060–1062.

Eaton, W.A.; Hoffrichter, J. *Methods in Enzymology* **1981**, *76*, 175.

Estabrook, R. *J. Biol. Chem.* **1956**, *223*, 781–794.

Garn, M.; Gisin, M.; Thommen, C.; Cevey, P. *Biotechnol. Bioeng.* **1989**, *34*, 423–428.

Gomy, C.; Jouan, M.; Quy Dao, N. *C.R. Acad. Sci. Paris* **1988**, *t. 306, Série II*, 417–422.

Guillou, M.; Tellier, C. *Anal. Chem.* **1988**, *60*, 2182–2185.

Halsey, S.A. *J. Inst. Brew.* **1985**, *91*, 306–312.

Harrison, D.E.F.; Chance, B. *Appl. Microbiol.* **1970**, *19*, 446–450.

Humphrey, A.E. *Chem. Eng.* **1974**, *81(26)*, 98–112.

Kelly, J.J.; Barlow, C.H.; Jinguji, T.M.; Callis, J.B. *Anal. Chem.* **1989**, *61*, 313–320.

Kovar, J. *J. Chromat.* **1985**, *333*, 389–403.

Kuehl, D.; Crocombe, R. *Appl. Spectrosc.* **1984**, *38*, 907–909.

Luli, G.W.; Schlasner, S.M.; Ordaz, D.E.; Mason, M.; Strohl, W.R. *Biotechnol. Techniques* **1987**, *4*, 225–230.

Martin, G.E.; Burgraff, J.M.; Dyer, R.H.; Buscemi, P.E. *J. Assoc. Off. Anal. Chem.* **1981**, *64*, 186–193.

Mayes, D.M.; Callis, J.B. *Appl. Spectrosc.* **1989**, *43*, 27–32.

Nagel, B.; Bayer, C.; Iske, U.; Glombitza, F. *Am. Biotech. Lab.* **1986**, *4(4)*, 12–17.

NIRS, Inc., 2431 Linden Lane, Silver Spring, Maryland 20910.

Norris, K.H. *Biotechnol. Bioeng.* **1983**, *25*, 603–607.

Norris, K.H. *NATO Adv. Study Ser. A* **1983**, *46*, 471–484.

Official Methods of Analysis of the Association of Analytical Chemists, Hormitz, W., Ed.; AOAC, Washington, DC, 1980.

Phelan, M.K.; Barlow, C.H.; Kelly, J.J.; Jinguji, T.M.; Callis J.B. *Anal. Chem.* **1989**, *61*, 1419–1424.

Reynolds, L.; Johnson, C.; Ishimaru, A. *Appl. Optics* **1976**, *15*, 2059–2067.

Scheper, T.; Schugerl, K. *J. Biotechnol.* **1986**, *3*, 221–229.

Scherer, J.R. In *Advances in Infrared and Raman Spectroscopy*, Clark, R.J.H.; Hester, R.E., Eds.; Heyden: London, 1979, Vol. 5; Issue 3, pp 149–216.

Schugerl, K. *Anal. Chim. Acta* **1988**, *213*, 1–9.

Sharaf, M.A.; Illman, D.L.; Kowalski, B.R. In *Chemometrics*, Elving, P.J.; Winefordner, J.D.; Kolthoff, I.M., Eds., John Wiley & Sons: New York, NY, 1986, Vol. 82; 254.

Walters, B.S.; Nielsen, T.J.; Arnold, M.A. *Talanta* **1988**, *35*, 151–155.

Wang, D.I.C.; Cooney, C.L.; Demain, A.L.; Dunnill, P.; Humphrey, A.E.; Lilly, M.D. In *Fermentation and Enzyme Technoloqy*, John Wiley & Sons: New York, 1979; p 79.

Weyer, L.G. *Appl. Spectrosc. Rev.* **1985**, *21*, 1–43.

Worsfold, P.J.; Ruzicka, J.; Hansen, E.H. *Analyst* **1981**, *106*, 1309–1317.

Zabriskie, D.W.; Humphrey, A. *Appl. Environ. Microbiol.* **1978**, *35*, 337–343.

In Situ On-line Optical Fiber Sensor for Fluorescence Monitoring in Bioreactors

John J. Horvath and Christopher J. Spangler, Center for Chemical Technology, National Institute of Standards and Technology, Gaithersburg, MD 20899

This work examines the fluorescence properties of natural cell components such as tryptophan, pyridoxine, NADH, and riboflavin in batch fermentations of S. cerevisiae. Measurements were made using both an on-line external flow-through cuvette loop and an in situ optical fiber/rod sensor. A good linear correlation was found between the dry mass of yeast and in situ fluorescence measurements of pyridoxine, tryptophan, and NADH. The sensitivity of pyridoxine and tryptophan fluorescence, for yeast cell mass measurement were approximately six and two times greater than NADH fluorescence, respectively. The pyridoxine and NADH signal were linear with the dry mass of yeast, while the tryptophan signal showed self-absorption effects starting at a dry mass of 0.5 g/l in the cuvette measurements. No deviations from linearity were observed with the in situ optical fiber/rod.

Various attempts in the past three decades have been made to measure on line the intrinsic fluorescence of a fermentation broth and relate the measured intensity to the cell mass and the metabolic state of the cells (Duysens and Amesz, 1957; Chance, 1954; Chance et al., 1964; Harrison and Harmes, 1972; Armiger et al., 1984). Most of the reported measurements have used varieties of commercial fluorometers in which the excitation and emission wavelengths are set to specified values by optical filter elements. Usually the excitation wavelength is chosen to be around 360 nm while the emission is recorded in a spectral region around 450 nm. This combination is best for detecting the molecule NADH. A commercial instrument designed to measure NADH fluorescence was introduced by BioChem Technology (Fluoro Measure™) in the 1970's followed later by Ingold's Fluorosensor™. These instruments use a mercury lamp with suitable filters for the excitation source, and detected the NADH fluorescence near 460 nm. Success of these attempts has been varied. In some cases, a good correlation between cell mass and the fluorescence signal was reported ((Ristroph, et al., 1977), or a causal relation established between the fluorescence signal and oxygen deprivation (Harrison and Chance, 1970). NADH fluorescence is only a good indicator of cell density when temperature, pH, and dissolved oxygen conditions are held constant and the energy substrate source is in excess. On the other hand, NADH culture fluorescence is a very rapid and good indicator of metabolic switches, energy substrate limiting conditions, and oxygen limitation (Einsele and Purhar, 1980).

Over the years roughly a half-dozen groups of investigators have actively pursued the use of on-line fluorometric measurements for monitoring bioreactors. Significant among these are Armiger, et al. (Armiger et al., 1986), Meyer, et al. (Meyer et al., 1984), Zabriskie and Humphrey (Zabriskie and Humphrey, 1978), Srinivas and Mutharasan (Srinivas and Mutharasan, 1987), Scheper, et al. (Scheper et al., 1987), and Horvath and Semerjian (Horvath and Semerjian, 1986). Their work has essentially involved the monitoring of NADH culture fluorescence with the exception of the last work which used the intrinsic fluorescence of yeast (tryptophan). Recently, Humphrey, et al. (Humphrey et al., 1989) have noted that other components of microbial cultures strongly fluoresce in the 300 to 550 nm region, overlapping the NADH fluorescence. These compounds include the aromatic amino acid, tryptophan and the B vitamins, pyridoxine and riboflavin. Humphrey and his colleagues have speculated on whether fluorometric monitoring of either the uptake or production of one of these compounds might be a better indicator of cell density and growth than monitoring NADH fluorescence.

In this study we examined the possibility of using the internic fluorescence of yeast cells for monitoring fermentations. Yeast is fluorescent due to the tryptophan which is part of the protein structures of the cell (Undenfriend, 1962; Konev, 1967). Absorption spectra of a typical protein, serum albumin, show a maximum at 280 nm, the absorbance maximum of tryptophan. In proteins containing all three aromatic amino acids tryptophan, phenylalanine and tyrosine only the fluorescence of tryptophan is observed (Teale, 1960).

In this paper we will present our techniques of using multiple excitation wavelengths and optimized varied emissions to observe the fluorescence signals from naturally occurring fluorescent species in a S. cerevisiae fermentation. The measurements were carried out in a bypass loop through which some of the fluid from a fermentor was continuously pumped to a cuvette located in a SLM fluorescence spectrometer before returning to the fermentor. Measurements were also made with an optical fiber/rod probe which was inserted into the fermentor. The optical rod was sterilized with the fermentation vessel and nutrient (medium). All recorded spectra from the fermentor were analyzed by comparing them to the spectra of individual fluorescing species which were shown to be present from our previous measurements.

Experimental Methods

Fluorometry: An SLM 8000C scanning spectrofluorometer, manufactured by SLM

Aminco Instruments, Inc.[*], was used in this study. This instrument used a 450 watt Xenon lamp as the light source and a double monochromator to select the excitation wavelength. The double monochromator was used to reduce scattering errors that occur when measuring turbid samples such as fermentation broths. The spectral bandpass was set at 4 nm, the typical scan duration was 3-5 minutes (depending on scan width), and the scan speed was 0.5 nm/s. The fluorescence signal of the whole culture broth was measured in a flow-through cuvette, with 1 cm path length, at a 90° angle (Figure 1). The signal was reported in units of relative intensity referenced to the signal of a reference cell containing a fluorescence quantum counter (rhodamine B in ethylene glycol), so all spectra are on the same scale. Spectra can be compared, i.e. an intensity of 10 on one figure is the same 10 on another, and all are self consistent. The experimental apparatus for the optical fiber/rod measurements is shown in Figure 2. For these measurements, the sample compartment of the SLM was removed and one leg of a bifurcated optical fiber bundle (Fiberguide Industries) was attached to the excitation monochromator and the other leg was connected to the emission monochromator. The common leg of the fiber bundle was removably attached to an unclad optical rod (Heraeus Amersil, Inc.) above the fermentor head plate, allowing the optical rod, which penetrated to within 4 cm of the bottom of the fermentor, to be autoclaved in situ with the vessel and nutrients (medium). The excitation light and the fluorescence light was transmitted by the distal end of the optical rod. Both the fiber bundle and optical rod were made out of

---

[*]Certain commercial equipment, instruments, and materials are identified in this paper in order to specify adequately the experimental procedure. In no case does such identification imply recommendation or endorsement by the NIST nor does it imply that the material or equipment is necessarily the best available for the purpose.

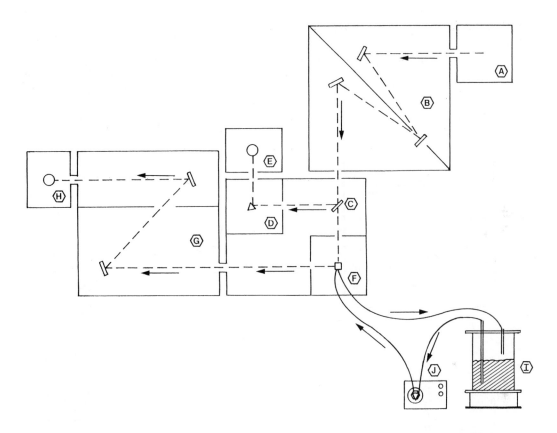

Fig. 1. Schematic diagram of the flow-through cuvette fluorescence experiment. Optical paths for the fluorescence measurements and flow paths for fermentation broth are indicated. A - 450 W Xenon Arc Lamp, B - Double Monochromator, C - Beam Splitter, D - Reference Cell, E - Reference PMT, F - Sample Cell, G - Single Monochromator, H - Emission PMT, I - Fermentor Vessel, J - Peristaltic Pump.

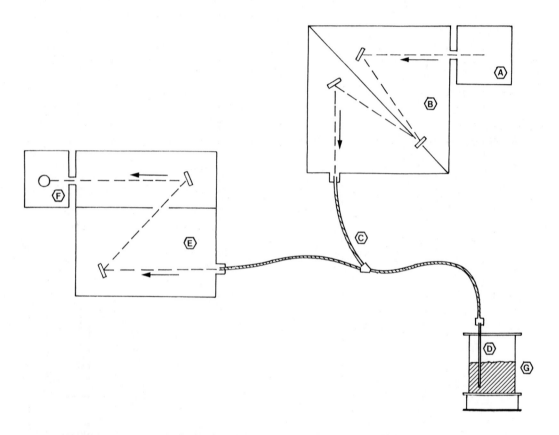

Fig. 2. Schematic diagram of the optical fiber/rod fluorescence experiments. A - 450 W Xenon Arc Lamp, B - Double Monochromator, C - Bifurcated Fiber bundle, D - Fused Silica Optical Rod, E - Single Monochromator, F - Emission PMT, G - Fermentor Vessel.

fused silica and the optical rod had a diameter of one-quarter inch. There was no reference channel as in the cuvette measurements but a blank background noise spectrum was measured at the beginning of each fermentation experiment.

For both experiments excitation wavelengths for yeast (tryptophan contained in cell proteins), pyridoxine, NADH and riboflavin were 290 nm, 320 nm, 345, nm, and 390 nm, respectively. The fluorescence spectra were integrated over the ranges of 320-360 nm, 380-420 nm, 440-480 nm and 510-550 for yeast, pyridoxine, NADH and riboflavin, respectively.

Microorganism: The yeast, Saccharomyces cerevisiae RTY110, containing the plasmid pRB58, was used as the model organism in this study. This plasmid-containing yeast secrets invertase and will grow on sucrose, although not as well as on glucose. It was for these reasons that this particular strain was picked. We believed that with this system it would be possible to follow, fluorometrically, the production of invertase. However, our early studies failed to detect any fluorescence that could be attributed to invertase when compared to our standard solution of 0.1% invertase. Chemical assays of several fermentations indicated that the plasmid was not producing invertase in very large quantities. In fact, the amount produced at the end of a fermentation was approximately 20 times below our detection limits for invertase. However, we observed fluorescence from yeast cultivations due to two different molecules contained in the cell, tryptophan and NADH. Two fluorescent compounds produced by the cells were identified as the B vitamins pyridoxine (B_6) and riboflavin (B_2). This gave us four independent fluorescent species to use for monitoring the progress of the fermentations. Yeast dry cell mass was calculated by filtering a known volume (usually 20 ml) of well-mixed fermentation broth through a .45 um filter. Careful microscopic examination of samples of the broth revealed that no foreign matter (e.g. precipitates) was responsible for any significant portion of the

dry weight. The filter and cells were then vacuum dried at room temperature overnight. The mass difference between the filter with and without cells was used to determine the dry cell mass.

Media: For the first set of studies, a complex medium was used for the growth and fluorescence studies. This medium contained 6.7 g/l of Bacto-yeast nitrogen base without amino acids (Difco) and 10 g/l of glucose, supplemented with 0.3 g/l of leucine. Unfortunately, this medium demonstrated a large broad-banded fluorescence signal over the region of interest, 300-500 nm. To have this large, unspecific, fluorescence background was undesirable for our studies. To avoid this background fluorescence problems we developed a synthetic, very low fluorescing media with composition as shown in Table 1. This medium was found to display essentially no fluorescence at all the excitation wavelengths and the absorption spectra were flat at zero absorbance units from 200 to 600 nm.

Table 1: Non-fluorescent Medium

Component	Concentration
Glucose	10.0 g/l
$(NH_4)_2SO_4$	5.0 g/l
KH_2PO_4	1.0 g/l
$MgCl_2$	0.5 g/l
$CaCl_2$	1.0 g/l
Inositol	2,000 ug/l
Niacin	400 ug/l
p-amino benzoic acid	200 ug/l
Thiamine	400 ug/l
Pantothenic acid	400 ug/l
$ZnSO_4$	400 ug/l
$FeSO_4$	200 ug/l
$CuSO_4$	40 ug/l
Biotin	4 ug/l

Growth Conditions: All fermentations were carried out using the Bioflo II fermentation system with a 3.3 L vessel containing 2.8 L of medium. The temperature and pH were microprocessor-controlled at 32°C and 6.0, respectively. The Bioflo II controlled the pH of the fermentation by addition of NaOH or HCl solutions as required. The dissolved oxygen (D.O.) electrode was calibrated at 0% with N_2 and 100% with 2.0 l/m air, at an agitation rate of 300 rpm. When using an exterior loop for the flow-through cuvette system, the dissolved oxygen level was maintained at 90% of saturation by adjusting the aeration and agitation rates. However, with the optical rod measurements, we observed inconsistent fluorescence data, which we were later able to attribute to changes in local environment due to agitation and aeration fluctuations. These are further discussed in the Results section. To eliminate these problems, we chose to keep the agitation and aeration rates constant at 300 rpm and 2.0 l/m, respectively, and to allow the D.O. level to respond to cell metabolism changes.

A nutrient-starved condition was created to allow observation of one of the yeast cell's metabolic constituents, NADH. This condition was characterized by a total absence of glucose in the fermentation broth.

This was ascertained by the use of a quantitative enzymatic test for glucose (Sigma Diagnostics -- Glucose (Trinder)). The effects of oxygen deprivation were studied by replacement of oxygen by nitrogen at the same flow rate. These techniques are similar to that of other researchers (Duyens and Amesz, 1957; Siano and Mutharasan, 1989).

Results

Flow-through Cuvette

The medium in Table 1 was found to be excellent for the growth of the yeast, and showed minimal background fluorescence. The next step was to determine the fluorescent species contained and produced in the fermentation. Our procedure was to compare spectra of pure molecules in water with the observed fluorescence in the fermentation broth. The species are shown in Figure 3, which is an overlay of spectra from three individual molecules in water with the yeast spectrum. The species are yeast at a concentration of 0.43 g/l (approximately

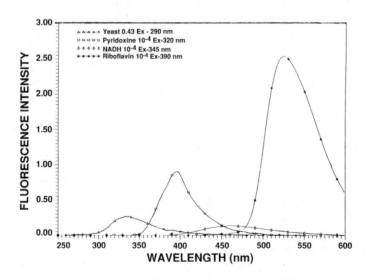

Fig. 3. Fluorescence spectra of species observed in the fermentation studies in this work. The yeast spectrum was obtained from the isolated solid component of the fermentation and resuspended in water (0.43 g/l). The other three species were made from stock aqueous solutions diluted to 10^{-4} m.

one-tenth the cell mass concentration at the end of the fermentation), pyridoxine, NADH and riboflavin. The latter three species were all shown at a concentration of 1 x 10^{-4} M in water. During a fermention the actual concentrations will differ widely. For example, right after the inoculum was added there was no measurable fluorescence from the yeast. After 50 hours, near the end of a batch fermentation, we observed a large fluorescence signal for a cell mass of approximately 4.5 g/l. In Figure 3, we see a fluorescence spectrum from 0.43 g/l dry mass yeast, a concentration that occurs at a fermentation time of 20 hours, one quarter of the total fermentation time. The total amount of yeast and the three other components constantly changed and never occurred at equal molar fractions. Figure 3 shows the respective spectral ranges and possible overlaps and interferences of the spectra of our four species. The actual magnitudes of the signals will be determined by the fermentation conditions.

The time evolution of a S. cerevisiae fermentation fluorescence spectrum is shown in Figure 4. This figure shows overlapping spectra taken at various times from zero hour (before inoculation) to almost 25 hr

(still in the exponential growth phase). In this figure, all spectra were obtained using an excitation wavelength of 290 nm. Several interesting features should be pointed out. In the first spectrum, at 0.0 hr, we observed a flat baseline from 320 to 500 nm with a rising baseline from 320 to 300 due to the scattered excitation light (290 nm). In fact, the strongest fluorescing species in Figure 3 is riboflavin, which showed the weakest fluorescence signal in a fermentation due to its low concentration. After 2 1/2 hr the yeast (tryptophan) signal appeared with a broad peak centered between 332-340 nm tailing off to the red past 400 nm. This spectrum continued to increase in intensity until around 8 hr when we observed the first signs of another peak centered at 392-408 nm, the combined signals of pyridoxine and NADH. These two peaks, yeast and pyridoxine/NADH continue to increase as shown by the 24:49 hr spectrum. Both peaks have increased greatly from the time of inoculation.

A better way to represent these data is shown in Figure 5, which is a fluorescence growth profile for the fermentation of S. cerevisiae. In this figure, we have three optical measurements (yeast, pyridoxine, and

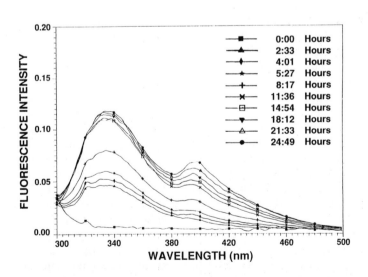

Fig. 4. Fluorescence emission spectra for excitation at 290 nm. The spectra were taken on line at indicated times during a fermentation of S. cerevisiae.

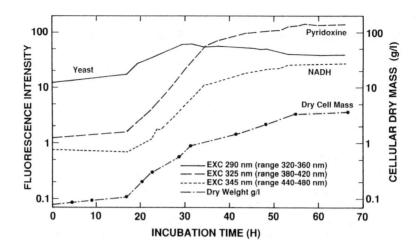

Fig. 5. Growth profile for a <u>S. cerevisiae</u> fermentation. Integrated fluorescence intensities taken with listed excitation (EXC) and emission wavelength range for each species (determined from pure solution measurements).

NADH) and one physical measurement (dry yeast cell mass) vs. fermentation time. The fluorescence of species was measured using a specific excitation wavelength and emission wavelength range. The wavelengths used are shown in Figure 5. In the dry cell mass curve, there are three of the four phases of a standard fermentation: (1). lag phase, 0 - 18 hrs; (2). exponential growth phase, 18 - 55 hr; (3). stationary phase, 55 - 68 hr. The fermentation was stopped before the fourth phase, the death phase, in which the cell mass starts to decline, was reached. The fluorescence signal characteristic of yeast followed the dry mass curve quite well until about 30 hr, at a cell mass of 0.5 g/l, at which the fluorescence signal started to decrease, while the cell mass increased. The yeast fluorescence signal continued to decrease as the fermentation continued. We attributed this decrease in the signal to both pre- and post-filter effects due to the high concentration and size (5-10 μm) of the yeast contained in the cuvette. The yeast cells are highly scattering and in a 1 cm square cuvette the actual fluorescent volume is small allowing the effects of scattering to reduce the excitation and emission beams resulting in a loss of signal. Any increase in fluorescence due to increased cell mass is counteracted by the increased scattering. However, in the fluorescence of pyridoxine and NADH we observed no effects of scattering due to their small size and concentration. As can be seen in Figure 5, the pyridoxine and NADH fluorescence signals both track the cell mass very well with the pyridoxine signal having an order of magnitude greater signal strength.

The pyridoxine signal appeared to be very sensitive to the cell concentration. For example, it increased nearly 40-fold during the 20 to 40 hour growth period, while the dry mass only increased 10-fold. This suggests that pyridoxine may be a very sensitive means of monitoring the cell concentration, at least in this yeast fermentation. A better way of indicating the relative sensitivity of pyridoxine fluorescence is shown in Figure 6, a plot of yeast dry mass vs. fluorescence intensity of the four cell growth parameters. We observed once again that the yeast leveled off out at a cell mass of ~0.5 g/l, the riboflavin showed no change at all with increasing cell mass, and the NADH correlated with cell mass but at relatively low signal strength. Pyridoxine exhibited a large linear range of fluorescence

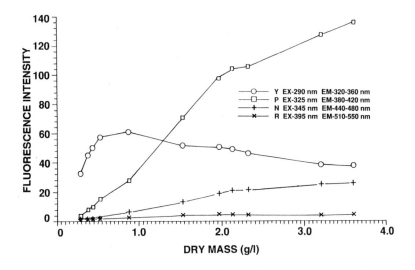

Fig. 6. Integrated fluorescence intensity individually optimized for the four fermentation species as a function of dry cell mass. Excitation wavelengths (EX) and integrated wavelength range (EM) for Y(yeast, P(pyridoxine), N(NADH) and R(riboflavin).

intensity versus cell mass. At the end of the fermentation, we can compare signal strength to cell mass to obtain a measure of sensitivity of each fluorescent species to cell mass, from which we obtain: yeast 11.1, pyridoxine 38.3, NADH 6.1, and riboflavin 0.8 in units of fluorescence signal per yeast cell mass.

Optical Fiber/Rod Measurements

The schematic diagram of the optical fiber/rod measurement system is shown in Figure 2. Figure 7 shows a plot of fluorescence intensity vs. concentration of pyridoxine using the optical fiber/rod fluorescence measurement system; a sensitivity of less than 10^{-7} M pyridoxine, is apparent. Figure 8 shows the fluorescence intensity of 5×10^{-5} M pyridoxine as a function of optical rod height from the bottom of the fermentor, and indicates increasing intensity up to a height of 4 cm, where the acceptance cone of the rod remains constant.

Discussions with other workers led us to investigate the effect of agitation and aeration rates on a constant signal (Rao, 1990). The data presented in Figure 9 indicates the way in which the signal from a 5×10^{-5} M pyridoxine sample varies as a function of agitation speed and aeration rates for a fixed optical rod position in the fermentor. Figure 9 indicates that for agitation rates between 300 and 800 rpm varying the aeration rate will not significantly change the observed signal. Figure 10 shows a growth profile of a yeast fermentation similar to Figure 5. An additional plot shows the dissolved oxygen concentration. These fermentation experiments with the optical fiber/rod were performed at constant agitation rate (300 rpm) and air flow (2.5 l/m).

The yeast pyridoxine and NADH fluorescence signals reflect the yeast mass growth. The pyridoxine signal is an order of magnitude greater than the NADH signal with a better signal-to-noise ratio. All three fluorescence signals correlate with the dry mass concentration, although there are differences between the individual fluorescence curves, the overall curves follow the dry mass growth curve.

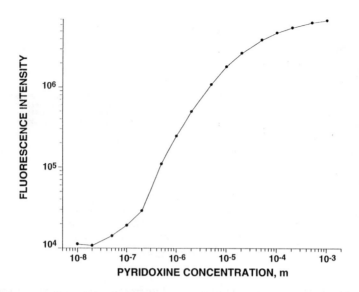

Fig. 7. Calibration curve for pyridoxine fluorescence using the optical fiber/rod. Integrated fluorescence intensity is plotted vs. pyridoxine concentration demonstrating good sensitivity and large dynamic range. Excitation wavelength (325 nm) and integrated fluorescence (380 - 420 nm).

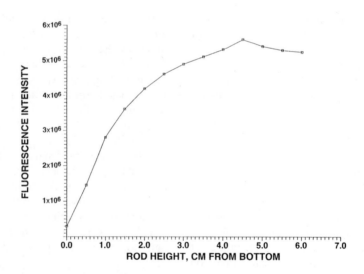

Fig. 8. Plot of integrated fluorescence intensity (380 - 420 nm) vs. height of optical fiber/rod from bottom of fermentor. A 5×10^{-5} M solution of pyridoxine was used as the test fluid.

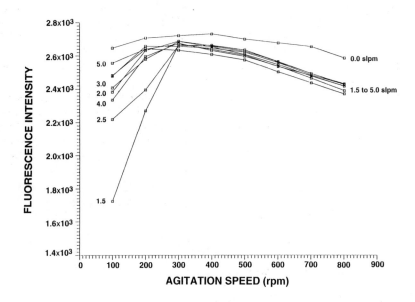

Fig. 9. Effect of agitation speed and aeration rates on the observed integrated fluorescence signal (380 - 420 nm) from the optical fiber/rod at a fixed position (4 cm above bottom) in a solution of 5×10^{-5}M pyridoxine.

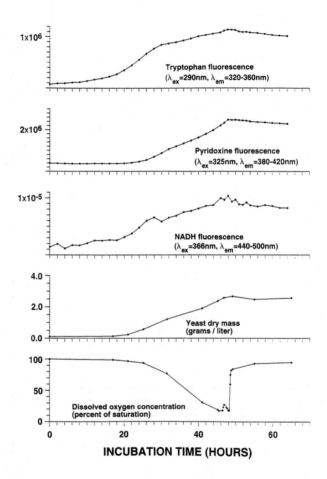

Fig. 10. Growth profiles for a <u>S. cerevisiae</u> fermentation obtained with the optical fiber/rod.

The optical rod collects fluorescence in the backscatter configuration while the cuvette measures in the right angle geometry. By measuring the fluorescence in situ with the optical rod, the linear range of cell mass measurement was greatly increased over external measurement. The other measured species, pyridoxine and NADH, were also monitored with better sensitivity than in the cuvette system. The dissolved oxygen concentration is a measure of yeast metabolic activity. There was little oxygen uptake when the yeast were dormant, but as exponential growth started, more oxygen was consumed until the yeast reached the stationary phase and the dissolved oxygen increased.

Figure 11 illustrates how the metabolic state of the yeast can be monitored by the fluorescence of NADH. Initially, the dissolved oxygen content was at 100% for glucose-starved yeast cells, which did not contain much NADH. As glucose was added, an immediate drop in dissolved oxygen occurred along with a large increase in NADH fluorescence as the yeast consume glucose. As the glucose was depleted the dissolved oxygen level rose to 100% and the NADH signal slowly decreased. Figure 12

illustrates the effects of glucose addition and oxygen deprivation to a carbon-starved yeast culture. Once again, upon addition of glucose we observed a rapid decrease of dissolved oxygen and an increase of NADH fluorescence. When we replaced the air by nitrogen we saw a large increase in NADH signal as the dissolved oxygen level dropped to 0%.

Discussion

Flow-through Cuvette

The results presented here indicate that cell and cell product fluorescence can be used to monitor cell mass in a non-fluorescent culture medium. The data show that measurement of the pyridoxine fluorescence may be a more sensitive indicator of cell mass than NADH fluorescence. The fluorescence from yeast was also stronger than that of NADH but only had a linear fluorescence response up to a cell mass of 0.5 g/l due to inner filter effects when measured in an external flow-through cuvette with a path length of 1 cm.

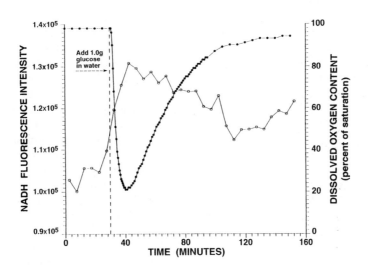

Fig. 11. Effects of addition of glucose to a carbon-starved fermentation on oxygen concentration and NADH fluorescence.

111

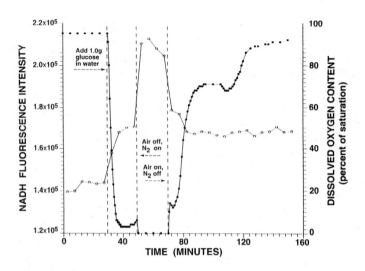

Fig. 12. Effects of glucose addition and oxygen deprivation on a carbon starved yeast culture.

Studies are underway to simultaneously measure the fluorescence and absorption spectra to correct the fluorescence signal for the effects of scattering and absorption in order to obtain a corrected yeast cell mass in a cuvette.

Figure 3 shows substantial spectral overlap between the four species studied. However, the concentration of riboflavin in a fermentation and its fluorescence intensity was negligible compared to the others and could be neglected. Researchers have found that excitation at 345 nm for NADH yields excitation of pyridoxine causing errors in the measurement of NADH fluorescence (Horvath et al., 1989). The commonly used excitation wavelength for NADH (366 nm) (Armiger et al., 1986; Zabriskie and Humphrey, 1978) will excite pyridoxine to a lesser degree than NADH, possibly reducing errors in the cell mass measurement by NADH fluorescence. NADH and pyridoxine fluorescence correlate well with yeast concentration, however, only on a relative basis. An absolute value for yeast cells/l cannot be calculated from the fluorescence due to uncertainties in the measurement volume, quantum efficiencies, and quenching cross sections of the observed species.

The protein fluorescence from yeast cells was found to track the dry cell mass accurately only at low cell densities (<0.5 g/l). At high cell densities, the effects of pre- and post-filtering caused the fluorescence signal to decrease as cell mass increased. The scattering and absorption by the yeast cells will filter out both the excitation light and the fluorescence emission. This could possibly be corrected for by measuring the absorption spectrum and the fluorescence at the same time. The pyridoxine signal appears to be the best indicator for measuring cell mass. It was the most sensitive, and had the largest dynamic range. Because of its low concentrations, there are no optical problems such as with the yeast signal at cell masses above 0.5 g/l. The spectral interferences from NADH affecting pyridoxine can be almost eliminated by exciting at 325 nm and observing the fluorescence between 380 and 420 nm as in this study.

112

Optical Fiber/Rod

The bifurcated optical fiber/rod fluorescence measurement system was shown to be a simple, rugged, and sensitive method for real time, in situ measurements in bioreactors. Our optical rod has undergone multiple steam sterilization cycles with no degradation or loss of sensitivity, allowing measurements without any chance of contamination. Figure 7 demonstrates the dynamic range of fluorescence which can be monitored with the optical fiber/rod: three orders of magnitude or more depending on the species of interest.

One surprising aspect of the optical rod was in its apparently large field of view, as is demonstrated in Figure 8. This figure shows the observed signal from the rod in a solution of 5×10^{-5} M pyridoxine as the probe tip was raised from the bottom of the fermentor, the fluorescence signal continued to increase until a height of approximately 4 cm where the signal appeared to level out. When the rod was at the bottom, the observed collection volume was very small, however as the rod was raised the collection volume was increased resulting in a larger fluorescence signal. Visual observation of the illuminated cone of light indicated the interaction length to be approximately 1 1/2-2 cm, however upon raising the rod further above the bottom, the signal increased until it became constant at 4 cm. This indicates that the optical rod observed a much larger volume than that expected from visual inspection.

Another question about the use of the optical fiber/rod concerned the way in which the signal might vary with changes in the local environment (e.g., agitation speed, aeration). The effects of agitation speed and aeration rate on the fluorescence signal of a 5×10^{-5} M pyridoxine solution at a rod height of 4 cm in the fermentor is shown in Figure 9. At zero l/m air, the fluorescence stayed within 4% of the average value with agitation rates of 100 to 800 rpm. This deviation is caused by the entrainment of air by the rotor at higher rpm. As aeration rate increased from 1.5 l/ to 5.0 l/m, there were large signal differences below 300 rpm agitation rate, but at 300 rpm and above the signals stayed within 5% of each other, all decreasing slightly with higher agitation rates. Above 300 rpm, the average of the five flows between 1.5 and 5 l/m lies 2% below the intensity of zero l/m and at 800 rpm the average signal of the five flows was 8% below the no air flow condition. These data indicated that large changes in agitation and aeration rate cause small changes in fluorescence signals above 300 rpm. These measurements were for a fixed optical rod position but they should be representative of the remainder of the reactor.

Figure 10 presents the fluorescence measurements, cell mass and DO concentration during a yeast fermentation obtained with the optical fiber/rod sensor. Upon comparing Figures 5 and 10 we see that the fluorescence signal from the yeast increases using the optical fiber/rod, throughout the dry mass curve in the fermentation, up to a dry cell mass of 2.68 g/l. In the flow-through cuvette the yeast fluorescence signal decreased from the effects of re-absorption and light scattering at a yeast dry mass of 0.5 g/l. The optical fiber/rod, which measures in the backscatter configuration with a variable path length can monitor the yeast fluorescence at higher optical densities than measurements in a 90° cuvette. At low concentrations and optical densities the path length will be long (4 cm), as optical density increases the pathlength shortens, maintaining light intensity and linear response at higher concentrations. The optical fiber/rod also collects the fluorescence from pyridoxine and NADH with excitation wavelengths from 290 nm to 366 nm and emission wavelengths from 320 nm to 500 nm, a broad coverage from the ultra-violet through the visible region of the spectrum.

The optical fiber/rod probe can also be used for monitoring the metabolic state of the yeast in situ. In Figure 11, the yeast were in a starved metabolic state at zero time due to lack of carbon source and were not using any oxygen, a state which is also indicated by a small level of NADH fluorescence. After 30 minutes, glucose was added and immediately the culture started to use oxygen and produced more NADH, as shown in the fluorescence signal rise. As the glucose was depleted, the culture returned to its resting state with no uptake of oxygen and a depressed NADH fluorescence signal. In Figure 12, we once again added glucose and the yeast culture rapidly used up most of the dissolved oxygen with an increase in NADH fluorescence. The oxygen was replaced by nitrogen, which moved the fermentation to an anaerobic state and significantly increased the amount of NADH present. These changes in metabolic states were instantly measured by the in situ optical fiber/rod fluorescence monitor.

The results presented in Figures 11 and 12 demonstrates the effects of oxygen and glucose deprivation on the observed fluorescence of NADH. When using NADH fluorescence for measurement of cell mass, the actual value of cell mass measured can be changed by the effects of oxygen concentration and/or carbon source. The actual cell mass present will not be changed instantaneously by carbon source or oxygen concentration changes, however the NADH fluorescence will, resulting in an inaccurate measure of cell mass. When using yeast fluorescence for measurement of cell mass it will be unaffected by changes in oxygen concentration or carbon source. Measurement of pyridoxine fluorescence is also not affected by changes in oxygen or carbon source (Horvath and Spangler, 1990).

Conclusions

A new optical fiber/rod sensor for in situ fluorescence measurements in bioreactors was described and tested. Fermentations of S. cerevisiae were studied with a on-line flow-through cell system and the optical fiber/rod for fluorescence measurements. It was established that the yeast fermentation broth had four strongly fluorescing species: pyridoxine, tryptophan, NADH, and riboflavin. Cell mass measurements obtained via NADH fluorescence were compared to

the new techniques of intrinsic yeast (tryptophan) fluorescence and fluorescence of pyridoxine produced by the yeast. All three measurements correlate well with yeast concentration. However in the cuvette measurements the yeast (tryptophan) signal was diminished by pre- and post filter effects at cell masses greater than 0.5 g/l. The optical fiber/rod probe was not affected by inner filter effects even at higher cell densities. Under some conditions yeast and pyridoxine fluorescence may be a better indicator of cell mass than NADH fluorescence. The signal from riboflavin was weak and was not correlated with the cell mass. The results indicate that for S. cerevisiae fermentations the fluorescence of pyridoxine can be a good indicator of cell mass up to a density of 4 g/l. The optical fiber/rod sensor showed superior performance compared to the flow-through cell system and was sterilized in situ with no loss of sensitivity. Fluorometers which are to be used for monitoring fermentations should have multiple excitation capability for the selective excitation of individual species. With proper care, fluorescence measurements can be used for quantitative measurements of individual species in fermentations.

References

Armiger, W.B., Forro, J.R., Maenner, G.F., and Zabriskie, D.W., "Analysis and Process Control of Fed-Batch Production of E. coli using Culture Florescence", *Proc. Biotech. '84*, Wash. D.C. **1984**.

Armiger, W.B., Forro, J.R., Montalvo, J.F. and Lee, J.F., *Chem. Eng. Commun.* **1986**, 45, 197-201.

Chance, B., Estabrook, R.W. and Ghosh, A., *Proc. Natl. Acad. Sci.* **1964**, 51, 1244-1451.

Chance, B., *Science* **1954**, 120, 767-775.

Duysens, L.N.M., and Amesz., J., *Biochim. Biophys. Acta* **1957**, 24, 19-26.

Einsele, A. and Purhar, E., *Acta Biotechnologica.* **1980**, 10, 33-37.

Harrison, D.E.F., and Chance, B., *Appl. Microbiol.* **1970**, 19, 446-450.

Harrison, D.E.F. and Harmes, C.S., *Proc. Biochem.* **1972**, 7, 13-16.

Horvath, J.J., and Semerjian, H.G., "LIF for Sensing in Bioreactors", paper presented at the Intl. Symp. on Biosensors, 192nd ACS National Meeting, Anaheim, CA, September, 1986; MBTD 21.

Horvath, J.J., Enriquez-Ortiz, A.B., Semerjian, H.G., "Online Monitoring of S. cerevisiae Fermentations using Multiple Excitation Wavelength Fluorescence Measurements" paper presented at 198th ACS National Meeting, Miami Beach, FL, September, 1989; MBTD 65.

Horvath, J.J. and Spangler, C.J., unpublished results, 1990.

Humphrey, A.E., Brown, K., Horvath, J.J. and Semerjian, H.G., *Bioproducts and Bioprocesses*, A Fiechter, T. Okada, and R. Tanner, eds., Springer-Verlag, **1989**, pp. 309-320.

Konev, S.V., *Fluorescence and Phosphorescence of Proteins and Nucleic Acids*, New York, Plenum Press, 1967.

Meyer, H.P., Beyeler, W. and Fiechter, A., *J. Biotechn.* **1984**, 1, 340-341.

Rao, G., Private Communication, Lehigh University, March 1990.

Ristroph, D.L., Watteeuw, C.M., Armiger, W.B., Humphrey, A.E., *J. Ferment. Technol.* **1977**, 55, 599-608.

Siano, S.A., and Mutharasan, R., *Biotech. Bioeng.* **1989**, 34, 660-670.

Scheper, T., Gebauer, A. and Schügerl, K., *The Chem. Eng. Journal* **1987**, 34, B7-B12.

Srinivas, S.P., and Mutharasan, R., *Biotechn. Letters* **1987**, 9, 139-142.

Teale, F.W.J., *Biochem. J.* **1960**, 76 381-388.

Undenfriend, S., *Fluorescence Assay in Biology and Medicine*, Academic Press, New York, 1962.

Zabriskie, D.W., and Humphrey, A.E., *Appl. Environ. Microbiol.* **1978**, 35, 337- 343.

Fiber Optic Biosensors Incorporating Sustained Release of Reagents

Tony Alex and **Henrik Pedersen**, Department of Chemical and Biochemical Engineering, **George H. Sigel, Jr.**, Fiber Optic Materials Research Program, Rutgers, The State University of New Jersey, P.O. Box 909, Piscataway, New Jersey 08855-0909

Enzyme based bioactive sensors combining miniature scale immobilized enzyme reactors, with optical signal transmission and detection capabilities are discussed. The sensors consist of a small bore nylon tube on which enzymes are immobilized that direct reactions leading to optically active end products. Intermediate reagents are provided to the reaction site via sustained release from a hollow fiber or ethylene vinyl acetate reservoir and optical fibers are used for the transmission of light. Two systems are investigated based on analytes that produce either hydrogen peroxide or analytes that cause a pH change in the reaction volume.

Since the first report of an enzyme electrode by Clarke and Lyons (1962), chemical sensors research has been continually expanding. A chemical sensor provides direct information about the chemical composition of its environment and usually consists of a physical transducer and a chemically selective layer. To date, a variety of transducers have been used such as electrodes (Guilbault, 1988; Turner, 1988), photodiodes (Aizawa *et al.*, 1984), fiber optics (Schultz and Mansouri, 1988), piezoelectric crystals (Maramatsu *et al.*, 1987) and field effect transistors (van der Shoot and Bergveld, 1987). These, in turn, have been coupled not only to selective biocatalysts such as immobilized enzymes (Guilbault, 1988; Turner, 1988; Aizawa *et al.*, 1984; Schultz and Mansouri, 1988; Maramatsu *et al.*,1987; van der Shoot, and Bergveld, 1987; Mascini *et al.*, 1983), but also to bacterial cells (Planchard *et al.*, 1988), plant and animal tissues (Uchiyama and Rechnitz, 1987), immunological substances (Bush and Rechnitz, 1987) and chemoreceptor structures (Belli and Rechnitz, 1986). Current developers of fiber optic sensors, in particular, are using emerging technologies that incorporate sophisticated biological and optical components. Examples of the applicability of such devices include sensors for the monitoring of diseases such as cardiac arhythmia (Gillette, 1985), diabetes (Mansouri and Schultz, 1984) and epilepsy (Wise, 1985), for anesthesia regulation (Konecny *et al.*, 1984) and blood gas monitoring (Dyson, 1983) used during surgical procedures and for "control" of artificial organs as replacements for the heart (Voorhees and Geddes, 1984), kidney (Sigdell, 1984) and pancreas (Shichiri *et al.*, 1985). None are currently available commercially, however. Nevertheless, the widespread availability of inexpensive, high quality optical fibers has led to a recent rapid rise in this technology which provides an alternative to electrodes for remote sensing.

The concept of a fiber optic sensor is, simply put, as follows: light from a source travels through an optical fiber to a transducer where the light is modulated before traveling via fiber optics to a detector. Fiber optic sensors have many advantages over the use of other sensor types such as electromagnetic immunity, physical separation of the sample and the detector, and the ability to easily multiplex many sensors to a central instrument (Seitz, 1984; Wolfbeis, 1987). Fiber optic sensors are easy to miniaturize which can lead to the development of very small, light and flexible instrumentation. Furthermore, analyses can be performed in real time and the coupling of small sensors for different analytes may allow for the simultaneous measurement of many different parameters.

The general design of our sensors employs a fiber or group of fibers that are connected to a miniature reactor. The reactor consists of a tube containing an enzyme or a coordinated set of enzymes covalently bound to the inner surface. Analytes diffuse into the tube, which is open at the probe end, and react with the enzyme and reagent in a controlled way to yield optically active compounds. These compounds are excited by light of a particular wavelength brought to the reactor by an optical fiber causing them to fluoresce. This signal is proportional to the analyte concentration and is transmitted to a detector communicating through the optical fibers. This paper looks at two related sensor prototypes that make use of this design.

Design of OTHER Fiber Optic Biosensors

The design of efficient transducers is important for the specificity and sensitivity of the sensor. We have employed enzymes in various configurations to meet these demands. In particular, the sensor typically contains an OTHER (open tubular heterogeneous enzyme reactor) which is a nylon tube on which an enzyme or enzymes are immobilized (Horvath and Solomon, 1972; Pedersen et al.,1981), and a reagent reservoir for sustained release of generally unavailable intermediate reagents into the reaction volume. In our case, the reservoir is a hollow fiber or an ethylene vinyl acetate matrix and one or more optical fibers are incorporated for the transmission of the optical signals. The arrangement is shown schematically in Fig. 1.

It is important to note some additional factors associated with this design. First, the tubular sleeve serves to protect the fiber tips from mechanical shocks and to eliminate stray light from the ends of the fibers. Second, the immobilization chemistry is performed on polymeric surfaces (nylon, for example) that afford much higher loadings than can be typically obtained with silica surfaces directly. Third, the bioactive and "optically active" components are separate and the probe tip could, in principle, be disposed or replaced in an easy and efficient manner. Fourth, the system design is able to accommodate a variety of sensing strategies. For example, the first sensor is suited to an analyte producing hydrogen peroxide upon reacting in the OTHER, which is a product of many oxidase enzymes, whereas the second sensor is suited to analytes that causes a pH shift to occur upon reaction in the OTHER. A wide range of analytes can be assayed using closely related procedures by simply substituting with an OTHER having a different enzyme. For example, cholesterol can be detected with the peroxidase sensor by replacing a glucose oxidase/peroxidase OTHER with a cholesterol oxidase/peroxidase OTHER. Finally, the volumes sampled by the sensor (contained in the OTHER sleeve) are on the order of 1-10 µl. Miniaturization is a key component of the design and is expected to be a factor not only in reducing sensor response times, but also in using the sensor in restricted environments where large sampling spaces are unavailable.

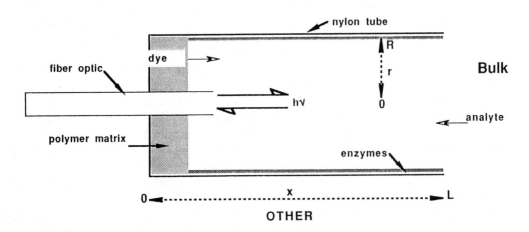

Fig. 1. Cross section of the biosensor.

Instrumentation

The general set-up used for the peroxide sensor incorporates a dual wavelength helium/cadmium laser (Liconix Model 4240N) for excitation. The laser can be adjusted to emit at two different wavelengths: 325 nm (10 mW) and 442 nm (40 mW). The 325 nm wavelength is of interest in the development of the glucose sensor. Spectral data was acquired with an EG&G (Princeton Applied Research, Princeton, NJ) OMA III system. Off line evaluation was done on a Perkin Elmer (Wilton, CT) model MPF-66 spectrofluorometer. UV-transmitting plastic clad silica fiber (PCS standard #33-PCS-2102) was obtained from EOTec (New Haven, CT).

Peroxide Sensor

In the development of the generic hydrogen peroxide sensor and its specific modification to glucose sensing, two enzymes were used, horseradish peroxidase (EC 1.11.1.7) and glucose oxidase (EC 1.1.3.4). Horseradish peroxidase catalyzes the oxidation by hydrogen peroxide of a number of substrates. The nonfluorescent intermediate p-hydroxyphenyl-acetic acid (HPA) is the substrate of choice for hydrogen peroxide determination since its oxidation, catalyzed by horseradish peroxidase, yields the highly fluorescent compound 6,6'-dihydroxy(1,1'-biphenyl) 3,3'-diacetic acid (DBDA). DBDA excites optimally in the ultraviolet (317 nm) and emits in the visible range (410 nm) of the light spectrum (Guilbault *et al.*, 1968). Excitation at 325 nm is also sufficiently strong to produce the fluorescent signal at 410 nm.

The conversion of non-fluorescent HPA to fluorescent DBDA can be utilized in the detection of any hydrogen peroxide producing system. In the case of glucose, a bienzyme system such as glucose oxidase/horseradish peroxidase will yield the final fluorescent signal. The first enzyme provides the specificity, whereas the second enzyme provides a common optical assay.

The fluorescent response of the sensor to glucose using a glucose oxidase/peroxidase OTHER, without the incorporation of a sustained release device, is shown in Fig. 2. A linear response was obtained for the glucose concentration range of 0 mM – 6 mM. The steady state data was collected by varying the glucose concentration in a well mixed vessel containing 0.1M phosphate buffer at pH 7 and 20 mM HPA. The exciting light from the laser at 325 nm was found to have negligible intensity fluctuations over several hours of operation and hence no internal reference for the excitation intensity was used.

pH Shift Sensor

A preliminary analysis of the pH shift sensor which is based on the pH sensitivity of the fluorescence at 511 nm of hydroxypyrene trisulfonic acid (HPTS) upon excitation at both 403 nm (acidic form) and 454 nm (basic form) was done. Such sensors have been developed previously; however, they have been primarily based on direct pH measurement or on pH measurement due to shifting chemical equilibria associated with solution hydration of a weak acid or base (Wolfbeis, 1986). A fiber optic pH sensor using this dye has been recently demonstrated (Luo and Walt, 1989). The pH shift sensor uses the pH sensor to detect analytes that cause a pH shift in the volume of the OTHER upon reaction with the enzyme. Again, the enzymatic step provides the specificity of the sensor and the second (chemical) reaction provides the common optical assay.

The chemical and enzymatic steps were both tested by conventional spectrofluorometry, as the required wavelengths were not available from the HeCd laser.

The relative fluorescence of HPTS in a solution of constant ionic strength is dependant upon the pH, the concentration of HPTS as well as on the excitation intensity. However, it has been demonstrated that the ratio of the fluorescence of the acidic and basic forms of the dye are independent of the concentration of the dye as well as the excitation intensity (Luo and Walt, 1989). This is verified in our system as shown in Table 1. Thus the signal is dependent only upon the pH if the ratio is monitored. This provides us with the flexibility of having to

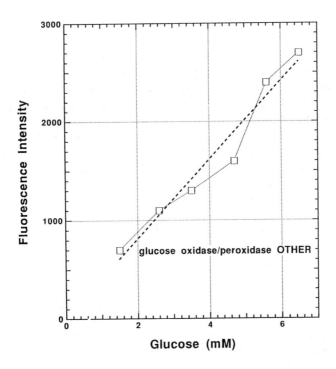

Fig. 2. Calibration curve for pH shift sensor with adenosine as the analyte. The signal is the ratio of the emission produces at 511 nm upon excitation at 403 nm and 454 nm in a solution assay in a spectrofluorometer. Data were collected in a 0.1 mM phosphate buffer at pH 7.

Table 1. Fluorescence ratio measurements at different pH and HPTS concentrations

pH	HPTS concentration (mM)	fluorescence due to 403 nm excitation	fluorescence due to 454 nm excitation	fluorescence ratio
7.00	10	62.84	27.81	2.26
7.00	20	125.0	54.66	2.28
7.00	30	233.6	102.9	2.27
7.00	40	290.0	128.0	2.27
7.33	10	41.66	32.87	1.27
7.33	20	97.37	75.44	1.29
7.33	30	143.1	111.8	1.28
7.33	40	183.5	143.1	1.28
7.70	10	34.28	49.91	1.46
7.70	20	77.46	113.3	1.46
7.70	30	119.2	176.4	1.48
7.70	40	156.2	231.9	1.48
8.30	10	35.71	60.53	1.70
8.30	20	67.21	113.6	1.69
8.30	30	105.4	179.1	1.70
8.30	40	139.5	236.7	1.70

obtain only a sustained release of the HPTS as opposed to a pure zero-order controlled release in the sensor volume.

Specifically, the analyte adenosine reacts to form ammonia and inosine in the presence of the enzyme adenosine deaminase (EC 3.5.4.4) immobilized on the OTHER. The ammonia formed causes the pH of the sensor volume to change with changes in the concentration of adenosine. This is further translated as a change in fluorescence signal, which is dependent only on the pH of the sensor volume.

A preliminary analysis of this system using solution assays has been done. Fig. 3 shows the change in intensity ratio of the fluorescence of HPTS due to the changes in pH caused by addition of adenosine. A linear response in the adenosine concentration range of 0 mM-3.5 mM is obtained in a 0.1mM phosphate buffer and a starting pH of 7. The limits of detection and the

sensitivity of the sensor to adenosine are determined by the buffer strength of the solution. A weak buffer will cause a greater signal shift per unit adenosine added to the system as compared to a stronger buffer. Hence, for every given buffer strength a particular range of adenosine concentrations can be measured.

Mathematical Modeling

In the peroxide sensor the final fluorescence signal arises from a particular sequence of steps. First, the analyte diffuses towards the active surface via the open end of the reactor tube. Simultaneously, the intermediate reagent diffuses into the reaction volume from the other end of the reactor. The active surface converts the analyte into an intermediary compound which reacts on the active surface with the released

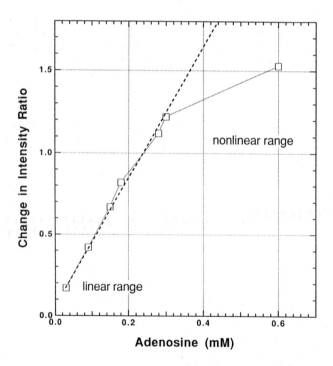

Fig. 3. Calibration curve for pH shift sensor with adenosine as the analyte. The signal is the ratio of the emission produced at 511 nm upon excitation at 403 nm and 454 nm in a solution assay. Data was collected in a 0.1 mM phosphate buffer at pH 7.

reagent to produce the fluorescent product. Subsequently the product diffuses into the sensor volume where it is excited with the light transmitted through the fiber and fluoresces. A simple lumped parameter model can be developed to describe these interactions and it has utility in the preliminary design of efficient sensors.

Model Equations

The governing equations for the four compounds are given below for a tube of radius R and length L. The equations describe transient, two-dimensional diffusion with a general sequence of surface reactions. In our case,

$$\frac{\partial C_i}{\partial t} = D_{i,r}\left[\frac{1}{r}\frac{\partial}{\partial r}\left(r\frac{\partial C_i}{\partial r}\right)\right] + D_{i,x}\frac{\partial^2 C_i}{\partial x^2} \qquad (1)$$

where the subscript i can correspond to glucose (i=1), hydrogen peroxide (i=2), HPA (i=3) or DBDA (i=4). C_i is the local, time dependent concentration of the compounds in the cylindrical volume occupied by the sensor tip as shown in Fig. 1.

The appropriate boundary conditions, for two reactions taking place, are generally written as

$$\frac{\partial C_i}{\partial r} = 0 \; ; \qquad\qquad r=0 \qquad (2a)$$

$$D_{i,r}\frac{\partial C_i}{\partial r} = \sum_{j=1}^{2}\nu_{ij}\kappa_j \; ; \qquad r=R \qquad (2b)$$

$$D_{i,x}\frac{\partial C_i}{\partial x} = 0 \; ; \qquad x=0, i\neq3 \qquad (2c1)$$

$$D_{i,x}\frac{\partial C_i}{\partial x} = f \; ; \qquad x=0, i=3 \qquad (2c2)$$

$$D_{i,x}\frac{\partial C_i}{\partial x} = -h_i(C_i-C_{i,b}) \; ; \qquad x=L \qquad (2d)$$

where $D_{i,r}$ and $D_{i,x}$ are radial and axial diffusion coefficients, respectively. The kinetic equation for each reaction is given as κ_j, ν_{ij} is the stoichiometric coefficient and f is the flux of reagent, where appropriate, into the volume. The flux is the sustained release rate of intermediate. The mass transfer coefficient is h_i and $C_{i,b}$ is the bulk concentration. For i=1, $C_{1,b}$ is the concentration whose value is sought.

Lumped Model

The average concentration in the sensor volume is given by the following equation

$$\langle C \rangle = \frac{2}{R^2 L} \int_0^L \int_0^R C \, rdrdx \qquad (3)$$

Integrating Eq. (1) and applying the boundary conditions given by Eqs (2a-d), we get the lumped model

$$\frac{d\langle C_i \rangle}{dt} = -(h_i/L)(\langle C_i \rangle - C_{i,b}) - (2/R)\sum_{j=1}^{2}\nu_{ij}\kappa_j - f/L \qquad (4)$$

The lumped model assumes that the wall and edge concentrations are replaced by the average values in the sensor volume. Appropriate corrections are accounted for in the parameters for the mass transfer coefficients, h_i, or the kinetic expressions.

Kinetic Expressions

These equations can be further reduced to dimensionless form by normalizing the species concentrations with the bulk analyte concentration $C_{1,b}$ and normalizing time using a retention time given by, $\tau = L/h_i$. This is then solved for species profiles in the reactor after substituting the appropriate kinetic rate expressions. The two reactions involve first the oxidation of glucose to produce hydrogen peroxide, which subsequently reacts with HPA to form DBDA in the second step. For simplicity, linear dependence of the rates on the concentrations are assumed, except for the intermediate which displays a saturation rate dependence. This linear dependence is reasonable since the reactor would most likely operate at high catalytic loadings. The expressions are then,

$$\kappa_1 = V_1 C_1 \qquad (5a)$$

$$\kappa_2 = V_2\frac{C_2 C_3}{K+C_3} \qquad (5b)$$

The model can be used to estimate, for example, the penetration depth of the analyte into the tube as a function of the enzyme loading on the inner surface. Response times and sensitivity can also be obtained from the model. These would define the appropriate L/R ratios and the minimum loading of enzyme for optimum operation of the tube.

Some representative results are depicted in Figs 4 and 5, which show the transient response of the different compounds in the tube after exposure to single and multiple changes in the bulk analyte (glucose) concentration. The actual response of the probe would be proportional to the DBDA curve.

Sustained Release of Intermediates

In order for the probe to truly be a sensor it is necessary to replenish the reagents in the sensor volume. The sensor designs do not require a fixed concentration of reagent in the sensor volume. This allows use of a method of reagent delivery that does not need be a zero order controlled release. We have found that entrapping the reagents in a hollow fiber membrane will afford us the concentration range we require, and continuously provide reagents in a nonlimiting fashion.

This hollow fiber trapped reagent can be located at the back of the OTHER and behind the optical fibers. Since it is contained within the reactor tube, its local concentration is maintained high with a relatively small mass flux of reagent. A similar procedure using a copolymer of ethylene vinyl acetate (EVA) as the trapping matrix was recently demonstrated by using pH-sensitive fluorescent dyes to passively monitor pH (Luo and Walt, 1989).

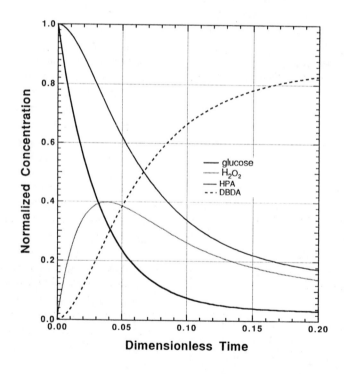

Fig. 4. Kinetics of the startup of the OTHER. The species concentrations are normalized against the bulk analyte concentration; $\tau =$ 1000 secs, L/R = 2, $V_1 = 10^{-4}$ cm/sec, $V_2 =$ 2.5 10^{-4} cm/sec. K is one half the bulk concentration.

122

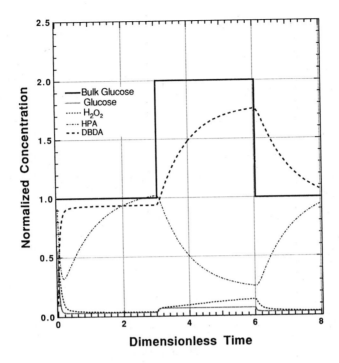

Fig. 5. Response of the sensor to changes in the bulk analyte concentration. The species concentrations are normalized against the bulk analyte concentration. Other parameters are the same as in Fig. 4.

Fig. 6 shows the cumulative release of HPA from a hollow fiber into water. The release profile suggests that HPA will be released for several days from the hollow fiber and the desired concentration range of the reagent in the sensor volume can be attained by using appropriate loadings of the reagent.

Summary

The previous section outlines approaches to employ immobilized enzymes with fiber optic waveguides. The advantages are summarized in the description of the sensor. The experiments used demonstrate the technology for a particular model system with extensions to closely related immobilized enzyme systems. Some of the key factors that contribute to our design are the heterogeneous loading of relatively large amounts of active enzyme, the incorporation of controlled release reagents, the small volumes, the shielding provided by a ruggedized tip, and the disposable nature of the final, bioactive portions of the sensor probe.

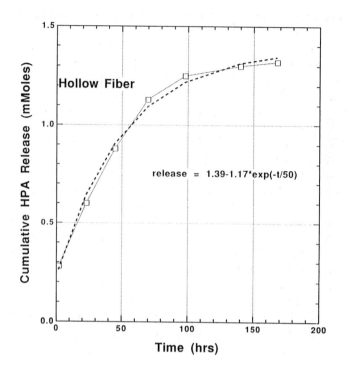

Fig. 6. Cumulative release of HPA from hollow fibers into water. Concentration determined by measuring absorbance at 285 nm.

References

Aizawa, M., Ikariyama, Y., Kuno, H., *Anal. Lett.,* **1984**, 17, 555.

Belli, S.L., Rechnitz, G.A., *Anal. Lett.*, **1986**, 19, 403.

Bush, D.L., Rechnitz, G.A., *Anal. Lett.,* **1987**, 20, 1781.

Clark, L.C., Lyons, C., *Ann. NY Acad. Sci.,* **1962**, 102, 29.

Dyson, C.W., Gehrich, J.L., Yafuso, M., Abstract, presented at the AmSECT Meeting, April 12, 1983.

Gillette, P., Implantable Sensors for Closed-Loop Prosthetic Systems, (W. H. Ko, ed) Futura Publishing Co., Mount Kisco, pp.251-256, 1985.

Guilbault, G.G., Brignac,P.J., Juneau,M., *Anal. Chem.,* **1968**, 40, 1256.

Guilbault, G.G., *Methods in Enzymology,* **1988**, 137, 14.

Horvath, Cs., Solomon, *B.A. Biotech. Bioeng.,* **1972**, 14, 885.

Konecny, E., Hattendorff, H.D., Leiss, M., Proceedings of the Symposium on Biosensors, A.R. Potvin and M. R. Neuman, eds., IEEE Service Center, Piscataway, NJ, pp.110-111, 1984.

Luo, S., Walt, D.A. *Anal. Chem.,* **1989**, 61, 174.

Mansouri, S., Schultz, J.S., Proceedings of the Symposium on Biosensors, A. R. Potvin, and M.R. Neuman, ed., IEEE Service Center, Piscataway, NJ, pp.112-115, 1984.

Maramatsu, H., Dicks, J.H., Tamiya, E., Karube, I., *Anal. Chem.*, **1987**, 59, 2760.

Mascini, M., Iannello, M., Palleschi, G., *Anal. Chim. Acta,* **1983**, 146, 135.

Pedersen, H., Horvath, Cs., in "Analytical Applications of Immobilized Enzymes and Cells (L. B. Wingard, Jr. E. K. Katchalski-Katzir, L. Goldstein) Academic Press, 1981.

Planchard, A., Mignot, L., Junter, G.A., *Sensors and Actuators*, **1988**, 14, 9.

Schultz, J.S., Mansouri, S., *Methods in Enzymology,* **1988**, 137, 349.

Seitz, R. W., *Anal. Chem.,* **1984**, 56, 16A.

Shichiri, M., Kawamori, R., Yamasaki, Y., Hakui, N., Asakawa, H., Abe, H., W. H. Ko, ed., Futura Publishing Co., Mount Kisco, NY, pp.197-210, 1985.

Sigdell, E.J., Report #5033, Biomedical Business International, Inc., Information Resources International, Inc., Tustin, CA, 1984.

Turner, A.P.F., *Methods in Enzymology,* **1988**, 137, 90.

Uchiyama, S., Rechnitz, G A., *Anal. Lett.,* **1987**, 20, 451.

van der Shoot, B.H., Bergveld, P., *Biosensors*, **1987**, 3, 161.

Voorhees III, W.D., Geddes, L.A., Proceedings of the Symposium on Biosensors, A. R. Potvin and M. R. Neuman, eds., IEEE Service Center, Piscataway, NJ, pp.20-23, 1984.

Wise, K.D., Implantable Sensors for Closed-Loop Prosthetic Systems, (W. H. Ko, ed) Futura Publishing Co., Mount Kisco, pp. 3-20, 1985.

Wolfbeis, O.S., *Pure and Applied Chem.,* **1987**, 59, 663.

Wolfbeis, O.S., Posch, H.E., *Analytica Chimica Acta* , **1986**, 185, 321.

Whole Cell Biosensors: A Brief Overview and Presentation of A Novel Neuron-based Approach

Devdatt L. Kurdikar, Rodney S. Skeen, Bernard J. Van Wie, Department of Chemical Engineering, Washington State University, Pullman, WA 99164-2710.
Charles D. Barnes, Simon J. Fung, Department of Veterinary and Comparative Anatomy, Pharmacology and Physiology, Washington State University, Pullman, WA 99164-6520.

This paper contains a discussion of the advantages and disadvantages of whole cell sensors with specific attention to a novel concept which will employ neuronal components as both the recognition elements and primary transducers in analyte quantitation. Preliminary results are presented on the effects of model analytes, serotonin and acetylcholine, on neurons from the central ganglia of the pond snail, *Limnea stagnalis*. With serotonin, a concentration dependent increase in firing frequency is shown as well as an increase in maximum firing frequency with temperatures up to 35°C. Also, the rate of desensitization is found to be faster at higher temperatures. Response to acetylcholine covers a narrower range and desensitization is more rapid.

Biosensors are generally defined as analytical devices that involve a biologically active or even living component as the primary recognition element in contact with the sample whose chemical activity or concentration is to be detected. Based on the type of element used in the sensing process, biosensors can broadly be classified into four categories: immobilized enzyme sensors, whole cell sensors, organelle sensors, and immunological sensors. The usual aim in biosensing is to produce an electronic signal that is proportional to the concentration or chemical activity of the species being tested. A high signal-to-noise ratio is preferred for easy detectability of the signal. The use of a biological component allows the detection of very low concentrations which is the principle advantage of biosensors.

Whole cell preparations have a few distinct advantages over other sensor types. First, there is no need to isolate and purify an enzyme for the specific compound to be measured. This can greatly reduce the cost of the sensor. Second, the biocatalytically active components found in whole cells usually have greater stability than the isolated enzyme. In addition, the possibility exists for regeneration of the biocatalytic activity by culturing cells in a growth medium. This potentially increases the lifetime of a whole cell based sensor. Finally, whole cells can be used in cases where enzymes for a particular compound haven't been isolated. The major disadvantage of these kinds of sensors is poor specificity due to the fact that cells may contain many types of enzymes that produce the same measurable by-product from different chemicals present in the sample to be analyzed.

This paper deals with whole cell sensors, giving specific attention to neuron-based biosensors.

1. Whole Cell Biosensors, A Brief Review

There has been rapid growth in the development of whole cell biosensors since the 1970s. In the past, immobilized bacteria, plant tissue and animal cells have been used for sensing, mostly in conjunction with electrochemical sensors. These devices use an immobilized layer of cells on the surface of the sensing membrane of the probe. The electrochemical sensor detects a metabolic by-product generated by the metabolism of the immobilized cells and the chemical to be measured. Many types of electrochemical devices, both amperometric and potentiometric, have been used in cellular sensors. The most common amperometric device is a dissolved oxygen probe. Examples of potentiometric devices are the ammonia, pH, and hydrogen sulfide sensors. Some recent novel approaches include work on neuron-based sensing (Skeen *et al.*, 1990a; Buch and Rechnitz, 1989a, 1989b) and on a microphysiometer to measure metabolic rates developed by scientists at Molecular Devices Corporation (Parce *et al.*, 1989).

Bacterial Biosensors

One of the earliest reported bacterial sensors was for determining the biological oxygen demand (BOD) of waste water (Karube *et al.*, 1977). It uses a layer of suspended bacteria around an oxygen sensor. When the sensor is immersed in a sample whose BOD is to be measured, the bacteria decomposes the organic material in the sample and at the same time consumes oxygen. This causes a decrease in the dissolved oxygen at the electrode and a current to flow between the anode and the cathode. The response of this sensor is slow, taking about 10 - 15 *min* before a steady state current is reached. The sensor gives reproducible results over a period of 10 days. Since then, Karube *et al.* (1989) have instigated the development of a microbial BOD sensor using thermophilic bacteria which allowed BOD measurements up to 50°C. This is advantageous over the previous devices, in that it allows BOD measurements in waste water with elevated temperatures, a situation commonly found in industrial applications.

Since the advent of BOD sensors, many bacterial sensors have been developed for various substances: sugars, alcohols, organic acids, antibiotics, and gases such as methane. A detailed review is given by Karube (1987). However, all of these are quite similar to the original idea. Another recent advance includes a bacterial electrode for sulfate developed by coupling a strain of the bacterium *Desulfovibrio desulfuricans* with a sulfide selective electrode (Kobos, 1986). Preliminary studies on this sensor report a lifetime of 10 days and a response time of 8 *min*. To make the response of the bacterial cell based sensors more specific, a novel selectivity enhancement scheme using enzyme and transport inhibitors has been proposed (Corcoran and Kobos, 1987). Recently, a hybrid disposable miniature L-lysine sensor has been reported (Suzuki *et al.*, 1990). This is a sensor consisting of an immobilized L-lysine decarboxylase and a miniature bacterial CO_2 sensor. The investigators report excellent selectivity to L-lysine, a response time of 1 - 3 *min* and a stable response after 25 repetitive operations.

Plant Tissue Based Biosensors

Plant tissue has also been shown to be good biocatalytic material for biosensors. The construction is like that of a bacterial sensor and examples include the urea and L-arginine sensors reported by Uchiyama and Rechnitz (1987) that use a potentiometric ammonia sensing electrode coupled with tissue portions from carnation and chrysanthemum flowers, respectively. Minced carnation petals give a highly specific response for urea. This sensor yields stable results for 10 days. However, immobilized chrysanthemum petals do not give a highly selective response to L-arginine and this sensor had a short usable life of only one day. These investigators also obtained a device sensitive to L-asparagine, by a modification which involved securing slices of the chrysanthemum receptacle rather than the flower petals to an ammonia gas sensitive electrode. This, too, had poor selectivity, but an improved lifetime of 18 days. Rechnitz (1988) has reported a variety of other plant tissue based biosensors such as the banana-based biosensor for the neurotransmitter dopamine, and, a fairly specific and sensitive mushroom-based sensor for tyrosine.

Mammalian Tissue Based Biosensors

One of the early reports on a mammalian tissue based sensor is that by Rechnitz *et al.* (1979). This sensor uses a thin slice of porcine kidney coupled with an ammonia gas sensing electrode. The response of this sensor to glutamine is highly specific and by using sodium azide as a preservative, a useful life of 28 days was reported. Since then a variety of mammalian sensors have been reported (Arnold and Rechnitz, 1987) such as the adenosine biosensor which uses immobilized mouse small intestinal mucosal cells at the surface of an ammonia gas-sensing probe, the guanine biosensor which uses a slice of rabbit liver with an ammonia gas-sensing probe, and a highly selective hydrogen peroxide biosensor which uses a slice of bovine liver immobilized on an oxygen sensitive probe.

A recent advance developed by scientists at Molecular Devices Corporation is the silicon microphysiometer (Parce *et al.*, 1989), which uses the Light Addressable Potentiometric Sensor (LAPS) developed by Hafeman *et al.* (1988) to measure local pH changes. The silicon microphysiometer measures, with high sensitivity, the metabolic responses of cells to physical and chemical stimuli by monitoring the acidity of the culture medium bathing the cells. Ligand-receptor interactions cause immediate changes in the cellular catabolic rates. The principle products of mammalian cell catabolic activity are lactate and CO_2, both of which are acidic in nature. Thus, the changes in the pH of the culture medium relate to the rate of catabolism. Recently, Owicki *et al.* (1990) used the silicon microphysiometer, with great success, to study changes in cellular metabolism rates caused by stimulation of β-adrenergic or muscarinic acetylcholine receptors.

Neuronal Devices

Neuronal devices use neurons as the primary sensing element and generate a digital output dependent on concentration of the chemical applied. The digital output is generated because, as the analyte interacts with the tissue it triggers changes in the transmembrane potential, either directly or through secondary messengers, which are then amplified by voltage dependent ionic channels to produce discrete all-or-none voltage spikes or action potential (AP) events. The large amplitude of the APs makes them easy to monitor, and it has been shown that the concentration of the detected chemical is encoded in

the frequency at which APs occur. Since the frequency information is digital in nature, the output of these devices are less prone to noise and baseline drift than the analog signals produced by most biosensing devices.

These devices have several advantages (Buch and Rechnitz, 1989c) over other whole cell biosensors, the most important one is their extremely short response time. This response is only limited by the time required to generate the action potential after the analyte-receptor binding event occurs, taking a matter of milliseconds. Another advantage is the degree of specificity that can be achieved because of the high selectivity of the particular receptors to a specific analyte. In addition, a broad analytical range, from 10^{-15} to 10^{-3} M has been reported for different analytes (Skeen *et al.*, 1990a; Buch and Rechnitz, 1989b). One disadvantage of neuronal sensors is that rinsing is necessary between two successive applications of different concentrations because of desensitization after prolonged exposure to analyte, making this approach unsuitable for certain continuous applications. Another disadvantage is one of cell viability, which depends on, among other things, the method that is chosen to extract electrical information from the cell. In our own research we have used intracellular recording to demonstrate sensing capabilities. This allows only short lifetimes, as it means impaling the cell with a microelectrode. A practical neuron-based sensor will need to have a reasonable lifetime since considerable effort would be involved in making this device, and thus extracellular recording, which is less traumatic to the cell, will be preferred. Such techniques have been demonstrated to be successful as shown below.

One of the early papers that reported a microfabricated electrode for extracellular recording was that by Gross (Gross, 1979; Gross *et al.*, 1977). The multi-microelectrode consists of an array of gold leads situated on a glass plate and deinsulated at their tips. This group recorded single unit extracellular activity from the central ganglia of the snail *Helix pomatia*, and reported that simultaneous single unit recording from small neurons is ensured if a recessed tip design is used and if a high resistance seal between the cell and electrical connection can be made. In addition, the possibility of extracellular recording of cultured mammalian neurons was first investigated by Gross *et al.* (1982). This allows the simultaneous long term stimulation of, and recording from, a large number of neurons. To do this, an array of photoetched gold plated conductors is deposited into the floor of a tissue culture chamber with a layer of insulation over it. Mouse spinal cord neurons are grown on the insulation layer of the multielectrode plate. With this approach it has been shown that spontaneous activity can be recorded continuously for a period of two days. Gross *et al.* (1985) also report the use of a different material for recording of spontaneous activity from monolayer neuronal culture. Advantages of this commercially available material, indium tin oxide (ITO), are its high transparency allowing unhindered observation of circuit components in monolayer cultures, easy adherability to glass, and good stability under warm culture medium. The disadvantages are high interface impedance and sensitivity of ITO to strong acidic and alkaline solutions.

Regehr *et al.* (1989) have recently reported a multielectrode array similar to that of Gross (1982, 1985). They used planar arrays of electrodes which had ITO conductive leads and insulation embedded in the bottom of a tissue culture dish. Large identified invertebrate neurons were grown on these culture dishes, and large seal resistances were obtained when a neuron completely covered an electrode. A signal to noise ratio of as high as 500:1 has been reported, and these multielectrode dishes were used to monitor spontaneous activity from *Helisoma* neurons for 13 days. Again, the critical step in these experiments is getting high seal resistance between the neuron and the electrode. The authors have suggested that in the case of smaller vertebrate neurons the seal can be formed by allowing non-neuronal cells to cover the exposed portions of the electrode. Another new method presented by Regehr *et al.* (1988) for long term recording and stimulating of cultured neurons is the "diving board electrode", a silicon microdevice, that can be glued into place at the bottom of a tissue culture dish. With this approach, a long flexible arm is used having at its end, a cup shaped structure with a gold strip between two insulating layers that is used to make electrical contact with the neuron. Connection to external electron-

ics is made by an insulated gold wire from the pedestal, out of the solution. In this manner a one-to-one connection is established for both stimulating and recording, and successful two way electrical connections have been made to several neurons in a single culture and maintained for up to four days.

The potential for using neuronal biosensors to detect analytes in solution has been demonstrated in the laboratories of G.A.Rechnitz. In 1986, a novel prototype biosensor based on the antennules of the blue crab, *Callinectes sapidus* was reported (Belli and Rechnitz, 1986). Essentially, a "pick up" probe electrode makes contact with a bundle of nerve fibers from an isolated antenna, and APs recorded from one or more neurons. A random low frequency firing is observed in the absence of an applied stimulus. When a chemical stimulus is introduced, signal activity is increased for the duration of the stimulus. A multi-unit response is obtained for certain amino acids while application of others yields no response indicating a receptor-based specificity. For example, this group reports that for a receptor-based response to glutamate, there is no response to glycine, alanine, proline or taurine, even at concentrations a hundred-fold more than the glutamate threshold.

Belli and Rechnitz (1988) further extended their previous study by testing a group of amino acids and closely related analogs to characterize the selectivity of amino acid receptors of the crab antennule. After applying a stimulus of L-glutamate and L-glutamine, the electrode was manipulated over several unidentified nerve fibers until some were found that gave a response. Data were collected for both single- and multi-unit cases. As was expected, the single-unit recordings offered the highest selectivity. Multi-unit cases demonstrated a variety of responses, since in these measurements a population of unidentified receptors was sampled, and the response observed was the summation of the individual responses of all of these receptors. The response time necessary to give significant data for analysis was reported to be 1 s for the single-unit case and 5-10 s for the multi-unit case. A lifetime of about 7 hours is reported suggesting that this could be increased to more than 48 hours by using a gentler micropipet technique. Recently, Buch and Rechnitz (1989a) reported further information on similar responses. For instance, glutamate was found to respond to a concentration range of six orders of magnitude and concentrations as low as 10^{-8} M. However, the multi-unit responses being non-specific, some axons responded to all 20 of the amino acids tested in this study. A response to $5'$-AMP and ADP was observed from 10^{-9} M to 10^{-2} M, while ATP gave no response in the same preparation. Buch and Rechnitz (1989b) also report sensitivity to excitatory amino acids such as kainic acid and quisqualic acid for concentrations as low as 10^{-15} M. However, this response does not show concentration dependency. A model explaining the wide sensitivity seen by neuron-based sensors was offered by Belli et al. (1987).

A major disadvantage of the above method is that it is difficult to determine a priori which region of the antennular nerve encodes the information produced by a particular chemoreceptive cell. Instead, one must search through an array of unidentified axons until a satisfactory response to an analyte is obtained. Hence, it may be difficult to construct a sensor for a predetermined analyte.

Work in our laboratories is aimed at furthering the field of neuron-based sensing by characterizing a model system which can provide repeatable and reproducible results. Such a system is necessary for the development of a practical sensor. Skeen et al. (1990a) have presented results using identified neurons from the snail Limnea stagnalis with serotonin, a neurotransmitter, as the model analyte. These authors demonstrate the graded and reversible increase in firing frequency, using the visceral ventral neurons VV1 and VV2, with increasing serotonin concentration over four orders of magnitude, from 10^{-6} to 10^{-3} M. However, to achieve repeatability of the maximum firing frequency response a concentration dependent rinse time is required. From these results, the authors suggest that a neuron-based sensor would operate best if solutions to be tested were introduced in short pulses and the time between pulses adjusted to compensate for varying serotonin concentrations. Since the magnitude of the maximum firing frequency varies from cell to cell, the authors also suggest that individual neurons need to be calibrated or responses of several cells averaged to give a reliable indicator of concentration.

Studies with the serotonin antagonist, methysergide, show that the changes in firing frequency are mediated through serotonin sensitive receptors. Finally, extended exposures of serotonin to the cells cause the response to diminish to a sub-optimal level after the maximum firing frequency is reached.

Since the neuron-based sensor may potentially be used in systems where temperature may not be easily controllable, such as in in vivo monitoring, the effects of temperature on such a system needs to be better understood. This paper presents the preliminary results of such a study currently underway in our laboratory.

2. Temperature Sensitivity of Neuronal Response to Neurotransmitters

Methods

To provide suitable neuronal preparation, the visceral ganglia were removed by dissection from an adult Limnea stagnalis snail and placed in a Sylgard 184 (Dow Corning, Midland, MI) lined 35 mm petri dish. The ganglia were pinned using a minute dissecting pin (Fine Science Tools, Belmont, CA) such that the giant visceral neurons VV1 and VV2 (Winlow and Benjamin, 1976) could be easily seen. The protective sheath of the ganglia was softened to allow easy microelectrode impalement by soaking in a solution of 0.2 % trypsin (Sigma type III) for 30 min.

The acetylcholine concentrations used were 10^{-3} - 10^{-4} M, and the serotonin concentrations were 10^{-6} - 10^{-3} M. For the experiments with acetylcholine, a new plug-flow chamber with a syringe pump (Sage Instruments, Cambridge, MA) was used. This chamber has complete solution turnover in 1 s at the 0.8 ml/min feed rate used in the experiments.

The equipment and experimental procedure used in these experiments were the same as those described by Skeen et al. (1990a). However, to briefly summarize the procedures, impalement of the VV1 and VV2 cells with a glass microelectrode, filled with 3 M KCl, was done under a dissecting microscope using a hydraulic micro-manipulator (Narishige

USA Inc., Greenvale, NY). Intracellular signals were conducted by a silver wire (A-M Systems Inc., Everett, WA) from the microelectrode to a Dagan 8700 amplifier (Dagan Corporation, Minneapolis, MN). The spontaneous activity from the impaled neuron was monitored continuously on an oscilloscope (Tektronix Corp., Beaverton, OR), stored on a modified VCR (Vetter Co., Rebersburg, PA), and sent to an on-line data analysis system. On-line data analysis was accomplished using a Das-16F analog-to-digital (A/D) converter (Metrabyte Corp., Taunton, MA) and an IBM-AT compatible microcomputer (Isotropic Computer Inc., Post Falls, ID). The frequency between the previous and the most recent AP was calculated and stored in the computer along with the time the recent AP occurred. The final analysis of the frequency versus time data was done off-line using in-house software which will be discussed in detail in a later paper. The firing frequency as a function of concentration, and AP characteristics for a cell as a function of temperature were monitored.

Results and Discussion

Figure 1 shows the response observed for increasing concentrations of serotonin at 32°C. In this case, serotonin was added in increasing concentrations from 10^{-6} M to 10^{-3} M followed by decreasing concentrations. As demonstrated by Skeen *et al.* (1990a, 1990b) at a lower temperature (28°C), here too a graded increase in firing frequency with concentration is observed. The response is repeatable when the required rinse times reported by Skeen *et al.* (1990a) are used. One also notes that the desensitization response for the extended application of 10^{-3} M is similar to what was observed in the previous studies.

Establishment of similar sensing behavior at temperatures different from those previously studied leads one to ask questions about subtle differences that are observed in the character of the response. To answer these questions, we first compare the global trends in behavior at the two temperatures. In Figure

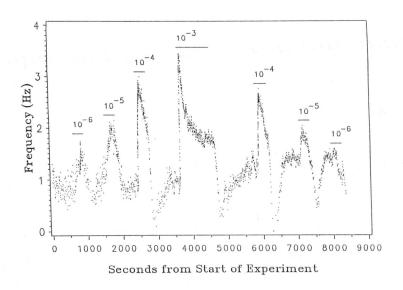

Fig. 1. Frequency spectrum of action potentials obtained when VV1 or VV2 neurons are exposed to 5 *min* applications of increasing and then decreasing concentrations of serotonin at 32°C. Each application was followed by a 10 - 25 *min* rinse with *Limnea* saline.

131

2, a graph is shown with maximum firing frequencies for both the increasing and decreasing concentrations of serotonin at 32°C. One should note that since the recording chamber acts like a well-mixed vessel, the data in the plot has been corrected, using an equation presented by Skeen *et al.* (1990a), to correspond to the actual concentration of the serotonin in the vessel when the maximum frequency occurs, which was usually around 3 *min* after flow from the stock solution was begun. The 32°C data in the figure is compared to the mean response and 95% confidence interval for 17 cells at 27 - 29°C. Although the 32°C results over the entire concentration range in the well-mixed recording chamber are preliminary, the observation of increased firing frequency at even slightly higher temperatures are consistent with those from a more detailed study underway in our laboratory on the effects of temperature in a plug-flow apparatus. One

should also note that one set of data presented by Skeen *et al.* at 27 - 29°C is in the range of that reported here for 32°C. This shows the variability one observes from one preparation to another, however, we expect the average maximum response to increase with higher temperatures. One also notes a slight decrease in maximum firing frequency by as much as 8% when the serotonin is added in decreasing concentrations. This decrease may indicate a small hysteresis effect, or alternatively may be attributed to insufficient rinse times after repeated exposures, irreversible alterations in behavior over prolonged experimentation or perhaps may simply be a result of experimental variability from one measurement to the next. Future work will be used to clarify this issue.

To confirm that temperature indeed is responsible for the generally higher responses suggested in Figure 2, other comparisons were

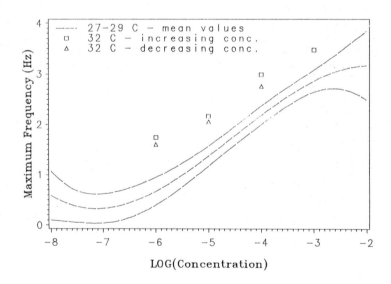

Fig. 2. Maximum frequency response data plotted against the logarithm of serotonin concentration at temperatures 27 - 29°C and 32°C. The middle dashed line is the mean maximum firing frequency observed at 28°C, while the outer lines are 95 % confidence limits on the mean. Squares are maximum firing frequency at 32°C with increasing concentrations, while triangles are at decreasing concentrations.

made. For example, Figure 3 shows, for a single cell at 10^{-4} M levels of serotonin and sufficient rinse times between applications, that the maximum firing frequency increases with temperature from 24°C to 35°C. However, at a higher temperature of 41°C, while the cell can be stimulated to respond by applying current in the nA range, it no longer generates spontaneous APs. In three other experiments this loss of activity was again observed occurring between 35 - 40°C depending on the period of time the cell was maintained at the elevated temperature. From Figure 3, an almost linear increase in maximum firing frequency is observed over 24 - 35°C. Hence, we can conclude that the average maximum firing frequency response at temperatures above 28°C for any given cell will certainly increase up to a point at which inactivation occurs.

Further verification of temperature dependence is offered on observation of the AP character, for a single cell, as a function of temperature. Differences in AP character represent alterations in either the voltage sensitive channel conductances, rates of enzymatic or metabolic activity, changes in receptor binding affinities or some combination thereof. Figure 4 shows the average of three APs for each of three temperatures, 24°C, 35°C and 41°C. The 41°C APs were generated on stimulation of the cell by passing a 100 ms current pulse. One observes that the peak minus the take-off value for the APs is 48 mV at 24°C, 46 mV at 35°C and 31 mV at 41°C. Thus, the peaks of the first two APs are very nearly the same while the third differs significantly. The half width of each AP is 5 ms at 24°C, 2.2 ms at 35°C and 2.1 ms at 41°C. In this case, the half width for the first is more than twice that of the other two. Preliminary conclusions are that while the AP heights do not vary significantly in the spontaneous firing range, the AP duration is significantly altered with temperature. Longer AP durations at lower temperatures can undoubtedly affect the time of recovery from an AP event, resulting in lower firing frequency.

In addition to the above, a comparison of the effects of sustained serotonin exposure on the firing frequency at two temperatures are shown in Figure 5. In both cases, the solution concentration was 10^{-3} M. The 32°C results are those from the experiment presented in Figure 1, while the 28°C results are those from another experiment. It is seen, that besides the higher peak response at 32°C, there is

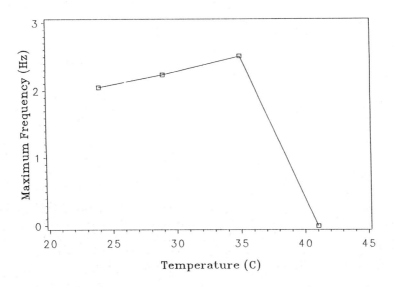

Fig. 3. Maximum firing frequency plotted against temperature for a single cell at a serotonin concentration of 10^{-4} M.

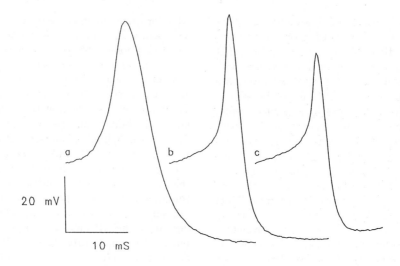

20 mV

10 mS

Fig. 4. The characteristics of the action potentials at (a) 24°C, (b) 35°C and (c) 41°C for a single cell.

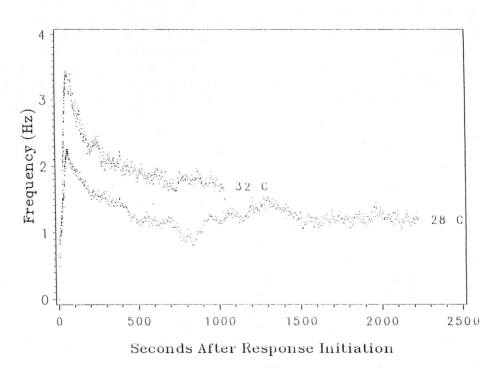

Seconds After Response Initiation

Fig. 5. Firing frequency plotted against time for an extended application of 10^{-3} M serotonin at 28°C and 32°C. The firing frequency increases and then decreases to a steady level due to desensitization, taking 125 s to fall to 80% of its maximum value at 32°C while taking 150 s to do the same at 28°C.

a more rapid decay in the firing frequency at a higher temperature than that observed for the lower temperature. Analysis of the data from Figure 5 shows that it takes 125 s for the firing frequency to fall to 80% of the maximum value at 32°C, while taking 150 s to do the same at 28°C. This trend has been repeatably confirmed in another separate set of ongoing experiments with a plug-flow chamber. As explained in our previous report (Skeen *et al.*, 1990a), the desensitization phenomenon makes this sensor unsuitable for certain continuous applications and would make it necessary for a sample to be introduced in short pulses to obtain an accurate indication of analyte concentration.

Preliminary studies on acetylcholine have also begun in our laboratory. A plug-flow chamber was used for recording in this experiment due to the rapid fall in the firing frequency when the cell was exposed to acetylcholine. A comparison of the serotonin and acetylcholine responses in the plug-flow chamber is shown in Figure 6. The circle in the figure near the serotonin curve indicates the point at which a rinse of *Limnea* saline was begun. No rinse step is shown for the acetylcholine. From the figure, it can be seen that the firing frequency for acetylcholine rapidly returns to baseline within 18 s of its peak, while that of serotonin would have shown desensitization to some intermediate value had an extended application been given. We also found that these particular cells responded to acetylcholine in the 10^{-4} to 10^{-3} M range which is a much narrower window of sensitivity than that for serotonin, and for practical purposes, one would, in general, select neurons that respond at lower concentrations. However, once found such neurons may still exhibit the rapid return to baseline, in which case, pulse application procedures should be used for sensing purposes.

3. Conclusions

The discussion presented in this paper reveals how neuronal biosensors offer distinct advantages over other whole cell preparations in terms of specificity, an extended analytical range and speed of response. A preliminary presentation of results on the effects of

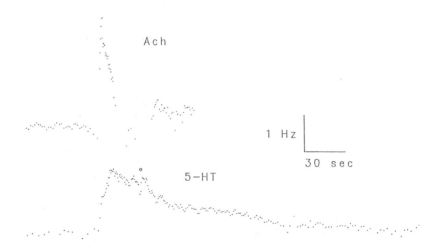

Fig. 6. A comparison of the firing frequency versus time profiles for applications of 10^{-4} M acetylcholine and serotonin. Due to the rapid decrease in the firing frequency of acetylcholine, a plug-flow chamber was used for this experiment. The responses are for two different cells.

temperature on neuronal response is also presented. It was found that an increase in temperature causes an increase in maximum firing frequency of the neurons. The general trend of this maximum firing frequency with concentration is the same as that observed at 28°C. An increase in temperature also causes desensitization to occur faster than that observed at 28°C. Efforts are now underway in our laboratory to fit results from a separate set of plug-flow experiments with a kinetic model for the desensitization process. Also, spontaneous AP generation disappears at temperatures between 35°C and 41°C. From this preliminary investigation of temperature effects, it can safely be concluded that temperature causes significant changes in the response of the system. However, as all of the above responses vary from cell to cell, more experiments of this type are being undertaken to extend and confirm these results. Other similar experiments will be done at lower temperatures. These studies are crucial since even small variations in the surrounding temperature appear to cause significant changes in neuronal responses.

When compared to serotonin, responses to acetylcholine show a much more rapid desensitization and a narrower range of sensitivity, namely 10^{-4} M to 10^{-3} M concentration.

Acknowledgments

This project was supported under the Microsensor Technology Program of the Washington Technology Center. Data analysis was aided by a computer workstation made available by an equipment grant from AT&T. Initiation of the biosensor effort in the Chemical Engineering Department at Washington State University and set up of the laboratory facility was aided by a NSF grant, no. ECE-8609910. An oscilloscope was donated by Tektronix Corporation. The authors thank Davis Bendezu for his technical assistance.

References

Arnold, G.A.; Rechnitz G.A. In *Biosensors: fundamentals and applications*; Turner, A.P.F.; Karube, I.; Wilson G.S.; Eds.; Oxford University Press: New York, 1987; pp 30-59.

Belli, S.L.; Rechnitz, G.A. *Anal. Lett.* **1986**, *19(3 & 4)*, 403.

Belli, S.L.; Buch, R.M.; Rechnitz, G.A. *Anal. Lett.* **1987**, *20(2)*, 327.

Belli, S.L.; Rechnitz, G.A. *Fresenius Z. Anal. Chem.* **1988**, *331*, 439.

Buch, R.M.; Rechnitz, G.A. *Biosensors* **1989a**, *4*, 215.

Buch, R.M.; Rechnitz, G.A. *Anal. Lett.* **1989b**, *22(13 & 14)*, 2685.

Buch, R.M., Rechnitz, G.A. *Anal. Chem.* **1989c**, *61(8)*, 533A.

Corcoran, C.A.; Kobos, R.K. *Biotechnol. Bioeng.* **1987**, *30*, 565.

Gross, G.W. *IEEE Trans. on Biomedical Engg.* **1979**, *BME-26(5)*, 273.

Gross, G.W.; Rieske, E.; Kreutzberg, G.W.; Meyer, A. *Neuroscience Lett.* **1977**, *6*, 101.

Gross, G.W.; Williams, A.N.; Lucas, J.H. *J. of Neuroscience Lett.* **1982**, *5*, 13.

Gross, G.W.; Wen, W.Y.; Lin, J.W. *J. of Neuroscience Methods* **1985**, *15*, 243.

Hafeman, D.G.; Parce, J.W.; McConnell, H.M. *Science* **1988**, *240*, 1182.

Karube, I.; Masunaga, T.; Mitsuda, S.; Suzuki, S. *Biotechnol. Bioeng.* **1977**, *19*, 1535.

Karube, I. In *Biosensors: fundamentals and applications*; Turner, A.P.F.; Karube, I.; Wilson, G.S., Eds.; Oxford University Press: New York, 1987; pp 13-29.

Karube, I.; Yokoyama, K.; Sode, K.; Eiichi, T. *Anal. Lett.* **1989**, *22(4)*, 791.

Kobos, R.K. *Anal. Lett.* **1986**, *19(3 & 4)*, 353.

Owicki, J.C.; Parce, J.W.; Kercso, K.M; Sigal, G.B.; Muir, V.C.; Venter, J.C.; Fraser, C.M.; McConnell, H.M. *Proc. natl. Acad. Sci. U.S.* **1990** (submitted)

Parce, J.W.; Owicki, J.C.; Kercso, K.M.; Sigal, G.B.; Wada, H.G.; Muir, V.C.; Bousse, L.J.; Ross, K.L.; Sikic, B.I.; McConnell, H.M. *Science* **1989**, *246*, 243.

Rechnitz, G.A. *C & EN* **1988**, *Sept 5*, 24.

Rechnitz, G.A.; Arnold, M.A.; Meyerhoff, M.E. *Nature* **1979**, *278*, 466.

Regehr, W.G.; Pine, J.; Rutledge, D.B. *IEEE Trans. on Biomedical Engg.* **1988**, *35(12)*, 1023.

Regehr, W.G.; Pine, J.; Cohan, C.J.; Mischke, M.D.; Tank, D.W. *J. of Neuroscience Methods* **1989**, *30*, 91.

Skeen, R.S.; Kisaalita, W.S.; Van Wie B.J.; Fung, S.J.; Barnes, C.D. *Biosensors and Bioelectronics* **1990a**, *5*, 491.

Skeen, R.S.; Kisaalita, W.S.; Van Wie B.J.; Barnes, C.D.; Fung, S.J. In: *Biosensors: fundamentals and applications*; Bowden, E.F.; Buck, R.D.; Hatfield, W.E.; Umana, M., Eds.; Marcel Dekker Inc.: New York, 1990b; pp 63-69.

Suzuki, H.; Tamiya, E.;Karube, I. *Anal. Chim. Acta* **1990**, *229*, 197.

Uchiyama, S.; Rechnitz, G.A. *Anal. Lett.* **1987**, *20*, 451.

Winlow, W.; Benjamin, P.R. In *Neurobiology of Invertebrates*; Salanki, J., Ed.; Akademiai Kaido: Budapest, Hungary, 1976; p 41.

Investigation of Immobilized-Cell Energetics and Metabolism using NMR Spectroscopy

Catherine A. Briasco, Genetics Institute, One Burtt Rd., Andover, MA 01810
Channing R. Robertson, Department of Chemical Engineering, Stanford University, Stanford, CA 94305

Nuclear magnetic resonance (NMR) spectroscopy is a non-invasive technique that can be applied to biological samples *in vivo* to identify and quantify selected molecules. We have evaluated a hollow-fiber reactor as a perfused sample container for NMR studies of microbial cells, with particular attention devoted to the effects of mass transfer limitation of the cells on the resulting ^{31}P NMR spectra. *E. coli* was immobilized and grown in hollow-fiber reactors. ^{35}S autoradiography indicated that protein-synthesizing cells occupied from 20% to almost 100% of the cell-containing region, depending on reactor design and operating parameters. ^{31}P NMR spectra of the cells, obtained during reactor operation, showed low volume-averaged concentrations of sugar phosphates, NTP (nucleoside triphosphates), and ratios of NTP/NDP in reactors with small volume fractions of growing cells. Intracellular pH was also depressed in these cells. In contrast, NMR spectra obtained from reactors without large regions of starved cells showed near-normal intracellular pH, metabolite concentrations, and NTP/NDP ratios. ^{31}P NMR of cells entrapped in a hollow-fiber reactor facilitates direct examination of cellular energetics, however the volume-averaged spectra are most unambiguously interpreted when the cell sample is as spatially homogeneous as possible with respect to the rate of cell metabolism.

NMR is a non-invasive spectroscopic technique that is becoming an increasingly popular and powerful tool in studies of *in vivo* cellular metabolism (Gadian, 1982; Burt, 1987). In its most common form, ^{13}C and/or ^{31}P NMR spectroscopy is used for the identification and quantification of specific cellular metabolites. One experimental limitation of NMR spectroscopy is that it is a relatively insensitive method, and therefore a concentrated cell sample is required to obtain a spectrum in a reasonably short time; for example, a sample of bacterial cells with approximately 10^{10} to 10^{11} cells/mL is required to obtain an NMR spectrum in a few minutes. In order to meet this requirement, many researchers prepare a highly concentrated cell slurry by centrifuging and resuspending free cell cultures. (See, for example, Axe and Bailey, 1987; Shanks and Bailey, 1990.)

A concentrated cell slurry is not an ideal cell sample, however, because the cells are not maintained in a sustained, non-starved metabolic state. Concentrations of nutrients and products are time-dependent, and thus the time period covered by each NMR spectrum must be as brief as possible, on the order of several minutes, to minimize averaging of the time-dependent metabolic state of the cells. The total experiment time, and hence the number of spectra that can be obtained in a given session, is also severely limited by degradation of the sample as nutrients are exhausted and inhibitory waste products accumulate.

A perfused cell sample, in which nutrients are continuously provided and waste products are continuously removed, can provide a greatly-improved system for NMR study of cell metabolism (Balaban, 1984). Constant perfusion with nutrient medium provides steady-state concentrations of nutrients and waste products within the cell sample. The desire to provide a steady-state cell sample for NMR studies has led several researchers to develop perfused, cell-containing devices that can be used within an NMR spectrometer. A variety of different sample geometries and operating strategies have been reported; examples of some of these are summarized in Table 1. In all such systems, the basic operating principle is that the cells are physically restrained, or immobilized, within the device,

Table 1. Summary of recent literature reports of perfused-cell devices suitable for use within an NMR spectrometer

Sample Type	Examples	Notes
microcarrier beads	Ugurbil *et al.* (1981)	mouse embryo fibroblasts
agarose gel threads	Foxall and Cohen (1983) Brindle and Krikler (1985)	*S. cerevisiae* *S. cerevisiae*
hollow fibers	Karczmar *et al.* (1983)	chick embryo fibroblasts, protozoa
hollow fiber reactor	Gonzalez-Mendez *et al.* (1982) Hrovat *et al.* (1985) Fernandez *et al.* (1988) Blute *et al.* (1988) Chresand *et al.* (1988) Drury *et al.* (1988) Heath *et al.* (1990)	CHO cells chick embryo fibroblasts hybridoma cells EAT cells: density measurements EAT cells: optimal fiber spacing EAT cells: O_2 transfer studies magnetic resonance flow imaging (cell-free bioreactor)
Reviews	Balaban (1984) Fernandez and Clark (1987) Gillies *et al.* (1989)	

Abbreviations: CHO: Chinese Hamster Ovary
EAT: Erlich Ascites Tumor

while fresh nutrient medium and/or oxygen are continuously added and spent medium is continuously withdrawn.

A critical factor in the design and operating of immobilized cell devices, however, is the relative balance between the reaction rate within the cell region and the rate of mass transport, either by convection or diffusion, into the cell region. For example, steep gradients in nutrient and product concentrations have been both predicted and experimentally observed for a wide variey of immobilized cell systems containing mammalian, mycelial, and bacterial cells (Karel *et al.*, 1985). Steep nutrient (or waste product) gradients can lead to the existence of regions of nutrient-starved (or product-inhibited) cells in the immobilized cell mass. Since high-resolution NMR produces a volume-averaged spectrum, the presence of a significant volume of metabolically inactive cells in the sample, either due to starvation or product inhibition, can lead to difficulties in interpreting the resulting spectra.

In this paper, we summarize our work in which hollow-fiber reactors were evaluated as perfused sample containers for cells during NMR studies. An important objective of the work was to characterize the extent of mass transfer limitations in immobilized cell/NMR devices, and to determine the influence of mass transfer limitations on the resulting NMR spectra. The model system used was *Escherichia coli* immobilization and growth in a NMR-compatible hollow fiber reactor, together with ^{31}P NMR spectroscopy of the high-density cell sample in the functioning reactors. Although virtually all methods of cell immobilization could be adapted for use in an NMR experiment, hollow fiber membrane reactors were chosen because of several unique features of their design and operation. The parallel arrangement of fibers in this reactor, resulting in steady laminar flow in the lumina at typical flow rates, permits relatively straightforward mathematical models of transport processes compared to flow past microspheres or gel beads. In addition, cells are physically entrapped in a HFMR, rather than chemically bonded to a surface as in some types of absorbed-cell reactors, and the lack of interaction between the cells and the hollow fiber membrane surfaces can eliminate experimental artifacts due to chemical attachment. *E. coli* was chosen as the model cell line to provide maximum flexibility in growth and operating conditions, since the objective of this work was to study reactors with a wide range of volumetric reaction rates, and hence exhibiting greatly different degrees of mass transfer limitation.

Materials and Methods

Strain: *E. coli B* (ATCC 11303) was obtained from the American Type Culture Collection and was maintained on nutrient agar plates at 4°C.

Nutrient Media: All nutrient media were prepared with reagent-grade chemicals and deionized water. The nutrient media used were: (PBM-1) phosphate-buffered, minimal medium M9 (Anderson, 1946) containing 23 mM glucose as carbon source; (PBM-2) similar to PBM-1 except containing 0.25 mM $MgSO_4 \cdot$ 7H_2O; and (MBM) 3-(N-morpholino) propane sulfonic acid (MOPS)-buffered medium as described by Neidhardt *et al.* (1974) except containing 70 mM MOPS, 2.0 mM K_2HPO_4 and 28 mM glucose. The phosphate-buffered media were sterilized by autoclaving, while the MOPS-buffered medium was filtered with a sterile Sterivex 0.2 μm filter (Millipore).

Hollow-Fiber Reactor Construction: A modified hollow-fiber reactor design (Fig. 1) was used to enable NMR spectroscopy of cells within the operating reactors. Reactors were constructed as described in Briasco *et al.* (1990a), using microporous, polypropylene hollow fibers (donated by Celanese Fibers Co., Charlotte, NC). The dimensions of the four reactors discussed in this paper are summarized in Table 2.

Reactor Inoculation and Operation: An anaerobic inoculation culture was prepared by inoculating a screw-cap bottle of sterile medium with 1% (v/v) of an overnight aerobic shaker culture. The anaerobic culture was incubated at either 37°C or 16°C, depending on subsequent

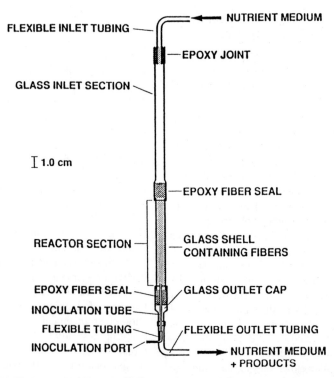

Fig. 1. Hollow-fiber reactor designed to allow operation inside the probe of a commercial NMR spectrometer. (Reproduced with permission from Briasco *et al.*, 1990a. Copyright 1990 John Wiley & Sons, Inc.)

140

Table 2.   **Reactor dimensions and nutrient media compositions**

| Reactor | Temperature (°C) | n | Reactor Shell | | | Reactor Feed Stream | | | |
			ID (cm)	OD (cm)	L_r (cm)	Medium Type	$S_{F,G}$ (mM)	pH	Q (ml/fiber·hr)
N1	37	240	0.67	0.90	8.0	PBM-1	23	7.2	6.25
S1	37	50	0.32	0.64	8.0	PBM-2	22	7.2	6.25
N2	16	360	0.67	0.90	8.0	MBM	27	8.2	3.33
S2	16	81	0.32	0.64	8.0	MBM	28	8.1	3.33

Note: All reactors contained fibers with OD=2.50×10^{-2} cm and ID=2.00×10^{-2} cm. Abbreviations: ID: inner diameter; OD: outer diameter.

SOURCE: Adapted from Briasco *et al.*, 1990b.

reactor operating temperature. Following sterilization, wetting and rinsing, the reactors were inoculated with an appropriate volume of exponential-phase anaerobic cells (Briasco *et al.*, 1990a). Reactors N1 and S1 were inoculated with cells equivalent to 6.4 gDM/L shell space, while N2 and S2 were inoculated at an initial cell density of approximately 90 gDM/L shell space.

^{35}S Autoradiography: Reactors S1 and S3 were labelled with ^{35}S to enable visualization of protein synthesis rate heterogeneities in the immobilized cell mass, following the method developed by Karel and Robertson (1989a). In order to avoid working with substantial quantities of radioactive medium, the labelled reactors were smaller, scaled versions of reactors examined with NMR. Model predictions (see section below) of the extent of mass transfer limitations were identical for an NMR reactor and its scaled version. Reactor S1 was scaled based on reactor N1, and S2 was based on reactor N2.

S1 and S3 (Table 2) were labelled with ^{35}S using a pulse-chase protocol. Reactor S1 was inoculated and fed with non-radioactive medium PBM-2 for 25 hours; this protocol was chosen to reproduce the conditions used for reactor N1. S1 was labelled immediately prior to the end of reactor operation with a two hour pulse of medium PBM-2 containing the same concentration of SO_4^{2-} but with a ^{35}S specific activity of approximately 5 Ci/mole. The pulse period was followed by a two hour chase period during which the reactor was perfused with unlabelled PBM-2 medium.

Reactor S3 was inoculated and perfused with unlabelled MBM medium for 30 hours, in order to reproduce the operating schedule of reactor N3. Immediately prior to the end of operation, the reactor was labelled with a 4 hour pulse of medium MBM-2 with a ^{35}S specific activity of approximately 5 Ci/mole followed by a 2 hour chase of unlabelled medium.

After the chase period, cells inside reactors S1 and S3 were perfused with a fixative of 20 g/L glutaraldehyde and 10 g/L paraformaldehyde in buffer. Reactors were then perfused with buffer and cut into 0.5 cm long sections, which were dehydrated by submersion for 2 hr each in progressively more concentrated ethanol solutions. The sections were then placed in low viscosity epoxy resin (Oliveira *et al.* 1983) overnight and the resin was polymerized at 70°C for 24 hours. Cross-sections (1 to 3 μm thick) were affixed to glass slides. The slides were coated with photographic emulsion sensitive to β-particle emission, then exposed and developed according to the procedure described and characterized by Karel *et al.* (1989a). Following development, the cross-sections were observed and photographed using bright field light microscopy with a Zeiss microscope equipped with a 4×5 Polaroid camera.

^{31}P NMR Spectroscopy: The metabolism of cells immobilized in reactors N1 and N2 was examined with ^{31}P NMR. Reactors were inserted into a Nicolet NT-300 spectrometer (General Electric, Fremont, CA) as described in Briasco *et al.* (1990a) Reactors were perfused with nutrient medium during data acquisition, and reactor temperature was controlled by heating or cooling the inlet medium. ^{31}P NMR spectra were recorded at 121.47 MHz with quadrature phase detection and ±10,000 Hz sweep width in 4K data blocks, without proton decoupling and without a deuterium lock. Pulse lengths were chosen for a tip angle of approximately 35° with 0.6 s delay between pulses. The summed FID's were processed by correcting the baseline with software provided by the spectrometer manufacturer, followed by exponential multiplication by 10 to 15 Hz and fast Fourier transformation. Chemical shifts are reported referenced to 0.5 M methylene diphosphonic acid, which was sealed within a capillary tube among the reactor fibers. Estimates of absolute metabolite concentrations were obtained by correcting for peak saturation using a spectrum acquired with a 7 s delay between pulses.

Mathematical Model: Mass transport in a hollow-fiber reactor is supplied primarily by diffusion of substrates through the fiber walls and into the immobilized cell layer. At the same time, substrates are consumed by the cells. If the overall rate of nutrient diffusion into the cell layer is significantly slower than the rate of nutrient consumption by the cells, than at some radial distance beyond the fiber wall the concentration of the limiting nutrient(s) will be exhaused. Beyond this limit, the cells are starved for the limiting nutrient. (A similar analysis applies when cell growth becomes limited by acid inhibition rather than nutrient starvation: as a first approximation, the limiting nutrient can be taken to be the basic form of the buffer component. In all studies described here, the nutrient medium composition was adjusted such that cell growth was predicted to be limited by glucose starvation rather than by organic acid inhibition.)

A model of reaction and diffusion in the immobilized cell layer was used to predict the relative importance of such mass transfer effects in the two experimentally-characterized reactor designs. Although numerous mathematical models of immobilized-cell growth have been described (Karel *et al.*, 1985), the practical use of many of these is restricted by an inability to independently estimate the values of required parameters. For this work, the relatively straightforward model described by Karel and Robertson (1989b) was adopted. This model assumes that each hollow fiber supplies a surrounding cylindrical volume in the reactor by

diffusion of nutrients across the fiber wall. (Convection of nutrients across the membrane walls is assumed to be negligible. This assumption, which implies that the contribution of Starling flow to mass transport is negligible, should be valid for these reactors, since the large number of fibers and short reactor lengths result in very small axial pressure drops.)

The model assumes that kinetics of immobilized-cell growth are equivalent to those for free-cell growth. Cell growth is described by reaction rate kinetics approximated as zeroth order in the limiting nutrient. The specific details of the application of this model to the hollow-fiber reactors described in this paper are presented elsewhere (Briasco *et al.*, 1990b). For purposes of comparison with experimental results, a summary of the model predictions are presented in Table 3. Kinetic parameters such as the growth rate, μ, and yield coefficients, $Y_{X/G}$ and $Y_{H+/G}$, were measured in batch cell cultures. Model results are presented in terms of ϕ, a Thiele modulus representing the ratio of the rates of reaction to diffusion in the cell mass; d_L, the predicted dimensional depth of the growing cell layer surrounding each fiber; and η, the fraction of the shell space volume that is occupied by non-starved cells.

Model predictions for the two reactor configurations are shown in Table 3. The value of η predicted for the first design (N1 and S1) is 0.34, indicating that 66% of the immobilized cell mass is expected to be starved. This is in sharp contrast to the expected η for N2/S2 of 1.0; all cells in this configuration are predicted to be adequately supplied with nutrients. The two configurations differed in the number of fibers and the concentration of cells, both of which contributed slightly to the different predicted effectiveness factors. By far the major influence, though, resulted from the difference in operating temperature. Reactors N2 and S2 were maintained at 16°C, while N1 and S1 were operated at 37°C. Since the growth rate of bacterial cells is extremely sensitive to temperature, this difference accounted for a change in the specific growth rate by a factor of seven. Taken together, the changes in reactor dimensions and operating conditions indicated in Table 3 resulted in values of ϕ^2 that differed by almost an order of magnitude.

Table 3.   Parameters and model predictions for anaerobic *E. coli* B grown in hollow-fiber reactors at 37°C and 16°C

		Reactors N1 and S1	Reactors N2 and S2	Units
Transport/Reaction				
temperature	T	310	289	K
cell concentration*	X	350	180	g/L
specific growth rate	μ	0.63	0.09	hr^{-1}
yield coefficients	$Y_{X/G}$	31	25	g DM/mol glu
	$Y_{H+/G}$	2.5	2.3	mol H$^+$/mol glu
Model Results				
squared Thiele modulus	ϕ^2	7.4	0.79	---
non-starved cell layer	d_L	37	140	μm
effectiveness factor	η	0.34	1.0	---

* Note: For N1 and S1, cell density was estimated from electron micrographs (Briasco *et al.*, 1990b) and data for *E. coli* free-cell excluded volumes (Stewart and Robertson, 1989). For N2 and S2, cell density (X) was estimated from inoculum density (X_o), cell growth rate (μ) and time prior to NMR spectroscopy or ^{35}S labelling (t), from $X = X_o \exp(\mu\, t)$.

SOURCE: Adapted from Briasco *et al.*, 1990b.

Results and Discussion

^{35}S Autoradiography

Autoradiographs taken from reactors S1 and S2 are shown in Fig. 2. Methylene blue staining showed that the interfiber area in both cross-sections was entirely filled with bacteria. The photograph taken from S1 contains silver grains associated with cells occupying a band measuring between 25 and 30 μm from the fiber wall. Thus, ^{35}S incorporation, and hence protein synthesis, occurred only in narrow zones surrounding each fiber in the reactor. The model results for this reactor predicted a growing cell layer depth of 37 μm. In practice, the estimate of this layer measured from autoradiographs is an upper limit of the actual value, due in part to cell layer convection during labelling and to imperfect resolution in the autoradiographs. These effects are discussed elsewhere (Karel, 1987; Karel and Robertson, 1989a). Nevertheless, considering the absence of adjustable parameters in the model, the agreement between experimental results and model predictions is very good.

The autoradiograph from reactor S2 (Fig. 2) shows a uniform distribution of silver grains; no radial dependence is evident. Thus there is no evidence of non-uniform protein synthesis in this reactor, at least within the resolution limits of the method. This supports the model prediction of $\eta=1$ for reactor S2 and hence reactor N2.

^{31}P NMR Spectroscopy

A ^{31}P NMR spectrum obtained from reactor N2 is shown in Fig. 3. The assignment of peak identities was determined from spectra measured for solutions of the commercially-available compounds and by comparison with previously-reported spectra of *E. coli* (Ugurbil *et al.*, 1978). The S-P peak, which includes contributions from both glucose-6-phosphate and fructose-1,6-biphosphate, and the NTP$_\beta$ peak are indicative of the volume-averaged energetic state of the cells within the reactor. In addition, the position of the P$_i$ peak is sensitive to local pH (Moon and Richards, 1973), and this enables estimation of the intracellular pH from NMR spectra. The spectrum shown here contains two P$_i$ peaks. The leftmost peak can be assigned a pH of 8.0 or greater, and is due to nutrient medium (pH 8.1) flowing through the lumina of the reactor. The other P$_i$ peak arises from intracellular phosphate, and thus can be used to

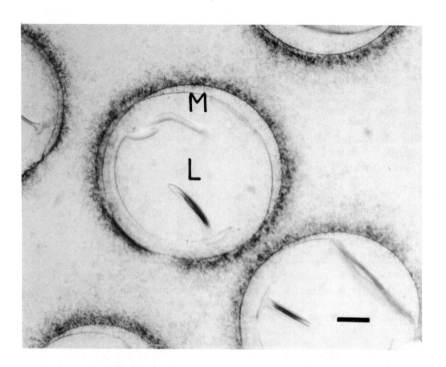

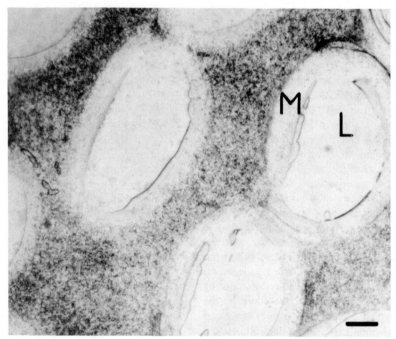

Fig. 2. Autoradiographs of *E. coli* B in hollow-fiber reactors S1 (**top**) and S2 (**bottom**); bar=50 μm. Reactors were subjected to a ³⁵S pulse-chase labelling protocol as described in the text. The black specks in the photographs are film grains exposed by β-particle emission from ³⁵S incorporated into cellular amino acids and proteins. Labels "**M**" and "**L**" indicate fiber membranes and lumina, respectively. (Reproduced with permission from Briasco *et al.*, 1990b. Copyright 1990 John Wiley & Sons, Inc.)

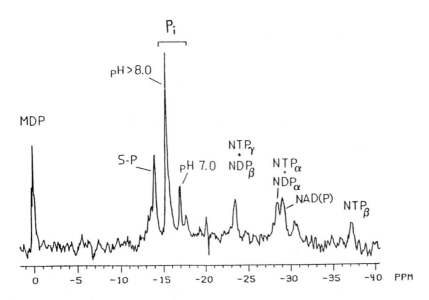

Fig. 3. ^{31}P NMR spectrum of anaerobic *E. coli* in reactor N2, recorded in 30 minutes beginning 8 hours after reactor inoculation. Abbreviations: MDP, methylene diphosphonic acid (chemical shift reference); S-P, sugar phosphates; P_i, inorganic orthophosphate; NDP and NTP, nucleoside di- and triphosphates; NAD(P), nicotinamide adenine dinucleotide and nicotinamide adenine dinucleotide phosphate. (Reproduced with permission from Briasco *et al.*, 1990b. Copyright 1990 John Wiley & Sons, Inc.)

estimate the internal *E. coli* pH of 7.0. This is a lower pH value than expected for energy-sufficient *E. coli*. Respiring *E. coli* have been reported to maintain a constant intracellular pH of 7.8 (Kashket, 1982) while the internal pH of anaerobic *E. coli* was found to reach a maximum of 7.5 soon after addition of glucose to the cell suspension (Ugurbil *et al.*, 1978). A possible explanation for the lower pH value observed here is the inherent inaccuracies in measurements of *absolute* pH values using NMR. As discussed by Roberts *et al.* (1981), the calibration of pH against chemical shift of P_i is sensitive to a number of factors such as concentration of free Mg^{++}, ionic strength of the medium, and concentrations of NTP and organic acids. Therefore, only an estimate of absolute pH can be obtained, and the NMR-measured intracellular pH of 7.0 cannot be considered markedly different from the previously reported value of a maximum of 7.5 for the intracellular pH of anaerobic *E. coli*.

In contrast with Fig. 3, a spectrum recorded from the cells within N1 (Fig. 4) exhibits extremely small peaks for both NTP_β and S-P. In addition, the P_i peak assigned to intracellular phosphate indicates a pH of less than 5.5, which is much more acidic than the value obtained from reactor N2. Qualitatively, these changes are consistent with the model prediction of a lower effectiveness factor, and hence a greater fraction of starved cells, in reactor N1 compared to reactor N2.

Further comparisons between reactors N1 and N2 can be obtained by integration of the NMR peaks to obtain both relative and absolute metabolite concentrations. Values for the relative concentrations of three different phosphorus-containing molecules are presented in Table 4 for three data sets. The time periods in the table are the number of hours after reactor inoculation. Individual spectra in these periods were averaged to obtain concentration estimates, which are presented here relative to the reference concentration of MDP. Two data sets, namely N2 (30-34 hours) and N1 (25-28 hours), were obtained shortly before termination of the reactor experiments and electron micrographs later showed that essentially all of the shell space in both reactors was completely filled with cells (Briasco *et al.*, 1990b). These two data sets can thus be taken as relative concentrations for the same volume of cells. The concentrations of all three metabolites are

145

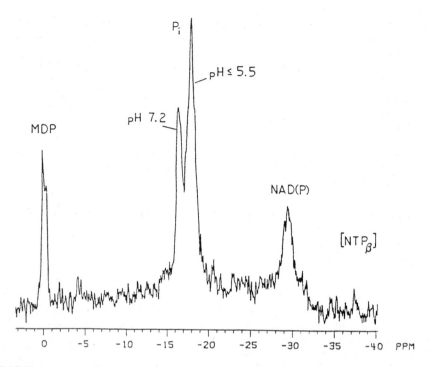

Fig. 4. ^{31}P NMR spectrum of anaerobic *E. coli* in reactor N1, recorded in 30 minutes beginning 24 hours after reactor inoculation. Abbreviations are given in Fig. 3. (Reproduced with permission from Briasco *et al.*, 1990b. Copyright 1990 John Wiley & Sons, Inc.)

greatest in reactor N2, consistent with the prediction of the larger value of η for this reactor. Although concentrations of sugar phosphate and NTP in N1 represent only 7% and 20%, respectively, of the values for N2, the concentration of NAD(P) in N1 is fully 70% of the N2 value. This observation is consistent with the roles of these metabolites: both sugar phosphate and NTP represent high-energy storage molecules and are rapidly degraded when cells are energy-starved. NAD(P) is used in the cell as a source of reducing power, and the NMR spectra show both the oxidized and reduced form. It is reasonable to expect that energy-starved cells could still possess a significant pool of this compound.

The data in Table 4 also confirm the assumption that cells are actively growing inside the reactor. Comparison of the N2 data for two time periods shows increases for all three compounds. Even more significantly, all three metabolites increase by a factor of two, within the accuracy of the data. This is additional evidence that substantial regions of energy-starved cells are absent in this reactor. If

such regions were present, then by the above comparison of reactors N1 and N2 we would expect a disproportionately larger increase in NAD(P) concentration.

In addition to relative concentrations, values for absolute concentrations can be obtained in those cases where the cell concentration is known. Unfortunately, it is virtually impossible to obtain reliable measurements of cell concentration in hollow-fiber reactors due to the difficulty of removing a representative sample from the sealed shell side of the reactor. An approximate cell concentration can be obtained from electron micrograph-derived cell number densities and previously-reported values for the free-cell volume of *E. coli* B (Stewart and Robertson, 1989); this calculation provided an estimate of 350 g DM/L for *E. coli*. With this value, absolute concentrations of NTP, NAD, and intracellular phosphate were calculated from NMR spectra of reactor N2 during the last 4 hours of operation. These estimates are shown in Table 5 together with past reports of concentrations in *E. coli* and *S. typhimurium*, a bacterium similar in many respects to *E. coli*.

Table 4. Metabolite concentrations measured from NMR spectra of reactors N1 and N2

Metabolite	Reactor N1	Reactor N2	
	25-28 hours	6-10 hours	30-34 hours
Sugar-P	7.2 ± 1.5	54 ± 12	100 ± 7.5
NAD(P)	69 ± 10	38 ± 14	100 ± 16
NTP	20 ± 3.2	48 ± 9.6	100 ± 15

Note: Concentrations are normalized to the largest values observed, which were measured for reactor N2 during the second NMR period. Values were corrected for different shell volumes in the reactors by dividing by the fraction of space available to cells.

SOURCE: Reprinted with permission from Briasco *et al.*, 1990b. Copyright 1990 John Wiley & Sons, Inc.

Table 5. Absolute intracellular metabolite concentrations measured from NMR spectra of reactor N2 and comparison values from literature reports

Metabolite	Reactor N2	Literature	Conditions and Source
NTP	5.1 ± 0.8	4.35	*S. typhimurium*; see note **b**; Bochner and Ames, 1982
NAD+NADP	13 ± 2	4.2 ± 0.9	*E. coli* B/r NF790; Andersen and von Meyenburg, 1977
P_i	27 ± 7	36 ± 4	anaerobic *E. coli* ML308-225, pH 6.0; Kashket, 1983
		44 ± 4	anaerobic, *E. coli* ML308-225, pH 7.0; Kashket, 1983
NTP/NDP	3.2 ± 0.4	--	
ATP/ADP	--	9.0	aerobic *E. coli* K12, 43°C; Kahru *et al.*, 1987
	--	3.7	anaerobic *E. coli* ML308-225, pH 7.0; Kashket, 1983
	--	3.1	anaerobic *E. coli* ML308-225, pH 6.0; Kashket, 1983

Note a: Values are mean ± standard deviation. Concentrations of NTP, NAD, and P_i are in units of mmol/L of cells.

Note b: The value reported by Bochner and Ames, in units of mmol per liter of intracellular water, has been corrected here for comparison purposes by assuming values of 2.33 ml intracellular water/gDM (Neuhard and Nygaard, 1987) and 350 gDM/L for reactor *E. coli* cells (Stewart and Robertson, 1989).

SOURCE: Reprinted with permission from Briasco *et al.*, 1990b. Copyright 1990 John Wiley & Sons, Inc.

The NMR-measured concentrations of NTP and intracellular inorganic phosphate are in remarkably good agreement with the literature values, considering the uncertainty in the reactor content of cells. The concentration of NAD measured for the reactor cells, however, is approximately three times larger than the previously-reported value for *E. coli*. A possible explanation for this discrepancy is that some other phosphorus-containing compound is contributing to the resonance observed at -29 to -30 ppm. Past reports have not attempted to assign absolute concentration values to the NAD(P) resonance, so the significance of this result is difficult to determine; it is, however, worthy of further investigation.

Another indicator of overall energetic state in the immobilized-cell layer is obtained from the ratio of NTP/NDP concentrations. This quantity was measured to be 3.2 ± 0.4 for cells in reactor N2, as shown in Table 5. Comparison of the measured value with past reports is difficult, however, due to the fact that researchers have usually reported data for the concentration ratio ATP/ADP. For purposes of this discussion, literature reports of ATP/ADP can be regarded as upper bounds to NTP/NDP, since the ratios CTP/CDP and UTP/UDP have been found to be lower than the ratios ATP/ADP and GTP/GDP (Bochner and Ames, 1982). A value of 9.0 for the ratio ATP/ADP was calculated from the data of Kahru *et al.* (1987) for aerobic *E. coli* K12. This value, significantly larger than that obtained from reactor N2, is for *aerobic* cells at 37°C. The effect of anaerobiosis is shown by comparison with Kashket's data for anaerobic *E. coli* ML308-225: the value of ATP/ADP for anaerobic cells appears to be much lower than that for aerobic cells (Kashket, 1983). In fact, the value of NTP/NDP measured for the immobilized cells is in good agreement with the values of 3.7 and 3.1 measured by Kashket. In addition, Kahru *et al.* (1987) demonstrated that a decrease in growth temperature also decreases the ATP/ADP ratio, at least for aerobic *E. coli* K12 cells. Considering the fact that the cells in reactor N2 were grown at 16°C under anaerobic conditions, the measured NTP/NDP ratio is in good agreement with literature reports, and is therefore also indicative of a kinetically-limited reactor, that is, one in which non-starved, nutrient supplied cells occupy essentially 100% of the reactor shell space.

Conclusions

The one-dimensional mathematical model of reaction and diffusion in the hollow-fiber reactors predicted growing cell volumes that were in good agreement with experimental results. ^{35}S autoradiography allowed estimation of the growing cell layer in the reactor characterized by a low effectiveness factor, and did not show evidence of starved cell regions in the case of the reactor predicted to have an effectiveness factor of unity. ^{31}P NMR provided information on intracellular pH and on the presence and concentration of molecules that are indicative of the energetic status of the cells, such as sugar phosphates and nucleoside triphosphates. As expected, the reactor characterized by a small effectiveness factor contained lower concentrations of these metabolites, and exhibited a severely depressed intracellular pH, when compared to the kinetically-limited reactor. Therefore, the extent of mass transport limitations in the immobilized cell layer must be evaluated when using this type of device for NMR studies of cell metabolism. While the focus in this work was on a specific type of immobilized-cell system, namely the hollow-fiber reactor, mass transport limitations of dense cell samples can occur in other perfused cell devices. The mathematical modeling and experimental techniques demonstraed in this work allow an improved description of mass transfer effects in immobilized-cell systems in general, and can be used with other NMR-compatible perfused cell devices in order to better understand the influence of mass transfer limitations on the resulting NMR spectra.

Nomenclature

d_L	dimensional depth of growing cell layer [m]
L_r	axial length of reactor [m]
n	number of fibers [-]
Q	flow rate per fiber [m^3 s^{-1}]
t	time [s]
T	absolute temperature [K]
X	cell concentration [g m^{-3}]
X_o	initial cell concentration [g m^{-3}]
$Y_{X/G}$	yield coefficient: ratio of cell mass

production to glucose consumption [g mol^{-1}]

$Y_{H+/G}$ yield coefficient: ratio of acid production to glucose consumption [mol mol^{-1}]

η effectiveness factor [-]

μ specific growth rate of cells [s^{-1}]

ϕ Thiele modulus [-]

$S_{F,G}$ concentration of glucose in reactor feed stream [mol m^{-3}]

$S_{F,B}$ concentration of buffer in reactor feed stream [mol m^{-3}]

Acknowledgments

We thank Professor Steven Karel (Princeton University) for advice and comments on this work. We also thank Fran Thomas of the Department of Biological Sciences at Stanford for preparing thin sections of the reactors, and Lois Durham of the Stanford Chemistry Department NMR Laboratory for technical assistance. The NMR spectrometer was purchased with NSF Grant CHE-81-09064. C. Briasco acknowledges financial support from an NSF Graduate Fellowship. Funding for this research was provided by NSF Grant ECE-86-13227 and by the Monsanto Co., St. Louis, MO.

References

Andersen, K.B.; von Meyenburg, K. *J. Biol. Chem.* **1977,** *252,* 4151-4156.

Anderson, E.H. *Proc. Natl. Acad. Sci. USA* **1946,** *32,* 120-128.

Axe, D.D.; Bailey, J.E. *Biotechnol. Lett.* **1987,** *9,* 83-88.

Balaban, R.S. *Am. J. Physiol.* **1984,** *246 (Cell Physiol. 15),* C10-C19.

Blute, T.; Gillies, R.J.; Dale, B.E. 1988. *Biotechnol. Progress* **1988,** *4,* 202-209.

Bochner, B.R.; Ames, B.N. *J. Biol. Chem.* **1982,** *257,* 9759-9769.

Briasco, C.A. *Investigation of immobilized microbial cell energetics using nuclear magnetic resonance spectroscopy;* Ph.D. Thesis, **1988,** Stanford University, Stanford, CA.

Briasco, C.A.; Ross, D.A.; Robertson, C.R. *Biotechnol. Bioeng.* **1990a,** 879-886.

Briasco, C.A.; Karel, S.F.; Robertson, C.R. *Biotechnol. Bioeng.* **1990b,** 887-901.

Brindle, K; Krikler, S. *Biochim. Biophys. Acta* **1985,** *847,* 285-292.

Burt, C.T. *Phosphorus NMR in Biology;* CRC Press: Boca Raton, FL, **1987.**

Chresand, T.J.; Gillies, R.J.; Dale, B.E. *Biotechnol. Bioeng.* **1988,** *32,* 983-992.

Drury, D.; Dale, B.E.; Gillies, R.J. *Biotechnol. Bioeng.* **1988,** *32,* 966-974.

Fernandez, E.J.; Clark, D.S. *Enzyme Microb. Technol.* **1987,** *9,* 259-271.

Fernandez, E.J.; Mancuso, A.; Clark, D.S. *Biotechnol. Progress* **1988,** *4,* 173-183.

Foxall, D.L.; Cohen, J.S. *J. Mag. Res.* **1983,** *52,* 346-349.

Gadian, D.G. *Nuclear magnetic resonance and its applications to living systems;* Oxford University Press: New York, NY **1982.**

Gillies, R.J.; MacKenzie, N.E.; Dale, B.E. *Bio/Technology* **1989,** *7,* 50-54.

Gonzalez-Mendez, R.; Wemmer, E.; Hahn, G.; Wade-Jardetzky, N.; Jardetzky, O. *Biochim. Biophys. Acta* **1982,** *720,* 274-280.

Heath, C.A.; Belfort, G.; Hammer, B.E.; Mirer, S.D.; Pimbley, J.M. *AIChE Journal* **1990,** *36,* 547-558.

Hrovat, M.I.; Wade, C.G.; Hawkes, S.P. *J. Mag. Res.* **1985,** *61,* 409-417.

Kahru, A.; Paalme, T.; Vilu, R. *FEMS Microbiol. Lett.* **1987,** *41,* 305-308.

Karczmar, G.S.; Koretsky, A.P.; Bissell, M.J.; Klein, M.P.; Weiner, M.W. *J. Mag. Res.* **1983,** *53,* 123-128.

Karel, S.F. *Reaction and diffusion in immobilized living bacteria: characterization with radioisotope labelling and autoradiography;* Ph.D. Thesis, **1987,** Stanford University, Stanford, CA.

Karel, S.F.; Libicki, S.B.; Robertson, C.R. *Chem. Eng. Sci.* **1985,** *40,* 1321-1354.

Karel, S.F.; Robertson, C.R. *Biotechnol. Bioeng.* **1989a,** *34,* 320-336.

Karel, S.F.; Robertson, C.R. *Biotechnol. Bioeng.* **1989b,** *34,* 337-356.

Kashket, E.R. *Biochemistry* **1982,** *22,* 5534-5538.

Kashket, E.R. *FEBS Lett.* **1983,** *154,* 343-346.

Moon, R.B.; Richards, J.H. *J. Biol. Chem.* **1973,** *248,* 7276-7278.

Neidhardt, R.C.; Bloch, P.L.; Smith, D.F. *J. Bacteriol.* **1974,** *119,* 736-747.

Neuhard, J.; Nygaard, P. In *Escherichia coli and Salmonella typhimurium: Cellular and molecular biology;* Neidhardt, F.C., Ed.; Amer. Soc. for Microbiol.: Washington, D.C., **1987,** Vol. 1; pp. 445-473.

Oliveira, L.; Burns, A.; Bisalputra, T.; Yang, K.-C. *J. Microscopy* **1983,** *132,* 195-202.

Roberts, J.K.M.; Wade-Jardetzky, N.; Jardetzky, O. *Biochemistry* **1981,** *20,* 5389-5394.

Shanks, J.V.; Bailey, J.E. *Biotechnol. Bioeng.* **1990,** *35,* 395-407.

Stewart, P.S.; Robertson, C.R. 1989. *Appl. Microbiol. Biotechnol.* **1989,** *30,* 34-40.

Ugurbil, K.; Rottenberg, H.; Glynn, P.; Shulman, R.G. *Proc. Natl. Acad. Sci. USA* **1978,** *75,* 2244-2248.

Ugurbil, K.; Guernsey, D.L.; Brown, T.R.; Glynn, P.; Tobkes, N.; Edelman, I.S. *Proc. Natl. Acad. Sci., USA* **1981,** *78,* 4843-4847.

Part III
Cell-Culture Systems

THE IDEAL REACTOR CATALYST synthesizes one product and no biomass. In cell bioreactors, this desideratum translates into living cells that neither divide nor die and that secrete their product into a liquid phase. Starting with prokaryotes and moving through the fungal, plant, and animal kingdoms, systems are described in which bacterial cells obey certain rules in producing inclusion bodies, yeast cells glycosylate and secrete desired proteins, fungal systems produce desired products (including proteins) under specified culture conditions, and plant cells respond to elicitors after growth has stopped and secrete products into an extraction phase. New developments are emerging slowly in protein production in animal cell culture. A thorough examination of specific phenomena such as glycosylation and antibody assembly is needed (and provided here) before significant additional progress can be made.

The Formation of β-Lactamase Inclusion Bodies in *Escherichia coli* and Comparison with *In Vitro* Studies

Gregory A. Bowden, Pascal Valax and George Georgiou, Department of Chemical Engineering, University of Texas at Austin, Austin, TX 78712

The high level expression of β-lactamase in *Escherichia coli* results in the accumulation of the protein in a misfolded, aggregated form (inclusion bodies) in the periplasmic space. The cellular and environmental factors that affect the aggregation of β-lactamase in the cell were investigated. The formation of inclusion bodies can be inhibited by growing the cells in the presence of non-metabolizable sugars such as sucrose and raffinose. It was shown that the addition of sugars inhibits the intracellular aggregation of β-lactamase by directly influencing the folding kinetics rather than indirectly by altering the rate of protein synthesis or secretion. The formation of inclusion bodies is also affected by the signal sequence which directs export of the protein in the periplasmic space. Furthermore, when the protein was expressed without a functional signal sequence it accumulated almost entirely in aggregated form within the cytoplasm. Since the investigation of protein folding and aggregation within the cell is technically difficult, refolding studies *in vitro* were used to complement *in vivo* observations. It was observed that higher recoveries of the correctly folded protein are obtained under conditions that approximate the pH and redox potential of the periplasmic space. Similarly, the presence of sucrose in the refolding buffer inhibits protein aggregation.

Proteins important for therapeutic, agricultural and other industrial applications can be produced in large quantities by expressing them in bacteria such as *Escherichia coli*. However, heterologous proteins, or even native proteins expressed at a high level, are often folded improperly and are sequestered in a misfolded form into intracellular insoluble protein aggregates known as inclusion bodies. Presumably, intracellular protein aggregation depends on the kinetics of polypeptide folding which are, in turn, influenced by the amino acid sequence and the intracellular conditions. A greater understanding of the relationship between protein aggregation and the folding environment within the cell can lead to the optimization of fermentation conditions and facilitate the choice of host and expression vector system for the efficient production of commercially important polypeptides.

Because of experimental difficulties, there is very little information on the folding and aggregation of proteins *in vivo*. However, some valuable analogies can be drawn from *in vitro* studies. In the presence of high concentrations of denaturants such as guanidine hydrochloride or urea, proteins lose their native three-dimensional structure and attain a random coil conformation (Tanford, 1968). The subsequent removal of the denaturant either by dialysis or dilution allows the polypeptide to fold back to the biologically active conformation. At high protein concentrations, the recovery of biological activity is less than 100%. Irreversibility in protein refolding arises mainly from the association of partially folded polypeptide chains which leads to the formation of aggregates. Studies by Goldberg, Rudolph and Jaenicke and more recently Brems and coworkers have shed light on the mechanism of protein aggregation *in vitro* (London and Goldberg, 1974; Jaenicke and Rudolph, 1983; Brems et al.,

1988). Aggregation is thought to occur from the intermolecular association of exposed hydrophobic surfaces which are normally confined to the core regions of the native protein. These hydrophobic domains may be transiently exposed to the solvent as a result of the formation of kinetic intermediates in the folding pathway. The intermediate (or intermediates) responsible for aggregation can have one of two fates: either continue on the folding pathway and give rise to the native conformation or self-associate to form aggregates. The former process is first order with respect to protein concentration whereas the latter follows second or higher order kinetics. For this reason higher protein concentrations tend to favor aggregation.

The concentration of the folding intermediate which is prone to aggregation is dependent on the folding kinetics. *In vitro*, temperature, pH and redox potential exert a significant effect on the folding pathway of the protein and, thus, on aggregation. Similarly, *in vivo*, the formation of inclusion bodies has been shown to be affected by the growth temperature (Schein and Noteborn, 1988; Takagi et al., 1988) and pH (Kopetski et al., 1989). In addition, cellular components have been shown to affect the folding pathway. In *E. coli*, the chaperonins GroEL, secB, trigger factor, and DnaK are proteins known to influence the folding (or unfolding) of proteins (Rothman, 1989). They are involved in a variety of cellular functions including maintenance of a protein in a relatively unfolded configuration suitable for secretion (GroEL, GroES, secB, trigger factor, DnaK; Crooke and Wickner, 1987; Phillips and Silhavy, 1990; Collier et al., 1988), in assisting the assembly of subunits into globular proteins (GroEL; Goloubinoff et al., 1989) and for protecting the organism from the effects of heat stress (DnaK, GroEL; Schlesinger, 1990).

2181–2/92/0152$06.00/0 © 1992 American Chemical Society

A convenient model for the study of inclusion body formation *in vivo* is the monomeric, secreted protein, R_{TEM} β-lactamase. The overexpression of β-lactamase in *E. coli* results in the formation of inclusion bodies in the periplasmic space (Georgiou et al., 1986). Due to the permeability of the outer cell membrane of *E. coli* to lower molecular weight compounds present in the growth media, the composition of the periplasmic space can be altered. This property is valuable in studying the influence of the cellular folding environment on protein aggregation. In this report we demonstrate that studies of the folding of β-lactamase *in vivo*, as well as *in vitro,* can provide valuable insight on the formation of inclusion bodies in *E. coli*.

MATERIALS AND METHODS

Bacterial strain and plasmids.

Escherichia coli strain RB791 (LacIq$_{L8}$) was used for all experiments (Brent and Ptashne, 1981). The plasmids pTac11 (4.6 kb) expressing native β-lactamase and pJG108 (9.6 kb) expressing a fusion between the ompA signal sequence and the native mature β-lactamase have been described earlier (Amann et al., 1983; Ghrayeb et al., 1984). The plasmid pKN has been constructed previously (Georgiou et al., 1988) by inserting the HindIII-BamHI fragment containing the neomycin resistance gene from pNEO (Pharmacia) into pTac11. The plasmid pTG2 containing the β-lactamase gene with a -20-1 deletion in the signal sequence was obtained from J. Knowles (Plückthun and Knowles, 1987). The plasmid pGB1 was constructed by cutting pKN and pTG2 (-20-1 deletion) with AatII and SalI (unique sites on both plasmids). The small fragment from pKN containing the *tac* promoter and the neomycin resistance gene and the large fragment from pTG2 (-20-1 deletion) containing the mutant β-lactamase were ligated and transformed into *E. coli* RB791. Cells harboring pGB1 were selected by resistance to neomycin sulfate.

Growth conditions and cell fractionation.

Except for experiments involving protein labelling with ^{35}S-methionine, cells were grown at 37⁰C in 125 ml Erlenmeyer flasks containing 25 ml of M9 minimal salts media, 0.2% fructose, 0.2% casein amino acids and 50 μg/ml ampicillin or neomycin sulfate. All sugars were sterilized by filtration and added to cultures at the concentrations indicated in the text. Unless otherwise stated, β-lactamase synthesis was induced with isopropylthiogalactoside (IPTG) to a final concentration of 0.1 mM when the cultures reached mid-exponential phase at an optical density (O.D.$_{600}$) between 0.35 and 0.4.

Cultures were fractionated into soluble and insoluble protein fractions as follows: 1 ml samples from overnight cultures were lysed in a French press at 20,000 pounds per square inch. The lysates were centrifuged at 12,000xg for 10 minutes. The supernatants were saved as the total soluble fraction

and the pellets were washed with 50 mM potassium phosphate, pH 6.5, and centrifuged as before. The insoluble fractions were solubilized in SDS electrophoresis buffer and resolved by SDS-PAGE.

Analytical procedures.

SDS-polyacrylamide gel electrophoresis (SDS-PAGE) was performed on 15% polyacrylamide gels (Laemmli, 1970). The method for the immunological detection of β-lactamase following electrophoresis and transfer to nitrocellulose filters has been described elsewhere (Towbin et al., 1979). β-lactamase enzymatic activities were measured from the hydrolysis of penicillin-G (Samuni, 1975). Protein concentrations were determined by the Bradford method (Bradford, 1976) using the Bio-Rad assay reagent with bovine serum albumin as the standard.

Pulse-chase experiments.

The use of ^{35}S-methionine incorporation in the measurement of protein synthesis rates and translocation kinetics are described in detail elsewhere (Bowden and Georgiou, 1990).

In vitro β-lactamase refolding experiments.

R_{TEM} β-lactamase was obtained and purified as described previously (Valax and Georgiou, 1990). For the refolding studies, known amounts of β-lactamase were lyophilized and dissolved in 50 mM potassium phosphate buffer, pH 7.0, containing 2 M guanidine hydrochloride (GuHCl) and 5 mM dithiothreitol. The samples were dialyzed against the same buffer at conditions described in the text for three hours (room temperature) in a PIERCE microdialyzer model 500 apparatus. The buffer was then changed and the protein allowed to refold for three more hours. In all experiments, the final GuHCl concentration was 0.02 M. The samples were subsequently centrifuged at 12,000xg for 20 minutes at 4°C and the activity remaining in the supernatant was measured. The pellets were washed in 50 mM potassium phosphate, pH 7.0 and resuspended in the same buffer. The suspension was then centrifuged as above and the activity of the supernatant was determined.

RESULTS

β-Lactamase Aggregation in *Escherichia coli*

The influence of protein synthesis rate on aggregation.

To examine the effects of high levels of protein synthesis on the formation of β-lactamase inclusion bodies, expression was directed by a strong promoter (*tac*). The rate of protein synthesis from

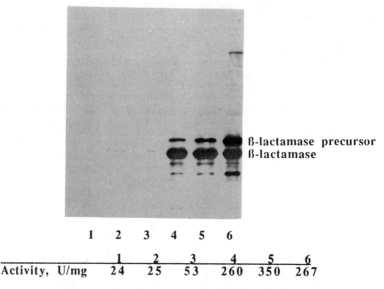

	1	2	3	4	5	6

ß-lactamase precursor
ß-lactamase

Activity, U/mg	1	2	3	4	5	6
	24	25	53	260	350	267

Fig. 1: Distribution of β-lactamase in the soluble and insoluble fractions of RB791(pTac11) cultures. Cell lysates from RB791(pTac11) grown in minimal media with increasing concentrations of IPTG were separated into soluble and insoluble fractions as described under "Materials and Methods". The insoluble fractions were solubilized by boiling in SDS electrophoresis buffer and a volume corresponding to exactly 20 μl of culture was loaded into each lane of a 15% polyacrylamide gel. β-lactamase was detected by immunoblotting. Lanes 1 to 6: insoluble fractions from cultures grown with 0, 0.0001, 0.005, 0.05, 0.1 and 1.0 mM IPTG, respectively. The positions of the mature β-lactamase and β-lactamase precursor bands are indicated. The specific activities of the β-lactamase in the soluble protein fractions corresponding to each lane are also shown. Adapted from Bowden and Georgiou, 1990.

the *tac* promoter depends on the concentration and time of addition of IPTG. The influence of the inducer concentration on the synthesis of β-lactamase in *E. coli* RB791(pTac11) cultures was investigated. As shown in Fig. 1, the appearance of a substantial amount of insoluble β-lactamase was observed in cultures grown in the presence of 0.05 mM IPTG. Previous studies have shown that the insoluble β-lactamase is due to protein sequestered in inclusion bodies in the periplasmic space (between the inner and outer membranes) of *E. coli* (Georgiou et al., 1986). The maximum amount of soluble β-lactamase is observed in the presence of 0.1 mM. The onset of inclusion body formation occurs at an IPTG concentration between 0.02 and 0.03 mM (data not shown). From Fig. 1 it is evident that the ratio of soluble to aggregated β-lactamase varies depending on the total amount of protein produced. These results represent the accumulation of protein in overnight cultures. The kinetics of inclusion body formation are shown in Fig. 2. Analysis of the distribution of the β-lactamase into the insoluble and soluble protein fractions by SDS-polyacrylamide gel electrophoresis and by measuring the specific activity, respectively, indicated that most of the insoluble β-lactamase is produced within the first

four hours after induction (Fig. 2). Subsequently, the majority of the newly synthesized β-lactamase accumulates in the soluble protein fraction.

To examine the relation between protein aggregation and the synthesis rate in RB791(pTac11), the rate of incorporation of ^{35}S-labelled methionine into β-lactamase and trichloroacetic acid-precipitable material was measured. The maximum synthesis rate was observed one hour after induction (Table 1). Subsequently, the synthesis rate decreased as the growth rate of the culture declined. However, the rate of β-lactamase synthesis does not necessarily reflect the rate of accumulation in the periplasmic space. Under certain conditions a fraction of the newly synthesized protein in the cytoplasm may be prevented from export through the inner membrane. Separate experiments were conducted to determine the rate of protein secretion (Bowden and Georgiou, 1990). From the combined rates of total protein synthesis and transport it was calculated that the maximum rate of β-lactamase accumulation in the periplasmic space was 11.9 cpm/10^5cells/sec. No protein aggregation could be detected until the rate of formation of the mature (i.e. the periplasmic) form of β-lactamase exceeded 2.5% of the total protein

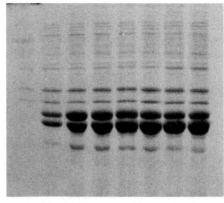

β-lactamase precursor
β-lactamase

	1	2	3	4	5	6	7	8
Activity, U/mg	27	141	220	301	320	341	348	344

Fig. 2: SDS-PAGE analysis of the production of insoluble β-lactamase over a 14 hour post-induction period from a RB791(pTac11) culture induced with 0.1 mM IPTG. Samples were prepared as described in Fig. 1 except that each lane was loaded with 100 μl of media. Lanes 1 to 8 are the insoluble fractions at 0, 2, 4, 6, 8, 10, 12 and 14 hours after induction. The specific activities of the β-lactamase in the soluble protein fractions corresponding to each lane are also shown.

Table 1: Rate of incorporation of ^{35}S-methionine into trichloroacetic acid precipitate and β-lactamase. A detailed description of the procedure is provided elsewhere (Bowden and Georgiou 1990). The rates of radioactivity incorporation were determined from least square fits (r≥0.98 for all experiments). The numbers in the "% β-lactamase synthesis" column represent the ratio of β-lactamase radioactivity over the total acid precipitate corrected for the difference in the methionine content between β-lactamase and the average *E. coli* protein. *The rate for the uninduced RB791(pJG108) culture was measured one hour after the time it would have been induced

PLASMID	[IPTG], mM	INDUCTION TIME, hrs	TOTAL, cpm/10^5cells/sec	β-LACT., cpm/10^5cells/sec	% β-LACT. SYNTHESIS
pTac11	0.1	1	246	34.0	10.7
pTac11	0.1	2	147	25.2	13.3
pTac11	0.02	1	239	7.3	2.4
pJG108	0	*	357	3.5	0.8
pJG108	0.1	1	281	6.8	1.9
pJG108	0.1	2	304	16.0	4.1
pJG108	0.1	3	110	7.5	5.3

Adapted from Bowden and Georgiou, 1990.

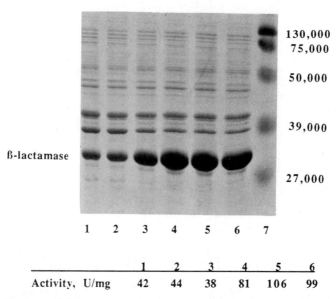

β-lactamase

	1	2	3	4	5	6
Activity, U/mg	42	44	38	81	106	99

Fig. 3: Distribution of β-lactamase in the soluble and insoluble fractions of RB791(pJG108) cultures. The culture samples were fractionated as described earlier. Each lane was loaded with 50 μl of culture media. Lanes 1 to 6: insoluble fractions from cultures grown with 0, 0.0001, 0.005, 0.05, 0.1 and 1.0 mM IPTG, respectively. Lane 7: molecular weight standards. The specific activities of the β-lactamase in the soluble protein fractions corresponding to each lane are also shown. Adapted from Bowden and Georgiou, 1990.

synthesis. These results clearly establish the significance of protein expression on aggregation.

Effect of signal sequence on β-lactamase aggregation in the periplasmic space.

The rate of accumulation of mature β-lactamase in the periplasmic space is partially dependent on the translocation kinetics of the cytoplasmic precursor of the protein. In turn, the kinetics of secretion are influenced by the signal sequence. It was of interest to see if the signal sequence affects the aggregation of the mature β-lactamase. Inouye and coworkers have constructed a fusion between the signal sequence of the *E. coli* outer membrane protein, ompA, and the mature sequence of β-lactamase (Ghrayeb et al., 1984). Cleavage of the signal sequence from the ompA-β-lactamase precursor upon secretion results in the formation of mature β-lactamase in which the N-terminal amino acid is identical to the native sequence. The ompA-β-lactamase protein is expressed from the *lpp* promoter from the *E. coli* lipoprotein gene and the operator region of the lactose operon (pJG108). Similar to the plasmid pTac11, the rate of protein synthesis can be controlled by the addition of IPTG. The formation of β-lactamase inclusion bodies in the periplasmic space of induced cultures of *E. coli* cells containing pJG108 is much more pronounced compared to cells expressing the protein with its wild type signal sequence. As shown in Fig. 3, β-lactamase appeared in the insoluble protein fraction

irrespective of the concentration of the inducer. Inclusion bodies were also observed in uninduced cultures. In contrast to the results obtained with RB791(pTac11) where both the precursor and the mature form of the protein are present in the insoluble fraction, only a very small amount of the precursor was detected in immunoblots of the insoluble fractions of cells expressing ompA-β-lactamase (data not shown). The kinetics of β-lactamase precursor were determined from radioactive pulse-chase experiments. The ompA-β-lactamase precursor was processed into the mature, secreted form of the protein at a rate (half-life of 45 seconds) identical to that of the native protein even in the fully induced cultures in which protein expression is the highest (data not shown). From these results it can be concluded that the effect of the ompA signal sequence on the aggregation of β-lactamase in the periplasmic space is not due to a change in the kinetics of the secretion step.

The relation between the protein synthesis rate and accumulation of aggregated mature β-lactamase in the periplasmic space of RB791(pJG108) cultures was investigated. As shown in Table 1, the specific rate of incorporation of ^{35}S-methionine into the TCA precipitate was higher in RB791(pJG108) compared to RB791(pTac11), which reflected slight differences in the growth rates. Since the ompA-β-lactamase precursor does not accumulate in the cytoplasm, the appearance of the mature protein in the periplasmic space is determined solely by the rate of protein

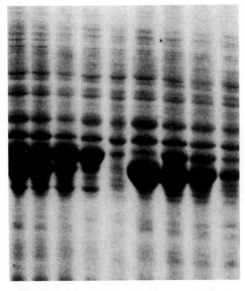

β-lactamase precursor
β-lactamase

1 2 3 4 5 6 7 8 9

Fig. 4: SDS-PAGE analysis of the insoluble protein fractions from cultures grown in the presence of varying concentrations of sucrose and raffinose and induced with 0.1 mM IPTG. Lanes 1 to 5: insoluble fractions from cultures grown with increasing concentrations of sucrose: 0, 0.15, 0.3, 0.45 and 0.6 M, respectively. Lanes 6 to 9: insoluble fractions from cultures grown with increasing concentrations of raffinose: 0.075, 0.15, 0.225 and 0.3 M, respectively. Each lane was loaded with 260 μl of media to accentuate the decrease in the amount of insoluble protein. Adapted from Bowden and Georgiou, 1988.

synthesis. For cultures induced with 0.1 mM IPTG, the maximum rates of mature β-lactamase accumulation in the periplasmic space for cells harboring pTac11 and those containing pJG108 were similar. However, the amount of soluble β-lactamase, as measured by the specific activity, was three-fold higher in RB791(pTac11) compared to RB791(pJG108) (compare Fig. 1 and 3). Similarly, in uninduced cultures of RB791(pJG108) where aggregation was extensive (Fig. 3), the β-lactamase synthesis rate was less than half of RB791(pTac11) cells induced with 0.02 mM IPTG (see Table 1) although in the latter all the protein was soluble. Consequently, the difference in the extent of aggregation between cells containing pJG108 and pTac11 is neither due to the secretion kinetics nor due to higher protein synthesis rates. Therefore, it can be concluded that the signal sequence exerts a direct effect on the aggregation of the mature protein. Most likely, the signal sequence affects a folding step which precedes the cleavage of the signal sequence by signal peptidase I.

Inhibition of β-lactamase aggregation by the additions of sugars.

Unlike the inner membrane, the outer membrane of gram-negative bacteria is readily permeable to hydrophilic solutes with a molecular mass less than 600 daltons (Decad and Nikaido, 1976). Small molecules equilibrate within the periplasmic space but since they cannot enter the cytoplasm they do not perturb most metabolic functions. Therefore, addition of small solutes may be used to control protein aggregation in the periplasmic space without affecting either cell growth or protein synthesis. For example, sucrose readily diffuses through the outer membrane but it is excluded from the cytoplasm. It is not metabolized by *E. coli* and does not affect growth or β-galactosidase synthesis even at relatively high concentrations (Richey et al., 1987). Furthermore, the role of sucrose and other sugars in stabilizing proteins *in vitro* is well established (Timasheff and Arakawa, 1989). When sucrose (M.W.=342 g/mole) was added to the culture medium of RB791(pTac11) cells induced with 0.1mM IPTG, a large increase in the specific activity of β-lactamase was observed. Analysis of the insoluble protein fractions by SDS-PAGE (Fig. 4, lanes 1-5) showed that the amount of aggregated mature β-lactamase was drastically reduced as the sucrose concentration was increased. The accumulation of precursor in the insoluble cell fraction was unaffected by the presence of sugars in the growth medium except at the higher concentration (0.6 M). However, under these conditions, the growth rate of the culture was significantly reduced. A maximum three-fold increase in the amount of soluble β-lactamase was observed in cultures supplemented with 0.4 M sucrose. Similar results were obtained with raffinose, a non-metabolizable trisaccharide (M.W. = 595 g/mole; Fig. 4). In the presence of 0.225 M raffinose the production of soluble β-lactamase was almost four-fold higher relative to the culture grown without raffinose. As with cultures grown in the presence of sucrose, addition of raffinose inhibited the formation of aggregated protein (Fig. 4, lanes 6-9). The addition of raffinose also prevented the accumulation of the unprocessed (cytoplasmic) β-lactamase precursor. A similar effect was noted in induced RB791(pTac11) cultures supplemented with varying concentrations of the monosaccharide, sorbose (M.W. = 180 g/mole) (data not shown). With these three sugars at the same concentration, the extent of the inhibition of β-lactamase aggregation decreased in the order: raffinose > sucrose > sorbose. Similar results were obtained with cultures overexpressing ompA-β-lactamase grown in media containing sucrose. In RB791(pJG108) cultures induced with 0.1 mM IPTG in media containing 0.4 M sucrose, nearly a ten-fold increase in the β-lactamase specific activity in the soluble fraction was observed (860 U/mg) compared to cells grown in media without sucrose (90 U/mg).

The addition of sugars changes the osmotic pressure of the growth medium. For example, 0.3 M sucrose increases the osmotic pressure of the growth medium from 220 to 480 mOs/kg H_2O. Changes in osmotic pressure can affect the growth rate, cell dimensions and the expression of certain genes. To determine the effect of the osmotic pressure on protein aggregation, RB791(pTac11) cells were grown in media with 0.3 M sucrose or the osmotically equivalent concentration of NaCl (0.185 M). The addition of NaCl did not affect protein aggregation or the specific activity of the soluble β-lactamase. Therefore, it can be concluded that the inhibition of inclusion body formation in the presence of sucrose is not due to the increased osmotic pressure.

No effect on the growth rate was observed in induced RB791(pTac11) cultures grown in media with low concentrations of sugars. However, at higher concentrations, both sucrose and raffinose caused a more significant reduction in the growth rate. The kinetics of protein synthesis and precursor translocation were measured in order to determine whether the inhibition of inclusion body formation was due to changes in the rate of accumulation of β-lactamase in the periplasmic space. No significant change in β-lactamase expression or the translocation kinetics could be detected in cultures grown in the presence of 0.15 M raffinose. In addition, the change in the volume of the periplasmic space is small in media containing 0.15 M raffinose (385 mOs/kg H_2O) (Richey et al., 1987), thus, the concentration of the folding protein was not altered. Since the addition of 0.15 M raffinose inhibits protein aggregation but does not affect the cell growth rate or β-lactamase synthesis, it must be altering the kinetics of protein folding and/or aggregation in the cell (Bowden and Georgiou,

1990). These results demonstrate that subtle manipulation of the growth environment can be employed to enhance the production of soluble recombinant proteins.

The influence of polyethylene glycol on β-lactamase aggregation.

Sugars and other polyhydroxylated compounds enhance the thermodynamic stability of proteins and prevent self-association *in vitro*. According to the widely accepted thermodynamic model of Timasheff and coworkers, this phenomenon is related to the exclusion of the co-solvent (i.e. sugar) molecules from the protein which affects the structure of water around the protein surface and makes the folded conformation energetically more favorable (Timasheff and Arakawa, 1989). On the other hand, compounds that bind to the protein are usually destabilizing. For example, polyethylene glycols (PEG's) of varying molecular weights have been shown to lower the thermal transition temperature of several proteins such as ribonuclease, chymotrypsinogen and β-lactoglobulin (Lee and Lee, 1987). As described above, protein stabilizing agents such as sugars prevent the aggregation of β-lactamase within *E. coli*. Based on these results it can be expected that a protein destabilizing agent will have the opposite effect and enhance inclusion body formation. This hypothesis was tested by examining the formation of inclusion bodies in *E. coli* cells grown in the presence of polyethylene glycols. As with sugars, polyethylene glycols of molecular weight up to 1,000 g/mole can pass through the outer membrane and equilibrate within the periplasmic space (Decad and Nikaido, 1976). The addition of 0.03 M PEG of molecular weight 200 to 1,000 g/mole had no effect on the cell growth rate, β-lactamase production or inclusion body formation. However, higher concentrations of PEG (0.2 M) resulted in lower accumulation of β-lactamase both in the soluble and the insoluble cell fractions indicating that protein synthesis is inhibited. All the PEG's tested (molecular weight range 200 to 1,000 g/mole) caused severe growth inhibition when added to cultures at a concentration of 0.3 M. These experiments underline the difficulties associated with manipulating the growth environment in order to affect protein folding.

Aggregation of β-lactamase in the cytoplasm of *Escherichia coli*.

The cytoplasm of *E. coli* differs significantly from the environment in the periplasmic space. The pH of the cytoplasm is about 7 to 7.2 whereas the pH of the periplasmic space varies depending on the external conditions. Furthermore, the cytoplasmic environment is reducing and prevents the formation of disulfide bonds which are essential for the stability and proper folding of secreted proteins. In addition, the composition of the cytoplasm differs from the periplasmic space. Many of the contaminants such as ribosomal components, plasmid DNA and a number of cytoplasmic proteins detected in

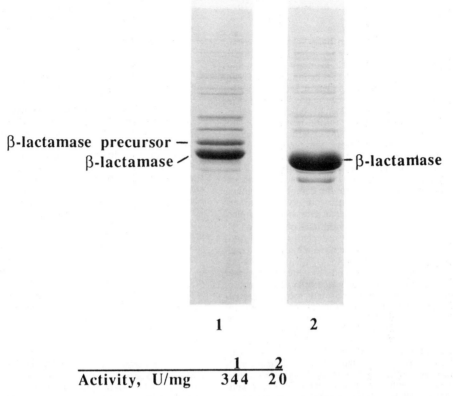

β-lactamase precursor —
β-lactamase ⁄

— β-lactamase

1   2

	1	2
Activity, U/mg	344	20

Fig. 5: Distribution of β-lactamase in the soluble and insoluble fractions of RB791(pKN) and RB791(pGB1) cultures induced with 0.1 mM IPTG. The culture samples were fractionated as described earlier. Each lane was loaded with 25 µl of media. Lane 1: insoluble protein fraction from RB791(pKN). Lane 2: insoluble protein fraction from RB791(pGB1). The positions of the β-lactamase and β-lactamase precursor are indicated. The specific activities of the β-lactamase in the soluble protein fractions corresponding to both lanes are also shown.

cytoplasmic inclusion bodies (Hart et al., 1990; Hartley and Kane, 1988) cannot be an integral part of aggregates formed in the periplasm because of the topological constraint placed by the inner membrane. The influence of these contaminants on the aggregation reaction(s) has not yet been determined.

Environmental differences between the cytoplasm and the periplasmic space can influence protein folding and, thus, affect protein aggregation. To investigate the effect of cellular location on folding and aggregation, a β-lactamase gene in which most of the signal sequence has been deleted (Plückthun and Knowles, 1987) was expressed from a strong *tac* promoter. In plasmid pGB1 (see Materials and Methods), the mature β-lactamase sequence is preceded by the codons of the first three amino acids of the signal sequence. Previous studies by Plückthun and Knowles have shown that this polypeptide folds properly within the cytoplasm giving rise to fully active β-lactamase. Analysis of the insoluble protein fraction from RB791(pGB1) cells induced with 0.1 mM IPTG by SDS-PAGE

indicated a single band corresponding to cytoplasmic β-lactamase (Fig. 5, lane 2). The sequence of the first fifteen amino acids on the N-terminal end of the protein from the inclusion bodies was determined by gas phase sequencing. The N-terminal amino acid was methionine followed by arginine, isoleucine and the sequence of the mature β-lactamase. Comparing the amount of insoluble cytoplasmic β-lactamase to the amount of aggregated precursor and mature β-lactamase from RB791(pKN) cells induced with 0.1 mM IPTG indicates that the extent of aggregation is significantly greater in the cytoplasm. For example, although the total level of β-lactamase in cells containing pGB1 and pKN was the same, the soluble protein in induced RB791(pGB1) cultures was twenty-fold lower compared to RB791(pKN) cells which produce the wild type polypeptide with the complete signal sequence. Further studies will be required to determine if the increased protein aggregation is due to lack of disulfide bond formation, the influence of cytoplasmic components, pH or the presence of the three remaining amino acids from the signal sequence on the N-terminus.

Folding and Aggregation of β-lactamase *in vitro*

Studies of inclusion body formation within the cell suffer from some inherent technical limitations: it is not possible to monitor the folding transition *in vivo* and the complexity of the intracellular environment makes it very difficult to delineate the relation between the physiological state of the cell and protein folding. At present, these issues can be addressed rather tentatively, by drawing analogies with *in vitro* experiments in which the protein is refolded from denaturant solutions. For these experiments, purified β-lactamase was denatured in phosphate buffer containing 2 M guanidine hydrochloride. This concentration of denaturant was sufficient for the complete unfolding of the protein as monitored by difference spectroscopy and fluorescence emission. Samples containing different concentrations of unfolded β-lactamase were then dialyzed against a 100-fold larger volume of buffer to remove the denaturant and allow the protein to refold. The efficiency of refolding was determined by measuring the enzymatic activity of β-lactamase and comparing it to the native protein. As expected, the recovery of enzymatic activity is a function of the protein concentration. In experiments conducted in phosphate buffer at pH 7.0, refolding was fully reversible (i.e. 100% activity) at protein concentrations up to 10 mg/ml. About 10% lower activity recovery accompanied by the formation of protein aggregates was observed with the same protein concentration but at pH 6.0 (Valax and Georgiou, 1990). Interestingly, the periplasmic space of cells growing in media of normal pH is acidic and could enhance the formation of inclusion bodies.

A decrease in the recovery of folded protein was observed when the β-lactamase was first unfolded under denaturing conditions in the presence of 5 mM dithiothreitol (DTT) and refolded in an oxidizing environment. When the protein is unfolded in the presence of 5 mM DTT, the single disulfide bond in β-lactamase is reduced. Refolding was performed by dialysis under oxidizing conditions to allow pairing of the cysteines. Starting with the reduced protein, the recovery of enzymatically active β-lactamase was about 20-40% lower compared to the refolding of the oxidized β-lactamase. The decrease in the refolding yield was related to the formation of visible aggregates.

At moderate protein concentrations, the aggregation of β-lactamase during oxidative refolding could be inhibited by the addition of sucrose (Fig. 6). In direct analogy with *in vivo* studies, the inhibition of aggregation was dependent on the sucrose concentration. For example, at a protein concentration of 12 mg/ml, the activity recovered after refolding was 60% in the absence of sucrose, 70% with 0.15 M and 85% with 0.3 M sucrose. Complete recovery of activity was obtained in the presence of 0.6 M sucrose. Thus, sucrose has the same effect on aggregation *in vitro* as it has on the formation of inclusion bodies. These results provide further evidence that the mechanism of inhibition of aggregation within the cell is due to the changes in the folding kinetics of the protein and not to physiological differences caused by the presence of sucrose in the growth media.

Conclusions

The formation of intracellular β-lactamase inclusion bodies in the periplasmic space of *Escherichia coli* was shown to be dependent on the rate of protein accumulation, the folding kinetics of the protein, and the intracellular environment. The fraction of the β-lactamase that formed inclusion bodies varied with the expression level and the signal peptide used for the secretion of the protein. Higher expression levels resulted in a greater proportion of the β-lactamase appearing in the aggregated form as expected since the aggregation reaction (second order

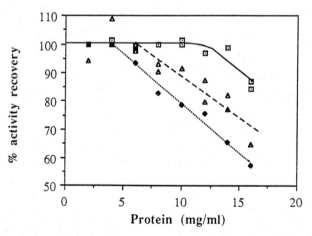

Fig. 6: Effect of sucrose concentration on the renaturation of β-lactamase. Samples with different concentrations of protein were unfolded in 2.0 M GuHCl, 5 mM Dithiothreitol and renatured by dialysis in potassium phosphate buffer, pH 6.0 (0.02 M final GuHCl concentration) with 0 M (◆), 0.15 M (△) and (▫) 0.3 M sucrose.

or greater) is favored by high concentrations of the folding protein. In addition, the ompA signal sequence was shown to enhance the aggregation of the mature β-lactamase when compared to the native protein. Whether the influence on the folding of a secreted protein in the periplasmic space by the signal sequence is specific to β-lactamase or a general phenomena has not been determined.

Our ability to manipulate protein folding and aggregation by altering intracellular conditions provides fundamental information on the nature of inclusion body formation as well as insight into the choice of fermentation conditions and the host/vector system. Both *in vitro* and *in vivo* studies found that the presence of non-metabolizable sugars (sucrose and raffinose) in the folding environment gives a significantly higher yield of protein in an active, soluble form. However, changing the folding environment *in vivo* by the addition of certain compounds to the growth media can lead to difficulties since many of them adversely affect cell viability. Fundamentally, the effect of the sugars on inclusion body formation shows that altering the folding kinetics of a protein has a drastic effect on its fate in the cell. Environmental differences in the intracellular compartments of *E. coli* were also shown to affect β-lactamase aggregation. Without a functional signal sequence, β-lactamase aggregated more readily in the cytoplasm than the native protein in the periplasmic space. Future studies will be focused on the cellular conditions responsible for this difference in the extent of aggregation.

References

Amann, E.; Brosius, J.; Ptashne, M., *Gene* **1983**, *25*, 167-178.

Bowden, G. A.; Georgiou, G., *Biotechnol. Prog.* **1988**, *4*, 97-101.

Bowden, G. A.; Georgiou, G., *J. Biol. Chem.* **1990**, *265*, 16760-16766.

Bradford, M. M., *Anal. Biochem.* **1976**, *72*, 248-252.

Brems, D. N.; Plaisted, S. M.; Havel, H. A.; Tomich, C. S. C., *Proc. Natl. Acad. Sci (USA)* **1988**, *85*, 3367-3371.

Brent, R.; Ptashne, M., *Proc. Natl. Acad. Sci. (USA)* **1981**, *78*, 4204

Collier, D. N.; Bankaitis, V. A.; Weiss, J. B.; Bassford, P. J., *Cell* **1988**, *53*, 273-283.

Crooke, E.; Wickner, W., *Proc. Natl. Acad. Sci. (USA)* **1987**, *84*, 5216-5220.

Decad, G. M.; Nikaido, H., *J. Bacteriol.* **1976**, *128*, 325-336.

Georgiou, G.; Shuler, M. L.; Wilson, D. B., *Biotech. Bioeng.* **1988**, *32*, 741-748.

Georgiou, G.; Telford, J. N.; Shuler, M. L.; Wilson, D. B., *Appl. Environ. Microbiol.* **1986**, *52*, 1157-1161.

Ghrayeb, J.; Kimura, H.; Takahara, M.; Hsiung, H.; Masui, Y.; Inouye, M., *EMBO J.* **1984**, *3*, 2437-2442.

Goloubinoff, P.; Gatenby, A. A.; Lorimer, G. H., *Nature* **1989**, *337*, 44-47.

Hart, R. A.; Rinas, U.; Bailey, J. E., *J. Biol. Chem.* **1990**, *265*, 12728-12733.

Hartley, D. L.; Kane, J. F., *Biochem. Soc. Trans.* **1988**, *16*, 101-102.

Jaenicke, R.; Rudolph, R., in *Biological Oxidations*, Sund, H.; Ullrich, V., Eds.; Springer Verlag: Berlin, **1983**, 63-90.

Kopetski, E. G.; Schumacher, G.; Buckel, P., *Mol. Gen. Genet.* **1989**, *216*, 149-155.

Laemmli, U. K., *Nature* **1970**, *227*, 680-685.

Lee, L. L.-Y.; Lee, J. C., *Biochemistry* **1987**, *26*, 7813-7819.

London, J. C.; Goldberg, M. E., *Eur. J. Biochem.* **1974**, *47*, 409-415.

Phillips, G. J.; Silhavy, T. J., *Nature* **1990**, *344*, 882-884.

Plückthun, A.; Knowles, J. R., *J. Biol. Chem.* **1987**, *262*, 3951-3957.

Richey, B.; Cayley, D. S.; Mossing, M. C.; Kolka, C.; Anderson, C. F.; Farrar, T. C.; Record, M. T., *J. Biol. Chem.* **1987**, *262*, 7157-7164.

Rothman, J. E., *Cell* **1989**, *59*, 591-601.

Samuni, A., *Anal. Biochem.* **1975**, *63*, 17

Schein, C. H.; Noteborn, M. H. M., *Bio/Technology* **1988**, *6*, 291-294.

Schlesinger, M. J., *J. Biol. Chem.* **1990**, *265*, 12111-12114.

Takagi, H.; Morinaga, Y.; Tsuchiya, M.; Ikemura, H.; Inouye, M., *Bio/Technology* **1988**, *6*, 948-950.

Tanford, C., *Adv. Prot. Chem.* **1968**, *23*, 122-283.

Timasheff, S. N.; Arakawa, T., in *Protein Structure: A Practical Approach*, Creighton, T. E., Ed., IRL Press: Oxford, **1989**, 331-345.

Towbin, H.; Staehelin, T.; Gordon, J., *Proc. Natl. Acad. Sci. (USA)* **1979**, *76*, 4350-4354.

Valax, P. P.; Georgiou, G., in *Protein Refolding*; ACS Symposium Series 470, Georgiou, G.; De-Bernardez-Clark, E., Eds., American Chemical Society: Washington, D.C., **1991**, pp 47–109.

Secretion of Heterologous Proteins and Peptides by Pichia pastoris

Robert S. Siegel, The Salk Institute Biotechnology/Industrial Associates, Inc. (SIBIA), P.O. Box 85200, San Diego, CA 92186

A methylotrophic yeast, Pichia pastoris has been developed as an extremely stable and scaleable recombinant protein expression system. Heterologous products have been secreted at concentrations in the gram per liter range. Secreted glycoproteins are not hyperglycosylated. Proteolysis can be reduced by modification of fermentation activity. In pilot plant-scale production, 27 grams of recombinant human epidermal growth factor were purified from one 250-liter fermentation run.

The methylotrophic yeasts are capable of growing on methanol as a sole carbon and energy source. Since the first discovery of one of these yeasts in 1969 (Ogata et al., 1969) much research has been directed to elucidating the biochemical pathways involved in methanol utilization (Anthony, 1982; Veenhuis et al., 1983) and the genetic (Sibirny et al., 1987) and physiological factors (Giuseppin et al., 1988) involved in regulation of methanol metabolism. During these investigations it was noted that alcohol oxidase, the enzyme responsible for the oxidation which is the first step in methanol metabolism, is produced in great abundance in methanol grown cells, but not in cells grown on other substrates such as glucose. In another line of research, methylotrophic yeasts were used in the development of a low cost process for the production of single cell protein. A process was developed for continuous culture of the methylotrophic yeast at cell densities exceeding 100 g/L (Wegner, 1981; Wegner, 1983).

The two complementary factors of high-productivity culture methodology and carbon-source regulation of a highly abundant protein led to the investigation of the methylotrophic yeasts as expression vectors for heterologous proteins.

Scientists at SIBIA conducted a research program to develop a transformation system for P. pastoris (Cregg et al., 1985), isolate methanol-regulated genes from P. pastoris (Ellis et al., 1985), and demonstrate the expression of the heterologous gene β-galactosidase under the control of a methanol-regulated promoter (Tschopp et al., 1987).

General Characteristics of the P. pastoris Expression System

The first reported expression of a commercially significant protein, hepatitis B surface antigen (HBsAg), in P. pastoris (Cregg et al., 1987) demonstrated several significant aspects of heterologous protein production in this host. In contrast to the common plasmid-based expression systems used in E. coli and S. cerevisiae, the expression of HBsAg in P. pastoris is based on the site-directed integration of a single copy of the recombinant gene into the host chromosome. Chromosomal integration enhances the inherent stability of the recombinant with respect to plasmid-based systems. Additionally, the use of a selective medium in the fermentation process is designed to foster strain stability. The transforming vector converts a histidine auxotrophic host to full autotrophy; the fermentation maintains continuous selection by employing a defined mineral salts medium without histidine. To further enhance stability, the bulk of the cell mass is first grown under repressing conditions where no product is made, thus avoiding selection for non-producers. The extreme stability of this strain was demonstrated by the isolation on non-selective plates of 25 colonies at the end of a production run generating 14 kg of dry cells. Each of these colonies was able to generate a culture which produced HBsAg as well as the original seed culture. This stability was also demonstrated in the constant level of specific HBsAg expression from low cell-density culture at 100-ml volume to high cell-density production in a 240-liter volume.

Expression of Secreted Proteins by P. pastoris

Secretion of proteins and peptides is directed by the presence of a signal sequence which guides the nascent polypeptide into the translocation machinery of the endoplasmic reticulum. Secretion of interferon by S. cerevisiae directed by its native signal was reported in 1982 (Hitzeman et al., 1982); in 1984, two groups reported use of the S. cerevisiae α-mating factor leader sequence to direct secretion of heterologous proteins from that organism (Brake et al., 1984; Bitter et al., 1984).

Soon after the demonstration that P. pastoris was suitable for cytoplasmic expression of heterologous proteins, it was reported that this yeast is also able to secrete extraordinarily high quantities of heterologous protein (Tschopp et al., 1987). Invertase, a S. cerevisiae enzyme, was accumulated to a concentration of 2 g/L in the extracellular broth of a transformed P. pastoris strain. Invertase was secreted and fully

processed from its own native signal sequence. Bovine lysozyme, a mammalian protein, was also efficiently secreted in a correctly processed mature form by recombinant P. pastoris using the native lysozyme signal sequence (Digan et al., 1988; Digan et al., 1989). Epidermal growth factor (EGF) is a small peptide derived from a very large precursor molecule. Secretion of human EGF (hEGF) from P. pastoris (Siegel et al., in preparation) was achieved in a similar fashion to that reported for EGF secretion by S. cerevisiae (Bitter et al., 1984), i.e., by use of a synthetic EGF gene fused to an α-mating factor leader. However, the resulting production of 500 mg/L to greater than 1 g/L of hEGF peptides was two orders of magnitude greater than the reported production in S. cerevisiae culture. In addition to the three proteins discussed above, other examples of proteins successfully produced by P. pastoris are shown in Table 1 (adapted from Thill, 1990).

The high yield of secreted invertase was obtained with a recombinant strain which contained only a single integrated copy of the invertase gene in the host genome (Tschopp et al., 1987), and the bovine lysozyme-producing strains also contained only a single copy of the integrated gene (Digan et al., 1989). On the other hand, the EGF-producing strains contained either two, four or six copies of an EGF expression cassette integrated into the host genome and expression level was strongly correlated with copy number (Siegel et al., in preparation).

Glycosylation by P. pastoris

Passage through the secretory pathway exposes the transiting protein to enzymatic machinery involved in secondary modifications of secreted proteins. The attachment of oligosaccharide residues to specific

Table 1. Heterologous Proteins Produced in P. pastoris

Secreted Proteins

Product	Concentration (Mgs/liter)
Bovine Lysozyme	300
Human Lysozyme	700
Human Serum Albumin	1500
Human Epidermal Growth Factor	500
Human Insulin-like Growth Factor-I	500
V_1 domain of Receptor CD4	100
Invertase	2500
Human Interleukin-2	1000
Aprotinin Analog	800

Cytoplasmic Proteins

Product	Concentration (Mgs/liter)
Superoxide Dismutase	750
Tumor Necrosis Factor	8000
Human Interleukin-2	4000
Hepatitis B Surface Antigen	300
HIV-1 gag	100

(Adapted from Thill, 1990)

sites on the protein may be the most striking of these secondary modifications, in some cases the mass of the sugar residues is greater than that of the base polypeptide in the final glycoprotein. The pattern of glycosylation is species-specific, and within multicellular organisms can be cell-type-specific (Goochee and Monica, 1990). Within the yeasts, the glycosylation pattern of S. cerevisiae has been the most studied, and many secretory mutants have been isolated which affect glycosylation. Invertase secreted by S. cerevisiae typically is characterized by variable length chains that may contain more than 50 mannose residues (Grinna and Tschopp, 1989). The mutation sec 18 blocks this hypermannosylation and gives N-linked oligosaccharides containing 8 mannose residues (Grinna and Tschopp, 1989). The invertase produced by P. pastoris appears similar to that from the sec 18 mutant of S. cerevisiae (Digan

et al., 1988; Grinna and Tschopp, 1989). Quantitative analysis of sugar moities of invertase secreted from P. pastoris show that greater than 85% of the oligosaccharide chains contain from 8 to 14 mannose residues. Most of the remaining chains contain more than 30 mannose residues, but these chains are still found to be shorter than typical oligosaccharide chains of S. cerevisiae (Grinna and Tschopp, 1989). The greater homogeneity of glycosylation of proteins by P. pastoris may be important in regulated applications of these products where good characterization is essential.

Proteolysis of Secreted Peptides

Fermentation broth is a potentially damaging environment for secreted heterologous peptides and proteins. For example, growth hormone releasing factor in the fermentation broth

164

of a recombinant S. cerevisiae strain was so severely degraded that proteolysis was the major limitation on production (Siegel and Brierley, 1990). Proteolysis has also been reported to occur in fermentation broth of recombinant P. pastoris strains. A carboxypeptidase activity was observed to sequentially remove C-terminal residues from hEGF, reducing the secreted hEGF (1-52) in length to hEGF (1-48) (Siegel et al., in preparation). However, it was noted that the EGF could be maintained in the (1-52) form by conducting the fermentation at pH 3 rather than pH 5 (Siegel et al., in preparation). This alteration of fermentation acidity was observed to be particularly useful in production of the T-cell receptor protein (Brierley et al., 1990). In fermentations conducted at pH 5, the entire product was nicked in the polypeptide chain in such a manner that the molecule appeared normal on non-reduced polyacrylamide gels, but fell apart on reduced gels. However, when the fermentation was conducted at low pH, most of the receptor protein remained intact at the end of the run. Thus, control of fermentation pH seems to have some generality as a means of controlling the proteolytic degradation of secreted proteins in recombinant P. pastoris fermentations due to decreased activity of the proteases at low pH.

Pilot Plant Production of Recombinant Human Epidermal Growth Factor

A pilot plant process has been described for the production of hEGF secreted by recombinant P. pastoris (Siegel et al., in preparation). In a 250-liter fermentor, run time was 80 h from inoculation to harvest. Oxygen transfer capacity reduced yield to 350 mg/L at the 250-liter scale compared to 500 mg/L at the 15-liter scale. In the 250-liter fermentor, a methanol feed rate of 1.2 l/h at 150-liter broth volume with an oxygen requirement of 33 mmole oxygen/g methanol allows calculation of an oxygen transfer rate of 210 mmole oxygen $l^{-1}h^{-1}$; this rate is one-third less than the rate of 330 mmole oxygen $l^{-1}h^{-1}$ at the 15-liter scale. In the earlier reported scale-up of cytoplasmic HBsAg no oxygen limitation was encountered and no oxygen consumption data was reported (Cregg et al., 1987). Secretion may have some effect on oxygen transfer, especially if the secreted proteins increase foaming and use of antifoam. However, the strain expressing cytoplasmic HBsAg was defective in methanol utilization and thus methanol was metabolized more slowly, generating a lower oxygen requirement than that of the EGF-secreting strain which had a normal methanol metabolism. Although the methanol utilization mutation eliminated oxygen limitation by reducing the oxygen demand, the HBsAg fermentation run time was more than 200 h compared to 80 h for the EGF fermentation. The EGF fermentation run time could be reduced to less than 60 h by the use of a seed fermentor; in the reported process the 250-liter fermentor was inoculated with only 1.5 liters of a low cell density shake flask culture.

The EGF recovery process (Siegel et al., in preparation) involved cell removal by centrifugation followed by an initial recovery step of adsorption onto a bulk C18 reverse phase resin. After desorption from the resin, human EGF was the major peptide component. After a second adsorption/desorption on a cation exchange resin to remove colored contaminants, the peptide purity was greater than 85%. The high purity allowed very high loading rates onto a preparative HPLC. HPLC fractions were pooled to give a final purity greater than 95%. Twenty-seven grams of lyophilized EGF were recovered from one run in a 250-liter fermentor.

Summary

The _Pichia_ _pastoris_ expression system has demonstrated the ability to secrete a number of biologically active heterologous proteins and peptides at commercially relevant concentrations. Although degradation of some secreted products has been observed, adjustment of the fermentation acidity has been shown to ameliorate the problem. The successful results of the pilot plant scale production of hEGF with the _Pichia_ _pastoris_ expression/secretion system, coupled with the range of products secreted at the one- and ten-liter fermentor scale suggest a general applicability of the _Pichia_ system to commercial production of a wide range of secreted polypeptides.

REFERENCES

Anthony, E. _The Biochemistry of Methylotrophs,_ Academic Press, N.Y., 1982.

Bitter, G.A., Chen, K.K., Banks, A.R., Lai, P.-H., _Proc. Natl. Acad. Sci. USA,_ 1984, _81,_ 5330-5334.

Brake, A.J., Merryweather, J.P., Coit, D.G., Herberlein, U.A., Masiarz, F.R., Mullenbach, G.T., Urdea, M.S., Valenzuela, P., Barr, P.J., _Proc. Natl. Acad. Sci. USA,_ 1984, _81,_ 4642-4646.

Brierley, R.A., Buckholz, R., Davis, G., Holtz, G., Odiorne, M., Siegel, R.S., Thill, G.P., presented at 199th National Meeting of American Chemical Society, Boston, MA. April 23-27, 1990.

Cregg, J.M., Barringer, K.J., Hessler, A.Y., Madden, K.R., _Mol. Cell. Biol.,_ 1985, _5,_ 3376-3385.

Cregg, J.M., Tschopp, J.F., Stillman, C., Siegel, R., Akong, M., Craig, W.S., Buckholz, R.G., Madden, K.R., Kellaris, P.A., Davis, G.R., Smiley, B.L., Cruze, J., Torregrossa, R., Velicelebi, G., Thill, G.P., _Bio/Technology,_ 1987, _5,_ 479-485.

Digan, M.E., Tschopp, J., Grinna, L., Lair, S.V., Craig, W.S., Velicelebi, G., Siegel, R., Davis, G.R., Thill, G.P., _Dev. Ind. Microbiol.,_ 1988, _29,_ 59-65.

Digan, M.E., Lair, S.V., Brierley, R.A., Siegel, R.S., Williams, M.E., Ellis, S.B., Kellaris, P.A., Provow, S.A., Craig, W.S., Velicelebi, G., Harpold, M.M., Thill, G.P., _Bio/Technology,_ 1989, 7, 160-164.

Ellis, S.B., Brust, P.F., Koutz, P.J., Waters, A.F., Harpold, M.M., Gingeras, T.R., _Mol. Cell. Biol.,_ 1985, _5,_ 1111-1121.

Giuseppin, M.L.F., van Eijk, H.M.J., Bes, B.C.M., _Biotech. Bioeng.,_ 1988, _32,_ 577-583.

Goochee, C.F., Monica, T., _Bio/Technology,_ 1990, _8,_ 421-427.

Grinna, L.S., Tschopp, J.F., _Yeast,_ 1989, _5,_ 107-115.

Hitzeman, R.A., Leung, D.W., Perry, L.J., Kohr, W.J., Hagie, F.E., Chen, C.Y., Lugovoy, J.M., Singh, A., Levine, H.L., Wetzel, R., Goeddel, D.V., _Rec. Adv. Yeast Mol. Biol.,_ 1982, _1,_ 173-190.

Ogata, K., Nishikawa, H., Ohsugi, M. _Agric. Biol. Chem.,_ 1969, _33,_ 1519-1522.

Sibirny, A.A., Titorenko, V.I., Efremov, B.D., Tolstorukov, I.I., _Yeast,_ 1987, _3,_ 233-241.

Siegel, R.S., Wondrack, L., Stillman, C., Davis, G.R., Hutton, J., Gross, R., Amarant, T., Nardi, R., Parikh, I., Thill, G.P., in preparation.

Siegel, R.S. Brierley, R.A., _Bio/Technology,_ 1990, _8,_ 639-643.

Thill, G.P., Presented at the 1990 Annual Meeting of the Society for Industrial

Microbiology, July 29-August 3, 1990 at Orlando, FL.

Tschopp, J.F., Brust, P.F., Cregg, J.M., Stillman, C.A., Gingeras, T.R., *Nucl. Acids Res.*, **1987**, *15*, 3859-3876.

Tschopp, J.F., Sverlow, G., Kosson, R., Craig, W., Grin-na, L., *Bio/Technology*, **1987**, *5*, 1305-1308.

Veenhuis, M., Van Dijken, J.P., Harder, W., *Adv. in Microbial. Physiol.*, **1983**, *24*, 1-81.

Wegner, E.H., U.S. Patent 4,414,329, 1981.

Wegner, E.H., U.S. Patent 4,617,274., 1983.

Streptomyces Bioprocessing: From Secondary Metabolites to Heterologous Proteins

Kimberlee K. Wallace Department of Chemical and Biochemical Engineering, University of Maryland Baltimore County, Baltimore, MD. 21228, USA.
Gregory F. Payne Department of Chemical and Biochemical Engineering and Center for Agricultural Biotechnology, University of Maryland Baltimore County, Baltimore, MD. 21228, USA.
Marilyn K. Speedie Department of Biomedicinal Chemistry, School of Pharmacy, University of Maryland at Baltimore, Baltimore, MD. 21201, USA.

Streptomyces are industrial microorganisms commonly used for producing secondary metabolites. Since the genetics of *Streptomyces* are becoming better characterized, it is possible to consider metabolically engineering these strains to improve yields and produce novel metabolites. In addition to secondary metabolite biosynthesis, *Streptomyces* have recently been considered for the production of foreign proteins. The advantage of using *Streptomyces* as a host microbe is its ability to secrete proteins. Progress in both areas, secondary metabolite production and heterologous protein expression, are reviewed.

Traditionally, *Streptomyces* have been exploited for the production of secondary metabolites since 60% of microbially produced antibiotics are derived from these organisms (Queener and Day, 1986). Recently, researchers have begun to employ *Streptomyces* as hosts for expression of heterologous proteins due to their ability to secrete proteins into the medium. Our objective is to review the physiological and genetic requirements associated with secondary metabolite biosynthesis and heterologous protein expression in *Streptomyces*.

Characterizing Secondary Metabolism in Terms of the Extracellular Environment

Commercial production of microbially derived secondary metabolites involves the use of complex carbon and nitrogen sources which permit the achievement of high cell densities prior to product biosynthesis. Along with strain improvement programs, improved media and cultivation techniques led to rapid increases in antibiotic productivity in the initial stages of fermentation development (Demain, 1973). In complex media, it is commonly observed that the onset of secondary metabolite production is associated with i) a decrease in growth rate; or ii) depletion of a key nutrient required for growth.

In a complex medium supporting rapid growth, production is non-growth associated such that product biosynthesis commences upon slowing or cessation of growth. In the production of the antibiotics chloramphenicol (Legator and Gottlieb, 1953; Malik and Vining, 1970) and candicidin (Liu et al., 1975) from *Streptomyces* and penicillin from *Penicillium* (Pirt and Righelato, 1967), cultivation in complex medium resulted in a rapid growth phase during which antibiotic production was absent. Liu et al. (1975) showed that the addition of phosphate to cultures actively producing candicidin stimulated growth and suppressed antibiotic production. Thus, induction of production in complex media is often correlated with a reduction in the growth rate. However, due to the undefined composition of the complex nutrient sources, determination of possible causal relationships between growth limitations and the onset of production is not possible.

To provide new insights into the environmental factors involved in antibiotic production, synthetic or defined culture media were developed. In contrast to complex medium, defined media support slower growth accompanied by growth associated production such that the production phase and the growth phase occur simultaneously (Pirt and Righelato, 1967; Haavik, 1974; Shapiro and Vining, 1983; Young et al., 1985; Lohr et al., 1989). Similar production patterns were observed in our model system: production of the antitumor antibiotic streptonigrin by *S. flocculus*. Cultivation in a complex medium yields non-growth associated production as shown in Figure 1a. From a plot of product versus cell concentration, as shown in Figure 1b, it is observed that during cell growth, streptonigrin production is suppressed as represented by the horizontal line. Although a period of concurrent growth and production is present, the majority of the antibiotic is biosynthesized upon cessation of growth as

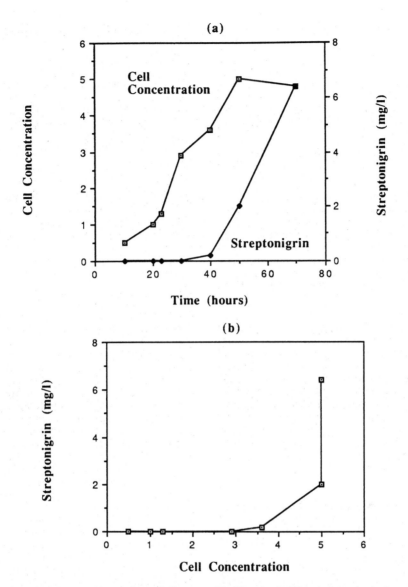

(a)

Cell Concentration

Cell Concentration

Streptonigrin (mg/l)

Streptonigrin

Time (hours)

(b)

Streptonigrin (mg/l)

Cell Concentration

Fig. 1. a) Fermentation profiles (adapted from Hartley and Speedie, 1984) of growth and streptonigrin production in complex medium containing 10 g/l glucose, 4 g/l beef extract, 4 g/l gelysate peptone, 1 g/l yeast extract, 2.5 g/l NaCl.   b) Differential plot of streptonigrin production as a function of cell concentration in complex medium.   Final point represents maximum streptonigrin titer obtained. Cell lysis was ignored in this plot.

shown by the vertical line in Figure 1b. In contrast, Figure 2a shows that growth in defined medium is much slower and production is growth associated.   This growth associated production is illustrated in Figure 2b by the linear relationship between product and biomass concentrations.

In addition to relating the onset of secondary metabolite biosynthesis to a reduction in the growth rate, it has been observed, in defined media, that the onset of production can be related to the deprivation of a key medium component.   The carbon, nitrogen and phosphorus sources are the media components most commonly studied.   In actinomycin (Gallo and Katz, 1972) and cephalosporin (Matsumura et al., 1978) biosynthesis, antibiotic production commences after glucose is depleted from the

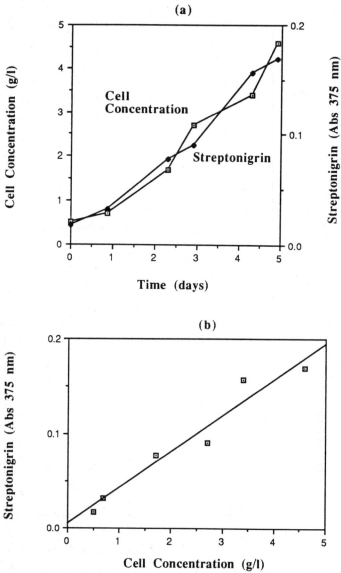

Fig. 2. a) Fermentation profiles of growth and streptonigrin production in defined medium containing 30 g/l glucose, 0.5 g/l MgSO₄ 7H₂O, 0.4 g/l CaSO₄ 2H₂O, 0.01 g/l FeSO₄ 7H₂O, 1.9 g/l NaCl, 1.0 g/l NH₄Cl, 75 mM potassium phosphate pH 7.2.   b) Differential plot of streptonigrin production as a function of cell concentration in a defined medium. Growth and production ceased simultaneously.   Cell concentration is represented by the squares and streptonigrin production is represented by the diamonds.

medium. Cultivation of S. *venezuelae* with a combination of nitrogen sources delayed production until depletion of both nutrients (Shapiro and Vining, 1984).   Again, generalizations which appear to hold for cultivation in complex medium are often invalid when cells are cultured in defined medium. Cultivation of S. *flocculus* in defined medium resulted in cessation of both growth and streptonigrin biosynthesis upon depletion of ammonium, the growth limiting nutrient, from the medium. Similarly, Young et al. (1985) observed in lincomycin production that depletion of the growth limiting nutrient resulted in cessation of both growth and antibiotic production.

Although limitation of the availability of nutrients is often required for the onset of production, the nature of the limitation may or may not be important. Whereas, thienamycin production was induced only in response to phosphate limitation, *S. cattleya* produced cephamycin C in response to phosphate, carbon or nitrogen limitations (Lilley et al., 1981). Doull and Vining (1990) observed that induction of actinorhodin production required a reduction in the growth rate, but the role of nutrient limitation was not clear.

A third observation in the study of secondary metabolism is that increased concentrations of nutrients required for growth often adversely affect product biosynthesis. Increased nitrogen levels decreased the titers of erythromycin (Flores and Sanchez, 1985) cephalosporin (Aharonowitz and Demain, 1973; Brana et al., 1985), and novobiocin (Kominek, 1972). However, Shapiro and Vining (1985) noted that the decrease in chloramphenicol production due to increased nitrogen levels could not be correlated with the extracellular ammonium concentration . We observed that although streptonigrin biosynthesis requires the availability of nitrogen, increased initial nitrogen levels appeared to reduce the specific productivity (Wallace et al., 1990). Again, our results could not be correlated to the extracellular ammonium concentration. Similarly, increased phosphate concentrations decreased the titers of candicidin (Liu et al., 1975), streptomycin (Shirato and Nagatsu, 1965; Miller and Walker, 1970), tylosin (Vu-Trong et al., 1981), bacitracin (Haavik, 1974), nanaomycin (Masuma et al., 1990) cephalosporin (Aharonowitz and Demain, 1973) and lincomycin (Young et al., 1985). Biosynthesis of cephalosporin (Aharonowitz and Demain, 1978), actinomycin (Gallo and Katz, 1972), and chloramphenicol (Bhatnagar et al., 1988) production were sensitive to increased concentrations of carbon.

In systems which are sensitive to increased nutrient levels, enhanced production can be obtained by limiting the availability of the suppressive nutrient. Providing carbon regulated systems with a low concentration of glucose and an alternate slowly utilized carbohydrate source results in a growth phase during which glucose consumption occurs and a production phase in which the alternate carbohydrate is metabolized (Pirt and Righelato, 1967; Gallo and Katz, 1972). Reduction of the ability of chloramphenicol producing cultures to assimilate glucose, by the use of a glucose analogue (methyl α-glucoside), resulted in increased antibiotic biosynthesis (Vining and Shapiro, 1984). In nitrogen regulated systems, decreasing the concentration of soluble nitrogen by the use

of ammonium trapping agents resulted in enhanced production of tylosin (Masuma et al., 1983), cephalosporin (Brana et al., 1985), and leucomycin (Omura et al., 1980). Similarly, we observed that supplying nitrogen slowly to the culture by the use of ammonium trapping agents or ammonium feeding resulted in increased streptonigrin production (Wallace et al., 1990).

In conclusion, although much research has been done, our qualitative understanding of how cultures respond to the extracellular environment is limited. Due to the lack of knowledge of the complex biosynthetic requirements of production, accurate quantification is impossible. Thus, process development remains empirical.

Characterizing Secondary Metabolism in Terms of the Intracellular Environment

We believe that before secondary metabolite production can be quantitatively characterized, it will be necessary to better understand the key intracellular requirements, both genetic and physiological, for biosynthesis. Genetically, product biosynthesis requires the availability of the biosynthetic enzymes. Physiologically, an adequate supply of precursors and energy sources must be present to prevent biosynthetic limitations.

Genetically, the production of secondary metabolites requires the expression of enzymes within the secondary metabolic pathways. Because the pathways for secondary metabolite biosynthesis are generally unknown, examination of how the cells regulate these pathways generally requires some assumptions. By extrapolation of knowledge from branched primary metabolic pathways (e.g. amino acids), it has been suggested that the first enzyme in the branch from primary to secondary metabolism may be the rate limitation to production. Thus, it is common for the expression of the pathway enzymes to be studied by measuring activity of the first enzyme branching primary and secondary metabolism.

In the production of secondary metabolites, it is seen that the pathway enzymes are not constitutively expressed but rather are induced or derepressed prior to antibiotic production (Jones and Westlake, 1974; Neway and Gaucher, 1981; Gil et al., 1985). Following induction, the secondary metabolic enzymes are often observed to be rapidly inactivated resulting in limitations to continued secondary metabolite biosynthesis (Friebel and Demain, 1977; Neway and Gaucher, 1981). Similarly, we observed that tryptophan C-methyltransferase, the first enzyme in the streptonigrin biosynthetic

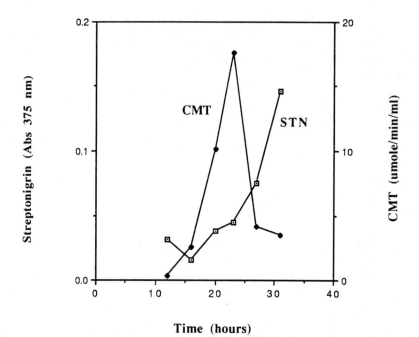

Time (hours)

Fig. 3. Profiles of tryptophan C-methyltransferase activity (CMT), the first enzyme in the streptonigrin biosynthetic pathway, and streptonigrin (STN) production illustrating induction and inactivation of enzyme activity in defined medium containing 30 g/l glucose, 0.5 g/l $MgSO_4$ $7H_2O$, 0.4 g/l $CaSO_4$ $2H_2O$, 0.01 g/l $FeSO_4$ $7H_2O$, 1.9 g/l NaCl, 1.0 g/l NH_4Cl, 75 mM potassium phosphate pH 7.2.

pathway, is induced prior to antibiotic biosynthesis and inactivated rapidly after reaching peak activity (Figure 3). Currently, we are attempting to determine the specific rate limitation to streptonigrin production. Although the rate limitation will likely change with time, we are considering three possible enzyme limitations to secondary metabolism: i) repression of enzyme synthesis by nutritional conditions as noted in chloramphenicol (Bhatnagar et al., 1988) and actinomycin (Gallo and Katz, 1972) production; ii) low expression of pathway enzymes as seen in cephalosporin production (Zhang et al., 1989); iii) enzyme inactivation as observed in patulin (Neway and Gaucher, 1981) and gramicidin S (Agathos and Demain, 1983) synthesis. As seen in Figure 3, the rate limitations may change. For instance, prior to induction, it is likely that expression of pathway enzymes limits production while enzyme inactivation may play a more important role at later times.

To relate the extracellular environment (e.g. nutrient deprivation) to an intracellular response (e.g. induction of secondary metabolic enzymes), the presence of intermediate messengers has been examined. Examination of these messengers has been done by analogy with better characterized systems. In *E. coli*, it is observed that increased levels of cAMP are required for induction of the enzymes of the *lac* operon. Although it is not known if cAMP is directly involved in antibiotic biosynthesis, Ragan and Vining (1978) observed a decline in cAMP levels prior to streptomycin biosynthesis. In addition to studying cAMP, others have attempted to relate the availability of other adenylate compounds to secondary metabolite biosynthesis. Vu-Trong et al. (1980, 1981, 1982) observed that tylosin production is independent of the energy charge but inversely correlates with the total intracellular adenylate pool. However, Martin (1976) observed an inverse correlation between candicidin production and both the intracellular ATP level and the energy charge. Thus, generalizations concerning the effect of individual or combinations of adenylates have not been (and may not be) established for secondary metabolite production.

In an attempt to correlate production levels with nitrogen availability, two different intracellular messengers have been examined. Since increased ammonium often suppresses antibiotic biosynthesis, a correlation between product biosynthesis and the activity of ammonium assimilating enzymes was sought. In *Streptomyces*, as in *E. coli*, ammonia assimilation at low ammonium levels is believed to occur by the high affinity glutamine synthetase-glutamate synthase enzymes. Although decreased ammonium availability stimulates ammonium assimilating enzyme activity, it has been difficult to interpret these results with respect to secondary metabolite biosynthesis (Brana et al., 1986; Flores and Sanchez, 1989). Also, since the onset of production of some secondary metabolites has been observed to occur following nutrient deprivation, correlations between product biosynthesis and intracellular mediators associated with nutritional downshift were examined. With *E. coli*, it is seen that nitrogen deprivation often invokes a stringent response as characterized by an increase in the intracellular ppGpp pool. Although the role of ppGpp in secondary metabolism is not known, Ochi (1987) suggests that stringent-type response may be a factor in streptomycin production since the onset of antibiotic biosynthesis and an increase in ppGpp levels occur simultaneously.

It has often been observed that secondary metabolism and sporulation occur simultaneously. With *Bacillus*, considerable efforts have been devoted to understanding the molecular details of this developmental regulation (see Smith et al., 1989). With *Bacillus*, it appears that the specificity of gene transcription is regulated by the use of multiple sigma factors for RNA polymerase. *Streptomyces* also have been shown to have different RNA polymerases based on differences in sigma factors (Westpheling et al., 1985; Buttner et al., 1988). At least one class of these sigma factors has been shown to have significant homology with *E. coli* and *Bacillus* sigma factors (Buttner, 1990).

Although the relationship between secondary metabolite biosynthesis and sporulation in *Streptomyces* is not clear, a bioregulator which appears to be unique to sporulation and secondary metabolite production in *Streptomyces* is known. Khokhlov and Tovarova (1979) observed that streptomycin producing cultures biosynthesized A-factor while non-producers lacked this compound. Addition of A-factor to non-producing and non-sporulating strains induced both streptomycin production and sporulation. More recently, Miyake et al.(1990) reported an A-factor binding protein and suggested that A-factor binding to this protein was required for derepression of streptomycin production and sporulation.

In addition to gene expression, the cells have physiological requirements for an adequate supply of primary metabolic precursors and for metabolic energy. For *de novo* biosynthesis, several metabolic pathways must be operative. Since secondary metabolites are derived from key primary metabolic intermediates (e.g acetyl CoA,

amino acids), the necessary primary metabolic pathways must be functioning to provide a sufficient precursor supply. Apparently precursor availability limited bicozamycin biosynthesis since Ochi et al. (1988) observed increased production by exogenous additions of leucine and isoleucine, both precursors of the antibiotic. Although this area is not well studied, precursor availability may be a limitation to secondary metabolite biosynthesis (Drew and Demain, 1977).

In addition to precursor availability, cells require energy for biosynthesis. This energy is often supplied by carbohydrates which can be consumed for three processes: growth, product biosynthesis and maintenance as described by

$$-q_s = \frac{1}{Y_{xs}} q_x + \frac{1}{Y_{ps}} q_p + m \qquad (1)$$

where $-q_s$, q_x and q_p are the rates of sugar consumption, growth and product biosynthesis per unit cell mass, Y_{xs} and Y_{ps} are the yields which include both the carbon and energy requirements for growth and product synthesis respectively, and m is the maintenance energy. Typically, maintenance coefficients of 0.022-0.028 g sugar/g cell-hr are observed (Pirt and Righelato, 1967; Mou and Cooney, 1976).

During secondary metabolite production phases, it has been observed that the maintenance energy requirement is substantial with 70% of the carbohydrate being consumed to meet this maintenance need (Cooney and Acevedo, 1977; Heinjen et al., 1979). To illustrate the importance of maintenance energy requirements consider production of the antibiotic cycloheximide. Since this antibiotic is produced by non-growing cultures in complex media, Equation 1 can be simplified to

$$-q_s = \frac{1}{Y_{ps}} q_p + m \qquad (2)$$

where the maintenance energy for this culture equals 0.028 g/g-hr (0.067 g/g-day) (Payne and Wang, 1989). Figure 4 shows that supplying glucose below the maintenance requirement results in a low observed product yield ($-q_p/q_s$) and low specific productivity, indicating that inadequate supplies of the carbon and energy source are unfavorable for biosynthesis. When glucose supply rates were above the maintenance requirement, cycloheximide production was observed. Above a feed rate of 0.9 g sugar/g cell-day, cellular metabolism is limiting such that increased productivity does not parallel increased

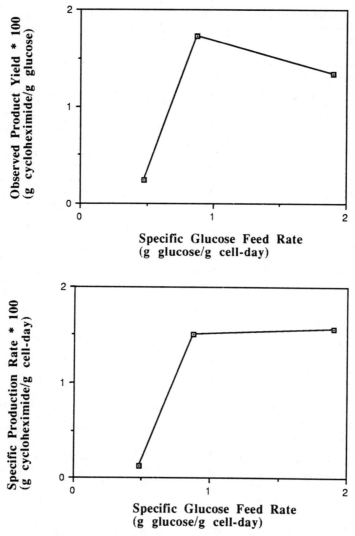

Fig. 4. Observed cycloheximide yield (-qp/qs) and specific cycloheximide production rates calculated during the production phase for glucose fed cultures (Payne, 1984).

supply, and glucose accumulates in the medium. Similarly, Pirt and Righelato (1967) observed that penicillin biosynthesis was suppressed until the availability of carbohydrate exceeded the maintenance requirement of the culture. Thus, carbohydrate availability may limit secondary product biosynthesis by failing to meet the high maintenance requirements of the system.

In summary, our current knowledge of secondary metabolism is limited. It is not clear what external stimuli or what internal messengers are involved in the induction of the pathways. Thus, empirical approaches are currently required

for medium design and process optimization. We believe that a better fundamental understanding is required before rational approaches to process optimization can be used. Firstly, with knowledge of intracellular rate limitations, it should be possible to genetically construct more efficient strains and to develop more efficient process strategies. Secondly, better understanding of the genetics and regulation of secondary metabolite producers may allow metabolic engineering strategies to be employed to redirect the cell's metabolism to obtain novel products.

The Future: Genetic Engineering for Secondary Metabolism

Because of the limited understanding of secondary metabolism in both metabolic and genetic terms, there has been limited impact of recombinant DNA technology on the commercial production of *Streptomyces* secondary metabolites. However, recent advances, as reviewed by Hutchinson et al. (1989) and Chater (1990), suggest that breakthroughs are on the horizon (also see Fayerman, 1986). Also, it was recently reported that introducing additional copies of cloned biosynthetic genes for valinomycin synthesis into the producing organism resulted in yield increases which were proportional to gene dosage (Perkins et al., 1990).

Before rational metabolic engineering approaches can be attempted, the genes for secondary metabolism must be identified. Fortunately, the genes coding for specific secondary metabolic enzymes are often clustered together (either on the chromosome or on plasmids). Two general techniques for identifying these gene clusters are evolving. The first technique relies on the observation that, in many cases, genes coding for antibiotic resistance are often linked to the biosynthetic genes (see Martin and Liras, 1989). By shotgun cloning DNA fragments into *E. coli* and selecting for antibiotic resistance in the *E. coli* transformants, it is possible to identify regions on the chromosome which code for antibiotic biosynthesis. A second technique which can be used to identify polyketide (Malpartida et al., 1987) and ß-lactam (Shiffman et al., 1988) structural genes uses the observation that genes for certain enzymes in these pathways show significant homology between different organisms. Thus, hybridization probes for polyketide and isopenicillin N synthase genes are used to identify regions of DNA responsible for these secondary metabolic pathways.

Once available, the genes coding for secondary metabolic products can be used to metabolically engineer production. Three approaches for the genetic engineering of secondary metabolism will be considered. First, it is possible to clone entire pathways into new organisms This was done by Malpartida and Hopwood (1984) who cloned genetic information for actinorhodin synthesis from *S. coelicolor* into *S. parvulus*. Stanzak et al. (1986) obtained erythromycin production in *S. lividans* by adding the biosynthetic genes from *Saccharopolyspora erythrae* (formerly *Streptomyces erythreus*) (for review see Donadio et al., 1989). It is also possible to clone only portions of a secondary metabolic pathway. Otten et al. (1990) obtained ε-rhodomycinone (a key intermediate in daunorubicin biosynthesis) production in a *S. lividans* which was transformed with genes from a daunorubicin producing *S. peucetius*.

A second genetic approach is to combine genetic information from various secondary metabolite producers to obtain novel hybrid metabolites. Pioneering work in this area was done by Hopwood's group who obtained novel polyketides (Hopwood et al. 1985) by adding genes for actinorhodin production into *S. violaceruber* which produces a related antibiotic, granaticin (see Hopwood et al. 1986). Omura et al. (1986) produced hybrid medermhodins by adding genes for actinorhodin synthesis into a medermycin producing *Streptomyces*. Epp et al. (1989) reported that a hybrid spiramycin (isovalerylspiramycin) was produced by a genetically altered *S. ambofaciens*. This cell was altered by the addition of a gene (*car*E) which presumably codes for an isovaleryl transferase enzyme in the carbomycin producing *S. thermotolerans*. Despite these encouraging results, Hutchinson et al. (1989) caution that failures in producing hybrid antibiotics could occur due to "enzyme specificity, metabolite channeling, and regulation of gene expression."

A third genetic approach is to clone regulatory genes into the host to stimulate secondary metabolite biosynthesis (see Champness et al., 1989). Although an interesting approach, knowledge of regulatory genes in *Streptomyces* is limited. However, Hopwood et al. (1986) observed increased actinorhodin production when *act* II genes were cloned into *S. coelicolor*. Addition of the presumed regulatory gene, *str* R, to *S. griseus* resulted in a 5-7 fold increase in streptomycin production (Ohnuki , 1986). Finally, cloning into *S. lividans* of the *afs*B gene, which is presumed to be regulatory for A-factor synthesis (Horinouchi et al., 1986), was observed to result in increased pigment production (Omer et al., 1988).

In conclusion, it appears that recent advances in the genetics of *Streptomyces* should begin to show rewards in terms of improving productivity of existing processes and for the production of novel metabolites.

Streptomyces for Heterologous Protein Production

Recently *Streptomyces lividans* has been considered as a host for foreign protein production. The motivation for using this gram positive bacteria is that *S. lividans* can excrete proteins into the medium, and this host has low levels of extracellular proteases. It should also be noted that although *Streptomyces* are known

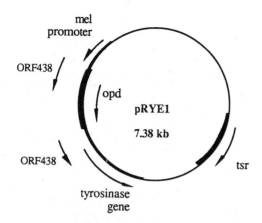

mel
promoter

ORF438

opd

pRYE1

7.38 kb

ORF438

tsr

tyrosinase
gene

Fig. 5. Derivative of the pIJ702 plasmid, commonly used to transform *S. lividans*. The pRYE1 plasmid shown here contains the organophosphate degradation gene (*opd*) inserted into the *mel* region of the pIJ702 plasmid (Steiert et al., 1989).

to be able to add sugar moieties to antibiotics, there have been no systematic studies to examine the ability of *Streptomyces* to glycosylate proteins. Typically, *S. lividans* is transformed using the multicopy plasmid pIJ702 (Katz et al.,1983) which codes for thiostrepton resistance and for the enzyme tyrosinase which is responsible for melanin production. As shown in Figure 5, foreign genes are inserted into the *mel* region of the pIJ702 plasmid, and transformants can be readily identified by selecting for thiostrepton resistance and loss of melanin producing ability. Table 1 lists foreign genes which have been cloned into and expressed by *S. lividans*. Fortunately, this host-plasmid system has been observed to be stable even in the absence of thiostrepton selection (e.g. see Payne et al., 1990).

Currently, expression of foreign proteins by *S. lividans* has been low, with few laboratories able to achieve extracellular levels in excess of 100 mg/l. Both genetic and physiological approaches have been used to enhance expression levels. Genetic approaches to improve production have focused on the promoter region and on secretion mechanisms (e.g. see Koller et al., 1989; von Heijne and Abrahmsen, 1989).

Recently, it is being better appreciated that physiological factors may also be critical to

foreign protein production. Erpicum et al. (1990) reported 10 to 20- fold increases in foreign protein expression by using medium containing glucose and ammonium. We observed that fed-batch operation could also stimulate production (Payne et al., 1990). Surprisingly, Table 2 shows that production was improved by feeding both the carbon (e.g. glucose) and nitrogen (e.g. tryptone) sources. In addition to increasing total production, glucose and tryptone feeding increases specific production (amount of enzyme produced per cell) and the specific activity of the extracellular protein (amount of extracellular enzyme/total extracellular protein). Currently, we believe that these improvements resulted from more efficient nutrient consumption due to redirection of the cells' metabolism away from acid production. Despite these improvements however, the foreign protein (parathion hydrolase in this case) was produced at levels less than 100 mg/l and appears to represent less than 20% of the extracellular protein.

In summary, *S. lividans* appears to be a promising host for foreign protein synthesis due to its ability to excrete proteins into the medium. By developing better genetic constructions and providing the culture with appropriate nutritional conditions, it should be possible to substantially improve heterologous protein production.

Conclusion

In conclusion, bioprocess improvements in the production of either secondary metabolites or heterologous proteins will require an understanding of both the physiology and genetics of the culture. Although the physiology of secondary metabolism in *Streptomyces* has been extensively studied, the genetics are not well characterized. Thus, critical details concerning the induction processes are unknown; therefore, process improvements have required extensive experimentation. Due to recent advances in the genetics of *Streptomyces* secondary metabolism, recombinant DNA technology may soon be used for process improvements as well as for production of novel metabolites. In contrast, the genetics required for heterologous protein production are becoming well characterized; however, the effect of culture physiology on production is often ignored. Our studies have shown that considerable improvements can be obtained by better matching fermentation operation with culture physiology.

Table 1: Expression of foreign genes in *S. lividans* using pIJ702
(Does not include enzymes involved in secondary metabolism)

<u>Procaryotic Genes</u>

Tyrosinase from *S. antibioticus*	Katz et al. (1983)
Agarase from *S. coelicolor*	Kendall and Cullum (1984)
Cholesterol oxidase from *Streptomyces*	Murooka et al. (1986)
β-lactamase from *S. albus* from *S. cacaoi*	Dehottay et al. (1986) Lenzini et al. (1987)
DD-peptidase from *Streptomyces* R61	Duez et al. (1987)
Cellulase from *Thermomonospora fusca*	Ghangas and Wilson (1987)
Endoglucanase from *Thermomonospora fusca*	Ghanges and Wilson (1988)
Parathion hydrolase from *Flavobacterium*	Steiert et al. (1989)
Tendamistat (α-amylase inhibitor) from *S. tendae*	Koller and Reiss (1989)
Esterase from *S. scabies*	Schottel et al. (1989)
Subtilisin inhibitor from *S. albogriseolus*	Obata et al. (1989)
Bacterial peptide pheromone from *Enterococcus faecalis*	Taguchi et al. (1989)
Lignin peroxidase from *S. viridosporous*	Wang et al. (1990)

<u>Eucaryotic Genes</u>

Bovine Growth Hormone	Gray et al. (1984)
Human Interleukin 2	Munoz et al. (1985)
Human Interferon α2	Pulido et al. (1986)
Monkey Proinsulin	Koller et al. (1989)
Sweet-tasting plant protein, thaumatin	Illingsworth et al. (1989)
CD-4 Receptor	Brawner et al. (1990)

Table 2: Summary of improvements in parathion hydrolase production

Experiment	Peak Extracellular Activity (Units/ml)	Activity per cell (units/mg)	Activity per extracellular protein (units/mg)
Batch flask	0.5-3.0	0.2-0.7	5.4
Fermentor			
Batch	4.5	0.7	ND
Glucose-fed	8.8	1.3	ND
Glucose-tryptone fed	25.7	2.6	17.8

Adapted from Payne et al, 1990.

ND = Not determined

Acknowledgements

The authors would like to acknowledge the financial support of the Systems Research Center at the University of Maryland and the American Society of Pharmacognosy.

References

Agathos, S. N.; Demain, A. L. In *Foundations of Biochemical Engineering: Kinetics and Thermodynamics in Biological Systems*; Blanch, H. W.; Papoutsakis, E. T.; Stephanopoulis, G. Eds.; American Chemical Society: Washington, DC., 1983, pp. 53-67.

Aharonowitz, Y.; Demain, A. L. *Antimicrob. Agents Chemother.* **1978**, *14*, 159-164.

Aharonowitz, Y,; Demain, A. L. *Arch. Microbiol.* **1973**, *115*, 169-173.

Bhatnagar, R. K.; Doull, J. L.; Vining, L. C. *Can. J. Microbiol* . **1988**, *34*, 1217-1223.

Brana, A. F.; Wolfe, S.; Demain, A. L. *Arch. Microbiol.* **1986**, *146*, 46-51.

Brana, A. F.; Wolfe, S.; Demain, A. L. *Can. J. Microbiol.* **1985**, *31*, 736-743.

Brawner, M.; Taylor, D.; Fornwald, J. Abstract presented at UCLA Symposia on Molecular and Cellular Biology, **1990**.

Buttner, M. J.; Chater, K. F.; Bibb, M. J. *J. Bacteriol.* **1990**, *172*, 3367-3378.

Buttner, M. J.; Smith, A. M.; Bibb, M. J. *Cell.* **1988**, *52*, 599-607.

Champness, W.; Adamidis, T.; Riggle, P. In *Genetics and Molecular Biology of Industrial Microorganisms*; Herschberger, C. L.; Queener, S. W.; Hegeman, G. Eds.; American Society for Microbiology: Washington, DC., 1989, pp. 53-59.

Chater, K. F. *Biotechnol.* **1990**, *8*, 115-121.

Cooney, C. L.; Acevedo, F. *Biotechnol. Bioeng.* **1977**, *19*, 1449-1462.

Dehottay, P.; Dusart, J.; Duez, C.; Lenzini, M.V.; Martial, J.A.; Frere, J-M.; Ghuysen, J-M.; Kieser, T. *Gene.* **1986**, *42*, 31-36.

Demain, A. L. *Adv. Appl. Microbiol.* **1973**, *16*, 177-202.

Donadio, S.; Tuan, J. S.; Staver, M. J.; Weber, J. M.; Paulus, T. J.; Maine, G. T.; Leung, J. O.; Dewitt, J. P.; Vara, J. A.; Wang, Y.; Hutchinson, C. R.; Katz, L. *In Genetics and Molecular Biology of Industrial Microorganisms*; Herschberger, C. L.; Queener, S. W.; Hegeman, G. Eds.; American Society for Microbiology: Washington, DC., 1989, pp. 53-59.

Doull, J. L.; Vining, L. C. *Appl. Microbiol. Biotechnol.* **1990**, *32*, 449-454.

Drew, S. W.; Demain, A. L. *Ann. Rev. Microbiol.* **1977**, *31*, 343-356.

Duez, C.; Piron-Fraipont, C.; Joris, B.; Dusart, J.; Urdea, M.S.; Martial, J.A.; Frere, J.M. *Eur. J. Biochem.* **1987**, *162*, 509-518.

Epp, J. K.; Huber, M. L.; Turner, J. R.; Schoner, B. E. In *Genetics and Molecular Biology of Industrial Microorganisms*; Herschberger, C. L.; Queener, S. W.; Hegeman, G. Eds.; American Society for Microbiology: Washington, DC., 1989, pp. 35-39.

Erpicum, T.; Granier, B.; Delcour, M.; Lenzini,

V. M.; Nguyen-Disteche, M.; Dusart, J.;
Frere, J.M.*Biotechnol. Bioeng.* **1990**,
35, 719-726.

Fayerman, J.T. *Biotechnol* . **1986**, *4*, 786-789.

Flores, M. E.; Sanchez, S. J. *Gen. Appl.
Microbiol.* **1989**, *35*, 203-211.

Flores, M. E.; Sanchez, S.J. *FEMS Microbiol.
Lett.* **1985**, *26*, 191-194.

Friebel, T. E.; Demain, A. L. *J. Bacteriol.*
1977, *130*, 1010-1016.

Gallo, M.; Katz, E. *J. Bacteriol.* **1972**, *109*,
659-667.

Ghangas, G.S.; Wilson, D.B. *Appl. and
Environmental Microbiol.* **1988**, *54*,
2521-2526.

Ghangas, G.S.; Wilson, D.B. *Appl. and
Environmental Microbiol.* **1987**, *53*,
1470-1475.

Gil, J. A.; Naharro, G.; Villanueva, J. R.;
Martin, J. F. *J. Gen. Microbiol..* **1985**,
131, 1279-1287.

Gray, G.; Selzer, G.; Buell, G.; Shaw, P.;
Escanez, S.; Hofer, S.; Voegeli, P.;
Thompson, C.J. *Gene.* **1984**, *32*, 21-
30.

Hartley, D. L.; Speedie, M. K. *Biochem J.*
1984, *220*, 309-313.

Haavik, H. I. *J. Gen. Microbiol.* **1974**, *84*,
226-230.

Heinjen, J. J.; Roels, J. A.; Stouthamer, A. H.
Biotechnol. Bioeng. **1979**, *21*, 2175-
2201.

Hopwood, D. A.; Malpartida, F.; Chater, K. F.
*In Regulation of Secondary Metabolite
Formation*; Kleinkauf, H.; Dohren, H.;
Dorndauer, H.; Nesemann, G., Eds.;
Workshop Conferences Hoechst; VCH
Publishers; Deerfield Beach, Fl. 1986, Vol.
16; pp. 23-33.

Hopwood, D. A.; Malpartida, F.; Kieser, H. M.;
Ikeda, H.; Duncan, J.; Fujii, I.; Rudd,
B. A. M.; Floss, H. G.; Omura, S. *Nature.*
1985, *314*, 642-644.

Horinouchi, S.; Suzuki, H.; Beppu, T. *J.
Bacteriol.* **1986**, *168*, 257-269.

Hutchinson, C. R.; Borell, C. W.; Otten, S. L.;
Stutzman-Engwall, K. J.; Wang, Y. *J.
Med. Chem.* **1989**, *32*, 929-937.

Illingworth, C.; Larson, G.; Hellekant, G. *J.
Industrial Microbiol.* **1989**, *4*, 37-42.

Jones, A.; Westlake, D. W. S. *Can J. Microbiol.*
1974, *20*, 1599-1611.

Katz, E.; Thompson, C.J.; Hopwood, D.A. *J.
Gen. Microbiol.* **1983**, *129*, 2703-2714.

Kendall, K.; Cullum, J. *Gene.* **1984**, *29*, 315-
321.

Khokhlov, A. S.; Tovarova, I. I. In *Regulation
of Secondary Product and Plant
Hormone Metabolism*; Luckner, M.;
Schreiber, K. Eds.; Pergomon Press:
New York, NY., 1979, pp. 133-146.

Koller, K-P.; Riess, G. *J. Bacteriol.* **1989**, *171*,
4953-4957.

Koller, K-P.; Riess, G.; Sauber, K.; Uhlmann,
E.; Wallmeier, H. *Biotechnol.* **1989**, 7,
1055.

Kominek, L. A. *Antimicrob. Agents Chemother.*
1972, *1*, 123-134.

Legator, M.; Gottlieb, D. *Antibiot. Chemother.*
1953, *3*, 809-817.

Lenzini, M.V.; Nojima, S.; Dusart, J.; Ogaware,
H.; Dehottay, P.; Frere, J-M.; Ghuysen,
J-M. *J. Gen. Microbiol.* **1987**, *133*,
2915-2920.

Lilley, G.; Clark, A. E.; Lawrence, G. C. *J.
Chem. Tech. Biotechnol.* **1981**, *31*,
127-134.

Liu, C.; McDaniel, E.; Shaffner, C. P.
Antimicrob. Agents Chemother. **1975**,
7, 196-202.

Lohr, D.; Buschulte, T.; Gilles, E. *Appl.
Microbiol. Biotechnol.* **1989**, *32*, 274-
279.

Malik, V. S.; Vining, L. C. *Can. J. Microbiol.*
1970, *16*, 173-179.

Malpartida, F.; Hallam, S. E.; Kieser, H. M.;
Motamedi, H.; Hutchinson, C. R.; Butler,
M. J.; Sugden, D. A.; Warren, M.;
McKillop, C.; Bailey, C. R.; Humphreys,
G. O.; Hopwood, D. A. *Nature.* **1987**,
325, 818-821.

Malpartida, F.; Hopwood, D. A. *Nature.* **1984**,
309, 462-464.

Martin, J. F. In *Microbiology-1976*:
Schlessinger, D., Ed.; American Society
for Microbiology: Washington, DC,
1976, pp. 548-552.

Martin, J. F; Liras, P. *Annu. Rev. Microbiol.*
1989, *43*, 173-200.

Masuma, R.; Zhen, D.; Tanaka, Y.; Omura, S.
J. Antibiot. **1990**, *43*, 83-87.

Masuma, R.; Tanaka, Y.; Omura, S. *J. Ferment.
Technol.* **1983**, *61*, 607-614.

Matsumura, M.; Imanaka, T.; Yoshida, T.,
Taguchi, H. *J Ferment. Technol.* **1978**,
56, 345-353.

Miller, A. L.; Walker, J. B. *J. Bacteriol.* **1970**,
104, 8-12.

Miyake, K.; Kuzuyama, T.; Horinouchi, S.;
Beppu, T. *J. Bacteriol.* **1990**, *172*,
3003-3008.

Mou, D.; Cooney, C. L. *Biotechnol. Bioeng.*
1976, *18*, 1371-1392.

Munoz, A.; Perez-Aranda, A.; Barbero, J.L.
*Biochem. and Biophysical Research
Communications.* **1985**, *133*, 511-519.

Murooka, Y.; Ishizaki, T.; Nimi, O.; Maekawa,
N. *Appl. and Environmental Microbiol.*
1986, *52*, 1382-1385.

Neway, J.; Gaucher, G. M. *Can. J. Microbiol.*
1981, *27*, 206-215.

Obata, S.; Furukubo, S.; Kumagai, I.;

Takahashi, H.; Miura, K. *J. Biochem.* **1989**, *105*, 372-376.

Ochi, K. *J. Bacteriol.* **1987**, *169*, 3608-3616.

Ochi, K.; Tsurumi, Y.; Shigematsu, N.; Iwami, M.; Umehara, K.; Okuhara, M. *J. Antibiot.* **1988**, *41*, 1106-1115.

Ohnuki, T.; Imanaka, T.; Aiba, S. *J. Bacteriol.* **1986**, *164*, 85-94.

Omer, C. A.; Stein, D., Cohen, S. N. *J. Bacteriol.* **1988**, *170*, 2174-2184.

Omura, S.; Ikeda, H.; Malpartida, F.; Kieser, H. M.; Hopwood, D. A. *Antimicrob. Agents Chemother.* **1986**, *29*, 13-19.

Omura, S.; Tanaka, Y.; Tanaka, H.; Takahashi, Y., Iwai, Y. *J Antibiot.* **1980**, *33*, 1568-1569.

Otten, S. L.; Stutzmann-Engwall, K. J.; Hutchinson, R. *J. Bacteriol.* **1990**, *172*, 3427-3434.

Payne, G. F. PhD. Thesis, University of Michigan, 1984.

Payne, G. F.; Delacruz, N.; Coppella, S. J. *Appl. Microbiol. Biotechnol.* **1990**, *33*, 395-400.

Payne, G. F.; Wang, H. Y. *Arch. Microbiol.* **1989**, *151*, 331-335.

Perkins, J. B.; Guterman, S. K.; Howitt, C. L.; Williams, V. E.; Pero, J. *J. Bacteriol.* **1990**, *172*, 3108-3116.

Pirt, S. J.; Righelato, R. C. *Appl. Microbiol.* **1967**, *15*, 1284-1290.

Pulido, D.; Vara, J.A.; Jimenez, A. *Gene.* **1986**, *45*, 167-174.

Queener, S. W.; Day, L. E. *The Bacteria: A Treatise on Structure and Function: Antibiotic Producing Streptomyces.* Academic Press: New York, New York, *Vol.9*, 1986 p. xviii.

Ragan, C. M.; Vining, L. C. *Can. J. Microbiol.* **1978**, *24*, 1012-1015.

Shapiro, S.; Vining, L. C. *Can. J. Microbiol.* **1985**, *31*, 119-123.

Shapiro, S.; Vining, L. C. *Can. J. Microbiol.* **1984**, *30*, 798-804

Shapiro, S.; Vining, L. C. *Can J. Microbiol.* **1983**, *29*, 1706-1714.

Shirato, S.; Nagatsu, C. *Appl. Microbiol.* **1965**, *13*, 669-672.

Shiffman, D.; Mevarech, M.; Jensen, S. E.; Cohen, G.; Aharonowitz, Y. *Mol..Gen . Genet..* **1988**, *214*, 562-569.

Schottel, J. L.; Willard, J. M. A.; Raymer, G. In *Genetics and Molecular Biology of Industrial Microorganisms*; Herschberger, C. L.; Queener, S. W.; Hegeman, G. Eds.; American Society for Microbiology: Washington, DC., 1989, pp. 13-118.

Smith, I.; Slepecky, R. A.; Setlow, P., *Regulation of Procaryotic Development*; American Society for Microbiology: Washington, DC, 1989.

Stanzak, R.; Matsushima, R. H.; Baltz, R. H.; Rao, R. N. *Biotechnol.* **1986**, *4*, 229-232.

Steiert, J. S.; Pogell, B. M.; Speedie, M. K.; Laredo, J. *Biotechnol..* **1989**, *7*, 65-68.

Taguchi, S.; Kumaga, I.; Nakayama, J.; Suzuki, A.; Miura, K. *Biotechnol.* **1989**, *7*, 1063-1066.

Vining, L. C.; Shapiro, S. *J. Antibiot.* **1984**, *37*, 74-76.

von Heijne, G.; Abrahmsen, L. *FEBS Letters* **1989**, *244*, 439-446.

Vu-Trong, K.; Gray, P. P. *Biotechnol. Bioeng.* **1982**, *24*, 1093-1103.

Vu-Trong, K.; Bhuwapathanapun, S.; Gray, P. P. *Antimicrob. Agents Chemother.* **1981**, *19*, 209-212.

Vu-Trong, K.; Bhuwapathanapun, S.; Gray, P. P. *Antimicrob. Agents Chemother.* **1980**, *17*, 519-525.

Wallace, K. K.; Payne, G. F.; Speedie, M. K. *J. Ind. Microbiol..* **1990**, *6*, 43-48.

Wang, Z.; Bleakley, B. H.; Crawford, D. L.; Hertel, G.; Rafii, F. *J. Biotechnol.* **1990**, *13*, 131-144.

Westpheling, J.; Ranes, M.; Losick, R. *Nature.* **1985**, *313*, 22-27.

Young, M. D.; Kempe, L. L.; Bader, F. G. *Biotechnol. Bioeng.* **1985**, *27*, 327-333.

Zhang, J.; Wolfe, S.; Demain, A. L. *Can. J. Microbiol.* **1989**, *35*, 399-402.

Strategy for the Production and Stabilization of Lignin Peroxidase from *Phanerochaete chrysosporium* in Air-lift and Stirred Tank Bioreactors

Austin H.C. Chen[1,3], Carlos G. Dosoretz[1,2,3], and Hans E. Grethlein[1,2,3], Department of Chemical Engineering[1] and Center for Microbial Ecology[2], Michigan State University, East Lansing, MI 48824, and Michigan Biotechnology Institute[3], Lansing, MI 48909

In the large scale production of lignin peroxidase in suspended cultures, high shear agitation has been reported to be a disadvantage. A comparative study of an air-lift bioreactor, which has the advantage of low shear agitation with low power, and a stirred tank bioreactor is presented. In general, there is a qualitative similarity in the behavior of the various enzymes formed in both bioreactors. Normally lignin peroxidase, a secondary metabolite, is maximum when the glucose is depleted, and rapidly decreases in activity after the maximum value is reached. This study showed that lignin peroxidase activity can be maintained at maximum level at least 6 days by fed-batch addition of glucose to insure a threshold level. We noticed a sequence of metabolites starting from the beginning of the fermentation; first a primary protease, then Mn-peroxidase, then lignin peroxidase, and finally a secondary protease. Each profile increases dramatically at the onset of the previous peak. The secondary protease activity can be controlled to a very low level by fed-batch addition of glucose. Apparently, inhibition of protease at secondary metabolism plays a major role in the stabilization of lignin peroxidase production. In considering the large scale production of lignin peroxidase in the future, air-lift bioreactor or other bioreactors without moving parts should have the better chance of success.

The potential applications of ligninase (lignin peroxidase) in detoxification of recalcitrant wastes (such as PCP and DDT, Mileski et al., 1988; Bumpus and Aust, 1987; Schreiner et al., 1988) and degradation of lignin (the second largest renewable resources in the world) and lignin model compounds (Ericksson et al., 1983; Buswell and Odier, 1987) have aroused the interest of many researchers in the last decade. One of the most common goals is the production of lignin peroxidase by white-rot fungus *Phanerochaete chrysosporium* in large scale bioreactors. Although the first lignin peroxidase was reported in 1983 (Tien and Kirk; 1983, Glenn et al., 1983), its production was limited to stationary cultures and was sensitive to agitation (Leisola and Fiechter, 1985). Only recently was it discovered that lignin peroxidase can be produced under agitated conditions in the presence of veratryl alcohol (Leisola and Fiechter, 1985, Jager et al., 1985). The production of lignin peroxidase in conventional stirred tank reactors has been difficult so far partly due to the sensitivity of fungus to mechanical agitation (Janshekar and Fiechter, 1988). Air-lift bioreactor, however, offers advantages of no mechanical impellers, mild agitation, high gas absorption efficiency, rapid mixing and no moving parts (Merchuk and Siegel, 1988). While, Leisola et al (1986) reported that 20 - 60 U/l lignin peroxidase activity was produced in a 1.5 liter air-lift reactor, Janshaker and Fiechter (1988) reported that the maximum ligninase activity produced in a 42 liter stirred tank reactor was 50 U/l. This study was motivated by the desire for a direct comparison of various extracellular enzymes produced in both air-lift and stirred tank bioreactors. The present paper intends to compare the time courses of various extracellular enzymes secreted by *P. chrysosporium*, the interaction between the sharp decrease of lignin peroxidase and the emergence of protease, and the strategy of stabilization of lignin peroxidase production in both air-lift and stirred tank bioreactors.

Materials and Methods

Microorganisms

Phanerochaete chrysosporium Burds wild-type BKM-F-1767 (ATCC 24725) was maintained at 37^0C on 2% malt agar slant (Jager et al., 1985). The organism was subcultured prior to experiments, incubated at 37^0C, and stored at 4^0C.

Culture Conditions and Media

All cultures were grown in a N-limited medium according to Tien and Kirk (1988) except dimethylsuccinate (DMS) was replaced by 20 mM acetate buffer (pH 4.5). The inoculum was grown in stationary culture initiated from fresh conidial suspension (5.0×10^5 spores per ml) by the method of Jager et al. After two days of stationary culture, the homogenized inoculum was incubated in three 2 liter flasks with 10 v/v% inoculation ratio at 37^0C and 175 rpm in an orbital shaker (G50, New Brunswick Scientific, NJ). When the pellets were formed after one day, all the cultures were transferred into either a 2 liter air-lift or a 2 liter stirred tank bioreactor (Starting time for this experiment). A few drops of antifoam (silicone emulsion; Dow corning FG-10) were added to eliminate the foam formed due to agitation. Continuous oxygenation was controlled at 0.2 l/l/min (0.2 liter of O_2 in 1 liter of culture per minute) for both reactors. The agitation rate for stirred tank reactor is 135 rpm with a marine impeller. No acid or base was added during this experiment. Samples were taken and analyzed daily.

Analytical Techniques

Reducing sugar was determined by the DNS (dinitrosalicylic acid) method using D-glucose as a standard according to Ghose (1987). **Protein content** was measured by the method of Bradford (1976) using bovine serum albumin fraction V as a standard and the Bio-Rad reagent. **Lignin peroxidase activity** was determined spectrophotometrically according to the procedure described by Tien and Kirk (1988). Enzyme activity was calculated as unit per liter (U/l) where one unit (U) was defined as 1 μM of veratryl alcohol oxidized to veratryl aldehyde in one minute. **Mn-peroxidase activity** was measured by the methods of Kuwahara *et al.* (1984) using phenol red as the substrate. The activity was expressed as the absorbance change at 610 nm (OD_{610}) in 3 minutes for 20 μl sample. **Protease activity** was determined as described by Dosoretz *et al.* (1990) using azocoll as the substrate. The activities were reported as unit per liter (U/l) where one unit (U) was defined as the amount of enzyme which catalyzes the release of azo dye and causes an absorbance change at 520 nm of 0.001 in one minute.

Results

The time courses of extracellular enzymes secreted from *Phanerochaete chrysosporium* (Fig. 1) in a 2 liter air-lift bioreactor suggested that a sequence of metabolites is produced starting from the beginning of the fermentation: first a protease, then Mn-peroxidase, then lignin peroxidase, and finally a secondary protease. Each profile increases dramatically at the onset of the previous peak. The peaks for protease, Mn-peroxidase, and lignin peroxidase activities occurred at 24, 72, and 116 hours respectively. Amazingly, the decay of the lignolytic enzyme (starting at 116 hours) was accompanied by the production of the secondary proteolytic activity.

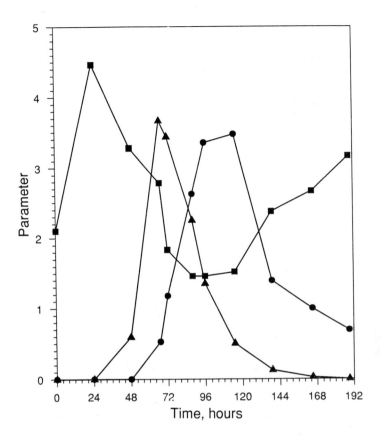

Fig. 1. Time course for the production of extracellular enzymes in a 2 liter air-lift bioreactor. ■ , Protease (x1); ▲ , Mn-peroxidase (x5); ●, Ligninase (x0.02).

The glucose consumption rate (Fig. 2) during the entire process was practically identical for both air-lift and stirred tank bioreactors. The glucose was added daily after 96 hours at the amount of 1.75 grams per liter of solution which was about the same as the glucose consumption rate shown in Fig. 2.

Apparently, a secondary protease was found after 116 hours (Fig. 3) for both air-lift and stirred tank bioreactors during the depletion of glucose. The higher secondary protease which appeared in air-lift reactor was accompanied by faster decay of lignin peroxidase activity (see Fig. 4). The effect of fed-batch glucose addition to protease profile was

significant. The results showed that the fed-batch addition of glucose after 96 hours significantly inhibited secondary protease (Fig. 3) and enhanced lignin peroxidase production (Fig. 4). The higher lignin peroxidase activity, which reached about 200 U/l, can be sustained for at least 6 days as long as a threshold level of glucose (around 2 mg/ml) remained.

Compared with the production of enzymes in an air-lift bioreactor, a maximum of 120 U/l lignin peroxidase activity was produced in a stirred tank reactor. While the lower lignin peroxidase activity in the stirred tank reactor may be due to the higher shear produced by the

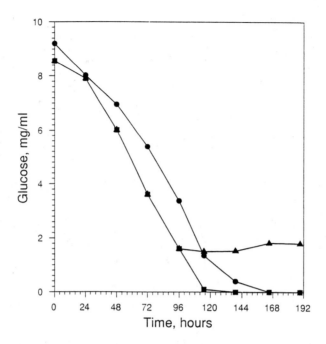

Fig. 2. Time course of glucose consumption in air-lift and stirred tank bioreactors with (w/) or without (w/o) fed-batch glucose addition. ■, Air-lift w/o addition; ▲, Air-lift w/ addition; ●, Stirred tank w/o addition.

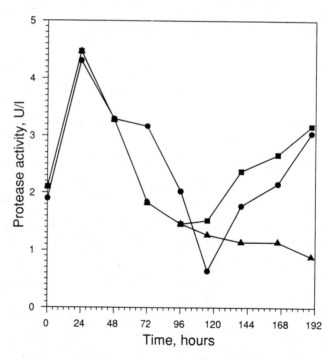

Fig. 3. Time course of protease activity in air-lift and stirred tank bioreactors with (w/) or without (w/o) fed-batch glucose addition. ■, Air-lift w/o addition; ▲, Air-lift w/ addition; ● , Stirred tank w/o addition.

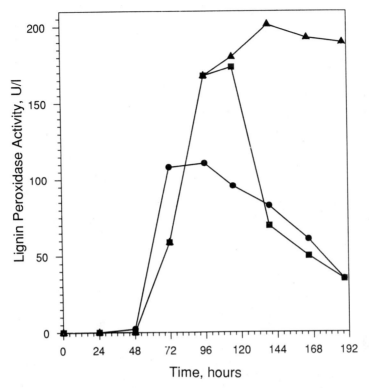

Fig. 4. Time course of lignin peroxidase activity in air-lift and stirred tank bioreactors with (w/) or without (w/o) fed-batch glucose addition. ■, Air-lift w/o addition; ▲, Air-lift w/ addition; ●, Stirred tank w/o addition.

internal agitated impellers, this activity was larger than the reported values (50 U/l maximum reported in C-limited medium in a 42 liter stirred tank reactor, Janshekar and Fiechter, 1988). The fact that the decay of lignin peroxidase activity is much slower in a stirred tank reactor than in a air-lift reactor is consistent with the lower secondary protease activity in the stirred tank bioreactor shown in Fig. 3.

Mn-peroxidase activity (Fig. 5), however, showed an opposite tendency when compared with the protease profiles. It peaked at 72 hours and continuously decayed after that to a very small level when there was no fed-batch glucose addition; however, it remained in a much higher level (about 60% of its maximum activity) when glucose was added. A higher Mn-peroxidase activity was found in a stirred tank compared with that in an air-lift bioreactor.

Discussion and conclusions

The production of extracellular enzymes in air-lift and stirred tank bioreactors is qualitatively similar but quantitatively different. A higher lignin peroxidase, lower Mn-peroxidase, and higher secondary protease activities were found in an air-lift than in a stirred tank bioreactor. This lower lignin peroxidase activity in stirred tank bioreactors suggests that mechanical agitation could be a major obstacle for the production of lignin peroxidase. It may also explain why conventional stirred tank reactors have not been considered widely in the study of lignin peroxidase production. Bioreactors without mechanical impellers should have a better chance of success for the large scale production of ligninase.

The primary protease that peaked at the glucose-sufficient condition (growing

185

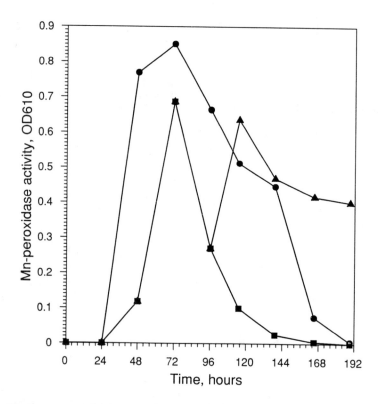

Fig. 5. Time course of Mn-peroxidase activity in air-lift and stirred tank bioreactors with (w/) or without (w/o) fed-batch glucose addition. ■, Air-lift w/o addition; ▲, Air-lift w/ addition; ●, Stirred tank w/o addition.

phase) is identical for both reactors. The secondary protease which peaked at glucose-depletion condition (idiophase), however, is higher in the air-lift than in the stirred tank reactor. This secondary protease appears at about the same time as the degradation of lignin peroxidase. This evidence and the apparent inverted relationship between the production of secondary protease and the decay of lignin peroxidase (after 116 hours) suggest that the stabilization of lignin peroxidase required the regulation of secondary protease. The stability of lignin peroxidase in air-lift bioreactor was found to be worse than that in stirred tank reactor due to the higher secondary protease. This study suggests a strategy for the stabilization of lignin peroxidase production. By fed-batch glucose addition at the time when lignin peroxidase activity reaches its maximum, one can simultaneously stabilize the lignin peroxidase and repress the protease activity during the period of the experiment. Since these results are promising, further studies of stabilization and production of lignin peroxidase in pilot plant scale are required.

Acknowledgments

The authors are grateful to the Research Excellence Funds from State of Michigan and the Michigan Biotechnology Institute for their support.

References

Bradford, M. M., *Anal. Biochem.*, **1976, 72**, 248.

Bumpus, J. A., Aust, S. D., *Appl. Environ. Microbiol.*, **1987, 53**, No 9, 2001.
Buswell, J. A., Odier, E., *CRC Critical Review in Biotechnology*, **1987, 6**, No 1, 1.

Dosoretz, C. G., Chen, H. C., Grethlein, H. E., *Appl. Environ. Microbiol.*, **1990, 56**, No 2, 395.

Ericksson, K. E., Johnsrud, S. C., Vallander, L., *Arch. Microbiol.*, **1983, 135**, 161.

Ghose, T. K., *Pure appl. Chem.*, **1987, 59**, 257.

Glenn, J. K., Morgan, M. A., Mayfield, M. B., Kuwahara, M., Gold, M. H., *Biochem. Biophys. Res. Commun.*, **1983, 114**, 1077.

Jager, A., Croan, S., Kirk, T. K., *Appl. Environ. Microbiol.*, **1985, 50**, No 5, 1274.

Janshekar, H., Fiechter, A., *J. Biotechnol.*, **1988, 8**, 97.

Kuwahara, M., Glenn, J. K., Morgan, M. A., Gold, M. H., *FEBS Lett.*, **1984, 169**, 247.

Leisola, M. S. A., Fiechter, A., *FEMS Microbiol. Letter*, **1985, 29**, 33.

Leisola, M., Troller, J., Fiechter, A., Linko, Y. Y., *Proc. the 3rd International Biotechnology on Pulp and Paper Industry*, **1986**, 46.

Merchuk, J. C., Siegel, M. H., *J. Chem. Tech. Biotechnol.*, **1988, 41**, 105.

Mileski, G. I., Bumpus, J. A., Jurek, M. A., Aust, S. D., *Appl. Environ. Microbiol.*, **1988, 54**, No 12, 2885.

Schreiner, R. P., Stevens, S. E., Tien, M., *Appl. Environ. Microbiol.*, **1988, 54**, No 7, 1858.

Tien, M., Kirk, T. K., *Science*, **1983, 221**, 661.

Tien, M., Kirk, T. K., *Meth. Enzymol.*, **1988, 161**, 238.

Bioprocessing of Higher Plant Cell Cultures

Henrik Pedersen and Sang Yo Byun, Department of Chemical and Biochemical Engineering, Rutgers University, P.O. Box 909, Piscataway, New Jersey 08855-0909
Chee-Kok Chin, Department of Horticulture, Rutgers University, P.O. Box 231, New Brunswick, New Jersey 08903-0231

The use of elicitor preparations to enhance secondary metabolite production is discussed. Various techniques for combining elicitation with precursor feeding and product extraction are demonstrated as means to control production and to allow for efficient bioprocessing of higher plant cells. Specific examples from poppy cultures producing benzophenanthridine alkaloids are highlighted.

Understanding the dynamic behavior of metabolite production in suspension cultures of higher plants is of obvious importance in the operation of bioreactors. Control over metabolic dynamics would certainly be a valuable asset in selecting strategies for enhancing productivity and moving closer towards commercial production of plant compounds. One way to regulate the dynamic behavior of secondary metabolites is to add so-called oligosaccharide elicitors to the medium. The oligosaccharide molecules modulate the metabolic pathways of a variety of secondary metabolites via action directed at the gene level. The products of these pathways may be important commercial specialty chemicals. For example, the benzophenanthridine alkaloids produced by california poppy (*Eschscholtzia californica*) cells following elicitation may have useful antibacterial action.

A preliminary experimental and theoretical analysis of the *Eschscholtzia* system has been recently developed (Byun et al., 1990a,b) with emphasis on precursor factors, elicitor timing and product extraction. In this paper, we demonstrate how bioreactor operation can be adjusted to manipulate and modify productivity in this cell line. Related examples from other systems are also described.

Bioprocessing Techniques and Options

The factors that regulate growth and product formation in plant cell suspension cultures can generally be divided into two major groups that are either associated with the biological aspects of the cells or with the engineering aspects of the reactor operation. They are not always exclusive, however, and should be thought of in this paper simply as a convenient tool for discussing process strategies. For instance, as shown schematically in Fig. 1A, simple bioprocessing of higher plant cell cultures is based on the scale-up of conditions that have been found suitable for growth of plant cells in suspension. The media components that make up the extracellular portion of the reactor contain nutrients and growth regulators derived from standard formulations such as MS media (Murashige and Skoog, 1962). There may be constitutive production of valuable secondary (2°) metabolites from intracellular primary (1°) metabolites under these conditions, but it is not generally the case. Rather, all the media precursors, such as the phosphate and carbohydrate source, or the modifiers added to the media as growth regulators, such as 2,4-dichlorophenoxyacetic acid (2,4-D) and indolyl-3-acetic acid (IAA), primarily have been adjusted to promote good cell growth.

Media modifications that are oriented toward the promotion of secondary metabolism are certainly possible, however, and we call this cell directed bioprocessing as shown in Fig 1B. Examples from the literature abound in this area and have been oriented to a great extent on increasing sucrose concentration or decreasing phosphate composition as well as adjusting the hormone levels (Knoblach and Berlin, 1980; Sahai and Shuler, 1984; Sugano et al., 1975). The result is usually to define appropriate conditions for the production of secondary metabolites and a "production media," or an "induction media," that is distinctly different than the growth media. This may further entail the need to switch culture conditions in the reactors from growth to production conditions or to operate staged reactors in a timely manner. Yamada and Fujita (1983) have demonstrated this reactor directed bioprocessing option in the case of pigment production by *Lithospermum erythrorhizon*. Another example of reactor directed bioprocessing is shown schematically in Fig. 1C that involves the use of extraction phases to promote metabolite production in the absence of any other media manipulations. This has been demonstrated with both liquid extraction phases (Bisson et al., 1983; Byun et al., 1990a) and

2181–2/92/0188$06.00/0 © 1992 American Chemical Society

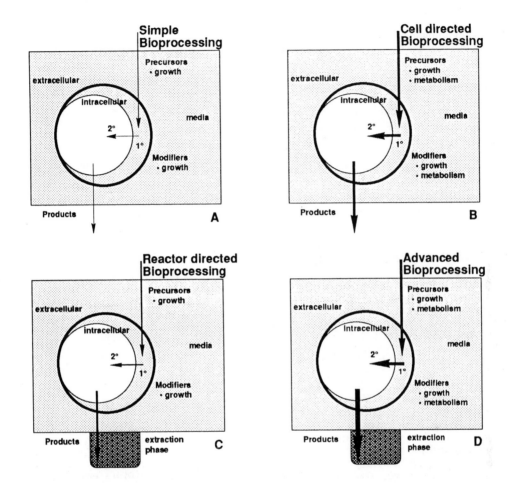

Fig. 1. Different bioprocessing options for the cultivation of cells of higher plants. The intracellular phase is depicted by the heavy circle and contains pools of primary and secondary metabolites (light circle). The thickness of the arrows represents the relative flux of material through the various compartments.

with solid extraction phases (Knoop and Beiderback, 1983; Payne et al., 1988). A coordinated scheme that brings in to play all of the different regulating factors falls under the heading of advanced bioprocessing as shown schematically in Fig. 1D. Here a concerted effort is made to promote metabolite production using factors that regulate cellular biosynthesis and to choose designs that favor further production advantages.

These different processing options can all be demonstrated in suspension cultures of california poppy and this is highlighted in the following sections. The major emphasis is on cell-directed phenomenom since these have appeared most often in the literature. In these examples, elicitors, that represent a unique type of cell-directed manipulation, play a central role. It should be kept in mind, however, that many other plant species and their associated secondary metabolites have similar features.

Cell Directed Bioprocessing

Suspension cultures of E. californica have been shown to produce the benzophenanthridine alkaloids sanguinarine, chelirubine, chelerythrine and macarpine (Berlin et al., 1983). Their biosynthesis has been well studied and some of the complex precursor relationships have been deduced (Takao et al., 1983). The pathway is summarized in Fig. 2 where the central intermediate (S)-reticuline is identified. This material is synthesized from two tyrosine precursors and is cyclized oxidatively to (S)-scoulerine by the berberine bridge enzyme, a key enzyme in the synthesis of many isoquinoline-type alkaloids (Rink and Böhm, 1975). The end products in poppy include sanguinarine that can be further reacted to give macarpine, the ultimate product in the pathway. Strategies for improving the alkaloid productivity of cell suspension cultures include the accelerated channeling of the primary metabolite tyrosine through the pathway and the enhanced expression or reactivity of the enzymes that make up the pathway, particularly the berberine bridge enzyme. Various bioprocessing options have been explored that are based on this cell-directed approach (Byun, 1989) and include the supplementation of the media with tyrosine, the use of elicitors that stimulate de novo enzyme synthesis (Byun et al., 1990a), and other culture condition and nutritional factors that may indirectly effect the movement of cellular metabolites down the alkaloid pathway. More than one of these strategies can be used concomitantly and, in particular, the employment of eliciting compounds has been shown to be quite effective (Collinge and Brodelius, 1989).

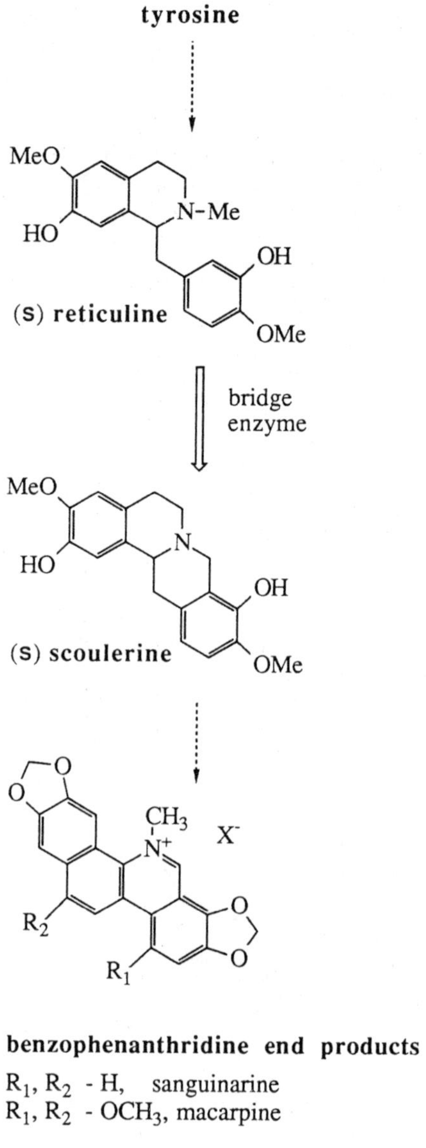

Fig. 2. Metabolic pathway in Eschscholtzia for the production of benzophenanthridine alkaloids.

Elicitation

The elicitation of secondary metabolites in a wide range of plants by a variety of chemical stimuli has been recently summarized by DiCosmo and Misawa (1985). The mechanism by which plants respond has not been completely elucidated, but some general steps have been put forth and some of the elicitor molecules have been identified (Templeton and Lamb, 1988). In particular, glucans isolated from the mycelia of a

190

variety of fungi can act as elicitors (Ayers *et al.*, 1976a,b; Funk *et al.*, 1987) and it has been shown that oligosaccharides prepared from *Collectotrichum lindemuthianum, Verticillium dahliae* and yeast extract (*Saccharomyces cerevisiae*) all induce the formation of benzophenanthridine alkaloids in california poppy (Byun, 1989) On the other hand, elicitors isolated from *Phytophthora megasperma* var. *sojae* do not stimulate a response that can be measured by enhanced alkaloid production. The maximum increase in total alkaloid levels was as much as 3.5-fold with both the yeast elicitor and with the *Verticillium dahliae* glucan elicitor. These results emphasize that a variety of microorganisms could possibly be used as a source of elicitors in suspension cultures of *E. californica*. The preparation of elicitor from yeast extract, however, is particularly straightforward and was therefore used as the sole source of the elicitor compounds reported here.

Dependence of Alkaloid Production on Elicitor Concentration

The elicitor concentration is a factor which strongly affects the intensity of the response as well as the distribution of products. Fig. 3 shows the relative accumulation of benzophenanthridine alkaloids in response to different elicitor

concentrations when the elicitor was introduced during the growth phase of the culture. The accumulation pattern of alkaloids measured after 16 hours as a function of the elicitor concentration is a complex phenomenon. The accumulation rate generally increases with elicitor levels at low concentrations, but was widely varying for the different alkaloids at high elicitor concentrations. The maximum accumulation of macarpine was observed at 60 μg of yeast elicitor per gram of fresh cell weight (μg YE/g FCW). This value, however, does not enhance sanguinarine concentration, which is a precursor of macarpine, relative to the unelicited level, at least at 16 hours after induction. Instead, as seen in Fig. 3, the maximum value of sanguinarine accumulation was obtained at 20 μg YE/g FCW. Other alkaloids in the pathway also have unique, optimal induction doses. Fig. 3 also demonstrates that "overloading" of elicitor has adverse effects. Inhibition by overdosed elicitor reduced the accumulation of total alkaloids and slightly decreased the cell growth (data not shown). At sufficiently low doses, however, the growth rate is unaffected or may even be slightly faster. Similar results with respect to cell growth have been reported for suspension cultures of parsley, *Petroselinum hortense*, (Hahlbrock *et al.*, 1981).

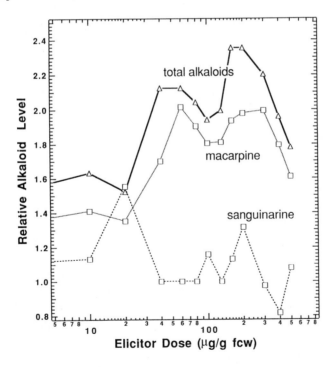

Fig. 3. Alkaloid production in california poppy as a function of the elicitor dose during the growth phase of the culture.

Elicitation at Different Growth Phases

The relationship between growth phase and formation of secondary metabolites in cultured plant cells is a topic of numerous publications (e.g., Drapeau *et al.*, 1986). In "normal" plant cell culture systems, the majority of compounds are formed in the stationary phase. The elicitor response has also proved to be dependent on the growth stage in many culture systems and, with few exceptions, most susceptible plant cell cultures respond to elicitation only during the so-called exponential growth phase. The particular point in the growth stage of a culture may further affect not only the quantitative response to elicitor treatment but also the production pattern. For instance, *Pythium* culture homogenate stimulated N-acetyltryptamine formation in 5-day-old *Catharanthus roseus* cultures; 10-day-old cells, in contrast, were shown to accumulate a wide spectrum of monoterpene indole alkaloids (Eilert *et al.*, 1986). Elicitor treatment at a time when a culture has already started to accumulate the inducible compounds does not necessarily enhance or accelerate accumulation and can even suppress already activated biosynthesis (Oba and Uritani, 1979). Further, Kombrink and Hahlbrock (1985) speculate that the formation of furanocoumarins in parsley cultures in the stationary phase might be a result of *auto*elicitation. This could occur through lysis of cells when endogenous elicitor derived from plant cell wall oligosaccharides might be released.

The influence of cultured cell growth stages on elicitor response is particularly obvious in california poppy when product accumulation following elicitation is investigated. This type of experiment is demonstrated in Fig. 4. The volumetric production of macarpine was highest when yeast elicitor was dosed at the 6th day from inoculation and product levels were measured after 48 hrs. This is also the case for the volumetric production of the total alkaloid. Elicitation effectively decreases the time required to accumulate product to a certain level when dosed just before or with the onset of rapid, inherent alkaloid production. Furthermore, the accumulation pattern of macarpine can be shifted from a typical non-growth associated pattern without elicitors to a growth associated profile with elicitation. As a result, a variety of controls can be brought into play by tuning the timing and dose of elicitor addition.

Additional Media Manipulations

Some of the normally ocurring media components can also be adjusted to promote product formation, including the growth

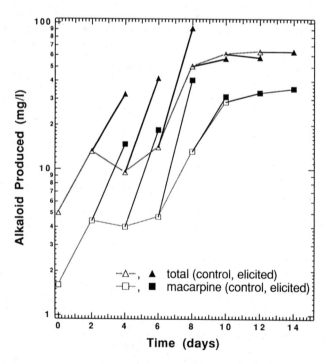

Fig. 4. Alkaloid production in california poppy as a function of the timing of elicitor addition.   The dose of elicitor was 60 µg/g FCW.

192

regulators (auxins and cytokinins), phosphate and nitrate levels, and the sucrose or other carbohydrate level. Addition of precursors or intermediates, that are not normally ocurring in the media, are also of interest. In addition, light and temperature can affect the production characteristics as well. Manipulations applied to these types of components can be done in concert with elicitation.

The concentration of growth regulators in the nutrient medium can affect expression of secondary metabolism in cultured cells quite dramatically. The mechanisms are indirect and poorly understood at the present time, however. In addition, only a few studies have dealt with the influence of this factor on cell response to elicitation. In early work with crude elicitor preparations, Haberlach et al. (1978) examined the effect of cytokinin and auxin concentration on resistance of Nicotiana tabacum callus cultures to the pathogenic fungi Phytophthora parasitica and observed that susceptibility or resistance was a function of the phytohormone balance. A high cytokinin ratio favored infestation and suppressed the hypersensitive response. In other experiments with suspension cultures of bean, the ratio of kinetin and auxin as well as the type of auxin markedly influenced Botrytis cineria homogenate-induced phaseollin accumulation (Dixon and Fuller, 1978). Cultures that were grown in the presence of 2,4-D accumulated lower levels of phaseollin than those grown with NAA and auxin.

The production of benzophenanthridine alkaloids in E. californica is promoted by an increase in the IAA level with and without elicitation (Byun, 1989). The highest alkaloid accumulation was observed at 5 μM of IAA. It was also found that IAA increased cell growth at concentrations lower than 10 μM, whereas it decreased dry cell weight at concentrations approaching 100 μM. The results are less dramatic than the effect of elicitation alone, however.

The influence of the sugar composition of the culture medium on elicitor-induced accumulation of secondary metabolites has generally been neglected. Sugars, and sucrose in particular, have, however, been regarded as important factors for secondary metabolite accumulation in the absence of elicitor substances. For example, in the production of Catharanthus alkaloids, several groups have demonstrated that high sucrose levels and modifications to the growth regulator composition can result in an enhanced alkaloid productivity (Carew and Kreuger, 1977; Zenk et al., 1977; Knobloch and Berlin, 1980). Such a production or induction media is usually used after a suitable growth phase in a distinct growth media since growth rates are often compromised under alkaloid producing conditions. This is also the case in E. californica

where cell growth is suppressed at high sucrose concentration. Alkaloid accumulation, however, is slightly enhanced. Increased sucrose levels also change the alkaloid accumulation pattern in that an increasing proportion of the alkaloids appear intracellularly with increasing sucrose concentration.

E. californica generally utilizes fructose over glucose, both of which are the products of a complete and rapid hydrolysis of sucrose. Elicitation, however, modifies the sucrose consumption whereby the sucrose is not completely degraded into fructose and glucose. Instead, constant levels of glucose and fructose are maintained throughout the culture period. Furthermore, elicited cells of E. californica utilize equal proportions of the hexoses in the media unlike the performance seen with nonelicited cells. These differences are not manifested in any remarkable changes in the alkaloid levels, however.

Light is a another factor often investigated in plant culture (Seibert and Kadkade, 1980). The production of secondary metabolites has been found to be both stimulated (Berl et al., 1986) and suppressed (van den Berg et al., 1988; Tabata et al., 1974) by exposing suspension cultures to different regimens of light and dark periods or by exposing the cultures to conditions characterized by light of a particular wavelength. Alteration in the pattern of product synthesis also can be seen for light- and dark-grown cultures. For instance, ajmalicine accumulates in dark grown Catharanthus cultures whereas serpentine accumulates by exposure to continuous light (Drapeau et al., 1987). In some cell lines, flavonoids and anthocyanins are not produced at all without light irradiation (Mantell and Smith, 1983). In addition, light is also involved in secretion of intracellular products into the medium (Kim et al., 1988). The effects of light on cell growth and alkaloid accumulation in elicited cells of E. californica elicitation have been measured over a culture time of 45 hours (Byun, 1989). The effect of light on cell growth during elicitation was not significant. The alkaloid accumulation, however, was greater under dark conditions and it was found that the total alkaloid production increased 12.7% and macarpine production increased 14.8% in 45 hrs compared to cells grown under identical conditions without elicitation. No differences in accumulation pattern of intra- and extracellular alkaloids were observed in this short period of time.

Precursor Effect with Elicitation

Based on knowledge of the biochemical pathways, it is possible to provide precursors or intermediates to cell suspension cultures in hopes of enhancing final product levels. This is not always an effective strategy, however. For instance, tobacco alkaloid precursors may actually lower the nicotine concentration in cell culture compared to controls (Miller *et al.*, 1983). In some cases, the precursor retards cell growth rates without a counterbalancing enhancement of the specific production rate. Feeding of tryptophan to cell cultures of *Cinchona*, for instance, did not result in any enhancement of the expected quinoline alkaloids, but severely limited growth (Wijnsma and Verpoorte, 1988). On the other hand, when only a few steps are involved between the intermediate and the desired product some success has been achieved. Biotransformation of both caffeine (Baumann and Frischknecht, 1982) and digitoxin derivatives (Rheinhard and Alfermann, 1980) have been noted.

Feeding of tyrosine, a precursor for secondary metabolism in E. californica, does not have an enhancing effect on alkaloid formation without elicitation. Instead, adverse effects were found that were manifested in suppressed cell growth (Byun, 1989). In precursor feeding experiments, it should be emphasized that the timing of feeding is very important. This is mainly due to the fact that different secondary metabolic activities are expressed at different growth phases and precursor feeding without any increase in metabolic activity is likely to result in insignificant or negative effects.

Precursor feeding experiments *with* elicitation, which induces certain key metabolic activities, may be quite effective, however. Experiments were therefore carried out with tyrosine and yeast elicitor in suspension cultures of *E. californica*. The results are shown in Fig. 5. In suspension cultures which had been cultivated for 3 days after inoculation, no change in cell growth was observed. An increase in alkaloid levels around tyrosine doses of 1 mg/g FCW is seen, but this is quickly replaced by rapidly decreasing levels at higher tyrosine concentrations. Precursor feeding together with elicitation on the 7th day from inoculation also shows that cell growth is not significantly reduced except at high tyrosine concentration. Alkaloid production, however, can be increased almost 80%. This enhancement is mainly due to an increase in intracellular accumulation of alkaloids. Again, above doses of 1 mg/g FCW the enhanced productivity is lost.

The biochemical mechanism underlying the variable response is not well understood. However, elicitation is known to induce L-tyrosine decarboxylase (TDC) activity in suspension cultured *E. californica* (Marques and Brodelius, 1988a,b). High TDC activity is observed as soon as 5 hours after elicitor addition. Increased tyrosine consumption is

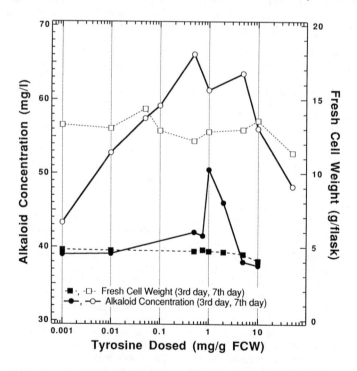

Fig. 5. Production of benzophenanthridine alkaloids in california poppy with tyrosine supplemented to the media and elicitor added on the 3rd day (solid symbols) or the 7th day (open symbols) after inoculation.

194

expected if TDC activity is induced, but getting the tyrosine to channel into the "activated" benzophenanthridine pathway is dependent, according to the data shown in Fig. 5, on both the time of addition of precursor and the amount of precursor added. In all cases, excessive amounts of tyrosine are detrimental. It should also be apparent that elicited cells could be used for enhancing simpler biotransformation reactions when such reactions can be directly coupled with induction of only a few key enzymes.

Reactor Directed Bioprocessing

Attention is now turned to the reactor side of the bioprocessing equation. Increasing productivity has been realized in a number of cases due to efficient bioreactor designs. For instance, as mentioned earlier in this paper if the growth and production media are distinctly different, sequential reactors may be used. The advantages of running such a scheme for the production of the naphthoquinone shikonin, in a semicontinuous fashion, has been convincingly demonstrated by Fujita *et al.* (1982) and Tabata and Fujita (1985). The yield of shikonin in large-scale operation reached 4 g/l in a 2 week culture. Other workers have explored the effects of reactor types and reactor hardware on large scale (20,000 liter) production of ginseng root (Furuya, 1988) and on the biotransformation of methyldigitoxin (Alfermann *et al.*, 1985). The latter group used a fed batch bioprocessing operation at the 200 liter scale to achieve production of 0.5 kg in 3 months.

It is also possible to consider bioprocessing options on a small scale. One possibility that is depicted in Fig. 1c is to carry out product separations in situ. Liquid-liquid extraction systems would be especially suited to batch type processing systems. Again, california poppy offers an example of this strategy.

Two-phase Culture
One type of extraction phase that can accumulate alkaloids in cultures of *E. californica* is composed of intermediate weight dimethyl siloxane polymers, a material that can often be found in silicone based antifoam fluids. It has many desirable properties that make it an "ideal" accumulation phase for suspension cultures of *E. californica* (Byun *et al.*, 1990b). It is very efficient in extracting benzophenanthridine alkaloids from the extracellular fluid phase without dramatically reducing cell growth. The partition coefficient for sanguinarine can be as large as 100, for instance. In this respect, it offers a large accumulation capacity separate from cellular or medium phases.

The biggest advantage of the polymer in suspension cultures of *E. californica*, however, is that it brings about a dramatic increase in the overall production of benzophenanthridine alkaloids. The reasons for this are not entirely clear, but the large accumulation capacity, that makes it possible to store correspondingly large amounts of alkaloids, probably lessens intracellular feedback inhibition on alkaloid production.

Two-Phase Airlift Fermentor Operation
Operations employing two-phase suspension cultures of higher plants in actual bioreactors are rare. We have done some limited work with 2 liter air-lift systems, however. The work reported here demonstrates the tehnique in straight batch processing systems, but the extension to fed-batch, semicontinuous or continuous sytems can also be envisioned.

In a simple, batch two-phase airlift fermentor, the suspension culture is grown in the reactor in the limited presence of the second phase. For example, the culture is run for the first half of the batch processing time without extraction phase and then exposed for the second half of the culture time to the continuous presence of the extraction phase (Byun, 1989). The total alkaloid concentration in the accumulation phase can be compared to the concentration in the cellular phase at the end of the run. The results of such an experiment are depicted in Fig. 6 by the curves connected with the open symbols. The alkaloid accumulation in the silicone fluid can also be easily observed by color changes. The color of the accumulation phase becomes darker, indicating the accumulation of alkaloids, whereas no significant color changes were observed in the distinct cellular phase. The net total alkaloid production of this system, compared to that without accumulation phase addition (data not shown), at the 15th day from inoculation was 1.8 times higher. It should be noted that this result was obtained without any optimization in the volume of accumulation phase or in the time of addition.

As is typically found in scale-up to airlift fermentor operation, the productivity of two-phase culture in the airlift fermentor was lower than that of a similar two-phase shake flask culture. This may be due to a decrease in oxygen transfer in the air lift fermentor or to stripping of volatiles and essential gases from the system (Kim *et al.*, 1990). Nevertheless, two-phase culture in airlift fermentors have some distinct advantages. The accumulation phase volume can be manipulated by layering the material in a quiescent region at the top of the fermentor without a corresponding change in the contact area between the two phases. On the other hand,

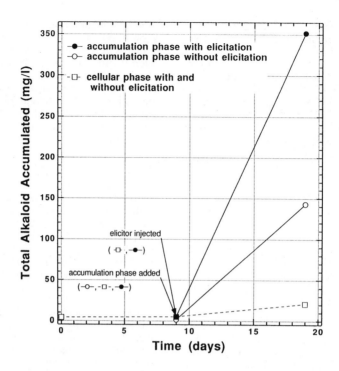

Fig. 6. Total alkaloid production in a two-phase, 2 liter air-lift fermentor with and without elicitation. Elicitor and/or accumulation phase were added on day 9 after inoculation.

the accumulation phase contact area can be adjusted by sparging the fluid into the fermentor without a corresponding change in the volume fraction. In shake flask cultures, it was found that volume fractions larger than 0.23 of accumulation phase cause a decrease in the cell growth (Byun *et al.*, 1990b).

The rapid and clean phase separation that is seen for the two phases is particularly applicable to *continuous* in situ recovery of alkaloids produced in an airlift fermentor. Finally, antifoam is not necessary in this system because the compounded silicone fluid is often used as an antifoam base and effectively eliminates foam generation. Optimization of the many operating conditions for this system can be expected to further increase productivity and may serve to remove obstacles to industrial application that are associated with low productivities.

Advanced Bioprocessing

As a final example, we briefly mention here a strategy that is derived from the obvious combination of the two previous types of bioprocessing steps – elicitation and extraction.

The further possibility of adjusting precursor levels in combination with these other steps is still in development.

Elicitation in a Two-phase Airlift Fermentor

The prior cell-directed bioprocessing results on alkaloid accumulation in shake flasks can also be extended to an air-lift fermentor and again illustrate that elicitation causes a significant increase in alkaloid production. In the two-phase airlift fermentor, elicitation can also further increase alkaloid production as shown in Fig. 6 by the curve connected with the solid symbols. The experiment was identical to the one shown in the figure for the direct use of the extraction phase, except that elicitor was also added to the media at the time of silicon polymer addition. Even without any optimization of the timing, elicitation was able to dramatically induce alkaloid formation in the two-phase airlift fermentor and encouraged further alkaloid production in *E. californica* cultures.

The combination of elicitation and two-phase culture has several advantages not acquired by the individual methods respectively. For example, it eliminates possible feed-back

196

inhibition by the relatively large amounts of secondary metabolites produced while it induces metabolic activities like transcriptional mRNA synthesis and the corresponding translation events. This can be done with independent control over the induction of the metabolic pathways and the regulation of those pathways by end products of the reactions. The processing steps are also relatively easy to implement and the transient responses that are seen are relatively rapid compared to usual plant cell culture time scales. Further work is in progress to optimize such a system.

CONCLUSION

This article has demonstrated some bioprocessing options that are suited to overcoming many of the limitations of secondary metabolite production in a particular plant cell culture system. The framework provided can reasonably be extended to many cell culture systems. It remains to be seen, however, whether some of the particular techniques employed are limited to cultures of california poppy. Nevertheless, we believe that these achievements are a strong encouragement for plant cell culture biotechnology and for successful commercial exploitation.

REFERENCES

Abeles, F.B.; Bosshart, R.; Forrence, L.E.; Habig, W.H. *Plant Physiol.* **1970**, *47*, 129-134.

Alfermann, A.W.; Spieler, H.; Reinhard, E. In *Primary and Secondary Metabolism of Plant Cell Cultures*; Neumann, K.H.; Barz, W.; Reinhard, E., Eds; Springer-Verlag: Berlin, 1985; pp. 316-322.

Anderson-Prouty, A.J.; Albersheim, P. *Plant Physiol.* **1975**, *56*, 286-291.

Ayers, A.R.; Ebel, E.; Valent, B.; Albersheim, P. *Plant Physiol.* **1976**, *57*, 760-765.

Ayers, A.R.; Jürgen, E.; Finelli, F; Berger, N.; Albersheim, P. *Plant Physiol.* **1976**, *57*, 751-759.

Baumann, T.W.; Frischknecht, P.M. In *Plant Tissue Culture*; Fujiwara, A., Ed.; Maruzen: Tokyo, 1982; pp. 365-366.

Beiderbeck, R. *Z. Pflanzenphysiol.* **1982**, *108*, 27-30.

Berlin, J.; Sieg, S.; Strack, D.; Bokern, M.; Hahms, H. *Plant Cell Tiss. Organ Cult.* **1986**, *5*, 163-174.

Berlin, J.; Forche, E.; Wray, V.; Hammer, J.; Hösel, W. *Z. Naturforsch.* **1983**, *38C*, 346-352.

Bisson, W.; Beiderbeck, R.; Reichling, J. *Planta Med.* **1983**, *47*, 164-168.

Butcher, D.N.; Conolly, J.D. *J. Exp. Bot.* **1971**, *22*, 314-322.

Byun, S.Y., *Ph.D. Thesis*, Rutgers, The State University of New Jersey, New Brunswick, NJ, 1989.

Byun, S.Y.; Pedersen, H.; Chin, C.K. *Ann. N.Y. Acad. Sci.* **1990a**, *589*, 54-66.

Byun, S.Y.; Pedersen, H.; Chin, C.K. *Phytochem.* **1990b**, *in press*.

Carew, D.P.; Krueger, R.J. *Lloydia* **1977**, *40*, 326-336.

Collinge, M.A.; Brodelius, P.E. *Phytochem.* **1989**, *28*, 1101-1104.

Cordell, G.A. In *Introduction to Alkaloids*; Geoffrey, A. Ed.; John Wiley & Sons: New York, NY, 1981; pp. 509-517.

Dicosmo, F.; Misawa, M. *Trends Biotechnol.* **1985**, *3*, 318-322.

Dixon, R.A.; Fuller, K.W. *Physiol. Plant Pathol.* **1978**, *12*, 279-288.

Drapeau, D.; Blanch, H.W.; Wilke, C.R. *Biotechnol. Bioeng.* **1986**, *28*, 1555-1563.

Drapeau, D.; Blanch, H.W.; Wilke, C.R. *Planta Med.* **1987**, *53*, 373-376.

Ebel, J.; Schmidt, W.E.; Loyal, R. *Arch. Biochem. Biophys.* **1984**, *232*, 240-248.

Eilert, U.; Constabel, F.; Kurz, W.G.W. *J. Plant Physiol.* **1986**, *126*, 11-12.

Eilert, U.; Kurtz, W.G.W.; Constabel, F. In *Plant Tissue and Cell Culture*; Green, C.E., Ed.; Alan R. Liss: New York, NY, 1987; pp. 213-219.

François, C.; Marshall, R.D.; Neuberger, A. *Biochem. J.* **1962**, *83*, 335-341.

Fujita, H.; Tabata, M.; Nishi, A.; Yamada, Y. In *Plant Tissue Culture*; Fujiwara, A., Ed.; Maruzen: Tokyo, 1982; pp. 399-400.

Funk, C.; Gügler, K.; Brodelius, P. *Phytochem.* **1987**, *26*, 401-405.

Furuya, T. In *Cell Culture and Somatic Cell Genetics of Plants*; Vasil, I.K., Ed.; Academic Press: San Diego, CA, 1988, Vol. 5; pp. 213-234.

Haberlach, G.T.; Budde, A.D.; Sequeira, L.; Helgeson, J.P. *Plant Physiol.* **1978**, *62*, 522-525.

Hahlbrock, K. In *Biochemistry of Plants*; Conn, E.E., Ed.; Academic Press: New York, NY, 1981, Vol. 7; pp. 425-455.

Hahn, M.G.; Albersheim, P. *Plant Physiol.* **1978**, *62*, 107-111.

Keen, N.T. *Science* **1975**, *187*, 187-188.

Kim, D.I.; Cho, G.H.; Pedersen, H.; Chin, C.K. *Biotechnol. Lett.* **1988**, *10*, 709-712.

Knoblach, K.H.; Berlin, J. *Z. Naturforsch.* **1980**, *35C*, 551-556.

Knoop, B.; Beiderbeck, R. *Z. Naturforsch.* **1983**, *38C*, 484-486.

Kombrink, E.; Hahlbrock, K. *Plant Cell Rep.* **1985**, *4*, 277-280.

Kuhn, D.N.; Chappell, J.; Boudet, A.;

Hahlbrock, K. *Proc. Natl. Acad. Sci. USA* **1984**, *81*, 1102-1106.

Mantell, S.H.; Smith, H. In *Plant Biotechnology*; Mantell, S.H.; Smith, H., Eds.; Cambridge Univ. Press: Cambridge, Great Britain, 1983; pp. 75-108.

Marques, I.A.; Brodelius, P.E. *Plant Physiol.* **1988a**, *88*, 46-51.

Marques, I.A.; Brodelius, P.E. *Plant Physiol.* **1988b**, *88*, 52-55.

Miller, R.D.; Collins, G.B.; Davis, D.L. *Crop Sci.*, **1983**, *23*, 561-565.

Murashige,T; Skoog, F. *Physiol. Plant.* **1962**, *15*, 473-497.

Oba, K. and I. Uritani, *Plant Cell Physiol.* **1979**, *20*, 819-826.

Payne, G.F.; Payne, N.N.; Shuler, M.L.; Asada, M. *Biotedchnol. Lett.* **1988**, *88*, 187-192.

Reinhard, E.; Alfermann, A.W. In *Advances in Biochemical Engineering;* A. Feichter, Ed.; Springer-Verlag: Berlin, 1980, Vol. 16; pp. 49-83.

Rink, E.; Böhm, H. *FEBS Lett.* **1975**, *49*, 396-399.

Sahai, O.P.; Shuler, M.L. *Biotechnol. Bioeng.* **1984**, *26*, 111-120.

Seibert, M.; Kadkade, P.G. In *Plant Tissue Culture as a Source of Biochemicals*; Staba, E.J., Ed.; CRC Press: Boca Raton, FL, 1980; pp. 123-141.

Sugano, N.; Iwata, R.; Nishi, A. *Phytochem.* **1975**, *14*, 1205-1207.

Tabata, M.; Mizukami,H.; Hiraoka, N.; Konoshima, M. *Phytochem.* **1974**, *13*, 927-932.

Tabata, M.; Fujita, Y. In *Biotechnology in Plant Science*; Zaitlin, M.; Day, P.; Hollaender, A., Eds; Academic Press: Orlando, FL, 1985; pp. 207-218.

Takao, N.; Kamigauchi, M.; Okada, M. *Helv. Chim. Acta* **1983**, *66*, 473-484.

Templeton, M.D.; Lamb, C.J., *Plant, Cell & Environ.* **1988**, *11*, 395-401.

Teuscher, E., *Pharmazie* **1973**, *28*, 6-18.

van den Berg, A.J.J.; Radema, M.H.; Rabadie, R.P., *Phytochem.* **1988**, *27*, 415-417.

Wijnsma, R.; Verpoorte, R. In *Cell Culture and Somatic Cell Genetics of Plants*; Vasil, I.K., Ed.; Academic Press: San Diego, CA, 1988, Vol. 5; pp. 335-355.

Yamada, Y.; Fujita,Y. In *Handbook of Plant Cell Culture*; Evans, D.E.; Sharp, W.R.; Ammirato, P.V.; Yamada, Y., Eds; Macmillan Press: New York, NY, 1983; pp. 502-512.

Zenk, M.H.; El-Shagi, H.; Arens, H.; Stöckigt, J.; Weiller, E.W.; Deus, B. In *Plant Tissue Culture and Its Biotechnological Applications*; Barz, W.H.; Reinhardt, E.; Zenk, M.H., Eds; Springer-Verlag: Berlin, 1977; pp. 27-43.

The Oligosaccharides of Glycoproteins: Factors Affecting their Synthesis and their Influence on Glycoprotein Properties

Charles F. Goochee#, Michael J. Gramer, Dana C. Andersen, Jennifer B. Bahr
Department of Chemical Engineering, Stanford University, Stanford, CA 94305-5025

and James R. Rasmussen#, Genzyme, 75 Kneeland St., Boston, MA 02111

The majority of proteins secreted by mammalian cells are glycoproteins. These proteins possess oligosaccharides covalently attached through an asparagine side chain ("asparagine-linked" or "N-linked") or through a threonine or serine side chain ("O-linked"). A given glycoprotein may contain only N-linked oligosaccharide chains, only O-linked oligosaccharide chains, or both.

The oligosaccharide structures of glycoproteins can have a profound effect on properties critical to the development of glycoprotein products for human therapeutic or diagnostic use, including clearance rate, antigenicity, immunogenicity, specific activity, solubility, resistance to thermal inactivation, and resistance to protease attack. In order to make rational decisions concerning the development of human therapeutic glycoproteins, it is important to understand how bioprocess factors influence oligosaccharide structure, and to understand the mechanisms that underlie the potential effects of oligosaccharide structure on glycoprotein properties.

The purpose of this review is to summarize the published data concerning these topics. Part 1 focuses on bioprocess factors that affect the outcome of glycoprotein oligosaccharide biosynthesis, including the effects of the host cell and cell culture environment. Part 2 is a short transitional section that summarizes data concerning the potential interaction of an oligosaccharide with the protein to which it is attached, a concept that is important to understanding the potential effect of oligosaccharides on glycoprotein properties. The effects of oligosaccharides on glycoprotein properties are summarized in Part 3, including discussion of molecular mechanisms thought to be responsible for these effects. Illustrative examples in this review are drawn primarily from the recent literature concerning generation of human therapeutic proteins using genetic engineering technology.

#corresponding authors

This review concludes with a summary discussion of the implications of the "glycosylation issue" for the synthesis of human therapeutic proteins.

PART 1. FACTORS AFFECTING THE SYNTHESIS OF GLYCOPROTEIN OLIGOSACCHARIDES

1.1 Synthesis of N-linked Oligosaccharides

The review article by Kornfeld and Kornfeld (1985) serves as an excellent, detailed introduction to the assembly of N-linked oligosaccharides. In summary, N-linked glycosylation begins with the synthesis of a lipid-linked oligosaccharide moiety ($Glc_3Man_9GlcNAc_2$-P-P-Dol), and its transfer en bloc to a nascent polypeptide chain in the endoplasmic reticulum (ER) (Fig. 1, reaction 1). Attachment occurs through asparagine, generally at the tripeptide recognition sequence Asn - X - Ser/Thr. A series of trimming reactions catalyzed by exoglycosidases in the ER (Fig. 1, reactions 2, 3 and 4) results in the "high mannose" oligosaccharide structure presented in Fig. 2A.

Further processing of N-linked oligosaccharides by mammalian cells continues in the compartments of the Golgi, where a sequence of exoglycosidase- and glycosyltransferase-catalyzed reactions generate high-mannose, hybrid and complex-type oligosaccharide structures (see Fig. 2, 3 and 4). One example of mammalian Golgi processing is represented in Fig. 1, reactions 5 through 11; the outcome of this pathway is the complex-type structure of Fig. 3A.

The glycosyltransferases responsible for N-linked oligosaccharide synthesis are very specific in the glycosidic bonds they construct; each utilizes a specific nucleotide sugar as a cosubstrate to generate a defined carbohydrate linkage (reviewed in Schachter et al., 1983;

2181–2/92/0199$11.50/0 © 1992 American Chemical Society

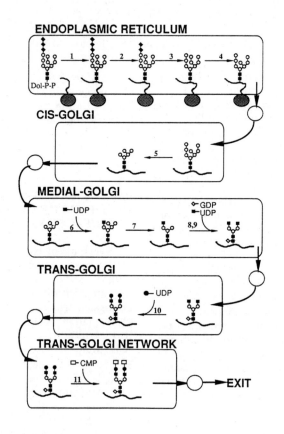

Figure 1. A Potential Pathway of Mammalian N-Linked Oligosaccharide Processing. The enzymes are:
    (1) oligosaccharyltransferase
    (2) α-glucosidase I
    (3) α-glucosidase II
    (4) ER α(1,2)mannosidase
    (5) Golgi α-mannosidase I
    (6) N-acetylglucosaminyltransferase I
    (7) Golgi α-mannosidase II
    (8) N-acetylglucosaminyltransferase II
    (9) α(1,6) fucosyltransferase
(10) β(1,4)galactosyltransferase
(11) α(2,3) sialyltransferase
The symbols are: ■, N-acetylglucosamine; O, mannose; ◆, glucose; ◊, fucose; ●, galactose; ▢, sialic acid. Dol-P-P is dolichyldiphosphate. The co-substrates for reactions 6,8,9,10 and 11 are the energized forms of the monosaccharides where: UDP is uridine diphosphate, GDP is guanosine diphosphate and CMP is cytidine monophosphate (Derived from figure 3 of Kornfeld and Kornfeld, 1985).

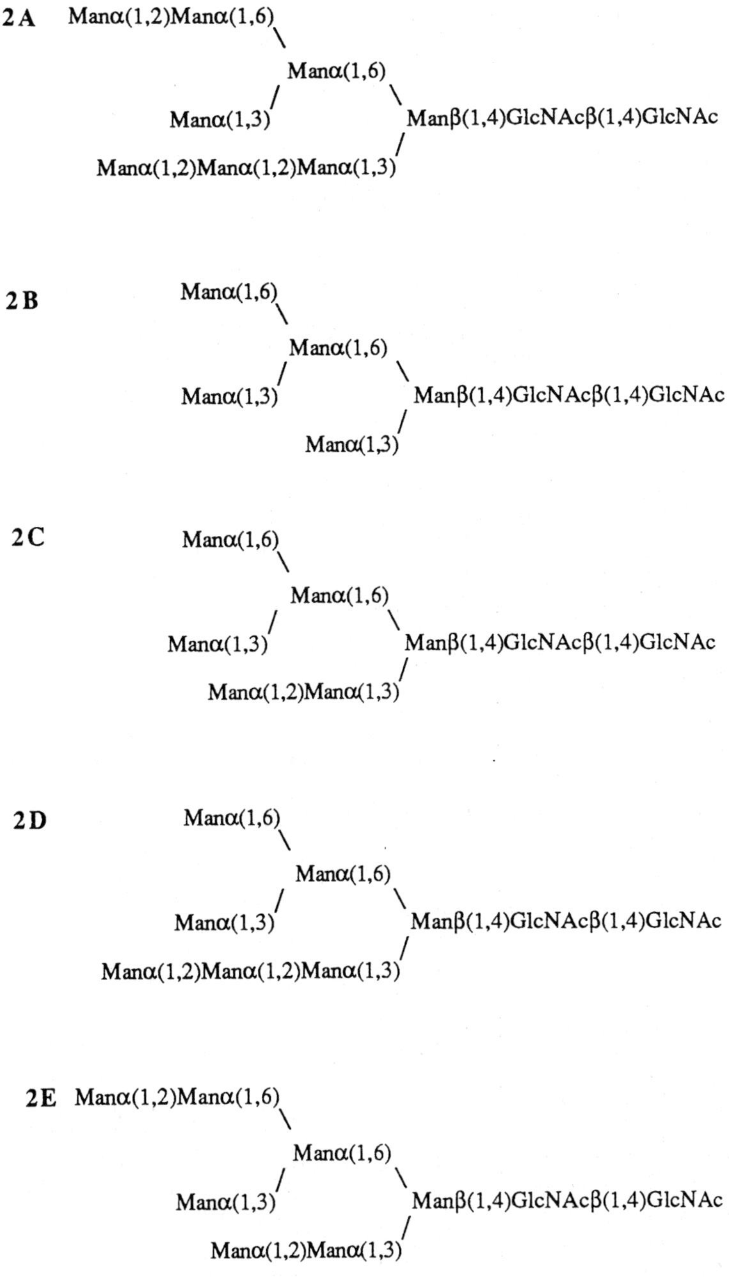

2 A   Manα(1,2)Manα(1,6)
                          \
                            Manα(1,6)
                          /   \
              Manα(1,3)   Manβ(1,4)GlcNAcβ(1,4)GlcNAc
        Manα(1,2)Manα(1,2)Manα(1,3) /

2 B   Manα(1,6)
                      \
                        Manα(1,6)
                      /   \
          Manα(1,3)   Manβ(1,4)GlcNAcβ(1,4)GlcNAc
                    Manα(1,3)   /

2 C   Manα(1,6)
                      \
                        Manα(1,6)
                      /   \
          Manα(1,3)   Manβ(1,4)GlcNAcβ(1,4)GlcNAc
              Manα(1,2)Manα(1,3) /

2 D   Manα(1,6)
                      \
                        Manα(1,6)
                      /   \
          Manα(1,3)   Manβ(1,4)GlcNAcβ(1,4)GlcNAc
        Manα(1,2)Manα(1,2)Manα(1,3) /

2 E   Manα(1,2)Manα(1,6)
                        \
                          Manα(1,6)
                        /   \
            Manα(1,3)   Manβ(1,4)GlcNAcβ(1,4)GlcNAc
              Manα(1,2)Manα(1,3)   /

Figure 2. Some Mammalian High Mannose-type N-linked Oligosaccharides.
(A) The $Man_8GlcNAc_2$ structure which is the potential end product of glycosidase activity in the ER.   (B-E) High mannose structures representing some potential products of Golgi processing. These structures were identified on t-PA from recombinant CHO cells (Spellman et al., 1989).

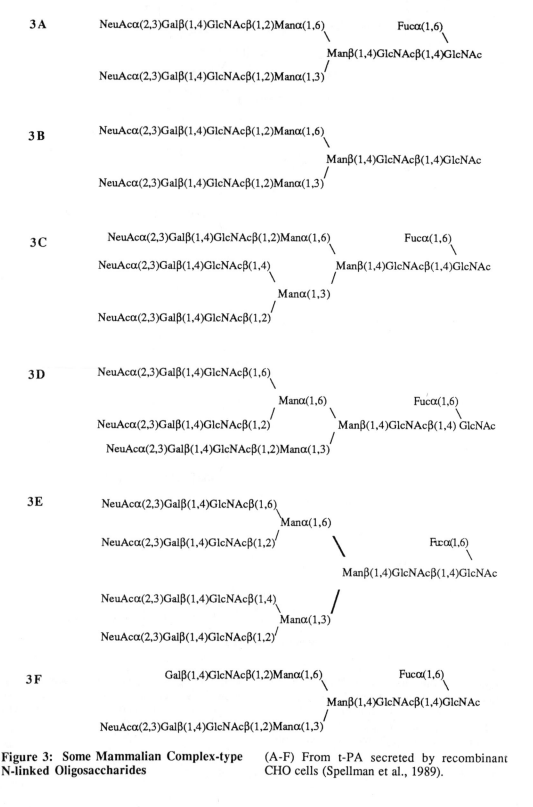

3A

NeuAcα(2,3)Galβ(1,4)GlcNAcβ(1,2)Manα(1,6)

Fucα(1,6)

Manβ(1,4)GlcNAcβ(1,4)GlcNAc

NeuAcα(2,3)Galβ(1,4)GlcNAcβ(1,2)Manα(1,3)

3B

NeuAcα(2,3)Galβ(1,4)GlcNAcβ(1,2)Manα(1,6)

Manβ(1,4)GlcNAcβ(1,4)GlcNAc

NeuAcα(2,3)Galβ(1,4)GlcNAcβ(1,2)Manα(1,3)

3C

NeuAcα(2,3)Galβ(1,4)GlcNAcβ(1,2)Manα(1,6)

Fucα(1,6)

NeuAcα(2,3)Galβ(1,4)GlcNAcβ(1,4)

Manβ(1,4)GlcNAcβ(1,4)GlcNAc

Manα(1,3)

NeuAcα(2,3)Galβ(1,4)GlcNAcβ(1,2)

3D

NeuAcα(2,3)Galβ(1,4)GlcNAcβ(1,6)

Manα(1,6)

Fucα(1,6)

NeuAcα(2,3)Galβ(1,4)GlcNAcβ(1,2)

Manβ(1,4)GlcNAcβ(1,4) GlcNAc

NeuAcα(2,3)Galβ(1,4)GlcNAcβ(1,2)Manα(1,3)

3E

NeuAcα(2,3)Galβ(1,4)GlcNAcβ(1,6)

Manα(1,6)

NeuAcα(2,3)Galβ(1,4)GlcNAcβ(1,2)

Fucα(1,6)

Manβ(1,4)GlcNAcβ(1,4)GlcNAc

NeuAcα(2,3)Galβ(1,4)GlcNAcβ(1,4)

Manα(1,3)

NeuAcα(2,3)Galβ(1,4)GlcNAcβ(1,2)

3F

Galβ(1,4)GlcNAcβ(1,2)Manα(1,6)

Fucα(1,6)

Manβ(1,4)GlcNAcβ(1,4)GlcNAc

NeuAcα(2,3)Galβ(1,4)GlcNAcβ(1,2)Manα(1,3)

Figure 3: Some Mammalian Complex-type N-linked Oligosaccharides (A-F) From t-PA secreted by recombinant CHO cells (Spellman et al., 1989).

3G

NeuAcα(2,6)Galβ(1,4)GlcNAcβ(1,2)Manα(1,6)

NeuAcα(2,6)Galβ(1,4)GlcNAcβ(1,4)

Fucα(1,6)

Manβ(1,4)GlcNAcβ(1,4)GlcNAc

Manα(1,3)

NeuAcα(2,3)Galβ(1,4)GlcNAcβ(1,2)

3H

Galβ(1,4)GlcNAcβ(1,2)Manα(1,6)

Manβ(1,4)GlcNAcβ(1,4)GlcNAc

Galβ(1,4)GlcNAcβ(1,3)Galβ(1,4)GlcNAcβ(1,2)Manα(1,3)

3I

Galα(1,3)Galβ(1,4)GlcNAcβ(1,2)Manα(1,6)

Manβ(1,4)GlcNAcβ(1,4)GlcNAc

Galα(1,3)Galβ(1,4)GlcNAcβ(1,2)Manα(1,3)

3J

Fucα(1,3)
|
Galβ(1,4)GlcNAcβ(1,2)Manα(1,6)

Fucα(1,6)

Manβ(1,4)GlcNAcβ(1,4)GlcNAc

Galβ(1,4)GlcNAcβ(1,2)Manα(1,3)
|
Fucα(1,3)

Figure 3: Some Mammalian Complex-type N-linked Oligosaccharides—Continued

(G) From human transferrin isolated from serum (Spik et al., 1985).
(H) From interferon-β1 isolated from human diploid fibroblasts (Kagawa et al., 1988).
(I) From interferon-β1 secreted by recombinant C127 cells (Kagawa et al., 1988).
(J) From human α-amylase isolated from human saliva (Tollefsen and Rosenblum, 1988).

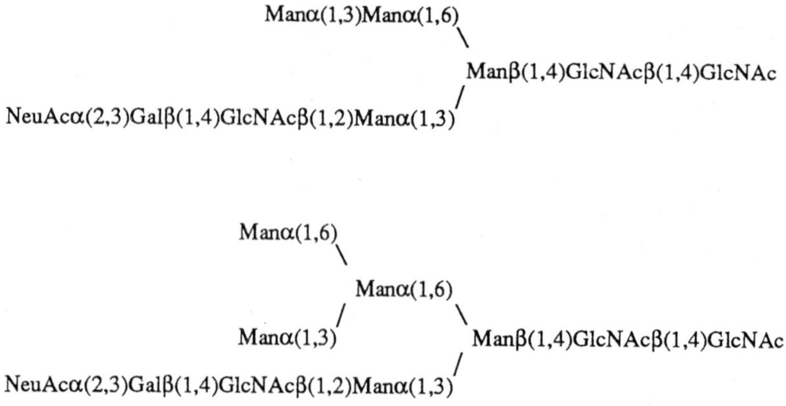

Manα(1,3)Manα(1,6)
\
Manβ(1,4)GlcNAcβ(1,4)GlcNAc
/
NeuAcα(2,3)Galβ(1,4)GlcNAcβ(1,2)Manα(1,3)

Manα(1,6)
\
Manα(1,6)
/ \
Manα(1,3) Manβ(1,4)GlcNAcβ(1,4)GlcNAc
/
NeuAcα(2,3)Galβ(1,4)GlcNAcβ(1,2)Manα(1,3)

Figure 4. Some Mammalian Hybrid-type N-linked Oligosaccharides

From t-PA secreted by recombinant CHO cells (Spellman et al., 1989)

Schachter, 1986). For example, N-acetylglucosaminyltransferase I (typically abbreviated GlcNAc Transferase I or GnT I) is the enzyme responsible for reaction 6 in Fig. 1. This enzyme utilizes a nucleotide sugar (UDP-GlcNAc) to add N-acetylglucosamine in a β(1,2) linkage to mannose. The enzyme is specific for the mannose bonded α(1,3) to the innermost mannose of the N-linked structure (Schachter et al., 1983).

It is estimated that 100 or more glycosyltransferases would be necessary to explain the known N- and O-linked oligosaccharide structures of mammalian glycoproteins and glycolipids (Paulson and Colley, 1989), although the synthesis of a specific class of oligosaccharide structures by a given cell type may involve only a small subset of these glycosyltransferases. For example, the N-linked oligosaccharides identified from erythropoietin (EPO) (Sasaki et al., 1988; Takeuchi et al., 1988), tissue-type plasminogen activator (t-PA) (Parekh et al., 1989b; Spellman et al., 1989) and interferon-β1 (Kagawa et al., 1988) synthesized by recombinant Chinese hamster ovary (CHO) cells would require the activity of just eight glycosyltransferases in the Golgi.

Yeast, insect, plant and mammalian cells share the features of N-linked oligosaccharide processing in the ER, including attachment of Glc$_3$Man$_9$GlcNAc$_2$-P-P-Dol (Fig. 1, reaction 1) and subsequent truncation to a Man$_8$GlcNAc$_2$ structure (product of reaction 4 in Fig. 1). However, oligosaccharide processing by these different cell types diverges in the Golgi apparatus. For example, plant cells produce complex-type oligosaccharides containing GlcNAc and Gal in linkages similar to those found in mammalian N-linked oligosaccharides, but plant-derived oligosaccharides do not possess sialic acid, a common constituent of mammalian complex-type oligosaccharides. Furthermore, plant-derived N-linked oligosaccharides frequently contain xylose, a monosaccharide not normally found in mammalian N-linked oligosaccharides (see Fig. 5) (reviewed in Kaushal et al., 1988).

N-linked oligosaccharide processing in the Golgi by yeast is very different from mammalian Golgi processing (reviewed in Kornfeld and Kornfeld, 1985; Kukuruzinska et al., 1987; Tanner and Lehle, 1987). In most strains of yeast, including *Saccharomyces cerevisiae*, the oligosaccharide chains are elongated in the Golgi through stepwise addition of mannose, leading to elaborate high mannose (mannan) structures sometimes containing more than 100 mannose monomers (see Fig. 6A). The elaborate nature of these yeast N-linked oligosaccharide structures is often reflected in the molecular weights of mammalian proteins produced in recombinant yeast. For example, the molecular weight of aglycosyl murine interferon-β (MuIFN-β) protein produced in recombinant *E. coli* is 19.5

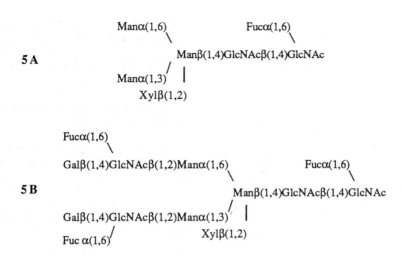

Mana(1,6)
\
Manβ(1,4)GlcNAcβ(1,4)GlcNAc
/
Mana(1,3)
Xylβ(1,2)

5 A

Fucα(1,6)
\

Fucα(1,6)
\
Galβ(1,4)GlcNAcβ(1,2)Mana(1,6)
\
Manβ(1,4)GlcNAcβ(1,4)GlcNAc
/
Galβ(1,4)GlcNAcβ(1,2)Mana(1,3)
Xylβ(1,2)
/
Fuc α(1,6)

5 B

Figure 5. N-linked Oligosaccharides from (Kaushal et al., 1988).
the Plant Enzyme Laccase

...
\
^Mana(1,6)
\
Mana(1,3)ManPO₄(6)   ^Mana(1,6)
\   \
Mana(1,3)Mana(1,2)Mana(1,2)Mana(1,6)
\ ⇐
Mana(1,2)Mana(1,6)
\
Mana(1,3)
/ \
Mana(1,3)Mana(1,2)Mana(1,2)   Manβ(1,4)GlcNAcβ(1,4)GlcNAc
/
Mana(1,3)Mana(1,2)Man*α(1,6)Mana(1,6)
/
Mana(1,3)

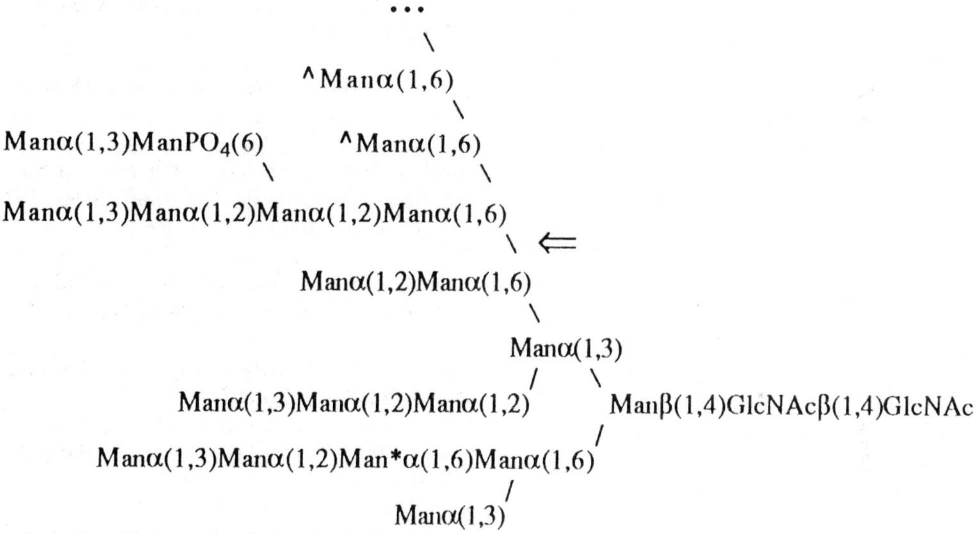

Figure 6. N-linked oligosaccharides from
Saccharomyces cerevisiae
(A) General Structure of N-linked
Oligosaccharides from Wild-type *S. cerevisiae*.
The structure below the arrow (⇐)
represents a core structure including 12
mannose units and a potential phosphorylation
site (*). Above the arrow (⇐) is the outer
chain which can be extended to over 10
repeats of Mana(1,6) as denoted by the
repeated dots (···). The first Mana(1,6) of the
outer chain is shown with an attached
Man₅PO₄ group. Each Mana(1,6) of the
outer chain may have such a group or a
truncated derivative, as indicated (^),

potentially raising the total number of
mannose to more than 50. An N-linked
glycosylation site of a glycoprotein produced
by wild-type *S. cerevisiae* may possess both
the core and outer chain, as shown above, or
just the core structure. Microheterogeneity of
oligosaccharide structure at a particular
glycosylation site is observed both in the core
and in the outer chain due to variability in both
the number of mannose linkages and the
degree of phosphorylation.
(B) The *S. cerevisiae mnn9* mutant is unable
to synthesize the outer chain. Therefore,
glycoproteins produced by that mutant
possess only the core structure below the
arrow (⇐) (Ballou, 1990).

205

kDa, while the apparent molecular weight of glycosylated MuIFN-β is 35 kDa when produced by murine T lymphocytes and is more than 100 kDa when produced in recombinant *S. cerevisiae* (Sedmak and Grossberg, 1989). As a second example, the molecular weight of aglycosyl HIV gp120 is 60 kDa, while the molecular weight of gp120 is 120 kDa when produced in mammalian cells (Leonard et al., 1990) and is up to 600 kDa when produced in *S. cerevisiae* (Hitzeman et al., 1990). Mutant *S. cerevisiae* strains permit synthesis of recombinant proteins with truncated N-linked, high mannose oligosaccharides (reviewed in Kukuruzinska et al., 1987; Tanner and Lehle, 1987; Ballou, 1990). For example, the *mnn9* mutant synthesizes a truncated $Man_{10}GlcNAc_2$ structure only slightly larger than a mammalian high-mannose structure (see Fig. 6B). Thus, the molecular weight of gp120 produced in the *mnn9* mutant is reduced to 120 kDa (Hitzeman et al., 1990).

The capability of insect cells for N-linked oligosaccharide processing in the Golgi is poorly understood. *Spodoptera frugiperda* Sf9 cells possess the glycosidases necessary to trim the $Glc_3Man_9GlcNAc_2$ precursor to $Man_3GlcNAc_2$ (see Fig. 7) (Kuroda et al., 1990). However, the capability of Sf9 for further processing to complex-type oligosaccharides is uncertain. Recombinant glycoproteins produced using Sf9 cells frequently have lower molecular weights than the corresponding native mammalian glycoproteins, suggesting limitations in Sf9 glycosyltransferase activities relative to mammalian cells. For example, the protein component of EPO has a molecular weight of 18.4 kDa (Lin et al., 1985; Jacobs et al., 1985); EPO isolated from human urine or produced by recombinant CHO cells has an apparent molecular weight near 35 kDa (Imai et al., 1990), while EPO produced using the Sf9/baculovirus expression system has a

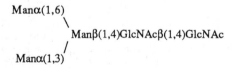

Figure 7. An N-linked Oligosaccharide from Insect Cells.
From influenza virus hemagglutinin expressed in *Spodoptera frugiperda* (Sf-9) cells (Kuroda et al., 1990).

molecular weight of 26.2 kDa (Quelle et al., 1989). Other examples illustrating lower molecular weights of Sf9-produced proteins include t-PA (Steiner et al., 1988), influenza virus hemagglutinin (Kuroda et al., 1990), human acid β-galactosidase (Itoh et al., 1990), and human interferon-β (Smith et al., 1983; Utsumi et al., 1989). Recent publications suggesting synthesis of N-linked complex-type oligosaccharide structures by Sf9 cells await further confirmation (Kuroda et al., 1990; Davidson et al., 1990).

The tripeptide sequence Asn - X - Ser/Thr is not a sufficient condition for covalent attachment of an N-linked oligosaccharide. Gavel and Heijne (1990) surveyed the locations of N-linked oligosaccharides in 147 glycoproteins and found that 10% of 465 of the Asn - X - Ser/Thr sites were not glycosylated, while Mononen and Karjalainen (1984) found that 30% of such sites in 105 proteins were not glycosylated. Several factors are believed to account for the 10 to 30% of unoccupied N-glycosylation sites (reviewed in Kornfeld and Kornfeld, 1985). Proline located at position X in the Asn - X - Ser/Thr - Y sequence almost always prohibits N-glycosylation; proline at position Y also lessens the probability of N-glycosylation (Bause, 1983; Roitsch and Lehle, 1989; Gavel and von Heijne, 1990). In addition, unoccupied sites are more likely to be found near the C-terminus of glycoproteins (Gavel and von Heijne, 1990). However, many other unoccupied N-glycosylation sites do not fit into these two catagories. The prevailing hypothesis is that unoccupied N-glycosylation sites often reflect a temporal competition between protein folding and the initiation of N-glycosylation. If the protein folds too rapidly upon entering the ER, an unglycosylated Asn - X - Ser/Thr sequon may become inaccessible as substrate for oligosaccharyltransferase (Fig. 1, reaction 1), either as a result of inappropriate protein tertiary structure or as the result of rapid sequon removal from the vicinity of oligosaccharyltransferase which is attached to the ER membrane (Sox and Hood, 1970; Gavel and von Heijne, 1990). The hypothesis that protein folding can inhibit initiation of N-glycosylation is supported by the results of Pless and Lennarz (1977), who demonstrated that enzyme-catalyzed transfer of lipid-linked oligosaccharide to α-lactalbumin in a cell-free system was not possible until the native α-lactalbumin protein conformation was disrupted (see also Kronquist and Lennarz, 1978).

Cell-type specific differences in N-glycosylation site occupancy between mammalian cells and yeast are suggested by recent studies involving heterologous expression of human interleukin-1β (hIL-1β). The single N-glycosylation site of hIL-1β at Asn-123 is unoccupied when this molecule is synthesized by human macrophages (March et al., 1985), but is N-glycosylated when synthesized by recombinant *S. cerevisiae* (Casagli et al., 1989; Livi et al., 1990).

Initiation of N-glycosylation can occur irregularly at a particular Asn - X - Ser/Thr site, a phenomenon termed "variable N-glycosylation site occupancy". For example, tissue-type plasminogen activator is secreted concurrently by a variety of mammalian cells in two distinct forms, type I and type II, which have identical protein structure but differ by the presence or absence of N-linked oligosaccharide at Asn-184 (Pohl et al., 1984; Parekh et al., 1989a; Spellman et al., 1989; Pfeiffer et al., 1989). Similarly, human granulocyte-macrophage colony-stimulating factor (hGM-CSF) is secreted by recombinant CHO cells (Donahue et al., 1986; Moonen et al., 1987) and human lymphocytes (Cebon et al., 1990) in several forms with apparent molecular weights ranging from 16 to 30 kDa due to variable site occupancy at its two N-glycosylation sites. Savvidou and coworkers (1984) demonstrated variable N-glycosylation site occupancy on a human myeloma light chain, and hypothesized that variable site occupancy is due to the same mechanism that has been invoked to explain the complete absence of oligosaccharide at some other Asn - X - Ser/Thr sites -- that is, the kinetic competition between N-linked glycosylation and folding as the nascent peptide is translocated into the ER.

In a few exceptional cases, N-linked glycosylation has been identified at sites other than Asn - X - Ser/Thr (reviewed in Gavel and von Heijne, 1990). For example, protein C is N-glycosylated at the sequence Asn - X - Cys (Miletich and Broze, 1990).

1.2 Synthesis of O-linked Oligosaccharides

The initial step in O-glycosylation by mammalian cells is the covalent attachment of N-acetylgalactosamine to serine or threonine via an α1 linkage. This reaction is catalyzed by the enzyme UDP-GalNAc:polypeptide N-acetylgalactosaminyltransferase (reaction 1 in Fig. 8) (reviewed in Sadler, 1984). No O-glycosylation sequon has been identified analogous to the Asn - X - Ser/Thr template required for N-glycosylation. In further contrast to N-glycosylation, no preformed, lipid-coupled oligosaccharide precursor is involved in the initiation of mammalian O-glycosylation. Nucleotide sugars (e.g. UDP-GalNAc and UDP-Gal) serve as the substrates for the first and all subsequent steps in O-linked processing.

The intracellular location for initiation of O-glycosylation in mammalian cells (Fig. 8, reaction 1) has been a source of controversy. Some reports suggest that initiation of O-glycosylation occurs in the ER or pre-Golgi compartments (e.g., Pathak et al., 1988; Tooze et al., 1988), while other reports indicate initiation in the Golgi (e.g. Deschuyteneer et al., 1988; Piller et al., 1990). Most of the evidence suggests that mammalian O-glycosylation follows the initiation of N-glycosylation, occurring either just before or just after the protein leaves the ER (Tooze et al., 1988). In an attempt to resolve these data, Carraway and Hull (1989) have hypothesized that the intracellular location for initiation of O-glycosylation may be dependent upon cell type and target protein.

Following the covalent attachment of GalNAc to serine or threonine, several different processing pathways are possible for mammalian O-linked oligosaccharides in the Golgi (reviewed by Sadler, 1984; Carraway and Hull, 1989). The most common pathway of O-glycosylation for cell-surface glycoproteins and plasma glycoproteins is outlined in Fig. 8. O-linked oligosaccharide structures from this pathway are evident in IL-2 from human lymphocytes and recombinant CHO, BHK, and Ltk⁻ cells (Conradt et al., 1985, 1989), EPO from recombinant CHO cells (Tsuda et al., 1990), immunoglobulin A from pooled human serum (Field et al., 1989), granulocyte colony-stimulating factor (G-CSF) from recombinant CHO cells (Oheda et al., 1988) and plasminogen from human, bovine and porcine serum (Marti et al., 1988).

Each enzyme involved in the pathway presented in Fig. 8 has a substrate specificity unique for O-linked structures. For example, the α(2,3)sialyltransferase and the α(2,6)sialyltransferase responsible for reactions 3 and 4 in Fig. 8 are distinct from the α(2,3)sialyltransferase and the α(2,6)sialyltransferase which act upon mammalian N-linked oligosaccharides (Gross et al., 1989).

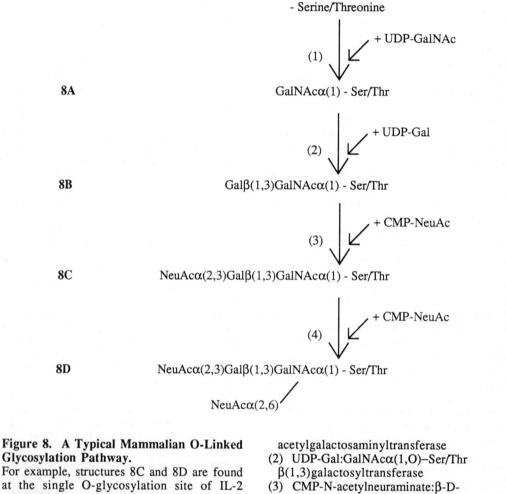

Figure 8. A Typical Mammalian O-Linked Glycosylation Pathway.
For example, structures 8C and 8D are found at the single O-glycosylation site of IL-2 secreted by recombinant CHO cells (Conradt et al., 1986).
Enzymes are:
(1) UDP-GalNAc:polypeptide N-acetylgalactosaminyltransferase
(2) UDP-Gal:GalNAcα(1,O)–Ser/Thr β(1,3)galactosyltransferase
(3) CMP-N-acetylneuraminate:β-D-galactoside α(2,3)sialyltransferase
(4) CMP-N-acetylneuraminate:α-D-N-acetyl-galactosaminide α(2,6)sialyltransferase

The biosynthesis of O-linked oligosaccharide structures in yeast is significantly different from mammalian O-glycosylation (reviewed in Tanner and Lehle, 1987; Kukuruzinska et al.., 1987). Yeast O-glycosylation begins in the ER with the covalent attachment of mannose to the recipient serine or threonine residue via a lipid intermediate (Dol-P-Man), an initiation step completely distinct from mammalian O-glycosylation. Up to four additional mannose residues are attached in the Golgi to yield a structure which has no analogue in mammalian systems [Manα(1,3)Manα(1,3)Manα(1,2)-Manα(1,2)Man-Ser/Thr] (Hard et al., 1989).

Yeast and mammalian cells do not have identical protein recognition sequences/structures for initiation of O-linked glycosylation. Recombinant hIGF-I expressed in *S. cerevisiae* is O-glycosylated at Thr-29 (Gellerfors et al., 1989; Elliott et al., 1990). In contrast, normal human serum hIGF-I is not O-glycosylated (Perdue, 1984; Clemmons, 1989; Gellerfors et al., 1989; Hard et al., 1989). The capability of insect cells for O-glycosylation is poorly understood. A recent report suggests that Sf9 insect cells are capable of performing the first two O-glycosylation steps outlined in Fig. 8 to produce pseudorabies virus gp50 containing GalNAc and Galβ(1,3)GalNAc, but unlike mammalian cells, they are unable to sialylate this structure (Thomsen et al., 1990). A difference between mammalian and insect cells in the initiation of O-linked glycosylation has been observed in one case; recombinant Sf9 cells secrete only nonglycosylated IL-2 (Smith et al., 1985), while cultured mammalian cells secrete a mixture of nonglycosylated and O-glycosylated IL-2 (Conradt et al., 1989).

Mammalian O-linked glycosylation occurs reproducibly at some potential Ser/Thr sites, while other potential Ser/Thr sites are never occupied or experience variable site occupancy (reviewed in Sadler, 1984; Carraway and Hull, 1989). For example, Ser-126 of EPO is O-glycosylated, while 16 other EPO Ser/Thr sites are never occupied (Lin et al., 1985; Jacobs et al., 1985; Sasaki et al., 1988; Tsuda et al., 1990). None of the more than 60 Ser/Thr sites of t-PA are O-glycosylated (Pennica et al., 1983; Pohl et al., 1984; Parekh et al., 1989a, Parekh et al., 1989b; Spellman et al., 1989; Pfeiffer et al., 1989). Variable site occupancy is evident at Thr-3 of IL-2 produced by human lymphocytes and recombinant CHO, mouse L and Ltk⁻ cells, while no other IL-2 Ser/Thr sites

are O-glycosylated (Conradt et al.,1985, 1986, 1989). These examples are representative of plasma glycoproteins, where only a small percentage of Ser/Thr sites are O-glycosylated. In contrast, the extracellular domains of some mammalian cell surface glycoproteins possess large numbers of O-linked oligosaccharides. For example, more than 80 O-linked oligosaccharides are found in the extracellular domain of leukosialin, representing glycosylation at approximately 85% of the possible Ser/Thr sites (Carlsson and Fukuda, 1986). Other examples of cell surface glycoproteins with large numbers of O-linked oligosaccharides include human erythrocyte glycophorins A, B and C (Blanchard et al., 1987) and the low density lipoprotein receptor (Cummings et al., 1983; Kozarsky et al., 1988).

Currently, it is not possible to accurately predict which Ser/Thr sites will be O-glycosylated. Experiments involving O-glycosylation of synthetic peptide substrates in a cell-free system indicate that the enzyme responsible for initiation of mammalian O-glycosylation [UDP-GalNAc:polypeptide N-acetylgalactosaminyltransferase] exhibits a high degree of amino acid sequence specificity, but examination of amino acid sequences of peptides and proteins in the vicinity of O-glycosylation attachment sites has not yet revealed any specific consensus sequence(s) (reviewed in Sadler, 1984). The presence of proline adjacent to serine or threonine appears to promote O-glycosylation in yeast (Lehle and Bause, 1984; Gellerfors et al., 1989) and mammalian cells (Young et al., 1979; Fiat et al., 1980). For example, the O-glycosylation sites of human interleukin-2 (hIL-2) produced by human lymphocytes (Robb et al., 1984) and human GM-CSF produced by recombinant CHO cells (Clark, 1988) are both located immediately adjacent to proline. In contrast, the lone O-glycosylation site of EPO produced by recombinant CHO cells is not located adjacent to a proline (Lin et al., 1985). Initiation of O-glycosylation probably occurs after the protein has assumed some measure of folding in the ER, suggesting that the peptide site for O-glycosylation must be available on the protein surface in an appropriate three-dimensional configuration (Sadler, 1984). Proline residues introduce β-turns in proteins, perhaps opening the protein structure to facilitate O-glycosylation. However, the presence of an adjacent proline is neither a necessary nor a sufficient condition for O-glycosylation. Secondary structure predictions based on amino acid sequences near O-glycosylated Ser/Thr fail to reveal any unique consensus secondary structural features (Sadler, 1984).

A new mammalian O-linked biosynthetic pathway involving glucose linked through serine (and containing xylose) has been reported recently in studies of human blood clotting factors VII and IX and protein Z (Nishimura et al., 1989). Factor IX contains only one form of this unusual O-linked structure (Xyl – Glc – Ser), while Factor VII and protein Z contain two forms (Xyl$_2$ – Glc – Ser and Xyl – Glc – Ser) (Nishimura et al., 1989).

1.3 Site Microheterogeneity of N-linked and O-linked Oligosaccharides

N-linked or O-linked processing usually leads to a distribution of oligosaccharide structures at a particular glycosylation site (termed site microheterogeneity). One source of microheterogeneity in N-linked oligosaccharide processing derives from branch points where two or more glycosyltransferases compete for a particular oligosaccharide structure as substrate (reviewed by Schachter et al., 1983; Schachter, 1986). For example, consider the competition of GlcNAc transferase III (GnT III) and GlcNAc transferase IV (GnT IV) for the oligosaccharide substrate indicated in Fig. 9A. If GnT III acts first, the resulting oligosaccharide structure (Fig. 9B) is no longer an acceptable substrate for GnT IV, and the oligosaccharide is committed to an oligosaccharide processing pathway leading to a bisected, biantennary complex-type structure. However, if GnT IV acts before GnT III , then the resulting structure (Fig. 9C) is committed to a processing pathway potentially leading to a bisected, triantennary complex-type structure. Microheterogeneity due to competing glycosyltransferases is also common in the latter stages of N-linked oligosaccharide processing -- for example, competing sialyltransferases with alternative linkage specificities (Fig. 3G).

Microheterogeneity is frequently observed due to variation in the presence or absence of terminal fucose or sialic acid (compare Fig. 3A with Fig. 3B and Fig. 3F). If variabilities in sialylation and fucosylation are discounted, the extent of site microheterogeneity due to other sources is often quite small. For example, the numerous oligosaccharide structures found at the lone N-linked glycosylation site of human urinary interferon-β1 from cultured human diploid fibroblasts include the fucosylation and

sialylation variants of just the four structures represented in Figures 3B (74%), 3C (8%), 3D (10%) and 3H (8%) (Kagawa et al., 1988).

Site microheterogeneity is also observed for O-linked oligosaccharides. For example, three oligosaccharide structures alternatively occur at the single O-glycosylation site of EPO produced in recombinant CHO cells. The major O-linked oligosaccharide of EPO from recombinant CHO cells is NeuAcα(2,3)-Galβ(1,3)[NeuAcα(2,6)]GalNAc (Fig. 8D), but the monosialylated and asialo versions (Fig. 8B and 8C) are also present (Sasaki et al., 1987, 1988). The mono- and di-sialylated versions of this same structure are found at the single O-glycosylation site in IL-2 produced by human lymphocytes and recombinant CHO, BHK and Ltk⁻ cells (Conradt, 1985, 1989), and at the single O-glycosylation site of G-CSF produced by recombinant CHO cells (Oheda et al., 1988; Kubota et al., 1990).

1.4 N-linked and O-linked Oligosaccharide Synthesis are Cell-type Dependent

Significant differences are observed between the N- and O-linked oligosaccharide structures from yeast, plant, and mammalian cells (as discussed in sections 1.1 and 1.2). Among mammalian cells, oligosaccharide processing is species dependent and cell-type dependent within a given species (reviewed in Kobata, 1984; Rademacher et al., 1988; Paulson, 1989; see also Parekh et al., 1987). The influence of cell type on glycosylation appears to be related primarily to the presence, concentration, kinetic characteristics, and compartmentalization of the individual glycosyltransferases and glycosidases (reviewed in Rademacher et al., 1988; Paulson and Colley, 1989; see also Paulson et al., 1989). This section will focus on the mechanisms underlying cell-type dependent glycosylation by cultured mammalian cells.

Cell-type dependent differences in oligosaccharide structure among mammalian cells are frequently attributable to differences in the presence of specific glycosyltransferase activities. For example, differences in branching structure are observed in the complex-type, N-linked oligosaccharides of t-PA produced by a human colon fibroblast (hcf) cell strain compared to those of t-PA produced by a Bowes melanoma cell line. These differences suggest that Bowes melanoma cells express GnT III and VII, but not GnT IV, while hcf cells express GnT IV, but not GnT III and VII (Parekh et al., 1989a).

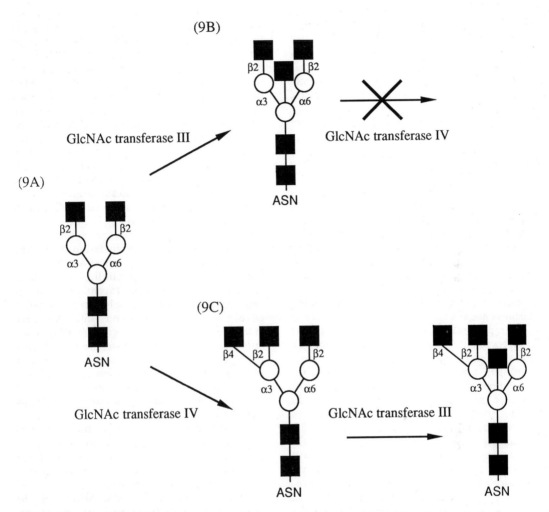

Figure 9. Microheterogeneity Can Result from Competing Glycosyltransferases. If GlcNAc transferase III acts first, then the resulting structure (9B) is not a substrate for GlcNAc transferase IV. The symbols are: ■, N-acetylglucosamine; O, mannose. (reviewed in Schacter, 1986)

211

The cell-type dependent presence or absence of α(1,3)galactosyltransferase provides another basis for differences in oligosaccharide processing among mammalian cells. This enzyme constructs the linkage galactose bonded α(1,3) to galactose [i.e., Galα(1,3)Gal] (see Fig. 3I). The Galα(1,3)Gal structure is found on the cell surfaces and some secreted proteins of New World monkeys and non-primate mammals (e.g. rodents, pigs, sheep, cows), but is not found on most cells and proteins of humans, Old World primates, or anthropoid apes (Galili et al., 1988; Thall and Galili 1990). For example, Galili and coworkers have detected the presence of the Galα(1,3)Gal epitope on the surface of SP/2 myeloma cells and on secreted mouse monoclonal antibodies (Galili et al., 1988; Thall and Galili 1990). This structure is also present on the N-linked oligosaccharides of recombinant interferon-β1 (Kagawa et al., 1988) and recombinant t-PA (Pfeiffer et al., 1989) synthesized by the mouse C127 cell line. In contrast, the Galα(1,3)Gal epitope is not found on the cell surface of many cultured human cells, including normal human fibroblasts and the HL60 and HeLa cell lines (Galili et al., 1988).

The presence of the Galα(1,3)Gal epitope in glycoproteins targeted for human therapeutic use is significant because one-half to one percent of circulating human IgG recognizes this epitope -- that is, approximately 100 μg of IgG per ml of human serum. (reviewed in Galili, 1988). The oligosaccharides of human interferon-β1 (Kagawa et al., 1988), EPO (Sasaki et al., 1987, 1988; Takeuchi et al., 1988) and t-PA (Spellman et al., 1989) produced in recombinant CHO cells do not possess the Galα(1,3)Gal structure. Likewise, the oligosaccharide structures of EPO produced by recombinant BHK cells possess no Galα(1,3)Gal epitopes (Tsuda et al., 1988). These data suggest that α(1,3)galactosyltransferase activity in these two cell types is diminished or absent. CHO cells do harbor DNA that hybridizes with a probe for a murine α(1,3)galactosyltransferase, but northern blot analysis reveals no evidence of α(1,3)galactosyltransferase message in CHO cells (Smith, D.F. et al., 1990).

Cell-type specific differences in sialylation and fucosylation have also been observed among mammalian cells. Two distinct sialic acid linkages, NeuAcα(2,3)Gal and NeuAcα(2,6)Gal, are found in N-linked glycoproteins isolated from human urine or plasma and from cultured human cell lines (see Fig. 3G). For example, EPO isolated from human urine contains about 60% of its sialylated structures with the NeuAcα(2,3)Gal linkage and 40% with the NeuAcα(2,6)Gal linkage (Takeuchi et al., 1988). Transferrin isolated from human serum and secreted by the human hepatoma cell line HepG2 also contains both sialic acid linkages (see Fig. 3G) (Spik et al., 1985; Campion et al., 1989). In contrast, recombinant CHO cells generate N-linked oligosaccharides containing only the NeuAc(α2,3)Gal linkage, while recombinant C127 cells generate N-linked oligosaccharides containing only the NeuAcα(2,6)Gal linkage, suggesting that only one of the two sialyltransferases is active in each cell type (Takeuchi et al., 1988). Human cells can possess an α(1,3)fucosyltransferase activity leading to generation of a Fucα(1,3)GlcNAc linkage (see Fig. 3J) (Tollefsen and Rosenblum, 1988). CHO cells apparently possess the α(1,3)fucosyltransferase gene, but it is not normally expressed (Campbell and Stanley, 1983). The feasibility of altering oligosaccharide processing capabilities through expression of exogenous glycosyltransferase genes has recently been demonstrated independently by several groups, who have introduced α(2,6)sialyltransferase (Lee et al., 1989), α(1,3)fucosyltransferase (Potvin et al., 1990) and α(1,3)galactosyltransferase (Smith, D.F. et al., 1990) into CHO cells.

While the examples noted above have focussed on oligosaccharide processing differences due to the presence or absence of specific glycosyltransferases, differences in oligosaccharide processing can also result from cell-specific differences in intracellular glycosyltransferase concentrations. For example, mammalian cells isolated from carcinomas, or transformed by oncogenic viruses or cellular oncogenes, frequently express increased N-linked oligosaccharide branching related to increased GlcNAc transferase V activity (Kobata, 1988; Yamashita et al., 1989; Palcic et al., 1990).

N-glycosylation site occupancy can also vary among mammalian cells. For example, cell-type dependent differences in N-glycosylation site occupancy are evident in studies of human interleukin-6 (hIL-6) synthesized by human monocytes, human fibroblasts, human endothelial cells and recombinant murine NIH/3T3 cells (Schiel et al., 1990), and in studies of hGM-CSF synthesized by recombinant CHO cells and recombinant COS-1 cells (Donahue et al., 1986).

Cell-type dependent differences in O-oligosaccharide processing have also been noted among mammalian cells. For example, significant differences are observed in the O-linked oligosaccharide structures of leukosialin synthesized by the K562 erythroid, HL-60 promyelocytic or HSB-2T lymphoid cell lines (Carlsson et al., 1986). Five O-linked structures are noted on leukosialin from K562 cells, including structures A, B, C and D of Fig. 8, plus the structure NeuAcα(2,6)GalNAc. In contrast, the four O-linked structures noted on leukosialin from HL-60 and HSB-2 cells include structure C in Fig. 8, plus three structures indicative of a cell-type specific β(1,6)GlcNAc glycosyltransferase activity: NeuAcα(2,3)-Galβ(1,4)GlcNAcβ(1,6)[Galβ(1,3)]GalNAc, Galβ(1,4)GlcNAcβ(1,6)[NeuAcα(2,3)-Galβ(1,3)]GalNAc, and NeuAcα(2,3)-Galβ(1,4)GlcNAcβ(1,6)-[NeuAcα(2,3)Galβ(1,3)]GalNAc.

Cell-type dependent differences in variable O-glycosylation site occupancy are also displayed among mammalian cells. For example, recombinant CHO cells secrete a ratio of glycosylated to nonglycosylated IL-2 on the order of 9:1, while recombinant BHK and Ltk⁻ cells and cultured human lymphocytes produce approximately equal proportions of glycosylated and nonglycosylated IL-2 (Conradt, 1985, 1986, 1989; Vita et al., 1990).

Alterations of cell-type dependent glycosylation can result from spontaneous mutations during serial cultivation of mammalian cells. A series of CHO clones has been isolated possessing a variety of mutations affecting N- and O-glycosylation (Stanley, 1984, 1987, 1989). Many of these mutants were isolated from CHO cultures which did not contain mutagens, a reflection of the genetic instability of CHO cells in culture (Stanley, 1983). Most CHO glycosylation mutants maintain a growth rate comparable to their parent population (Stanley, 1984, 1989). Mutation usually diminishes a glycosylation capability -- for example, the LEC 2 CHO mutant has reduced capacity for sialylation of both N- and O-linked sugars due to reduced ability to transport sialic acid into the Golgi (Stanley, 1984, 1989). However, a mutation may lead to a new glycosylation capability. For example, the LEC 10 and 11 CHO mutants express an α(1,3)fucosyltransferase activity not expressed in the parent CHO population (Campbell and Stanley, 1983; Stanley, 1987).

These results provide a note of caution in long-term studies employing cultured mammalian cells -- particularly studies employing mutagens to facilitate gene amplification. Takeuchi and colleagues (1989) have recently described a CHO glycosylation mutant cloned at random in conjunction with amplification of the EPO gene (to be discussed further in section 3.5).

1.5 Oligosaccharide Synthesis is Protein-Dependent

The protein exerts considerable influence upon its own oligosaccharide processing (reviewed in Kornfeld and Kornfeld, 1985; Yet et al., 1988). For example, consider the protein-dependent outcomes of oligosaccharide processing for human interferon-β1 and human EPO. Biantennary complex-type oligosaccharides comprise the major N-linked oligosaccharide structure for interferon-β1 isolated from human diploid fibroblasts (82% of N-linked structures), recombinant CHO cells (68%), recombinant C127 cells (59%) and recombinant PC8 cells (52%), with near absence of tetraantennary structures in all cases (Kagawa et al., 1988). In contrast, biantennary complex-type structures comprise 6% or less of the N-linked oligosaccharides of EPO from recombinant CHO cells, with tetraantennary complex-type structures comprising more than 80% of the N-linked structures (Takeuchi et al., 1988; Sasaki et al., 1988). Tetraantennary structures also comprise the major N-linked oligosaccharides of EPO isolated from human urine (approximately 70%) (Takeuchi et al., 1988; Tsuda et al., 1988) and EPO produced by recombinant BHK cells (73%) (Tsuda et al., 1988).

Oligosaccharide processing can differ between N-glycosylation sites on the same protein, as demonstrated by recent studies of the N-linked oligosaccharides of t-PA. Four potential N-glycosylation sites are apparent from the t-PA amino acid sequence, Asn-117, Asn-184, Asn-218 and Asn-448. The site at Asn-218 (with the amino acid sequence Asn - Pro - Ser) is never occupied (Pohl et al., 1984). The oligosaccharides at the other three N-glycosylation sites have been analyzed for t-PA secreted by Bowes melanoma cells (Parekh et al., 1989a), human intestinal fibroblasts (Parekh et al., 1989a), recombinant CHO cells (Spellman et al., 1989) and recombinant C127 cells (Pfeiffer et al., 1989). For t-PA from each of the four cell types, it was observed that the site at Asn-184 is subject to variable site occupancy, the oligosaccharide structures at Asn-117 are high-mannose, and the structures at Asn-184 and Asn-448 are complex-type (although high-mannose structures were also identified at Asn-448 for the Bowes melanoma cell line). These data clearly demonstrate that oligosaccharide processing is influenced by the local molecular environment at each N-glycosylation site. In fact, for proteins with multiple glycosylation sites, site-specific differences in oligosaccharide processing are the rule rather than the exception. Site-to-site differences may be extensive (as with t-PA), or may involve more subtle differences in branching and terminal processing, as observed at the three N-glycosylation sites of EPO (Sasaki et al., 1988).

Five independent mechanisms have been proposed to account for site-specific differences in oligosaccharide processing following initiation of N- and O-linked glycosylation. One possibility is that access to an oligosaccharide chain as substrate may be sterically hindered by the local protein structure (Hubbard, 1988; Lee et al., 1990).

Secondly, the interaction of oligosaccharide with the local protein surface may lead to a restricted oligosaccharide conformation which either promotes or inhibits its suitability as a substrate for a particular glycosyltransferase; that is, glycosyltransferase K_m's may be site-dependent (Carver et al., 1989). Carver and Cumming (1987) have employed this concept to explain site-dependent differences in oligosaccharide processing in the constant and variable regions of an immunoglobulin G (IgG); at the N-glycosylation site in the constant region, interaction of oligosaccharide with protein apparently results in an oligosaccharide conformation which can not be recognized as substrate by a particular glycosyltransferase (Carver and Cumming, 1987).

Third, interaction of the glycosyltransferase enzyme with the local protein structure may lead to site-specific oligosaccharide processing differences. This type of interaction has been demonstrated for N-acetylglucosaminylphosphotransferase, the glycosyltransferase which adds GlcNAc 1-phosphate to high mannose oligosaccharides as a first step in the mechanism for intracellular targeting of glycoproteins to lysosomes (Lang et al., 1984).

Fourth, it has been hypothesized that oligosaccharide processing at a given glycosylation site may be dependent upon events occurring at other glycosylation sites (Parekh et al., 1987). For example, the oligosaccharide at one glycosylation site may sterically hinder oligosaccharide processing at a second adjacent site; thus, carbohydrate-carbohydrate interactions in the hinge region of IgG may be responsible in part for an observed restriction of N-linked oligosaccharide processing in that region (Rademacher et al., 1988). An alternative possibility is that the oligosaccharide structure at one glycosylation site may influence the protein tertiary structure at a second site. Persaud and coworkers (1988) employed this hypothesis to explain why initiation of O-glycosylation at Thr-95 and Thr-98 of human myelin basic protein occurs sequentially rather than randomly.

Fifth, oligosaccharide processing may be influenced on a site-specific basis by the interaction of protein subunits to form oligomers. This mechanism follows from the fact that protein folding and subunit assembly occur in the ER, while most glycosyltransferase activity occurs later in the Golgi. For example, placental cells secrete the human chorionic gonadotropin (hCG) α-subunit alone or combined noncovalently with a β-subunit to form the hCG heterodimer. Most of the α-subunit secreted alone is O-glycosylated, while α-subunit secreted in conjunction with β-subunit is not O-glycosylated. This inhibition of α-subunit O-glycosylation could result from steric hindrance of the particular O-glycosylation site by the β-subunit, or from alteration of α-subunit tertiary structure as a result of β-subunit association, prohibiting the α-subunit from serving as a substrate for GalNAc transferase (reviewed in Baenziger and Green, 1988). As a second example, Dahm and Hart (1986) demonstrated effects of quaternary structure on N-linked glycosylation of β-subunits of the heterodimeric cell surface glycoproteins Mac-1 and LFA-1 secreted by the macrophage-like cell line P388D$_1$ (Dahm and Hart, 1986). In addition to potential effects of oligomeric association on protein tertiary structure and substrate accessibility, Dahm and Hart raised the possibility that oligomeric association may lead to altered intracellular routing resulting in differential exposure to oligosaccharide processing enzymes.

Although the effects of local protein structure on oligosaccharide processing are extensive, these results do not diminish the importance of cell type in determining the outcome of oligosaccharide processing. The array of potential oligosaccharide processing reactions is defined by the enzyme repertoire of the cell, but the final outcome of oligosaccharide processing may be constrained to varying degrees by the local protein structure (Hubbard, 1988; Yet et al., 1988). For example, the full repertoire of known CHO cell glycosyltransferase activities for N-glycosylation are evident in the highly branched oligosaccharide structures of recombinant EPO (Takeuchi et al., 1988), while the predominance of biantennary N-linked oligosaccharide structures for CHO-produced interferon-β1 suggests protein-specific constraint of CHO cell oligosaccharide processing (Kagawa et al., 1988).

1.6 N-linked and O-linked Glycosylation are Affected by the Cell Culture Environment

The outcome of N-linked glycosylation is influenced by cell culture variables such as glucose concentration, ammonium ion concentration and the hormonal content of the medium (reviewed in Goochee and Monica, 1990). Among the potential mechanisms to explain such effects are: 1) depletion of the cellular energy state, 2) disruption of the local ER and Golgi environment, 3) interference with vesicle trafficking, and 4) modulation of glycosidase and glycosyltransferase activities. Control of mRNA transcription rate and/or stability is presumed to be responsible for many of the effects of hormones on oligosaccharide processing. Recently, this hypothesis has been confirmed in experiments demonstrating increases in specific glycosyltransferase mRNA and corresponding increases in their respective enzymes in response to cellular stimulation by dexamethasone (Wang et al., 1989) or retinoic acid (Cummings and Mattox, 1988; Larsen et al., 1989).

To our knowledge, only one example of an effect of the cell culture environment on O-linked protein glycosylation has been published. In that study, significant changes in the O-linked structures of the cell surface protein leukosialin accompanied the activation of human T lymphocytes by anti-CD3 antibodies and IL-2 (Piller et al., 1988). Both N- and O-glycosylation occur as the result of sequential actions of enzymes in several intracellular compartments. It would not be surprising to find that O-linked glycosylation is influenced by many of the same environmental variables that affect N-linked glycosylation.

PART 2. GLYCOPROTEIN OLIGOSACCHARIDE CONFORMATION

In section 1.5 (mechanism 2), we saw that differences in oligosaccharide processing on a site-specific basis might potentially be due to a restricted oligosaccharide conformation caused by interaction of the oligosaccharide with the local protein surface. In that same section (mechanism 4), it was suggested that changes in protein tertiary structure might result due to interaction of oligosaccharide with protein. In subsequent sections (Part 3), we will see that some effects of oligosaccharides on glycoprotein properties are potentially explained on the basis of oligosaccharide-protein interactions. In this transitional section, we will first review the published data concerning oligosaccharide conformation, then discuss the mechanisms for oligosaccharide interaction with protein surface and the proof that such interactions do exist.

Considerable progress has been made over the past decade in understanding the tertiary structure of free oligosaccharides in solution (reviewed in Homans et al., 1987; Carver et al., 1989; Meyer, 1990). The overall oligosaccharide solution conformation at a given moment is determined by the orientation of the constituent monosaccharides to one another, expressed in terms of the interglycosidic torsion angles, ϕ and ψ. NMR provides a powerful technique for experimentally determining these angles. However, NMR results may be misleading for an interglycosidic linkage exhibiting considerable conformational flexibility; under these circumstances, the NMR "average" structure may only represent a transitional conformation which is infrequently achieved. Computation of conformational potential energy as a function of torsional angles provides the best available method for validating an NMR-deduced conformation; that is, if a single potential energy minimum is calculated for a particular glycosidic linkage, this minimum should be consistent with the NMR-derived conformation. If the potential energy calculations reveal multiple local energy minima, then several interpretations of the NMR results are possible, including the existence of considerable flexibility about a particular glycosidic linkage (reviewed in Homans et al., 1987).

NMR and computational techniques have led to a consensus that some interglycosidic linkages exhibit considerable conformational flexibility -- for example, the Manα(1,6)Man linkage. The existence of this linkage in the inner structure of N-linked oligosaccharides (see Fig. 2 and 3) suggests conformational flexibility for one antenna of N-linked oligosaccharide structures (reviewed in Homans et al., 1987; Carver et al., 1989). The solution conformations have also been examined for the NeuAcα(2,3)Galβ and NeuAcα(2,6)Galβ linkages which terminate many complex-type glycoproteins (see Fig. 3G). Sialic acid linked α(2,3) to galactose prefers to assume a conformation extended away from the oligosaccharide chain, while sialic acid bonded α(2,6) prefers to fold back on the oligosaccharide chain (Berman, 1984; Breg et al., 1989). NMR data and computations have been interpreted as indicating considerable flexibility for both sialic acid linkages about their preferred conformations (Breg et al., 1989).

Much less is known about the conformation of oligosaccharides when they are N-linked or O-linked to protein. One possibility is that the oligosaccharide may interact minimally with the protein, protrude into the solvent away from the protein and exhibit conformations similar to those found in free solution. Alternatively, the oligosaccharide may associate with the protein at one or more locations on the protein surface, with resulting constraints on oligosaccharide conformation.

High affinity interactions of protein and saccharide have been documented in several cases: plant and animal lectins which bind carbohydrate ligands (Spohr et al., 1985; Hindsgaul et al., 1985); antibodies specific for carbohydrate structures (Lemieux et al., 1988); and enzymes that have carbohydrate as substrate (reviewed in Quiocho, 1989; Quiocho, 1986). Quiocho and colleagues have used X-ray crystallography to study the protein conformational changes associated with the binding of L-arabinose-binding protein and maltose-binding protein with their sugar substrates (reviewed in Quiocho, 1989; Quiocho, 1986). Their analysis reveals that protein-saccharide associations are stabilized by extensive hydrogen bonding of saccharide hydroxyl groups with nitrogen and oxygen atoms of amino acid side chains and by hydrogen bonding of the saccharide ring oxygen with an amino acid amine group. Van der Waals forces also contribute significantly to protein-carbohydrate interactions. Some of these attractive van der Waals forces result from close association of saccharide hydrophobic patches with aromatic tryptophan and phenylalanine side chains (Quiocho, 1989).

The concept of hydrophobic interactions involving monosaccharides is somewhat counterintuitive. Hydroxyl groups and ring oxygens of monosaccharides do introduce significant polar character to oligosaccharide surfaces; approximately 70% of oligosaccharide surfaces are polar according to estimates by Vyas et al. (1988). However, the pyranose ring of monosaccharides also possesses hydrophobic regions which could interact with non-polar amino acid side chains (Yet and Wold, 1990; Vyas et al., 1988). The location and size of these monosaccharide hydrophobic patches are defined by the absence of hydroxyl groups in limited regions of the pyranose ring. For example, examination of the boat configuration of D-glucose reveals a hydrophobic patch on the β-face composed of C3, C5 and C6 and a second hydrophobic patch on the α-face composed of C2 and C4 (Quiocho, 1989). The methyl group of L-fucose provides another source for non-polar interaction between oligosaccharide and protein (Lemieux et al., 1988). Thus, oligosaccharides are potentially capable of hydrophobic interactions with appropriate amino acid side chains. Such interactions would be stabilized by van der Waals attractions as well as by entropic considerations -- that is, the increased entropy of water molecules associated with burial of hydrophobic groups on the protein surface (Dill, 1990).

Both hydrophobic and polar interactions have been implicated in protein/oligosaccharide contacts in glycoproteins. For example, Sutton and Phillips (1983) used X-ray crystallography to determine the three-dimensional structure of the Fc fragment of rabbit IgG, which contains one N-linked oligosaccharide at Asn-297 of each heavy chain. They found that the contacts between the two heavy chains and the N-linked oligosaccharides in the CH2 domain were primarily non-polar in nature, involving interactions of the oligosaccharides with Phe-243 and Pro-246 on both protein chains. A single polar contact was observed between one oligosaccharide and Thr-260 of one polypeptide chain. Both polar and nonpolar contacts were also noted between oligosaccharides and protein in the CH2 and CH3 domains of human IgG (Deisenhofer, 1981).

Ionic forces represent another mechanism for oligosaccharide/protein interactions in glycoproteins. Sialic acid has a negatively charged carboxyl group at neutral pH, permitting ionic interactions with amino acid side chains. The other monosaccharides of N- and O-linked oligosaccharides are uncharged at neutral pH. Some cell types possess the enzymatic capability for covalent attachment of sulfate (Hortin et al., 1986; Roux et al., 1988; Smith, P.L. et al., 1990) or phosphate (Lee and Nathans, 1988; Purchio et al., 1988) to oligosaccharides, providing additional bases for ionic oligosaccharide/protein interactions.

Ionic attractions may be involved in protein/oligosaccharide interactions for α1-acid glycoprotein (Li et al., 1983; Perkins et al., 1985). This glycoprotein possesses bi-, tri- and tetraantennary complex-type oligosaccharides at five N-linked sites. Perkins and coworkers divided the α1-acid glycoprotein into two fractions using lectin chromatography: a ConA binding fraction containing α1-acid glycoprotein enriched in biantennary oligosaccharides, and a ConA non-binding fraction enriched in α1-acid glycoprotein with tri- and tetraantennary oligosaccharides. The oligosaccharide conformations of the two α1-acid glycoprotein fractions were examined using neutron and X-ray scattering techniques (Li et al., 1983; Perkins et al., 1985). The oligosaccharides of the α1-acid glycoprotein fraction enriched in biantennary complex-type structures were bound to the protein surface at low, "physiological" NaCl concentrations, but became extended into free solution as the NaCl concentration was raised above 0.6 M. In contrast, the triantennary and tetraantennary oligosaccharides remained bound to the protein surface over a broad range of NaCl concentrations. These authors hypothesized that the primary basis of oligosaccharide binding to protein for α1-acid glycoprotein is ionic attraction of terminal sialic acid to the protein surface -- a mode of interaction that could be disrupted by increasing NaCl concentration. A biantennary oligosaccharide, carrying at most two sialic acid moieties, would be more easily displaced from the protein surface than a tetraantennary oligosaccharide possessing up to four sialic acid moieties (Perkins et al., 1985). The authors estimated that up to 50% of the surface of α1-acid glycoprotein is masked by oligosaccharide moieties (Perkins et al., 1985).

The potential for protein/oligosaccharide interactions involving hydrogen bonding, van der Waals forces and ionic attractions does not guarantee that glycoprotein oligosaccharides will associate with the protein surface. For example, neutron scattering, X-ray scattering and X-ray crystallography data for α1-antitrypsin suggest that its three N-linked oligosaccharides are freely extended into the aqueous medium under physiological conditions (Smith, K.R. et al., 1990).

In summary, the available data support the possibility that protein-oligosaccharide interactions could affect protein and oligosaccharide conformation. The existence of such interactions will be glycoprotein-dependent -- that is, dependent upon both oligosaccharide structure and protein surface structure.

PART 3. EFFECTS OF OLIGOSACCHARIDES ON GLYCOPROTEIN PROPERTIES

3.1 Effect of Oligosaccharides on Glycoprotein Solubility

The N-linked oligosaccharides of glycoproteins often promote protein solubility and inhibit protein aggregation. Enzymatic or chemical removal of most or all N-linked oligosaccharides decreases the solubility of EPO (Dordal et al., 1985), porcine RNase (Grafl et al., 1987), glucose oxidase (Takegawa et al., 1989), fibrinogen (Langer et al., 1988), and interferon-β (Conradt et al., 1987). Invertase with its full complement of N-linked oligosaccharides remains soluble in the unfolded protein state, while aglycosyl invertase aggregates in the unfolded state (Schulke and Schmid, 1988a, 1988b). Removal of the O-linked oligosaccharide from granulocyte colony-stimulating factor (G-CSF) results in increased protein aggregation (Oh-eda et al., 1990).

In contrast, the presence of sialylated oligosaccharides has been associated with decreased solubility for some immunoglobulins; removal of sialic acid with neuraminidase increases the solubility of some monoclonal IgM's and IgG's at low temperature (Tsai et al., 1977; Weber and Clem, 1981; Lawson et al., 1983) and at 37 °C (Lawson et al., 1983). For one of these "cryoimmunoglobulins", the reduced solubility is apparently related to the presence of sialylated, N-linked oligosaccharides in the variable region (Middaugh and Litman, 1987).

Protein solubility is a complex phenomenon which is only partially understood (reviewed in Arakawa and Timasheff, 1985). In general terms, protein solubility is promoted by factors which increase the affinity of the protein for its solvent, and reduced by factors which promote protein-protein attractive forces. For example, non-polar amino acid side chains on the protein surface promote protein aggregation via hydrophobic interactions (Arakawa and Timasheff, 1985).

Oligosaccharides may contribute to decreased glycoprotein hydrophobicity and increased glycoprotein solubility through masking of protein surface hydrophobic groups as well as through strong affinity of the oligosaccharides for water. In support of this argument, the removal of glycoprotein oligosaccharides invariably increases protein retention time in reverse phase chromatography (for example, see interferon-β: Utsumi et al., 1987; t-PA: Parekh et al., 1989a). In contrast, for some immunoglobulins, the charged sialic acid moieties apparently provide the basis for intermolecular contacts causing aggregation (Middaugh and Litman, 1987).

3.2 Effect of Oligosaccharide Structure on Glycoprotein Resistance to Protease Digestion

N-linked oligosaccharides frequently play a role in protecting a glycoprotein from extracellular proteolytic attack. For example, carbohydrate-depleted IgM becomes more susceptible to trypsin attack (Sibley and Wagner, 1981); carbohydrate-depleted yeast invertase is more susceptible to attack by subtilisin and pronase (Yamamoto et al., 1987); carbohydrate-depleted porcine pancreatic ribonuclease is more susceptible to attack by trypsin and subtilisin (Wang and Hirs, 1977); aglycosylated fibronectin from chicken embryo fibroblasts is more susceptible to pronase digestion (Olden et al., 1979); aglycosyl hCG is more susceptible to proteolysis by chymotrypsin (Merz, 1988); and carbohydrate-depleted β-N-acetylhexosaminidase from *Penicillium oxalicum* (Yamamoto et al., 1987) and glucoamylase (Takegawa et al., 1988) are more sensitive to proteolysis by pronase, subtilisin and trypsin. Furthermore, removal of just terminal sialic acid from EPO results in increased susceptibility to proteolysis by trypsin (Goldwasser et al., 1974).

The absence of O-oligosaccharides leads to increased susceptibility to proteolytic attack for some mammalian glycoprotein receptors, including LDL receptor (Kozarsky et al., 1988) and decay accelerating factor (Reddy et al., 1989).

The absence of N-oligosaccharides leads to increased intracellular proteolytic degradation of many glycoproteins, including ACTH-endorphin precursor (Loh and Gainer, 1979) and IgA (Taylor and Wall, 1988).

It has been proposed that oligosaccharides protect glycoproteins from proteolytic attack by masking potential cleavage sites on the protein surface (Montreuil, 1984; Yet and Wold, 1990). Removal of terminal sialic acid alone could result in increased exposure of protein sites to proteolytic attack, particularly if desialylation is accompanied by the transition of the oligosaccharide from a protein-bound conformational state to an unbound state (Montreuil, 1984). Yet and Wold (1990) proposed that hydrophobic interactions may be responsible for the masking effect in other cases; they hypothesized that the protective effect of oligosaccharides against protein attack by chymotrypsin is due to specific non-polar interactions between oligosaccharide moieties and protein Tyr, Phe and Leu side chains at the point of protease attack. As an alternative to these "masking hypotheses", Merz (1988) hypothesized that the increased susceptibility of aglycosyl hCG to chymotrypsin is due to a change in protein conformation upon deglycosylation which exposes new chymotrypsin proteolysis sites on the protein surface.

3.3 Effect of Oligosaccharides on Glycoprotein Resistance to Thermal Denaturation

In some cases, partial or complete removal of glycoprotein oligosaccharides leads to increased susceptibility to thermal denaturation. For example, enzymatic removal of sialic acid from EPO results in a significant increase in the rate of thermal inactivation at 70 °C (Goldwasser et al., 1974; Tsuda et al., 1990). Complete removal of N- and O-linked oligosaccharides results in a further increase in the rate of EPO thermal denaturation (Tsuda et al., 1990). Partial or complete removal of oligosaccharides has also been demonstrated to increase the rate of thermal inactivation for human thyroxine binding globulin (Grimaldi et al., 1985), glucoamylase (from *Rhizopus niveus*)(Takegawa et al., 1988) and yeast acid phosphatase (Barbaric et al., 1984.

Such results have been interpreted by some investigators to represent an effect of oligosaccharides on glycoprotein thermal stability. However, as noted by Grafl and colleagues (1987), "these results are equally well explained by a decreased solubility of the deglycosylated species, which leads to preferential irreversible aggregation in the thermally unfolded state."

A careful study by Puett (1973) concluded that the oligosaccharides contributed 0.3 to 0.4 kcal/mol to the thermal stability of bovine pancreatic ribonuclease, but several other studies have identified no effect of oligosaccharides on thermal stability for porcine RNase (Wang and Hirs, 1977; Grafl et al, 1987), yeast invertase (Schulke and Schmid, 1988a), recombinant human GM-CSF (Wingfield et al., 1988) and glucose oxidase (Kalisz et al., 1990).

3.4 Effect of Oligosaccharides on Glycoprotein Folding, Subunit Assembly, and Secretion

N- and O-linked oligosaccharides facilitate the protein folding, subunit assembly and secretion of some glycoproteins. These roles of N-linked oligosaccharides have been examined *in vivo* using N-glycosylation inhibitors such as the antibiotic tunicamycin, which blocks synthesis of the $Glc_3Man_9GlcNAc_2$-P-P-Dol precursor necessary for initiation of N-glycosylation (Fig. 1, reaction 1) (reviewed in Olden et al., 1982; Elbein, 1987). The consequences of the absence of N-oligosaccharide on secretion is very protein-dependent. In many cases, a protein lacking N-oligosaccharides is either not secreted or is secreted at a greatly reduced rate. Examples include IgM from some mouse plasmacytomas and hybridomas (Hickman and Kornfeld, 1978; Sidman, 1981; Yuan, 1982); α_1-protease inhibitor, α_2-macroglobulin and ceruloplasmin from human hepatoma cells (Bauer et al., 1985); t-PA and factor VIII from recombinant CHO cells (Dorner et al., 1987); and procollagen from human fibroblasts (Housley et al., 1980). Cell surface expression of some glycoprotein receptors is also inhibited by tunicamycin treatment. Examples include acetylcholine receptor in cultured muscle cells (Olden et al., 1982), recombinant CD4 receptors in CHO cells (Konig, et al., 1989) and transferrin receptors in A431 cells, a human epidermoid carcinoma cell line (Reckhow and Enns, 1988).

In contrast, the secretion of many glycoproteins is largely unaffected by the tunicamycin-induced absence of N-oligosaccharides. Examples include: recombinant hGM-CSF from CHO cells (Cebon et al., 1990); IgM from a mouse B cell lymphoma (Sibley and Wagner, 1981); IgG from cultured hybridomas and plasmacytomas (Hickman and Kornfeld, 1978; Sidman, 1981); transferrin from cultured rat hepatocytes (Edwards et al., 1979; Struck et al., 1978), and interferon from human leukocytes (Mizrahi et al., 1978). The absence of N-oligosaccharides does not affect the surface expression of acetylcholine receptors by chicken embryo muscle cells (Prives and Olden, 1980) or the asialoglycoprotein receptor by HepG2 cells (Breitfeld et al., 1984).

Relatively few studies have addressed the effect of O-glycosylation on glycoprotein secretion or cell surface expression. A chemical inhibitor of O-glycosylation analogous to tunicamycin has not been identified. The importance of O-glycosylation for intracellular folding, assembly and secretion of recombinant proteins has been primarily examined using the CHO mutant *ldlD*, which contains a conditional mutation preventing O-glycosylation (Kingsley et al., 1986; Kozarsky et al., 1988). Using this mutant it has been determined that secretion of recombinant human apolipoprotein E is unaffected by the absence of O-glycosylation (Zanni et al., 1989; Wernette-Hammond et al., 1989). The ldlD mutant has also been utilized to demonstrate that the synthesis, cell surface expression, and cleavage rate of transforming growth factor-α are unaffected by the absence of O-glycosylation (Teixida et al., 1990).

Recently, site directed mutagenesis has been utilized to explore the importance of individual N- and O-linked oligosaccharide sites to glycoprotein secretion. For example, point mutations to remove individually the three N-linked and one O-linked glycosylation sites of EPO suggest that two N-glycosylation sites (Asn-38 and Asn-83) and the single O-glycosylation site (Ser-126) are required for efficient EPO secretion from recombinant CHO cells (Dube et al., 1988).

Subunit assembly and secretion of oligomeric proteins is sometimes facilitated by the presence of N-linked oligosaccharides. Studies employing tunicamycin with cultured mouse pituitary cells have revealed that N-glycosylation is required for efficient *in vivo* assembly of the α- and β-subunits of thyrotropin (reviewed in Weintraub et al., 1985). In contrast, the assembly of IgG heavy and light chains and secretion of the resulting structure occurs efficiently in the absence of N-linked glycosylation (Hickman and Kornfeld,

1978; Sidman, 1981). Site-directed mutagenesis has revealed that N-glycosylation is required for efficient *in vivo* assembly and secretion of α- and β-subunits of hCG in recombinant CHO cells (Matzuk and Boime, 1988a, 1988b). The presence of N-linked oligosaccharides also affects the *in vivo* and *in vitro* assembly of yeast invertase dimers into higher oligomers and the equilibrium of di-, tetra-, and octomeric forms. Under hypoosmolar conditions, glycosylated invertase dimers assemble into tetramers and octomers, while aglycosyl invertase remains in dimeric form. In the presence of NaCl, aglycosyl invertase dimers assemble into octomers, but the equilibrium between oligomeric forms differs between the glycosylated and aglycosyl invertase (Esmon et al., 1987; Reddy et al., 1990).

O-linked oligosaccharides are less likely to affect *in vivo* subunit assembly of oligomeric proteins since subunit assembly in the ER probably precedes O-glycosylation in most cases. Thus, the absence of O-linked oligosaccharides does not affect intracellular assembly of the α- and β- subunits of recombinant hCG and secretion of the resulting dimer (Matzuk et al., 1987). In contrast, the presence of an O-oligosaccharide on the α-subunit of hCG does interfere with the *in vitro* assembly of the α- and β-subunits (reviewed in Sairam, 1989).

Glycoprotein monomers and oligomers whose secretion is inhibited in the absence of glycosylation frequently accumulate as insoluble aggregates in the ER (Hickman and Kornfeld, 1978; Dorner et al., 1987; Yeo et al., 1989). Some degree of correct protein tertiary and quaternary structure is required for exit from the ER. Misfolded proteins are retained in the ER, frequently in association with the ER-binding protein BiP (also known as GRP78) (reviewed in Pfeffer and Rothman, 1987; Lodish, 1988; Rose and Doms, 1988; Gething and Sambrook, 1989; Hurtley and Helenius, 1989). Recently, Dorner et al. (1987) observed association with BiP of two proteins (factor VIII and t-PA) whose secretion by recombinant CHO cells had been retarded in the absence of N-oligosaccharides. These data are consistent with the hypothesis that N- and O-oligosaccharides promote protein secretion for some proteins by assisting in the achievement of the proper tertiary or quaternary structure necessary for exit from the ER (Leavitt et al., 1977; Dorner et al., 1987).

In some cases, the presence of oligosaccharides may be a necessary condition for protein folding and maintenance of proper tertiary or quaternary structure. For example, Walsh and coworkers (1990) examined the effect of N-linked oligosaccharides on the secondary structure of human plasma β2-glycoprotein I. Changes in the circular dichroic spectra after enzymatic removal of 96% of the oligosaccharide moieties suggest that one-third of the β2-glycoprotein amino acids originally in random coil conformation assume β-turns after oligosaccharide removal.

For other proteins, the presence of oligosaccharides may be a necessary condition for proper protein folding, but not a necessary condition for maintenance of the folded state. For example, it has been hypothesized that N-linked oligosaccharides can promote formation of appropriate intramolecular disulfide bonds (Vidal et al., 1989) and inhibit formation of inappropriate intermolecular disulfide bonds (Machamer and Ross, 1988b). This hypothesis was employed to explain why many viral proteins form ER aggregates with inappropriate disulfide bonds when virally-infected cells are treated with tunicamycin or when specific N-linked glycosylation sites are deleted by site-directed mutagenesis (Machamer and Ross, 1988a, 1988b; Vidal et al., 1989). The data presented in these studies can also be explained by the hypothesis presented below (Ng et al., 1990).

In many cases, the presence of oligosaccharides may not be an absolute requirement for either protein folding or maintenance of the folded state, but the oligosaccharides may increase the probability of proper protein folding -- for example, by increasing the solubility of partially folded proteins and preventing their aggregation before the folded state is achieved (Leavitt et al., 1977). Studies involving the *in vitro* protein folding of porcine ribonuclease (Wang and Hirs, 1977; Grafl et al., 1987) have demonstrated that the presence of N-linked oligosaccharides is not required for achievement of the final folded state, but the rates of individual folding steps were up to three-fold higher for the glycosylated species (Grafl et al., 1987). The presence of N-linked oligosaccharides was also not an absolute requirement for protein folding and assembly of yeast invertase dimers (Chu et al., 1978; Schulke and Schmid, 1988b). However, glycosylation did "promote" invertase folding by increasing the solubility of the unfolded or partially folded proteins, thereby inhibiting irreversible aggregation (Schulke and Schmid, 1988b). The presence of N-linked oligosaccharides could play a comparable role *in vivo*, promoting protein secretion by inhibiting aggregation of partially-folded or fully-folded proteins in the ER (Leavitt et al., 1977; Dorner et al., 1987).

3.5 Effect of Oligosaccharides on Glycoprotein Specific Activity

The biological activity of a human therapeutic glycoprotein *in vivo* is dependent upon a variety of factors, including glycoprotein specific activity. Specific activity measurements are conventionally assessed using an *in vitro* assay. For example, the specific activity of EPO is frequently assessed *in vitro* by measuring ^{59}Fe incorporation into cultured rat bone marrow cells (e.g., see Takeuchi et al., 1989). The effect of N-linked oligosaccharides on specific activity is typically assessed following the partial or complete removal of oligosaccharide structures by enzyme or chemical treatment, or following synthesis of the protein in the presence of tunicamycin. Particular care is necessary to assure that specific activity measurements are not compromised by aggregation of aglycosyl proteins or by their increased susceptibility to protease attack (Gomi et al., 1983; Utsumi et al., 1989).

The oligosaccharides of glycoproteins frequently have a significant effect on specific activity (reviewed in Rademacher et al., 1988). Examples of increased specific activity in the absence of oligosaccharides include: aglycosyl prolactin has higher receptor binding affinity and greater ability to stimulate cell proliferation than its glycosylated counterpart (Markoff et al., 1988; Pellegrini et al., 1988); deglycosylated fibrinogen stimulates more rapid fibrin assembly that its glycosylated counterpart (Langer et al., 1988); and removal of sialic acid increased the specific activity of protein C (Yan et al., 1990). Examples of decreased specific activity in the absence of oligosaccharides include: removal of sialic acid abolishes factor IX clotting activity (Chavin and Weidner, 1984); removal of 95% of the N-linked oligosaccharides results in a 17% increase in K_m for glucose oxidase (Kalisz et al., 1990); deglycosylation of HIV-1 envelope glycoprotein gp120 reduces its binding affinity to CD4 receptors (Fennie and Lasky, 1989; Fenouillet

et al., 1990; Hitzman et al., 1990); and desialylation eliminates human α_1-acid glycoprotein immunomodulatory effects (Pos et al., 1990; Bennett and Schmid, 1980).

In contrast, it has been reported that the partial or complete absence of oligosaccharide moieties does not measurably affect the *in vitro* specific activity of many glycoprotein enzymes and several glycoprotein hormones, including glucoamylase (Takegawa et al., 1987), bovine RNase B and DNase A (Tarentino et al., 1974), yeast invertase (Tarentino et al., 1974; Yamamoto et al., 1987), IL-2 (Rosenberg et al., 1984; Doyle, M.V., 1985; Naruo et al., 1985), interleukin-5 (Tavernier et al., 1989; Tominaga et al., 1990) and macrophage colony-stimulating factor (type β) (Halenbeck et al., 1989).

A number of well documented cases are discussed in the following paragraphs which further illustrate involvement of oligosaccharide moieties of glycoproteins in the up- or down-regulation of glycoprotein specific activity.

The glycoprotein hormones. The N-linked oligosaccharides of the glycoprotein hormones lutropin (LH), follitropin (FSH), thyrotropin (TSH) and chorionic gonadotropin (CG) are required for bioactivity (reviewed in Weintraub et al., 1985; Baenziger and Green, 1988; Weintraub et al., 1989; Sairam, 1989). These hormones are heterodimers consisting of noncovalently associated α- and β-subunits. The α-subunits of the four hormones have identical amino acid sequence, while the β-subunits are unique peptides which confer biological specificity to the respective hormones. These hormones contain N-linked oligosaccharides on both the α- and β-subunits, and may additionally contain O-linked oligosaccharides on a hormone-dependent basis. Removal of the N-linked oligosaccharides from CG, LH, TSH or FSH does not inhibit receptor binding of these hormones. In fact, the binding affinity of LH and CG is apparently enhanced in the absence of N-linked oligosaccharides (Sairam, 1990). Although able to bind to their receptors, the hormones lacking N-linked oligosaccharides are unable to elicit an appropriate response from their target cells *in vitro*; that is, they are unable to activate adenylate cyclase to produce cyclic AMP as a second messenger (Sairam, 1990). In fact, the hormones lacking N-linked oligosaccharides serve as competitive inhibitors of the native glycosylated forms (Kalyan and Bahl, 1983; Sairam, 1990). The N-linked oligosaccharides of the α-subunit are critical for hormone signal transduction (Kaetzel et al., 1989; Sairam, 1989; Matzuk et al., 1989).

In contrast, the O-linked oligosaccharides appear to have no effect on glycoprotein hormone signal transduction. Human chorionic gonadotropin (hCG) contains four O-linked oligosaccharides on the β-subunit. Matsuk and colleagues (1990) have recently examined the relationship of O-glycosylation to specific activity using recombinant hCG produced without O-oligosaccharides using the *ldlD* mutant CHO cell line. They found that the presence or absence of O-oligosaccharides does not affect hCG binding to its receptor nor does it affect *in vitro* receptor signal transduction (as measured by intracellular cAMP accumulation and steroid production). In spite of having no effect on specific activity, the absence of O-linked oligosaccharides leads to a three-fold decline in hCG biological activity *in vivo*, presumably as a result of increased clearance rate (Matsuk et al., 1990).

IgE potentiating factor and IgE suppressive factor. A graphic example of the influence of oligosaccharide structure on specific activity comes from studies of IgE potentiating factor (IgE-pF) and IgE suppressive factor (IgE-sF), proteins capable of either stimulating or inhibiting the synthesis of IgE by lymphocytes. Recent evidence suggests that these glycoproteins have the same protein core structure, but that their potentiating or suppressive nature is determined by their respective oligosaccharide moieties (reviewed in Ishizaka, 1988).

The immunoglobulins. Oligosaccharide moieties affect the specific activity of immunoglobulins. Human and mouse IgG possess an N-linked oligosaccharide structure on each heavy chain in the constant region at Asn-297. The effectiveness of Fc receptor binding and complement activation by IgG is dependent upon the presence of oligosaccharide at that site. Reduction of these effector functions in deglycosylated IgG is thought to result from protein conformational changes in the C_H-2 domain (Nose and Wigzell, 1983; Leatherbarrow et al., 1985; Tsuchiya et al., 1989; Walker et al., 1989).

IgG and other immunoglobulins may also possess N-linked oligosaccharides in the hypervariable region of either the heavy or light chain (Sox and Hood, 1970; Spiegelberg et al., 1970; Savvidou et al., 1984; Taniguchi et al., 1985; Arvieux et al., 1986; Rademacher et al., 1986; Wallick et al., 1988). The frequency and location of these oligosaccharides is dependent upon the generation of Asn - X - Ser/Thr sites during genetic recombination.

Variable region glycosylation occurs frequently in human IgG. Abel et al. (1968) determined that 25% of 76 human myeloma-derived IgG's contained oligosaccharide within the variable regions of the heavy chain and/or the light chain (see also Spiegelberg et al., 1970). Rademacher and Dwek (1983) determined an average of 2.8 oligosaccharide chains per IgG in pooled, human IgG. Two of those chains are attached in the Fc region at Asn-297 of the heavy chains, while the remainder are in the hypervariable region on either the heavy or light chain, suggesting a prevalence of hypervariable region glycosylation consistent with the earlier Abel study. The existence of unoccupied Asn - X - Ser/Thr sites in the IgG variable region has also been noted (Sox and Hood, 1970; Spiegelberg et al., 1970).

Morrison and coworkers (Wallick et al., 1988) have demonstrated that glycosylation in the IgG variable region can affect antibody affinity for antigen. The binding affinity of a monoclonal antibody toward dextran was reduced 10-fold in the absence of variable region glycosylation. Morrison and coworkers thought the most likely explanation of this result was an effect of the oligosaccharide on the protein conformation of the IgG variable region.

An effect of variable region glycosylation on antibody affinity for antigen is also suggested by recent experiments probing the phenomenon of "nonprecipitating" antibodies (reviewed in Margni and Binaghi, 1988). Approximately 5 to 15% of IgG antibodies present in serum are unable to precipitate antigen. These nonprecipitating antibodies are unable to fix complement and serve as a competitive inhibitor of complement fixation by normal, "precipitating" antibodies. The antigen binding affinities of the two IgG binding sites are comparable for precipitating antibodies. Margni and Binaghi determined that the antigen binding affinities differ by as much as 100-fold at the two binding sites of non-precipitating antibodies; this binding asymmetry is apparently due to differences in variable region glycosylation between the two IgG Fab regions. In at least some cases, bivalent binding is restored following enzymatic removal of all variable region oligosaccharides. The Margni and Binaghi group has recently reported that approximately 10% of all circulating human IgG possess asymmetric glycosylation in the variable region (Borel et al., 1989). One potential basis for this asymmetry would be variable site occupancy at an Asn - X - Ser/Thr site in the heavy or light chain variable regions, a phenomenon demonstrated previously by Savvidou and coworkers (1981, 1984) .

Erythropoietin. Erythropoietin (EPO) contains three N-linked, complex-type oligosaccharides and one O-linked oligosaccharide (see references in Tsuda et al., 1990). Enzymatic removal of terminal sialic acid from EPO using neuraminidase results in increased specific activity as measured by the in vitro bone marrow culture assay (Briggs et al., 1974; Goto et al., 1988; Goldwasser et al., 1974; Tsuda et al., 1990; Takeuchi et al., 1990). For example, Tsuda and colleagues (1990) observed a two-fold increase in specific activity following removal of EPO sialic acid, which correlated with increased affinity of asialo-EPO for its receptor. Further enzymatic removal of galactose and N-acetylglucosamine from asialo-EPO to yield truncated, high mannose-type structures, such as those presented in Fig. 7, results in no further enhancement of specific activity beyond that obtained by desialylation (Tsuda et al., 1990; Takeuchi et al., 1990). The removal of all N- and O-linked oligosaccharides from EPO reportedly results in comparable or somewhat reduced specific activity in comparison with fully glycosylated EPO (Tsuda et al., 1990; Takeuchi et al., 1990). Dube and coworkers (1988) examined the effect of the individual EPO N- and O-glycosylation sites using site-directed mutagenesis and expression in recombinant BHK cells. Elimination of N-linked glycosylation sites at Asn-24 or Asn-38 reduces in vitro specific activity by 80% in comparison with fully glycosylated EPO; the effect on specific activity of eliminating the O-glycosylation site at Ser-126 and the N-glycosylation site at Asn-83 could not be determined due to inhibition of EPO secretion in those mutants (Dube et al., 1988).

Human urinary EPO and EPO isolated from recombinant CHO and BHK cells contain a predominance of tetraantennary complex-type N-linked oligosaccharides (Sasaki et al., 1987; Tsuda et al., 1988; Takeuchi et al., 1988). Takeuchi and colleagues (1989) have recently described isolation of EPO subfractions from a mutant recombinant CHO cell line that produces two distinct subsets of EPO glycoforms -- a subset enriched in biantennary complex-type N-linked oligosaccharides (EPO-bi) and a subset enriched in tetraantennary complex-type structures (EPO-tetra); the EPO-tetra is considered to be comparable to EPO previously isolated from recombinant CHO cells. EPO-bi demonstrates three-fold higher specific activity in comparison with EPO-tetra in the in vitro bone marrow culture assay (Takeuchi et al., 1989).

Tissue-type Plasminogen Activator. Significant differences have been observed between the specific activities of type I and type II t-PA measured in vitro using the clot lysis assay, apparently reflecting an effect of the N-linked oligosaccharide at Asn-184. This effect has been examined using endogenous t-PA isolated from Bowes melanoma cells (Einarsson et al., 1985; Wittwer et al., 1989) and human intestinal fibroblasts (Wittwer et al., 1989). For both cell types, type II t-PA has 20-30% greater specific activity than type I t-PA. This trend is also evident in comparison of the in vitro activities for type I and type II t-PA from recombinant CHO and C127 cells (Parekh et al., 1989b).

The trend of increased t-PA specific activity with decreased extent of glycosylation is further demonstrated in comparisons of the in vitro activity of aglycosyl t-PA to its glycosylated counterparts. Recombinant t-PA produced by CHO cells in the presence of tunicamycin has significantly greater fibrin binding and fibrinolytic potency than its glycosylated counterparts (Hansen et al., 1988). Endogenous t-PA produced by human colon fibroblasts or Bowes melanoma cells in the presence of tunicamycin demonstrates more than two-fold greater specific activity as measured by the in vitro clot lysis assay (Wittwer et al., 1989).

Species-specific differences in glycosylation can contribute to differences in t-PA specific activity. Wittwer and colleagues (1989) observed that type I and type II t-PA from Bowes melanoma cells each have approximately 30% greater specific activity than their counterparts isolated from human colon fibroblasts.

Human granulocyte-macrophage colony-stimulating factor. Human granulocyte-macrophage colony-stimulating factor (hGM-CSF) produced by mammalian cells potentially possesses two N-linked oligosaccharides at Asn-27 and Asn-37 and two O-linked oligosaccharides at Ser-7 and Ser-9 (Clark, 1988). Both N-glycosylation sites are subject to variable site occupancy. Fully glycosylated mammalian hGM-CSF has an apparent molecular weight in the vicinity of 30 kDa, with 14.5 kDa attributable to the protein (Moonen et al., 1987; Cebon et al., 1990). In general, the specific activity of hGM-CSF is inversely related to its extent of glycosylation. Aglycosyl hGM-CSF has 10- to 20-fold higher specific activity than fully glycosylated hGM-CSF (Moonen et al., 1987; Cebon et al., 1990), while hGM-CSF with one N-linked oligosaccharide has intermediate activity (Cebon et al., 1990). This effect correlates with a strong inverse relationship between the extent of glycosylation and hGM-CSF receptor binding affinity (Cebon et al., 1990).

Receptors. Many cell surface receptors are glycoproteins, including most of the hormone and cytokine receptors. Full receptor function requires appropriate receptor binding to ligand, followed by an intracellular response mediated by the receptor/ligand complex. Oligosaccharide moieties can affect *in vitro* receptor binding affinity and/or the intracellular response mechanism for some receptors. For example, enzymatic removal of the N-linked oligosaccharide moieties or expression of receptors in the presence of tunicamycin leads to loss of measurable receptor binding affinity for the murine insulin receptor (Ronnett and Lane, 1981), human interferon-γ receptor (Fisher et al., 1990) and the human EGF receptor (Gamou and Shimizu, 1988). Fully glycosylated human transferrin receptor has about 10-fold higher affinity for transferrin than does the nonglycosylated receptor (Hunt et al., 1989). Removal of the O-linked oligosaccharide from the mouse egg surface glycoprotein ZP3 abolishes its sperm receptor activity, while removal of N-linked oligosaccharide has no effect (Florman and Wasserman, 1985). Boege et al. (1988) observed that aglycosyl β2-adrenoreceptors retain their original ligand binding properties, but lose coupling efficiency -- that is, the ligand-bound aglycosyl receptors are unable to stimulate intracellular adenylate cyclase activity.

In contrast to the results presented above, complete inhibition of N-glycosylation by tunicamycin does not inhibit the binding affinity or other functional properties of the human asialoglycoprotein receptor (Breitfeld et al., 1984). In an odd twist, while the complete absence of N-linked oligosaccharides does not interfere with asialoglycoprotein receptor function, the selective removal of terminal sialic acid from the glycosylated receptor abolishes its ligand binding capacity, apparently because the asialo receptor binds to its own galactose-terminated oligosaccharides (Paulson et al., 1977; Stockert et al., 1977).

The presence of O-linked oligosaccharides affects the functionality of the LDL receptor indirectly by protecting the receptor from proteolytic attack following surface expression. Transfection of the gene for the LDL receptor into a CHO cell line deficient in O-glycosylation (*ldlD*) results in the expression of receptors that are non-functional due to rapid proteolytic degradation (Kozarsky et al., 1988).

Fisher and coworkers (1990) explored the mechanism responsible for loss of human interferon-γ receptor binding affinity in the absence of its N-linked oligosaccharides. Their results suggest that the N-linked oligosaccharides contribute to maintenance of the interferon-γ receptor protein tertiary structure that is necessary for effective ligand binding.

3.6 Effect of Oligosaccharides on Glycoprotein Conformation

The results presented in section 3.5 document effects of oligosaccharides on the specific activities of many glycoproteins. Several mechanisms could explain such results, including 1) steric hindrance of a functional site by the oligosaccharide, 2) direct participation of the oligosaccharide at the functional site of the glycoprotein and 3) an effect of the oligosaccharide on protein tertiary structure. This third mechanism is clearly plausible given the possibilities for interaction of oligosaccharide with the protein surface via hydrogen bonds, ionic attractions and hydrophobic attractions (see section 2). In this section, evidence for effects of oligosaccharide on glycoprotein secondary and tertiary structure will be reviewed.

The effect of oligosaccharides on glycoprotein secondary structure has been analyzed using methodologies such as circular dichroic (CD) spectroscopy, which probes polypeptide secondary structure, fluorescence spectroscopy, which probes the environment near aromatic residues, and proton and C-13 NMR, which probe the spatial relationships of hydrogen or carbon atoms. Several studies have detected differences in glycoprotein CD spectra as a result of deglycosylation. The CD spectrum of porcine pancreatic RNase exhibits a subtle but significant shift following deglycosylation, indicative of a change in secondary structure in the vicinity of at least two tyrosine side chains (Wang and Hirs, 1977). As noted in section 3.4, significant changes in the CD spectrum of human plasma β2-glycoprotein were observed after enzymatic removal of 96% of the oligosaccharide moieties, suggesting that one-third of the β2-glycoprotein amino acids originally in random coil conformation were present in β-turns after oligosaccharide removal (Walsh et al., 1990). Significant changes in CD spectra have also been observed in comparison of glycosylated and aglycosyl hCG (Merz, 1988). Changes in CD spectra and C-13 NMR spectra were noted in deglycosylated ovine submaxillary gland mucin (Gerken et al., 1989; Shogren et al., 1989). Intrinsic tryptophan fluorescence measurements suggest that glycosylation may affect protein conformations of human placental fibronectin (Zhu et al., 1990) and human thyroxine binding globulin (Grimaldi et al., 1985), while NMR data suggest effects of O-glycosylation on the conformation of human myelin basic protein (Persaud et al., 1988).

The effect of oligosaccharides on protein conformation is clearly protein-dependent. In contrast to the results presented above, CD spectroscopy has revealed no detectable differences in secondary structure in glycosylated and aglycosyl variants of human interferon–β (Utsumi et al., 1987), bovine RNase B and DNase A (Tarentino et al., 1974) and yeast invertase (Chu et al., 1978). No effect of glycosylation on the tertiary structure of human interferon–β is evident through NMR analysis (Utsumi et al., 1987).

3.7 Effect of Oligosaccharides on Glycoprotein Antigenicity

Oligosaccharide structures can serve as a basis for antibody recognition (reviewed in Feizi and Childs, 1987). Many mammalian circulating antibodies are targeted against specific oligosaccharide determinants. For example, approximately 1% of circulating human IgG is specific for the terminal Galα(1,3)Galβ(1,4)GlcNAc epitope (Galili et al., 1985) (see Fig. 3I). As a second example, carbohydrate structures form the basis of the A, B and O blood group determinants (Fig. 10)

(reviewed in Watkins, 1980; Clausen and Hakomori, 1989; Green, 1989). Individuals not expressing one of these blood group determinants normally have circulating antibodies that recognize it as antigen (reviewed in Green, 1989). As a last example, most humans have circulating antibodies against N-linked yeast mannan chains with the general form shown in Fig. 6A (Chew and Theus, 1967; Savolainen et al., 1990).

Oligosaccharides may also contribute indirectly to glycoprotein antigenicity (Alexander and Elder, 1984; Meager and Leist, 1986). For example, an oligosaccharide may inhibit access to an antigenic site by steric hindrance or by charge interactions contributed by sialic acid (Schauer, 1988). Alternatively, the antigenicity of a glycoprotein may be altered due to conformational differences in the protein as a result of oligosaccharide-protein interactions. Such subtle mechanisms are supported by the recent work of Gribben et al. (1990), who employed human anti-sera raised against *Saccharomyces cerevisiae*-produced GM-CSF to probe the antigenicity of recombinant CHO-produced GM-CSF. The anti-sera did not cross-react with CHO-produced GM-CSF containing O-linked oligosaccharides; however, the same anti-sera did cross-react with CHO-produced GM-CSF after enzymatic removal of the O-linked oligosaccharides. These results demonstrate

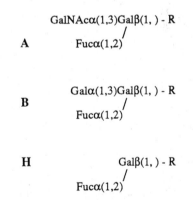

Figure 10. The Minimal Structures Representing the ABH Antigens of the Human ABO Blood Group System
R can contain linear or branched inner core saccharides linked to protein or lipid depending on the tissue expressing the antigen (reviewed in Clausen and Hakomori, 1989)

exposure of an antigenic site upon removal of O-linked oligosaccharides, but they do not distinguish between the steric hindrance and conformation mechanisms noted above. A protein conformational change has been suggested by several authors to explain the differences in antigenicity between glycosylated and aglycosyl forms of hCG (Hattori et al., 1988; Rebois and Liss, 1986; Sairam et al., 1988), ovine luteinizing hormone (Sairam et al., 1988) and Semliki forest virus glycoprotein (Kaluza et al., 1980).

3.8 Effect of Oligosaccharides on Glycoprotein Immunogenicity

While oligosaccharides clearly affect glycoprotein antigenicity, the effect of oligosaccharides on glycoprotein immunogenicity (ability to elicit an immune response) is less clear. Antibodies to oligosaccharide are elicited in animals when the oligosaccharide is conjugated to an immunogenic protein. For example, antibodies to plant-specific oligosaccharide structures similar to that shown in Fig. 5A were raised in rabbits immunized with plant glycoproteins (Kaladas et al, 1983). As an additional example, antibodies against a common O-linked yeast structure, Manα(1,3)Manα(1,2)-Manα(1,2)Man, were raised in rabbits injected with that oligosaccharide conjugated to the plant protein edestin (Zopf et al., 1978).

Little data is available concerning the possible immunogenicity of a glycoprotein with a native amino acid sequence but non-native oligosaccharide structure. Human EPO produced by recombinant CHO cells is the only recombinant glycoprotein for which extensive clinical trial data involving antibody development has been published. No antibodies against recombinant EPO have been detected in patients receiving this product, even after chronic administration up to 18 months (Canaud et al., 1990; Faulds and Sorkin 1989; Lim et al., 1989; Eschbach et al., 1987). Takeuchi and coworkers (1988) have reported the oligosaccharide structures from human urinary EPO and recombinant CHO-derived EPO to be essentially identical, with the exception that the CHO-produced material contains sialic acid only in an α2,3 linkage, while the human urinary material contains both the α2,3 and α2,6 linkages. The EPO example does not provide a satisfactory or conclusive answer to the larger question of the possible effect of non-native oligosaccharide structures

on immunogenicity. For example, it is possible that the EPO glycoforms from recombinant CHO simply represent a subset of the naturally occurring EPO glycoforms that contain sialic acid in both α2,3 and α2,6 linkages.

Nevertheless, the absence of antibodies to EPO from recombinant CHO is an important and encouraging result from a biotechnological standpoint. The detailed analyses of N- and O-linked oligosaccharides from CHO-produced t-PA (Spellman et al., 1989), interferon-β (Kagawa et al., 1988) and IL-2 (Conradt et al., 1989) have also revealed oligosaccharide structures similar to those found on their native human glycoprotein counterparts, although no corresponding data concerning immunogenicity has been published.

Proteins which are normally glycosylated can potentially have altered immunogenicity when administered in aglycosyl form due to the tendency of aglycosyl proteins to aggregate (section 3.1). Protein aggregation enhances immunogenicity through a mechanism that is poorly understood. For example, increased protein aggregation has been correlated with increased immunogenicity of clinically administered human growth hormone preparations (Moore and Leppert, 1980). In addition, deliberate aggregation of muscle creatine kinase (Man et al., 1989) and cytochrome c (Reichlin et al., 1970) has been demonstrated to significantly increase the immunogenicity of these proteins, eliciting antibodies that recognize the native, unaggregated proteins. In some cases, it may be possible to avoid the insolubility problems associated with aglycosyl proteins through a carefully chosen formulation buffer or through covalent attachment to the protein of an oligosaccharide substitute, such as polyethylene glycol. For example, site-specific attachment of polyethylene glycol to human recombinant E. coli-produced IL-2 at its lone O-linked glycosylation site increases the solubility while significantly reducing the immunogenicity of human IL-2 administered in mice (Katre, 1990).

Antibody development has been reported in patients receiving the E. coli-produced human glycoproteins GM-CSF (Thompson et al., 1989) and interferon beta-ser (Konrad et al., 1987; Sarna et al., 1986; Hawkins et al., 1985). However, these studies are inconclusive concerning the effect of missing oligosaccharides on immunogenicity, since these studies may be complicated by other factors -- for example, amino acid point mutations, differences in N-terminal protein processing and potential impurities. Resolution of this issue is further hindered because the glycosylated counterparts of the *E. coli*-produced proteins have not been available for clinical trials.

3.9 Effect of Oligosaccharides on Glycoprotein *in vivo* Circulatory Half-life

Oligosaccharides play a significant role in defining the *in vivo* glycoprotein clearance rate, a critical property in determining the efficacy of an injected therapeutic protein. High *in vitro* specific activity will be of little consequence if an injected protein is too rapidly eliminated from the circulatory system. For example, the specific activity of EPO is increased upon desialylation, but the *in vivo* activity is abolished due to rapid *in vivo* clearance (Briggs et al., 1974; Fukuda et al., 1989).

Several circulatory clearance mechanisms are associated with high affinity receptors recognizing terminal monosaccharides of glycoprotein oligosaccharides. Other clearance mechanisms are more indirectly related to carbohydrate structure. These direct and indirect mechanisms will be briefly reviewed in this section.

The asialoglycoprotein receptor found on hepatocytes binds glycoproteins exhibiting terminal galactose or N-acetylgalactosamine (GalNAc), including desialylated, complex-type N-linked oligosaccharides (e.g., Fig. 3F, 3H and 3I) and desialylated O-linked oligosaccharides (e.g., Fig. 8A and 8B) (reviewed in Weiss and Ashwell, 1989). It is presumed that this receptor serves a major role in the turnover of serum glycoproteins (Weiss and Ashwell, 1989). The functional receptor is an oligomer possessing multiple sites for Gal and GalNAc attachment. Binding of a glycoprotein to this receptor is enhanced by the presence of multiple terminal galactose and N-acetylgalactosamine moieties; for a given asialoglycoprotein, binding will increase with increasing numbers of oligosaccharide groups and with increased oligosaccharide branched structure. The receptor guides bound glycoprotein to lysosomes via endocytosis and the free receptor is then recycled to the cell surface (Weiss and Ashwell, 1989). The asialoglycoprotein receptor is thought to be primarily responsible for the rapid *in vivo* clearance noted for desialylated EPO (Fukuda et al., 1989). Hepatic clearance of EPO by the asialoglycoprotein receptor is apparently promoted by the presence of multiple N-acetyllactosamine repeats (see Fig. 3H) (Fukuda et al., 1989). The loss of *in vivo* biological activity of desialylated GM-CSF also apparently results from rapid clearance by the asialoglycoprotein receptor (Donahue et al., 1986).

A mannose receptor has been identified on the surface of several cell types, including liver endothelial cells and resident macrophage cells, especially those of the spleen, lung and liver (Kupffer cells) (reviewed in Ezekowitz and Stahl, 1988; Stahl, 1990). Oligosaccharides terminating in mannose are infrequently found on the surface of mammalian cells or on mammalian glycoproteins, but are abundant on the surfaces and proteins of lower organisms, such as yeast (see Fig. 6). The mannose receptors of liver endothelial cells and resident macrophages apparently represent a means for recognizing and eliminating cells and glycoproteins bearing high-mannose-type oligosaccharides (Stahl, 1990). The mannose receptor exhibits affinity for terminal monosaccharides according to the order Man $\cong$ Fuc>GlcNAc>Glu>>Gal. Because of the affinity of the receptor for fucose, the receptor is frequently referenced as the mannose/fucose receptor (Stahl, 1990). The mannose receptor is an oligomeric membrane glycoprotein with a molecular weight of 162 kDa. The receptor directs bound glycoprotein to lysosomes via endocytosis and is recycled to the cell surface. The mannose receptor is apparently responsible for circulatory clearance of human glycosylated α-amylase on the basis of its N-linked oligosaccharides possessing terminal fucose bonded α(1,3) to GlcNAc (see Fig. 3J) (Tollefsen and Rosenblum, 1988). Human t-PA is cleared through the mannose receptor of liver endothelial cells and Kupffer cells via the high-mannose oligosaccharide found on t-PA at Asn-117 (Owensby et al., 1988; Krause et al., 1990; Rijken et al., 1990; Smedsrod and Einarsson, 1990). Enzymatic removal of the high-mannose oligosaccharide at Asn-117 using Endo-H (Tanswell et al, 1989) or site-directed mutagenesis to eliminate the glycosylation site at Asn-117 (Hotchkiss et al., 1988) results in enhanced t-PA circulatory half-life. The mannose receptor will presumably represent a major clearance mechanism for glycoproteins produced in recombinant *Saccharomyces cerevisiae* and insect cells which possess terminal mannose or GlcNAc moieties (see Fig. 6 and 7).

Human proteins with molecular weights less than about 70 kDa are continuously removed from the circulation through the kidney. The filtration rate through the kidney glomerular tubules is sensitive to protein tertiary structure as well as molecular weight, and is inhibited by the presence of surface charge (reviewed in Kaniwar, 1984). This filtration system does not possess a sharp molecular weight cutoff. For example, Knauf et al. (1988) observed that the circulatory half-life of IL-2 increased steadily as the molecular weight was increased from 19.5 kDa to 72 kDa by conjugation with polyethylene glycol, and no further increases in half-life were observed for IL-2/PEG conjugates with higher molecular weights. For low molecular weight glycoproteins, the oligosaccharides can prolong glycoprotein circulatory half life by increasing size and surface charge -- the latter contributed by sialic acid (Gross et al., 1988, 1989). For example, α1-acid glycoprotein with its normal complement of five sialylated, complex-type oligosaccharides has a molecular weight in the range of 43 to 60 kDa. Thus, fully glycosylated α1-acid glycoprotein is cleared slowly from the rat circulatory system -- 97% remained in the circulatory system 10 minutes after injection. In contrast, aglycosyl α1-acid glycoprotein (molecular weight = 23 kDa) is cleared rapidly through the kidney -- 36% remained in the circulatory system 10 minutes after injection (Gross et al., 1988).

A protein may be cleared from the circulatory system simultaneously by independent mechanisms. An example of this paradigm is provided by the studies of α1-acid glycoprotein clearance introduced in previous paragraph. Gross and coworkers (1988, 1989) examined the clearance of α1-acid glycoprotein possessing high mannose oligosaccharides in place of its normal, sialylated complex-type oligosaccharides. They observed clearance via kidney filtration, as well as clearance in the spleen and liver via the mannose receptor -- 16% of the α1-acid glycoprotein remained in the circulatory system after 10 minutes. That is, the clearance rate of α1-acid glycoprotein with high mannose oligosaccharides was much faster than the clearance rate of either α1-acid glycoprotein with sialylated complex-type oligosaccharides or the aglycosyl α1-acid glycoprotein (Gross et al., 1988). The α1-acid glycoprotein example graphically illustrates the potential of oligosaccharide structure to either promote or limit glycoprotein circulatory half life.

Some carbohydrate-independent clearance mechanisms have been identified, but are not well-characterized. For example, in addition to clearance via the mannose receptor, t-PA is simultaneously cleared by hepatocytes via a second mechanism which is apparently carbohydrate-independent and t-PA-specific (Owensby et al., 1988; Krause et al., 1990; Rijken et al., 1990; Smedsrod and Einarsson, 1990). While low molecular weight aglycosyl proteins are cleared via kidney filtration, little is known about the clearance mechanisms of high molecular weight, aglycosyl proteins (Gross et al, 1989).

Binding to a circulating antibody provides another potential basis for rapid glycoprotein clearance. The formation of immune complexes of antigen and antibody leads to clearance of antigen via mechanisms involving the complement system and cellular receptors (reviewed in Schiefferli et al., 1986; Anderson, 1989; Baatrup, 1990). These clearance mechanisms have been postulated to explain the rapid clearance of therapeutic proteins during clinical trials where an immunogenic response was elicited (Vallbracht et al., 1981; Quesada et al., 1983; Trown et al., 1983; Gribben et al., 1990). However, more rapid protein clearance has not been reported in all cases where an an immunogenic response was observed to an injected therapeutic protein (Konrad et al., 1987). Some investigators have proposed that specific non-precipitating circulating antibodies may in some cases lead to enhanced circulatory half-life of the target protein, by serving as physiological carriers which inhibit normal clearance mechanisms such as kidney filtration (Bendtzen et al., 1990).

In a section 1.4, it was noted that mouse C127 cells generate glycoproteins possessing the oligosaccharide structure Galα(1,3)Gal, an epitope recognized by one-half to one percent of circulating human antibodies. The pharmacokinetics of recombinant, C127-derived human t-PA containing terminal Galα(1,3)Gal epitopes has been recently examined in chimpanzees, who also have natural circulating antibodies specific for that determinant (Tanigawara et al., 1990). As a comparison, the same study also examined clearance of CHO-derived human t-PA which does not contain the Galα(1,3)Gal epitope. The C127-derived t-PA was cleared no faster than the CHO-derived t-PA, in spite of the presence of serum antibodies against Galα(1,3)Gal epitopes. The interpretation of this result is difficult, since t-PA is being rapidly cleared under normal conditions by at least two simultaneous mechanisms. For example, it is possible that the presence $Gal\alpha(1,3)Gal$ did not contribute to t-PA clearance. It is also possible that t-PA clearance was mediated primarily by $Gal\alpha(1,3)Gal$, while normal clearance mechanisms were inhibited. A more conclusive test would be comparison of clearance rates utilizing a glycoprotein such as EPO which is cleared relatively slowly from the circulation in the absence of $Gal\alpha(1,3)Gal$.

DISCUSSION

The efficacy of a human therapeutic glycoprotein is dependent upon many properties that are potentially affected by oligosaccharide structure (reviewed in Part 3). Oligosaccharides play a prominent role in defining glycoprotein circulatory half-life. In addition, the oligosaccharides of glycoproteins usually enhance protein solubility, and often promote protein folding, secretion, resistance to protease attack and resistance to thermal inactivation. These latter properties are affected by oligosaccharides on a protein-specific basis due to the influence of oligosaccharides on protein surface chemistry and protein tertiary structure.

Specific activity is also frequently affected upon partial or complete removal of oligosaccharides, but is rarely abolished. In fact, the absence of oligosaccharides often enhances specific activity as measured *in vitro* (e.g. EPO and t-PA). Oligosaccharides may influence specific activity by affecting glycoprotein tertiary structure. Alternatively, oligosaccharides may promote or inhibit ligand-receptor interactions through a steric effect or through a charge effect related to the presence of sialic acid on the oligosaccharides of the ligand and/or the receptor.

Differences in the N-linked and O-linked oligosaccharide structures produced by mammalian, yeast, insect and plant cells are documented in Part 1. On the basis of these data, it is difficult to be enthusiastic about the use of yeast, insect and plant cells as hosts for the production of human therapeutic glycoproteins that require substantial circulatory residence time. These cells synthesize oligosaccharide structures with terminating mannose, GlcNAc and/or galactose moieties that should be recognized by the high affinity receptors of the circulatory clearance mechanisms associated with hepatocytes and resident macrophages. Additional immunogenic effects of glycoproteins possessing yeast-, insect- or plant-specific oligosaccharide structures are possible, although as yet unproven.

Differences in oligosaccharide processing among prospective mammalian host cells could also be significant. For example, the $Gal\alpha(1,3)Gal$ moiety determinant found on glycoproteins produced using the C127 mouse expression system could have a significant effect on glycoprotein circulatory half-life, given the prevalence in the human blood stream of antibodies directed against that epitope. The $Gal\alpha(1,3)Gal$ determinant is also likely to be found on proteins produced by transgenic cows, sheep and pigs, since those animals fall in the category of species who express $\alpha(1,3)$ galactosyltransferase.

Indeed, it is somewhat surprising that Chinese hamster ovary cells do not express $\alpha(1,3)$ galactosyltransferase. As noted in section 1.4, Chinese hamster ovary cells do harbor DNA that hybridizes with a probe for a murine $\alpha(1,3)$ galactosyltransferase, suggesting that a native CHO gene expressing $\alpha(1,3)$ galactosyltransferase exists, but is not active.

Detailed N-linked and O-linked oligosaccharide structures have been determined for several glycoproteins produced using recombinant CHO cells, including EPO, t-PA, interferon-$\beta 1$ and IL-2 . A pleasant surprise from these recent analyses has been the remarkable degree to which the oligosaccharide structures from the CHO-produced glycoproteins correspond to the structures of those same proteins isolated from human urine or produced using normal human diploid cells. Indeed, the correspondence is closer than would be expected if those same proteins were produced in recombinant continuous human cell lines of tumor origin. Chinese hamster ovary cells have emerged as the cell line of first choice for the synthesis of recombinant human therapeutic glycoproteins, although CHO cells do possess deficiencies that may limit their applicability in specific cases, such as limited capability for γ-carboxylation and inability for oligosaccharide sulfation.

Finally, it was stressed in Part 1 that oligosaccharide processing will be dependent upon the cell culture environment. Protein translation occurs on the basis of an mRNA template, assuring high fidelity of protein structure. In contrast, oligosaccharide processing occurs as a result of the sequential actions of competing enzymes in several different intracellular compartments. It is not surprising that the outcome of this set of reactions should vary with environmental conditions. In addition, due to the profound effect of oligosaccharide structure on glycoprotein surface chemistry, it is to be expected that oligosaccharide structure will have a significant effect on most downstream processing steps. A given downstream processing step may select for a particular set of glycoforms. Conversely, changes in a downstream processing procedure may change the set of purified glycoforms.

ACKNOWLEDGEMENTS

We gratefully acknowledge the thoughtful comments and editorial assistance of Timothy Hahn, Thomas Monica and Steven Williams.

This project was supported through a National Science Foundation Presidential Young Investigator Award to CG (EET-8857712), and through Merck and 3M Faculty Development Awards to CG.

REFERENCES

Abel, C.A.; Spielberg, H.L.; Grey, H.M. *Biochem.* **1968**, *7*, 1271-1278.

Alexander, S.; Elder, J.H. *Science* **1984**, 226, 1328-1330.

Anderson, C.L. *Clin. Immunol. Immunopath.* **1989**, *53*, S63-S71.

Arakawa, T.; Timasheff, S.N. *Methods in Enzymology* **1985**, *114*, 49-77.

Arvieux, J.; Willis, A.C.; Williams, A.F. *Mol. Immunol.* **1986**, *23*, 983-990.

Ashford, D.; Dwek, R.A.; Weply, J.K.; Amatayakul, S.; Homans, S.W.; Lis, H.; Taylor, G.N.; Sharon, N. Rademacher, T.W. *Eur.J. Biochem.* **1987**, 166, 311-320.

Baatrup, G. *Danish. Med. Bull.* **1989**, *36*, 443-463.

Ballou, C.E. *Methods in Enzymology* **1990**, 185, 440-469.

Baenziger, J.U.; Green, E.D. *Biochim. Biophys. Acta* **1988**, *947*, 287-306.

Barbaric, S.; Mrsa, V.; Ries, B.; Mildner, P. *Arch. Biochem. Biophys.* **1984**, *234*, 567-575.

Bauer, H.C.; Parent, J.B.; Olden, K. *Biochem. Biophys. Res. Comm..* **1985**, *128*, 368-375.

Bause, E. *Biochem. J.* **1983**, *209*, 331-336.

Bendtzen, K.; Svenson, M.; Jonsson, V.; Hippe, E. *Immunol. Today* **1990**, *11*, 167-169.

Bennett, M.; Schmid, K. *Proc. Natl. Acad. Sci.* **1980**, *77*, 6109-6113.

Berman, E. *Biochemistry* **1984**, *23*, 3754-3759.

Blanchard, D.; Dahr, W.; Hummel, M.; Latron, F.; Beyreuther, K.; Cartron, J.-P. *J. Biol. Chem.* **1987**, *262*, 5808-5811.

Boege, F.; Ward, M.; Jurss, R.; Hekman, M.; Helmreich, E.J.M. *J. Biol. Chem.* **1988**, *263*, 9040-9049.

Borel, I.M.; Gentile, T.; Angelucci, J.; Margni, R.A.; Binaghi, R.A. *Biochim. Biophys. Acta* **1989**, *990*, 162-164.

Breg, J.; Kroon-Batenburg, L.M.J.; Strecker, G.; Montreuil, J.; Vliegenthart, J.F.G. *Eur. J. Biochem.* **1989**, *178*, 727-739.

Breitfeld, P.P.; Rup, D.; Schwartz, A.L. *J. Biol. Chem.* **1984**, *259*, 10414-10421.

Briggs, D.W.; Fisher, J.W.; George, W.J. *Am. J. Physiol.* **1974**, *227*, 1385-1388.

Campbell, C.; Stanley, P. *Cell* **1983**, *35*, 303-309.

Campion, B.; Leger, D.; Wieruszeski, J.-M.; Montreuil, J.; Spik, G. *Eur. J. Biochem.* **1989** *184*, 405-413.

Canaud, B.; Polito-Bouloux, C.; Garred, L.J.; Rivory, J.-P., Donnadieu, P.; Taib, J.; Florence, P.; Mion, C. *Am. J. Kidney Dis.* **1990**, *15*, 169-175.

Carlsson, S.R.; Fukuda, M. *J. Biol. Chem.* **1986**, *261*, 12779-12786.

Carlsson, S.R.; Sasaki, H.; Fukuda, M. *J. Biol. Chem.* **1986**, *261*, 12787-12795.

Carraway, K.L.; Hull, S.R. *BioEssays* **1989**, *10*, 117-121.

Carver, J.P.; Cumming, D.A. *Pure & Appl. Chem.* **1987**, *59*, 1465-1476.

Carver, J.P.; Michnick, S.W.; Imberty, A.; Cumming, D.A. In *Carbohydrate Recognition in Cellular Function*; Ciba Foundation Symposium 145; John Wiley & Sons: Chichester, UK, **1989**; pp 6-26.

Casagli, M.C.; Borri, M.G.; Bigio, M.; Rossi, R.; Nucci, D.; Bossu, P.; Boraschi, D.; Antoni, G. *Biochem. Biophys. Res. Comm.* **1989**, *162*, 357-363.

Cebon, C.; Nicola, N.; Ward, M.; Gardner, I.; Dempsey, P.; Layton, J.; Duhrsen, U.; Burgess, A.W.; Nice, E.; Morstyn, G. *J. Biol. Chem.* **1990**, *265*, 4483-4491.

Chavin, S.I.; Weidner, S.M. *J. Biol. Chem.* **1984**, *259*, 3387-3390.

Chew, W.H.; Theus, T.L. *J. Immunol.* **1967**, *98*, 220-224.

Chu, R.K.; Trimble, R.B.; Maley, F. *J. Biol. Chem.* **1978**, *253*, 8691-8693.

Clark, S. C. *Int. J. Cell Cloning* **1988**, *6*, 365-377.

Clausen, H.; Hakomori, S-i.. *Vox Sang.* **1989**, *56*, 1-20.

Clemmons, D.R. *Brit. Med. Bull.* **1989**, *45*, 465-480.

Conradt, H.S.; Geyer, R.; Hoppe, J.; Grotjahn, L.; Plessing A.; Mohr, H. *Eur. J. Biochem.* **1985**, *153*, 255-261.

Conradt, H.S.; Ausmeier, M.; Dittmar, K.E.J.; Hauser, H.; Lindenmaier, W. *Carbohydrate Res.* **1986**, *149*, 443-450.

Conradt, H.S.; Egge, H.; Peter-Katalinic, J.; Reiser, W.; Siklosi, T.; Schaper, K. *J. Biol. Chem.* **1987**, *262*, 14600-14605.

Conradt, H.S.; Nimtz, M.; Dittmar, K.E.J.; Lindenmaier, W.; Hoppe, J.; Hauser, H. *J. Biol. Chem.* **1989**, *264*, 17368-17373.

Cummings, R.D.; Kornfeld, S.; Schneider, W.J.; Hobgood, K.K.; Tolleshaug, H.; Brown, M.S.; Goldstein, J.L. *J. Biol. Chem.* **1983**, *258*, 15261-15273.

Cummings, R.D.; Mattox, S.A. *J. Biol. Chem.* **1988**, *263*, 511-519.

Dahms, N.M.; Hart, G.W. *J. Biol. Chem.* **1986**, *261*, 13186-13196.

Davidson, D.J.; Fraser, M.J.; Castellino, F.J. *Biochemistry* **1990**, *29*, 5584-5590.

Deisenhofer, J. *Biochemistry* **1981**, *20*, 2361-2370.

Deschuyteneer, M.; Eckhardt, A.E.; Roth, J.; Hill, R.L. *J. Biol. Chem.* **1988**, *263*, 2452-2459.

Dill, K.A. *Biochemistry* **1990**, *29*, 7133-7155.

Do, S.-I.; Enns, C.; Cummings, R.D. *J. Biol. Chem.* **1990**, *265*, 114-125.

Donahue, R.E.; Wang, E.A.; Kaufman, R.J.; Foutch, L.; Leary, A.C.; Witek-Giannetti, J.S.; Metzger, M.; Hewick, R.M.; Steinbrink, D.R.; Shaw, G.; Kamen, R.; Clark, S.C. *Cold Spring Harbor Symposia on Quantitative Biology* **1986**, *51*, 685-692.

Dordal, M.S.; Wang, F.F.; Goldwasser, E. *Endocrinology* **1985**, *116*, 2293-2299.

Dorner, A.J.; Bole, D.G.; Kaufman, R.J. *J. Cell Biology* **1987**, *105*, 2665-2674.

Doyle, M.V.; Lee, M.T.; Fong, S. *J. Biological Response Modifiers* **1985**, *4*, 96-109.

Dube, S.; Fisher, J.W.; Powell, J.S. *J. Biol. Chem.* **1988**, *263*, 17516-17521.

Edwards, K.; Nagashima, M.; Dryburgh, H.; Wykes, A.; Schreiber, G. *FEBS Letters* **1979**, *100*, 269-272.

Einarsson, M.; Brandt, J.; Kaplan, L. *Biochim. Biophys. Acta* **1985**, *830*, 1-10.

231

Elbein, A.D. *Methods in Enzymology* **1987**, *138*, 661-709.

Elliott, S.; Fagin, K.D.; Narhi, L.O.; Miller, J.A.; Jones, M.; Koski, R.; Peters, M.; Hsieh, P.; Sachdev, R.; Rosenfeld, R.D.; Rohde, M.F.; Arakawa, T. *J. Protein Chemistry* **1990**, *9*, 95-104.

Eschbach, J.W.; Egrie, J.C.; Downing, M.R.; Browne, J.K.; Adamson, J.W. *N. Engl. J. Med.* **1987**, *316*, 73-78.

Esmon, P.C.; Esmon, B.E.; Schauer, I.E.; Taylor, A.; Schekman, R. *J. Biol. Chem.* **1987**, *262*, 4387-4394.

Ezekowitz, R.A.B.; Stahl, P.D. *J. Cell. Sci. Suppl.* **1988**, *9*, 121-133.

Faulds, D.; Sorkin, E.M. *Drugs* **1989**, *38*, 863-899.

Feizi, T.; Childs, R.A. *Biochem. J.* **1987**, *245*, 1-11.

Fennie, C.; Lasky, L.A. *J. Virol.* **1989**, *63*, 639-646.

Fenouillet, E.; Gluckman, J.C.; Bahraoui, E. *J. Virol.* **1990**, *64*, 2841-2848.

Fiat, A.-M.; Jolles, J.; Aubert, J.-P.; Laucheux-Lefebvre, M.-H.; Jolles, P. *Eur. J. Biochem.* **1980**, *111*, 333-339.

Field, M.C.; Dwek, R.A.; Edge, C.J.; Rademacher, T.W. *Biochem. Soc. Trans.* **1989**, *17*, 1034-1035.

Fisher, T.; Thomas, B.; Scheurich, P.; Pfizenmaier, K. *J. Biol. Chem.* **1990**, *265*, 1710-1717.

Florman, H.M.; Wassarman, P.M. *Cell* **1985**, *41*, 313-324.

Fournet, B.; Leroy, Y.; Wieruszeski, J.-M.; Montreuil, J.; Poretz, R.D.; Goldberg, R. *Eur. J. Biochem.* **1987**, *166*, 321-324.

Fukuda, M.N.; Sasaki, H.; Lopez, L.; Fukuda, M. *Blood* **1989**, *73*, 84-89.

Galili, U.; Macher, B.A.; Buehler, J.; Shohet, S.B. *J. Exp. Med.* **1985**, *162*, 573-582.

Galili, U. *Blood Cells* **1988**, *14*, 205-220.

Galili, U.; Shohet, S.B.; Kobrin, E.; Stults, C.L.M.; Macher, B.A. *J. Biol. Chem.* **1988**, *263*, 17755-17762.

Gamou, S.; Shimizu, N. *J. Biochem.* **1988**, *104*, 388-396.

Gavel, Y.; von Heijne, G. *Protein Engineering* **1990**, *3*, 433-442.

Gellerfors, P.; Axelsson, K.; Helander, A.; Johansson, S.; Kenne, L.; Lindqvist, S.; Pavlu, B.; Skottner, A.; Fryklund, L. *J. Biol. Chem.* **1989**, *264*, 11444-11449.

Gerken, T.A.; Butenhof, K.J.; Shogren, R. *Biochemistry* **1989**, *28*, 5536-5543.

Gething, M.-J.; Sambrook, J. *Biochem. Soc. Symp.* **1989**, *55*, 155-166.

Goldwasser, E.; Kung, C.K.-H.; Eliason, J. *J. Biol. Chem.* **1974**, *249*, 4202-4206.

Gomi, K.; Morimoto, M.; Nakamizo, N. *Gann.* **1983**, *74*, 737-742.

Goochee, C.F.; Monica, T. *Bio/Technology* **1990**, *8*, 421-427.

Goto, M.; Akai, K.; Murakami, A.; Hasimoto, C.; Tsuda, E.; Ueda, M.; Kawanishi, G.; Takahashi, N.; Ishimoto, A.; Chiba, H.; Saski, R. *Bio/Technology* **1988**, *6*, 67-71.

Grafl, R.; Lang, K.; Vogl, H.; Schmid, F.X. *J. Biol. Chem.* **1987**, *262*, 10624-10629.

Green, C. *FEMS Microb. Immunol.* **1989**, *47*, 321-330

Gribben, J.G.; Devereux, S.; Thomas, N.S.B.; Keim, M.; Jones, H.M.; Goldstone, A.H.; Linch, D.C. *Lancet* **1990**, *335*, 434-437.

Grimaldi, S.; Robbins, J.; Edelhoch, H. *Biochemistry* **1985**, *24*, 3771-3776.

Gross, V.; Heinrich, P.C.; vom Berg, D.; Steube, K.; Andus, T.; Tran-Thi, T.-A.; Decker, K.; Gerok, W. *Eur. J. Biochem.* **1988**, *173*, 653-659.

Gross, V.; Steube, K.; Tran-Thi, T.-A.; Gerok, W.; Heinrich, P.C. *Biochem. Soc. Trans.* **1989**, *17*, 21-23.

Gross, H.J.; Rose, U.; Krause, J.M.; Paulson, J.C.; Schmid, K.; Feeney, R.E.; Brossmer, R. *Biochemistry* **1989**, *28*, 7386-7392.

Halenbeck, R.; Kawasaki, E.; Wrin, J.; Koths, K. *Bio/Technology* **1989**, *7*, 710-715.

Hansen, L.; Blue, Y.; Barone, K.; Collen, D.; Larsen, G.R. *J. Biol. Chem.* **1988**, *263*, 15713-15719.

Hard, K.; Bitter, W.; Kamerling, J.P.; Vliegenthart, J.F.G. *FEBS Lett.* **1989**, *248*, 111-114

Hattori, M.-a.; Hachisu, T.; Shimohigashi, Y.; Wakabayashi, K. *Mol. Cell. Endocrinol.* **1988**, *57*, 17-23.

Hawkins, M.; Horning, S.; Konrad, M.; Anderson, S.; Sielaff, K.; Rosno, S.; Schiesel, J.; Davis, T.; DeMets, D.; Merigan, T.; Borden, E. *Cancer Res.* **1985**, *45*, 5914-5920.

Hickman, S.; Kornfeld, S. *J. Immunol.* **1978**, *121*, 990-996.

Hindsgaul, O.; Khare, D.P.; Bach, M.; Lemieux, R.U. *Can. J. Chem.* **1985**, *63*, 2653-2658.

Hitzman, R.A.; Chen, C.Y.; Dowbenko, D.J.; Renz, M.E.; Liu, C.; Pai, R.; Simpson, N.J.; Kohr, W.J.; Singh, A.; Chisholm, V.; Hamilton, R.; Chang, C.N. *Methods in Enzymology* **1990**, *185*, 421-440.

Homans, S.W.; Dwek, R.A.; Rademacher, T.W. *Biochemistry* **1987**, *26*, 6571-6578.

Hortin, G.; Green, E.D.; Baenziger, J.U.; Strauss, A.W. *Biochem. J.* **1986**, *235*, 407-414.

Hotchkiss, A.; Refino, C.J.; Leonard, C.K.; O'Connor, J.V.; Crowley, C.; McCabe, J.; Tate, K.; Nakamura, G.; Powers, D.; Levinson, A.; Mohler, M.; Spellman, M.W. *Thomb. Haem.* **1988**, *60*, 255-261.

Housley, T.J.; Rowland, F.N.; Ledger, P.W.; Kaplan, J.; Tanzer, M.L. *J. Biol. Chem.* **1980**, *255*, 121-128.

Hubbard, S.C. *J. Biol. Chem.* **1988**, *263*, 19303-19317.

Hunt, R.C.; Riegler, R.; Davis, A.A. *J. Biol. Chem.* **1989**, *264*, 9643-9648.

Hurtley, S.M.; Helenius, A. *Annu. Rev. Cell Biol.* **1989**, *5*, 277-307.

Ishizaka, K. *Ann. Rev. Immunol.* **1988**, *6*, 513-534.

Itoh, K.; Oshima, A.; Sakuraba, H.; Suzuki, Y. *Biochem. Biophys. Res. Comm.* **1990**, *167*, 746-753.

Jacobs, K.; Shoemaker, C.; Rudersdorf, R.; Neill, S.D.; Kaufman, R.J.; Mufson, A.; Seehra, J.; Jones, S.S.; Hewick, R.; Fritsch, E.F.; Kawakita, M.; Shimizu, T.; Miyake, T. *Nature* **1985**, *313*, 806-810.

Kaetzel, D.M.; Virgin, J.B.; Clay, C.M.; Nilson, J.H. *Molecular Endocrinology* **1989**, *3*, 1765-1774.

Kagawa, Y.; Takasaki, S.; Utsumi, J.; Hosoi, K.; Shimizu, H.; Kochibe, N.; Kobata, A. *J. Biol. Chem.* **1988**, *263*, 17508-17515.

Kaladas, P.M.; Goldberg, R.; Poretz, R.D. *Mol. Immunol.* **1983**, *20*, 727-735.

Kalisz, H.M.; Hecht, H.-J.; Schomburg, D.; Schmid, R.D. *J. Mol. Biol.* **1990**, *213*, 207-209.

Kaluza, G.; Rott, R.; Schwarz, R.T. *Virology* **1980**, *102*, 286-299.

Kalyan, N.K.; Bahl, O.P. *J. Biol. Chem.* **1983**, *258*, 67-74.

Kaniwar, Y.S. *Lab. Invest.* **1984**, *51*, 7-21.

Kaplan, E.H.; Rosen, S.T.; Norris, D.B.; Roenigk, H.H. Jr.; Saks, S.R.; Bunn, P.A. Jr. *J. Natl. Cancer Inst.* **1990**, *82*, 208-212.

Katre, N.V. *J. Immunol.* **1990**, *144*, 209-213.

Kaushal, G.P.; Szumilo, T.; Elbein, A.D. *Biochemistry of Plants* **1988**, *14*, 421-463.

Kingsley, D.M.; Kozarsky, K.F.; Hobbie, L.; Krieger, M. *Cell* **1986**, *44*, 749-759.

Knauf, M.J.; Bell, D.P.; Hirtzer, P.; Luo, Z.-P.; Young, J.D.; Katre, N.V. *J. Biol. Chem.* **1988**, *263*, 15064-15070.

Kobata, A. In *Biology of Carbohydrates*; Ginsburg, V.; Robbins, P.W., Eds.; John Wiley & Sons: New York, NY, **1984**, Vol. 2, pp 87-110..

Kobata, A. *J. Cell. Biochem.* **1988**, *37*, 79-90.

Konig, R.; Ashwell, G.; Hanover, J.A. *Proc. Natl. Acad. Sci.* **1989**, *86*, 9188-9192.

Konrad, M.W.; Childs, A.L.; Merigan, T.C.; Borden, E.C. *J. Clin. Immunol.* **1987**, *7*, 365-375.

Kornfeld, R.; Kornfeld, S. *Ann. Rev. Biochem.* **1985**, *54*, 631-664 .

Kozarsky, K.; Kingsley, D.; Krieger, M. *Proc. Natl. Acad. Sci.* **1988**, *85*, 4335-4339.

Krause, J.; Seydel, W.; Heinzel, G.; Tanswell, P. *Biochem. J.* **1990**, *267*, 647-652.

Kronquist, K.E.; Lennarz, W.J. *J. Supramolecular Structure* **1978**, *8*, 51-65.

Kubota, N.; Orita, T.; Hattori, K.; Oh-eda, M.; Ochi, N.; Yamazaki, T. *J. Biochem.* **1990**, *107*, 486-492.

Kukuruzinska, M.A.; Bergh, M.L.E.; Jackson, B.J. *Ann. Rev. Biochem.* **1987**, *56*, 915-944.

Kuroda, K.; Geyer, H.; Geyer, R.; Doerfler, W.; Klenk, H.-D. *Virology* **1990**, *174*, 418-429.

Lang, L.; Reitman, M.; Tang, J.; Roberts, R.M.; Kornfeld, S. *J. Biol. Chem.* **1984**, *259*, 14663-14671.

Langer, B.G.; Weisel, J.W.; Dinauer, P.A.; Nagaswami, C.; Bell, W.R. *J. Biol. Chem.* **1988**, *263*, 15056-15063.

Larsen, R.D.; Rajan, V.P.; Ruff, M.M.; Kukowska-Latallo, J.; Cummings, R.D.; Lowe, J.B. *Proc. Natl. Acad. Sci* **1989**, *86*, 8227-8231.

Lawson, E.Q.; Hedlund, B.E.; Ericson, M.E.; Mood, D.A.; Litman, G.W.; Middaugh, R. *Arch. Biochem. Biophys.* **1983**, *220*, 572-575.

Leatherbarrow, R.J.; Rademacher, T.W.; Dwek, R.A.; Woof, J.M.; Clark, A.; Burton, D.R.; Richardson, N.; Feinstein, A. *Mol. Immun.* **1985**, *22*, 407-415.

Leavitt, R.; Schlesinger, S.; Kornfeld, S. *J. Biol. Chem.* **1977**, *252*, 9018-9023.

Lee, E.U.; Roth, J.; Paulson, J.C. *J. Biol. Chem.* **1989**, *264*, 13848-13855.

Lee, S.-J.; Nathans, D. *J. Biol. Chem.* **1988**, *263*, 3521-3527.

Lee, S.-O.; Connolly, J.M.; Ramirez-Sota, D.; Poretz, R.D. *J. Biol. Chem.* **1990**, *265*, 5833-5839.

Lehle, L.; Bause, E. *Biochim. Biophys. Acta* **1984**, *799*, 246-251.

Lemieux, R.U.; Hindsgaul, O.; Bird, P.; Narasimhan, S.; Young, W.W. *Carbohydrate Res.* **1988**, *178*, 293-305.

Leonard, C.K.; Spellman, M.W.; Riddle, L.; Harris, R.J.; Thomas, J.N.; Gregory, T.J. *J. Biol. Chem.* **1990**, *265,* 10373-10382.

Li, Z.-Q.; Perkins, S.J.; Loucheux-Lefebvre, M.H. *Eur. J. Biochem.* **1983**, *130*, 275-279.

Lim, V.S.; Kirchner, P.T.; Fangman, J.; Richmond, J.; DeGowin, R.L. *Am. J. Kidney Dis.* **1989**, *14*, 496-506.

Lin, R.K.; Suggs, S.; Lin, C.-H.; Browne, J.K.; Smalling R.; Egrie, J.C.; Chen, K.K.; Fox, G.M.; Martin, F.; Stabinsky, Z.; Badrawi, S.M.; Lai, P.-H.; Goldwasser, E. *Proc. Natl. Acad. Sci.* **1985**, *82*, 7580-7584.

Livi, G.P.; Ferrara, A.; Roskin, R.; Simon, P.L.; Young, P.R. *Gene* **1990**, *88*, 297-301.

Lodish, H.F. *J. Biol. Chem.* **1988**, *263*, 2107-2110.

Loh, Y.P.; Gainer, H. *Endocrinol.* **1979**, *105*, 474-487.

Machamer, C.E.; Rose, J.K. *J. Biol. Chem.* **1988a**, *263,* 5948-5954.

Machamer, C.E.; Rose, J.K. *J. Biol. Chem.* **1988b**, *263,* 5955-5960.

Man, N.t.; Cartwright, A.J.; Andrews, K.M.; Morris, G.E. *J. Immunol. Methods* **1989**, *125*, 251-259.

March, C.J.; Mosley, B.; Larsen, A.; Cerretti, D.P.; Braedt, G.; Price, V.; Gillis, S.; Henney, C.S.; Kronheim, S.R.; Grabstein, K.; Conlon, P.J.; Hopp, T.P.; Cosman, D. *Nature* **1985**, *315*, 641-647.

Margni, R.A.; Binaghi, R.A. *Ann. Rev. Immunol.* **1988**, *6*, 535-554.

Mark, D.F.; Lu, S.D.; Creasey, A.A.; Yamamoto, R.; Lin, L.S. *Proc. Natl. Acad. Sci.* **1984**, *81*, 5662-5666.

Markoff, E.; Sigel, M.B.; Lacour, N.; Seavey, B.K.; Friesen, H.G.; Lewis, U.J. *Endocrin.* **1988**, *123*, 1303-1306.

Marti, T.; Schaller, J.; Rickli, E.E.; Schmid, K.; Kamerling, J.P.; Gerwig, G.J.; van Halbeek, H.; Vliegenthart, J.F.G. *Eur. J. Biochem.* **1988**, *173*, 57-63.

Matzuk, M.M.; Krieger, M.; Corless, C.L.; Boime, I. *Proc. Natl. Acad. Sci.* **1987**, *84*, 6354-6358.

Matzuk, M.M.; Boime, I. *J. Cell Biology* **1988a**, *106*, 1049-1059.

Matzuk, M.M.; Boime, I. *J. Biol. Chem.* **1988b**, *263*, 17106-17111.

Matzuk, M.M.; Keene, J.L.; Boime, I. *J. Biol. Chem.* **1989**, *264*, 2409-2414.

Matzuk, M.M.; Hsueh, A.J.W.; LaPolt, P.; Tsafriri, A.; Keene, J.L.; Boime, I. *Endocrinology* **1990**, *126*, 376-383.

Merz, W.E. *Bioch. Biophys. Res. Comm.* **1988**, *156*, 1271-1278.

Meyer, B. In *Carbohydrate Chemistry*; Thiem, J., Ed; from the series *Topics in Current Chemistry* **1990**, *154*, 142-208.

Middaugh, C.R.; Litman, G.W. *J. Biol. Chem.* **1987**, *262*, 3671-3673.

Miletich, J.P.; Broze, G.J. *J. Biol. Chem.* **1990**, *265*, 11397-11404.

Mizrahi, A.; O'Malley, J.A.; Carter, W.A.; Takatsuki, A.; Tamura, G.; Sulkowski, E. *J. Biol. Chem.* **1978**, *253*, 7612-7615.

Mononen, I.; Karjalainen, E. *Biochim. Biophys. Acta* **1984**, *788*, 364-367.

Montreuil, J. *Bio. Cell* **1984**, *51*, 115-132.

Moonen, P.; Mermod, J.-J.; Ernst, J. F.; Hirschi, M.; DeLamarter, J. F. *Proc. Natl. Acad. Sci.* **1987**, *84*, 4428-4431.

Moore, W.V.; Leppert, P. *J. Clin. Endocrin. Metab.* **1980**, *51*, 691-697.

Morrison T.G.; McQuain, C.O.; Simpson, D. *J. Virol.* **1978**, *28*, 368-374.

Nishimura, H.; Kawabata, S.-I.; Kisiel, W.; Hase, S.; Ikenaka, T.; Takao, T.; Shimonishi, Y.; Iwanga, S. *J. Biol. Chem.* **1989**, *264*, 20320-20325.

Ng, D.T.W.; Hiebert, S.W.; Lamb, R.A. *Mol. Cell. Biol.* **1990**, *10*, 1989-2001.

Nose, M.; Wigzell, H. *Proc. Natl. Acad. Sci. USA.* **1983**, *80*, 6632-6636.

Oheda, M.; Hase, S.; Ono, M.; Ikenaka, T. *J. Biochem.* **1988**, *103*, 544-546.

Oh-eda, M.; Hasegawa, M.; Hattori, K.; Kuboniwa, H.; Kojima, T.; Orita, T.; Tomonou, K.; Yamazaki, T.; Ochi, N. *J. Biol. Chem.* **1990**, *265*, 11432-11435.

Olden, K.; Parent, J.B.; White, S.L. *Biochim. Biophys. Acta* **1982**, *650*, 209-232.

Olden, K.; Pratt, R.M.; Yamada, K.M. *Proc. Natl. Acad. Sci.* **1979**, *76*, 3343-3347.

Owensby, D.A.; Sobel, B.E.; Schwartz, A.L. *J. Biol. Chem.* **1988**, *263*, 10587-10594.

Palcic, M.M.; Ripka, J.; Kaur, K.J.; Shoreibah, M.; Hindsgaul, O.; Pierce, M. *J. Biol. Chem.* **1990**, *265*, 6759-6769.

Parekh, R.B.; Tse, A.G.D.; Dwek, R.A.; Williams, A.F.; Rademacher, T.W. *EMBO J.* **1987**, *6*, 1233-1244.

Parekh, R.B.; Dwek, R.A.; Thomas, J.R.; Opdenakker, G.; Rademacher, T.W; Wittwer, A.J.; Howard, S.C.; Nelson, R.; Siegel, N.R.; Jennings, M.G.; Harakas, N.K.; Feder, J. *Biochem.* **1989a**, *28,* 7644-7662.

Parekh, R.B.; Dwek, R.A.; Rudd, P.M.; Thomas, J.R.; Rademacher, T.W; Warren, T.; Wun, T.-C.; Hebert, B.; Reitz, B.; Palmier, M.; Ramabhadran, T.; Tiemeier, D.C. *Biochem.* **1989b**, *28,* 7670-7679.

Pathak, R.K.; Merkle, R.K.; Cummings, R.D.; Goldstein, J.L.; Brown, M.S.; Anderson, R.G.W. *J. Cell Biol.* **1988**, *106*, 1831-1841.

Paulson, J.C.; Hill, R.L.; Tanabe, T.; Ashwell, G. *J. Biol. Chem.* **1977**, *252*, 8624-8628.

Paulson, J.C. *Trends in Biochemical Sciences* **1989**, *14*, 272-276.

Paulson, J.C.; Colley, K.J. *J. Biol. Chem.* **1989**, *264*, 17615-17618.

Paulson, J.C.; Weinstein, J.; Schauer, A. *J. Biol. Chem.* **1989**, *264*, 10931-10934.

Pellegrini, I.; Gunz, G.; Ronin, C.; Fenouillet, E.; Peyrat, J.-P.; Delori, P.; Jaquet, P. *Endocrin.* **1988**, *122*, 2667-2674.

Pennica, D.; Holmes, W.E.; Kohr, W.J.; Harkins, R.N.; Vehar, G.A.; Ward, C.A.; Bennett, W.F.; Yelverton, E.; Seeburg, P.H.; Heyneker, H.L.; Goeddel, D.V. *Nature* **1983**, *301*, 214-221.

Perdue, J.F. *Can. J. Biochem. Cell Biol.* **1984**, *62*, 1237-1245.

Perkins, S.J.; Kerckaert, J.-P.; Loucheux-Lefebvre, M.H. *Eur. J. Biochem.* **1985**, *147*, 525-531.

Persaud, R.; Fraser, P.; Wood, D.D.; Moscarello, M.A. *Biochim. Biophys. Acta* **1988**, *966*, 357-361.

Pfeffer, S.R.; Rothman, J.E. *Ann. Rev. Biochem.* **1987**, *56*, 829-852.

Pfeiffer, G.; Schmidt, M.; Strube, K.-H.; Geyer, R. *Eur. J. Biochem.* **1989**, *186*, 273-286.

Piller, F.; Piller, V.; Fox, R.I.; Fukuda, M. *J. Biol. Chem.* **1988**, *263,* 15146-15150.

Piller, V.; Piller, F.; Fukuda, M. *J. Biol. Chem.* **1990**, *265*, 9264-9271.

Pless, D.D.; Lennarz, W.J. *Proc. Natl. Acad. Sci.* **1977**, *74*, 134-138.

Pohl, G.; Kallstrom, M.; Bergsdorf, N.; Wallen, P.; Jornvall, H. *Biochemistry* **1984**, *23*, 3701-3707.

Pos, O.; Oostendorp, R.A.J.; van der Stelt, M.E.; Scheper, R.J.; van Dijk, W. *Inflammation* **1990**, *14*, 133-141.

Potvin, B.; Kumar, R.; Howard, D.R.; Stanley, P. *J. Biol. Chem.* **1990**, *265*, 1615-1622.

Prives, J.M.; Olden, K. *Proc. Natl. Acad. Sci.* **1980**, *77*, 5263-5267.

Puett, D. *J. Biol. Chem.* **1973**, *248,* 3566-3572.

Purchio, A.F.; Cooper, J.A.; Brunner, A.M.; Lioubin, M.N.; Gentry, L.E.; Kovacina, K.S.; Roth, R.A.; Marquardt, H. *J. Biol. Chem.* **1988**, *263*, 14211-14215.

Quelle, F.W.; Caslake, L.F.; Burkert, R.E.; Wojchowski, D.M. *Blood* **1989**, *74*, 652-657.

Quesada, J.R.; Gutterman, J.U. *J . Natl. Cancer Inst.* **1983,** *70*, 1041-1046.

Quiocho, F.A. *Ann. Rev. Biochem.* **1986**, *55*, 287-315.

Quiocho, F.A. *Pure & Appl. Chem.* **1989**, *61*, 1293-1306.

Rademacher, T.W.; Dwek, R.A. *Prog. Immun.* **1983**, *5*, 95-112.

Rademacher, T.W.; Homans, S.W.; Parekh, R.B.; Dwek, R.A. *Biochem. Soc. Symp.* **1986**, *51*, 131-148.

Rademacher, T.W.; Parekh, R.B.; Dwek, R.A. *Ann. Rev. Biochem.* **1988**, *57*, 785-838.

Rebois, R.V.; Liss, M.T. *J. Biol. Chem.* **1987**, *262*, 3891-3896.

Reckhow, C.L.; Enns, C.A. *J. Biol. Chem.* **1988**, *263*, 7297-7301.

Reddy, P.; Caras, I.; Krieger, M. *J. Biol. Chem.* **1989**, *264*, 17329-17336.

Reddy, A.V.; MacColl, R.; Maley, F. *Biochemistry* **1990**, *29*, 2482-2487.

Reichlin, M.; Nisonoff, A.; Margoliash, E. *J. Biol. Chem.* **1970**, *245*, 947-954.

Rijken, D.C.; Otter, M.; Kuiper, J.; van Berkel, T.J.C. *Thrombosis Res.* **1990**, *Supplement X*, 63-71.

Robb., R.J.; Kutny, R.M.; Panico, M.; Morris, H.R.; Chowdhry, V. *Proc. Natl. Acad. Sci.* **1984**, *81*, 6486-6490.

Roitsch, T.; Lehle, L. *Eur. J. Biochem.* **1989**, *181*, 525-529.

Ronnett, G.V.; Lane, M.D. *J. Biol. Chem.* **1981**, *256*, 4704-4707.

Rose, J.K.; Doms, R.W. *Ann. Rev. Cell Biol.* **1988**, *4*, 257-288.

Rosenberg, S.A.; Grimm, E.A.; McGrogan, M.; Doyle, M.; Kawasaki, E.; Koths, K.; Mark, D.F. *Science* **1984**, *223*, 1412-1415.

Roux, L.; Holojda, S.; Sundblad, G.; Freeze, H.H.; Varki, A. *J. Biol. Chem.* **1988**, *263*, 8879-8889.

Sadler, J.E. In *Biology of Carbohydrates*; Ginsburg, V.; Robbins, P.W., Eds; John Wiley & Sons: New York, NY, 1984, Vol. 2, pp 199-288.

Sairam, M.R. *FASEB J.* **1989**, *3*, 1915-1926.

Sairam, M.R. *Biochem. J.* **1990**, *265*, 667-674.

Sairam, M.R.; Linggen, J.; Bhargavi, G.N. *Bioscience Reports* **1988**, *8*, 271-278.

Sarna, G.; Pertcheck, M.; Figlin, R.; Ardalan, B. *Cancer Treat. Rep.* **1986**, *70*, 1365-1372.

Sasaki, H.; Bothner, B.; Dell, A.; Fukuda, M. *J. Biol. Chem.* **1987**, *262*, 12059-12076.

Sasaki, H.; Ochi, N.; Dell, A.; Fukuda, M. *Biochem.* **1988**, *27*, 8618-8626.

Savolainen, J.; Viander, M.; Koivikko, A. *Allergy* **1990**, *40*, 54-63.

Savvidou, G.; Klein, M.; Horne, C.; Hofmann, T.; Dorrington, K.J. *Molecul. Immunol.* **1981**, *18*, 793-805.

Savvidou, G.; Klein, M.; Grey, A.A.; Dorrington, K.J.; Carver, J.P. *Biochemistry* **1984**, *23*, 3736-3740.

Schachter, H.; Narasimhan, S.; Gleeson, P.; Vella, G. *Can. J. Biochem. Cell Biol.* **1983**, *61*, 1049-1066.

Schachter, H. *Biochem. Cell Biol.* **1986**, *64*, 163-181.

Schauer, R. *Adv. Exp. Med. Biol.* **1988**, *228*, 47-72.

Schiel, X.; Rose-John, S.; Dufhues, G.; Schooltink, H.; Gross, V.; Heinrich, P. *Eur. J. Immunol.* **1990**, *20*, 883-887.

Schifferli, J.A.; Ng, Y.C.; Peter, D.K. *New Engl. J. Med.* **1986**, *315*, 488-495.

Schulke, N.; Schmid, F.X. *J. Biol. Chem.* **1988a**, *263*, 8827-8831.

Schulke, N.; Schmid, F.X. *J. Biol. Chem.* **1988b**, *263*, 8832-8837.

Sedmak, J.J.; Grossberg, S.E. *J. Interferon Res.* **1989**, *9(suppl. 1)*, S61-S65.

Shogren, R.; Gerken, T.A.; Jentoft, N. *Biochemistry* **1989**, *28*, 5525-5536.

Sibley, C.H.; Wagner, R.A. *J. Immunol.* **1981**, *126*, 1868-1873.

Sidman, C. *J. Biol. Chem.* **1981**, *256*, 9374-9376.

Smedsrod, B.; Einarsson, M. *Thromb. Haem.* **1990**, *63*, 60-66.

Smith, D.F.; Larsen, R.D.; Mattox, S.; Lowe, J.B.; Cummings, R.D. *J. Biol. Chem.* **1990**, *265*, 6225-6234.

Smith, G.E.; Summers, M.D.; Fraser, M.J. *Mol. Cell. Biol.* **1983**, *3*, 2156-2165.

Smith, G.E.; Ju, G.; Ericson, B.L.; Moschera, J.; Lahm, H.-W.; Chizzonite, R.; Summers, M.D. *Proc. Natl. Acad. Sci.* **1985**, *82*, 8404-8408.

Smith, K.R.; Harrison, R.A.; Perkins, S.J. *Biochem. J.* **1990**, *267*, 203-212.

Smith, P.L.; Kaetzel, D.; Nilson, J.; Baenziger, J.U. *J. Biol. Chem.* **1990**, *265,* 874-881.

Sodora, D.L.; Cohen, G.H.; Eisenberg, R.J. *J. Virol.* **1989**, *63*, 5184-5193.

Sox, H.C.; Hood, L. *Proc. Natl. Acad. Sci.* **1970**, *66*, 975-982.

Spellman, M.W.; Basa, L.J.; Leonard, C.K.; Chakel, J.A.; O'Connor, J.V.; Wilson, S.; van Halbeek, H. *J. Biol. Chem.* **1989**, *264*, 14100-14111.

Spiegel, R.J.; Jacobs S.L.; Treuhaft, M.W. *J. Interferon. Res.* **1989**, *9 (Suppl. 1)*, S17-S24.

Spiegel, R.J.; Spicehandler, J.R.; Jacobs, S.L.; Oden, E.M. *Am. J. Med.* **1986**, *80*, 223-228.

Spiegelberg, H.L.; Abel, C.A.; Fishkin, B.G.; Grey, H.M. *Biochemistry* **1970**, *9*, 4217-4223.

Spik, G.; Debruyne, V.; Montreuil, J.; van Halbeek, H.; Vliegenthart, J.F.G. *FEBS Letters* **1985**, *183*, 65-69.

Spohr, U.; Hindsgaul, O.; Lemieux, R.U. *Can. J. Chem.* **1985**, *63*, 2644-2652.

Stahl, P.D. *Am. J. Respir. Cell Mol. Biol.* **1990**, *2*, 317-318.

Stanley, P. *Methods in Enzymology* **1983**, *96*, 157-184.

Stanley, P. *Ann. Rev. Genet.* **1984**, *18*, 525-552.

Stanley, P. *Methods in Enzymology* **1987**, *138*, 443-458.

Stanley, P. *Mol. Cell. Biol.* **1989**, *9*, 377-383.

Steiner, H.; Pohl, G.; Gunne, H.; Hellers, M.; Elhammer, A.; Hansson, L. *Gene* **1988**, *73*, 449-457.

Stockert, R.J.; Morell, A.G.; Scheinberg, I.H. *Science* **1977**, *197*, 667-668.

Struck, D.K.; Siuta, P.B.; Lane, M.D.; Lennarz, W.J. *J. Biol. Chem.* **1978**, *253*, 5332-5337.

Sutton, B.J.; Phillips, D.C. *Biochem. Soc. Trans.* **1983**, *11*, 130-132.

Takegawa, K.; Inami, M.; Yamamoto, K.; Kumagai, H.; Tochikura, T.; Mikami, B.; Morita, Y. *Biochim. Biophys. Acta* **1988**, *955*, 187-193.

Takegawa, K.; Fugiwara, K.; Iwahara, S.; Yamamoto, K.; Tochikura, T. *Biochem. Cell Biol.* **1989**, *67*, 460-464.

Takeuchi, M.; Takasaki, S.; Miyazaki, H.; Takashi, K.; Hoshi, S.; Kochibe, N.; Kobata, A. *J. Biol. Chem.* **1988**, *263*, 3657-3663.

Takeuchi, M.; Inoue, N.; Strickland, T.W.; Kubota, M.; Wada, M.; Shimizu, R.; Hoshi, S.; Kozutsumi, H.; Takasaki, S.; Kobata, A. *Proc. Natl. Acad. Sci. USA.* **1989**, *86*, 7819-7822.

Takeuchi, M.; Takashi, K.; Hoshi, S.; Shimada, M.; Kobata, A. *J. Biol. Chem.* **1990**, *265*, 12127-12130.

Tanigawara, Y.; Hori, R.; Okumura, K.; Tsuji, J.-i.; Shimizu, N.; Noma, S.; Suzuki, J.; Livingston, D.J.; Richards, S.M.; Keyes, L.D.; Couch, R.C.; Erickson, M.K. *Chem. Pharm. Bull.* **1990**, *38*, 517-522.

Taniguchi, T.; Mizuochi, T.; Beale, M.; Dwek, R. A.; Rademacher, T. W.; Kobata, A. *Biochem.* **1985**, *24,* 5551-5557.

Tanner, W.; Lehle, L. *Biochim. Biophys. Acta* **1987**, *906*, 81-99.

Tanswell, P.; Schluter, M.; Krause, J. *Fibrinolysis* **1989**, *3*, 79-84.

Tarentino, A.L.; Plummer, T.H.; Maley, F. *J. Biol. Chem.* **1974**, *249,* 818-824.

Tavernier, J.; Devos, R.; van der Heyden, J.; Hauquier, G.; Bauden, R.; Fache, I.; Kawashima, E.; Vandekerckhove, J.; Contreras, R.; Fiers, W. *DNA* **1989**, *8*, 491-501.

Taylor, A.K.; Wall, R. *Mol. Cell. Biol.* **1988**, *8*, 4197-4203.

Teixida, J.; Wong, S.T.; Lee, D.C.; Massague, J. *J. Biol. Chem.* **1990**, *265,* 6410-6415.

Thall, A.; Galili, U. *Biochemistry* **1990**, *29*, 3959-3965.

Thompson, J.A.; Lee, D.J.; Kidd, P.; Rubin, E.; Kaufman, J.; Bonnem, E.M.; Fefer, A. *J. Clin. Oncol.* **1989**, *7*, 629-637.

Thomsen, D.R.; Post, L.E.; Elhammer, A.P. *J. Cell. Biochem.* **1990**, *43*, 67-79.

Tollefsen, S.E.; Rosenblum, J.L. *Am. J. Physiol.* **1988**, *255*, G374-G381.

Tominaga, A.; Takahashi, T.; Kikuchi, Y.; Mita, S.; Naomi, S.; Harada, N.; Yamaguchi, N.; Takatsu, K. *J. Immunol.* **1990**, *144*, 1345-1352.

Tooze, S.A.; Tooze, J.; Warren, G. *J. Cell Biol.* **1988**, *106*, 1475-1487.

Trown, P.W.; Kramer, M.J.; Dennin, R.A.; Connell, E.V.; Palleroni, A.V.; Quesada, J.; Gutterman, J.U. *Lancet* **1983**, *1*, 81-87.

Tsai, C.M.; Zopf, D.A.; Yu, R.K.; Wistar, R.; Ginsburg, V. *Proc. Natl. Acad. Sci* **1977**, *74*, 4591-4594.

Tsuchiya, N.; Endo, T.; Matsuta, K.; Yoshinoya, S.; Aikawa, T.; Kosuge, E.; Takeuchi, F.; Miyamoto, T.; Kobata, A. *J. Rheumatol.* **1989**, *16*, 285-290.

Tsuda, E.; Goto, M.; Murakami, A.; Akai, K.; Ueda, M.; Kawanishi, G.; Takahashi, N.; Sasaki, R.; Chiba, H.; Ishihara, H.; Mori, M.; Tejima, S.; Endo, S.; Arata, Y. *Biochemistry* **1988**, *27*, 5646-5654.

Tsuda, E.; Kawanishi, G.; Ueda, M.; Masuda, S.; Sasaki, R. *Eur. J. Biochem.* **1990**, *188*, 405-411.

Utsumi J.; Yamazaki, S.; Hosoi, K.; Kimura, S.; Hanada, K.; Shimazu, T.; Shimizu, H. *J. Biochem.* **1987**, *101*, 1199-1208.

Utsumi, J.; Mizuno, Y.; Hosoi, K.; Okano, K.; Sawada, R.; Kajitani, M.; Sakai, I.; Naruto, M.; Shimizu, H. *Eur. J. Biochem.* **1989**, *181*, 545-553.

Vallbracht, A.; Treuner, J.; Flehmig, B.; Joesters, K.-E.; Niethammer, D. *Nature* **1981**, *289*, 496-497.

Vidal, S.; Mottet, G.; Kolakofsky, D.; Roux, L. *J. Virol.* **1989**, *63*, 892-900.

Vita, N.; Magazin, M.; Marchese, E.; Lupker, J.; Ferrara, P. *Lymphokine Research* **1990**, *9*, 67-79.

Vyas, N. K.; Vyas, M. N.; Quiocho, F. A. *Science* **1988**, *242*, 1290-1295.

Walker, M.R.; Lund, J.; Thompson, K.M.; Jefferis, R. *Biochem. J.* **1989**, *259*, 347-353.

Wallick, S.C.; Kabat, E.A.; Morrison, S.L. *J. Exp. Med.* **1988**, *168*, 1099-1109.

Walsh, M.T.; Watzlawick, H.; Putnam, F.W.; Schmid, K.; Brossmer, R. *Biochemistry* **1990**, *29*, 6250-6257.

Wang, F.-F.C.; Hirs, C.H.W. *J. Biol. Chem.* **1977**, *252*, 8358-8364.

Wang, X.C.; O'Hanlon, T.P.; Lau, J.T.Y. *J. Biol. Chem.* **1989**, *264*, 1854-1859.

Watkins, W.M. *Adv. Hum. Genet.* **1980**, *10*, 1-136.

Weber, R.J.; Clem, L.W. *J. Immunol.* **1981**, *127*, 300-305.

Weintraub, B.D.; Stannard, B.S.; Magner, J.A.; Ronin, C.; Taylor, T.; Joshi, L.; Constant, R.B.; Menezes-Ferreira, M.M.; Petrick, P.; Gesundheit, N. *Rec. Prog. Horm. Res.* **1985**, *41*, 577-606.

Weintraub, B.D.; Gesundheit, N.; Taylor, T.; Gyves, P.W. *Ann. N.Y. Acad. Sci.* **1989**, *553*, 205-213.

Weiss, P.; Ashwell, G. *Prog. Clin. Biol. Res.* **1989**, *300*, 169-184.

Wernette-Hammond, M.E.; Lauer, S.J.; Corsini, A.; Walker, D.; Taylor, J.M.; Rall, S.C. *J. Biol. Chem.* **1989**, *264*, 9094-9101.

Wingfield, P.; Graber, P.; Moonen, P.; Craig, S.; Pain, R.H. *Eur. J Biochem.* **1988**, *173*, 65-72.

Wittwer, A.J.; Howard, S.C.; Carr, L.S.; Harakas, N.K.; Feder, J.; Parekh, R.B.; Rudd, P.M.; Dwek, R.A.; Rademacher, T.W. *Biochem.* **1989**, *28*, 7662-7669.

Wittwer, A.J.; Howard, S.C. *Biochem.* **1990**, *29*, 4175-4180.

Yamamoto, K.; Takegawa, K.; Kumagai, H.; Tochikura, T. *Agric. Biol. Chem.* **1987**, *51*, 1481-1487.

Yamashita, K.; Koide, N.; Endo, T.; Iwaki, Y.; Kobata, A. *J. Biol. Chem.* **1989**, *264*, 2415-2423.

Yan, S.C.B.; Razzano, P.; Chao, Y.B.; Walls, J.D.; Berg, D.T.; McClure, D.B.; Grinnell, B.W. *Bio/Technology* **1990**, *8*, 655-661.

Yeo, T.-K.; Yeo, K.-T.; Olden, K. *Biochem. Biophys. Res. Comm.* **1989**, *160*, 1421-1428.

Yet, M.-G.; Shao, M.-C.; Wold, F. *FASEB J.* **1988**, *2*, 22-31.

Yet, M.-G.; Wold, F. *Arch. Bioch. Biophys.* **1990**, *278*, 356-364.

Yodoi, J.; Hirashima, M.; Ishizaka, K. *J. Immun.* **1980**, *125*, 1436-144.

Young, J.D.; Tsuchiya, D.; Sandlin, D.E.; Holroyde, M.J. *Biochemistry* **1979**, *18*, 4444-4448.

Yuan, D. *Mol. Immunol.* **1982**, *19*, 1149-1157.

Zanni, E.E.; Kouvatsi, A.; Hadzopoulou-Cladaras, M.; Krieger, M.; Zannis, V.I. *J. Biol. Chem.* **1989**, *264*, 9137-9140.

Zhu, B.C.-R.; Laine, R.A.; Barkley, M.D. *Eur. J. Biochem.* **1990**, *189*, 509-516.

Zopf, D.-A.; Tsai, C.M.; Ginsburg, V. *Arch. Biochem. Biophys.* **1978**, *185*, 61-71.

Modeling the Regulation of Antibody Synthesis and Assembly in a Murine Hybridoma

M.C. Flickinger[+], Department of Biochemistry, and Institute for Advanced Studies in Biological Process Technology, University of Minnesota, St. Paul, Minnesota 55108
T. Bibila, Department of Chemical Engineering and Materials Science, and Institute for Advanced Studies in Biological Process Technology, University of Minnesota, St. Paul, Minnesota 55108

Understanding the regulation of murine monoclonal antibody (MAb) synthesis, chain assembly and secretion, particularly under slow growth conditions, may be important in optimizing the productivity of large scale immunoglobulin production systems. An initial structured, unsegregated kinetic model of MAb synthesis and secretion is described based on intracellular balances of heavy (H) and light (L) chain mRNAs, the intracellular balances of H and L chains and H and L chain assembly. The model predicts a decrease in intracellular H and L mRNA levels with transition of the hybridoma cells from the exponential into the stationary phase of growth.

Enhancement of multichain protein synthesis and secretion productivity in mammalian cells cultivated *in vitro* would be a significant advance for cell culture technologies currently being used for production of therapeutic glycoproteins (Murakami, 1990). The ability to manipulate cellular secretion rate, however, depends on an understanding of the regulation of protein synthesis, assembly, modification and secretion. As biochemical knowledge of individual steps in this pathway accumulates, structured mathematical models become increasingly essential as a rapid, inexpensive method for describing as well as predicting optimization of the entire pathway leading to protein secretion.

Murine monoclonal antibody (MAb) synthesis, chain assembly and secretion (Fig.1) during both rapid and slow hybridoma growth is a potentially useful model for investigation of multichain glycoprotein synthesis in mammalian cells cultivated in vitro. In this system, the rate of MAb production appears not to be proportional to the rate of cell proliferation; rather it depends on the total number of cells in the culture (Tolbert *et al.*, 1985a, Long *et al.*, 1988, Velez *et al.*, 1986, Altschuler *et al.*, 1986, Dean *et al.*, 1987, Lee *et al.*, 1989). Hence,

many commercial systems have been developed to maintain cells at densities of greater than 10^7 cells/ml for prolonged periods of time in order to optimize glycoprotein secretion rate and yield (Fig. 2). Unfortunately, little biological information exists on the regulation of protein synthesis in *in vitro* culture under conditions of either rapid exponential growth or for prolonged periods under conditions of minimal cell proliferation. Attempts to investigate MAb secretion under conditions of very slow growth using physical methods of synchronization or by continuous culture at low dilution rates have often resulted in low cell viabilities (Castillo *et al.*, 1988) (Fig.3). Entrapment of cells in perfusion reactors, although useful for maintenance of cells with continuous perfusion of nutrients, does not facilitate access to the cells for determination of viability, secretion rate or analysis of pools of intracellular components of the pathway of protein synthesis.

In order to investigate the regulation of MAb synthesis by either rapidly growing or non-proliferating viable hybridoma cells, we have chosen to compare the intracellular levels of MAb heavy (H) and light (L) chain mRNAs, polypeptides and the kinetics of chain assembly during exponential growth in batch culture and

+To whom all correspondence should be addressed.

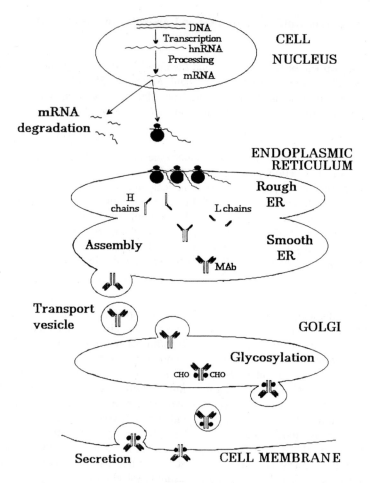

Fig. 1. Schematic diagram of antibody synthesis and secretion.

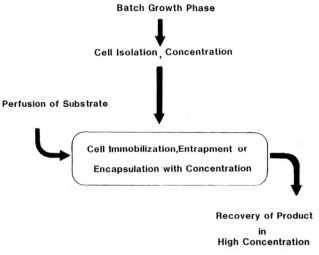

Fig. 2. Commercial in vitro MAb production methods to maintain high cell density and MAb secretion using cells in a slowly growing or non-growing state.

242

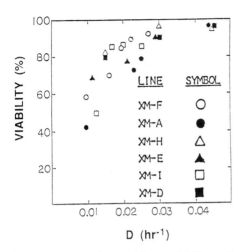

VIABILITY (%)

LINE	SYMBOL
XM-F	O
XM-A	●
XM-H	△
XM-E	▲
XM-I	□
XM-D	■

D (hr⁻¹)

Fig. 3. Decrease in hybridoma viability at low dilution rate in continuous culture (Castillo *et al.*, 1988). Reproduced by permission.

during stationary phase culture under conditions where cell population viability is greater than 90%. These two physiological conditions can be achieved by cultivation of cells in spinner flasks or using a total cell recycling reactor to increase cell density to >10^7 viable cells/ml and prolong cell viability for several hundred hours at a minimal growth rate (Flickinger *et al.*, 1990) (Fig. 4). Using this latter method, specific MAb secretion rate can be determined as a function of hybridoma growth rate while maintaining high cell viability by continuous perfusion of medium to a dense population of cells (1-3 x 10^7 viable cells/ml) in a homogeneous suspension (Flickinger *et al.*, 1990).

Initial experimental determination of variation in the levels of MAb H and L chain mRNAs and their half-lives has been carried out using spinner flask cultures in order to describe the regulation of MAb synthesis during batch culture. Heavy and light chain translation rates and assembly kinetics are currently being experimentally determined in batch culture. Currently, this experimental work is being extended to describe the regulation of MAb synthesis and secretion in systems other than batch culture using the total cell recycle reactor method (Flickinger *et al.*, 1990) (Figs. 5,6).

Materials and Methods

The cell line used in this study is a murine hybridoma designated 9.2.27 which secretes an IgG_{2a} monoclonal antibody against a 240 kD human melanoma surface glycoprotein (Morgan *et al.*, 1981). For batch experiments 9.2.27 hybridoma cells were grown in 1 liter spinner flasks (Bellco) in filter sterilized Dulbecco's Modified Eagle's Medium (DMEM), containing 4 mM L-glutamine, 4.5 g/l of glucose and supplemented with 3% fetal bovine serum (Gibco). Cultures were maintained at 37°C in a 5% CO_2 atmosphere. Viable cell concentration was determined by 0.4% erythrosine B vital staining and counting in the hemacytometer (Kruse and Patterson, 1973). Fig. 7 shows a typical growth curve in a 1 liter batch spinner culture. Cells grow exponentially with a maximum specific growth rate of 0.04 hr⁻¹, for 60 hours after inoculation. The cells then enter stationary phase which lasts for approximately 20 hours (60-80 hours) during which time cell viability is high (≥ 80%). After 80 hours the culture viability begins to decrease rapidly with the entrance of cells into the death phase. Monoclonal antibody (IgG_{2a}) concentration in the cell culture supernatant, was determined by using an ELISA assay (Lee *et al.*, 1986). Total cellular RNA was isolated using the guanidium isothiocyanate-phenol-chloroform extraction method (Chomzynski and Sacchi, 1987). Intracellular heavy and light chain mRNA levels were determined by dot blot hybridization of total cellular RNA to heavy and light chain cDNA probes radioactively labeled by nick translation (Bibila and Flickinger, 1991). Quantitation was made as previously described (Bibila and Flickinger, 1991). The half-lives of the mRNAs coding for the heavy and light immunoglobulin chains, were determined by following the decay of the mRNAs (decrease in intracellular content) after blocking transcription with either actinomycin D, an inhibitor of transcription elongation, at a final concentration of 3 mg/ml, or dichloro-benzimidazole riboside, (DRB), an inhibitor of transcription initiation, at a final concentration of 20 mg/ml (Bibila and Flickinger, 1991).

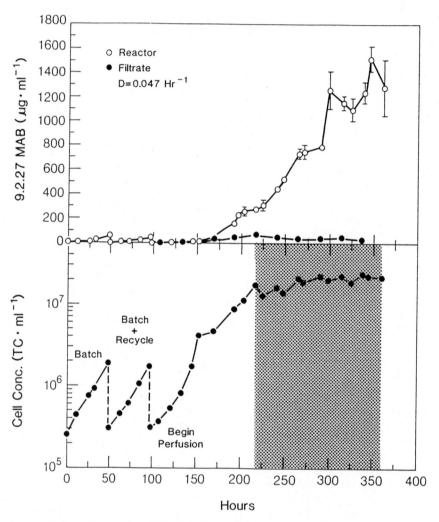

Fig. 4. Accumulation of a murine IgG$_{2a}$ MAb during continuous perfusion in a total recycle reactor. Shaded area indicates data at very slow growth rate where hybridoma population viability is greater than 90% viable cells (Reproduced with permission from Flickinger *et al.*, 1990, Copyright 1990, Springer-Verlag).

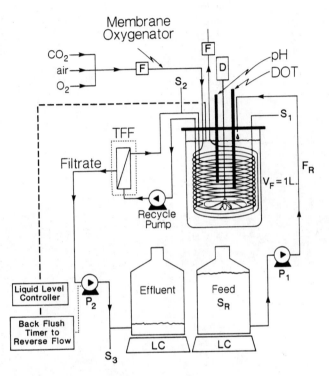

Fig. 5. Diagram of a 1 liter total cell recycle reactor with continuous perfusion of fresh medium to sustain hybridoma viability during very slow growth. (Reproduced with permission from Flickinger *et al.*, 1990, Copyright 1990, Springer-Verlag). F - filter, P - medium, recycle and filtrate pumps, D - drive motor, S - limiting substrate concentration, TFF - tangential flow filter, F_R - perfusion rate, LC - load cell.

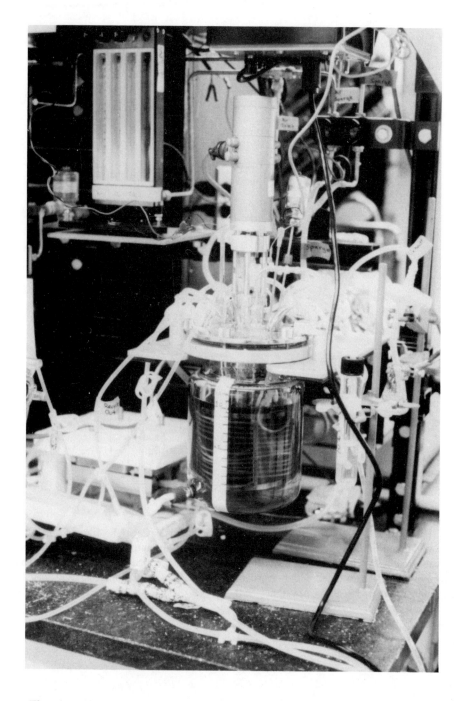

Fig. 6. The total cell recycle reactor with tangential flow recycle filter.

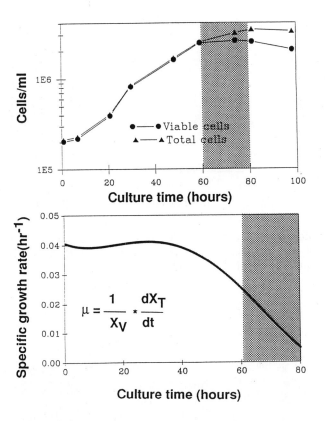

Fig. 7. Growth of the murine 9.2.27 hybridoma in a 1 liter spinner culture. (Reproduced with permission from Bibila and Flickinger, 1991, Copyright 1991, John Wiley & Sons).

Results

Levels of Total Cellular RNA and Heavy and Light Chain mRNAs During Exponential and Stationary Phase

One liter 9.2.27 spinner cultures were sampled at various stages of growth for determination of total cellular RNA and determination of heavy and light chain mRNA levels. Total cellular RNA content was found to decrease from about 2×10^{-5} µg/cell in exponentially growing 9.2.27 cells to about 0.5×10^{-5} µg/cell in cells in the late stationary phase. Total cellular RNA content appears to decrease almost linearly with decreasing specific growth rate (Fig. 8).

Intracellular light and heavy chain mRNAs appear to represent an increasingly higher percentage of total cellular RNA as cells enter from the exponential into the stationary phase (Fig. 9). However, the light and heavy chain mRNA content on a per cell basis appears to remain constant during the exponential phase and begins to decrease with time after entrance of the cells into the stationary phase (Fig. 10). The ratio of the light to heavy chain mRNA was found to be constant with time and equal to 1.5, indicating the presence of an excess of light chain mRNA, in agreement with data on myeloma cell lines (Schibler *et al.*, 1978).

Heavy and Light Chain mRNA Half-lives

Actinomycin D or DRB were added to cells at different stages of their growth; early exponential (30 hours of growth), middle exponential (45 hours of growth) or stationary

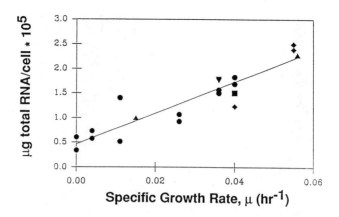

Fig. 8. Determination of total cellular RNA content as a function of hybridoma growth rate in 1 liter batch culture. (Reproduced with permission from Bibila and Flickinger, 1991, Copyright 1991, John Wiley & Sons).

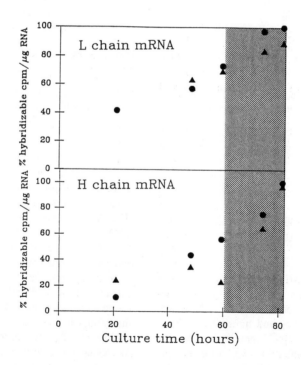

Fig. 9. Relative intracellular levels of heavy (H) and light (L) chain mRNAs per µg of total RNA during batch cultivation. Shaded area indicates data during slow growth at cell viabilities greater than 90%. (Adapted from Bibila and Flickinger, 1991, Copyright 1991, John Wiley & Sons).

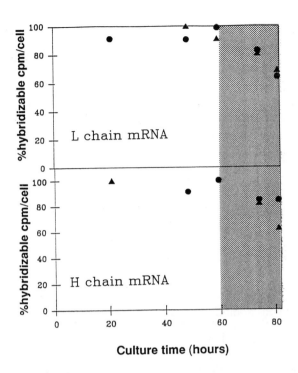

Culture time (hours)

Fig. 10. Relative intracellular levels of heavy (H) and light (L) chain mRNAs per cell during batch cultivation. Normalization was done by dividing all values by the maximum observed value of intracellular mRNA levels. (Adapted from Bibila and Flickinger, 1991, Copyright 1991, John Wiley & Sons).

phase (70 hours of growth), in order to determine the half-lives of the heavy and light chain mRNAs at each growth stage. Actinomycin D was found to seriously affect cell viability and morphology for exposure times longer than 3 hours; therefore, all experiments with actinomycin D were performed within 3 hours of treatment. The half-lives of the heavy and light chain mRNAs were determined at different stages of cell growth and found to decrease almost linearly with time as cells entered stationary phase. Half-lives range from about 5 hours in the middle exponential phase to about 2 hours in the stationary phase.

Although a trend of decreasing mRNA stability with entrance of the cells into the stationary phase can be observed on the basis of actinomycin D experiments, the measured absolute values for the mRNA half-lives are probably lower than the real values, due to the secondary toxic effects of actinomycin D on the cells (Revel *et al.*, 1964, Soeiro and Amos, 1966, Craig *et al.*, 1971, Singer and Penman,

1972). For this reason, mRNA half-lives were also determined using the transcriptional inihibitor, DRB. In the presence of DRB , viability of 9.2.27 cells remained high for 8-12 hours after addition. Fig. 11 shows the decay of light and heavy chain mRNAs after addition of DRB at 46 and 70 hours of growth. The half-life of the light chain mRNA was determined to be 14.4 ± 2.9 hours and 6.2±1.0 hours in the exponential and stationary phases respectively. The half-life of the heavy chain mRNA was determined to be 12.0±3.75 and 6.2±1.0 hours in the exponential and stationary phases respectively. The half-lives of the heavy and light chain mRNAs are similar at the same growth stage. Statistical analysis using a T test showed that they are equal with a 95% confidence limit. However, the values measured using DRB are much higher than when using actinomycin D. The stationary phase half-lives are shorter than during exponential growth, indicating again that half-life decreases with entrance of cells into stationary phase. This

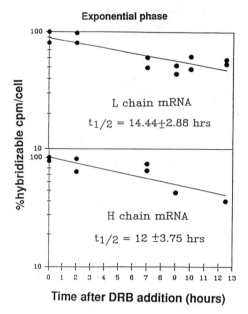

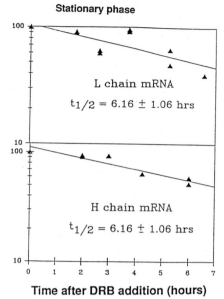

Fig. 11. Decay of heavy (H) and light (L) chain mRNAs after addition of dichloro-benzimidazole riboside (DRB). (Adapted from Bibila and Flickinger, 1991, Copyright 1991, John Wiley & Sons).

observed destabilization of the heavy and light chain mRNAs with decreasing specific growth rate would be in agreement with the observed trend for decrease in the intracellular levels of these mRNAs under the same conditions. These results suggest that the observed decrease of the heavy and light chain mRNA levels during the stationary phase of 9.2.27 hybridoma growth is most probably controlled at the post-transcriptional level, specifically at the level of mRNA decay.

An Initial Structured Kinetic Model

Intracellular heavy and light chain mRNA balances

The intracellular balances of heavy and light chain coding mRNAs can be represented by the following:

$$[Accumulation] = [synthesis] - [decay] - [dilution by growth] \quad (1)$$

The transcription (synthesis) rates of the heavy and the light chain mRNAs were kept constant with time during batch culture and equal to 3,000 mRNAs/cell/hr and 4,500 mRNAs/cell/hr respectively (Schibler et al., 1978). As an initial approximation, the decay rate of the heavy and light chain mRNAs was varied with time during batch culture, in order to vary the half-life of the mRNAs linearly from about 12 hours in the middle exponential phase to about 6 hours in the stationary phase, according to our experimental observations. The growth rate in the dilution by growth term was also varied with batch cultivation time from its maximum value in the exponential phase to about zero in the late stationary phase.

Intracellular balances of heavy and light chains

The intracellular balances of the heavy and light chains can be represented by the following:

$$[Accumulation]=$$
$$[translation]-[dilution \ by \ growth]-[consumption \ in \ Ab \ assembly](2)$$

The H and L chain mRNA translation rates were calculated as follows:

$$T_H, T_L =$$

$$(Ribosomes/mRNA)*(Ribonucleotides \ translated/ribosome/unit \ time)$$

$$(Translatable \ ribonucleotides/protein \ molecules)$$

During exponential growth of mouse plasmacytoma cells, H and L chain mRNAs have been found to be loaded with 18-20 and 6-8 ribosomes respectively (Stevens, 1974). Assuming an average ribosome velocity of 20 ribonucleotides/sec (Potter, 1972), and knowing the length of the H and L chain mRNAs coding (translated) sequences (Auffray and Rougeon, 1980) (1.35 and 0.75 kb respectively), an average value for the H and L chain translation rates can be calculated for exponentially growing cells as 17 and 12 chains/mRNA/min, respectively. The translation rates of both the H and the L chain mRNAs have been shown to decrease with entrance of cells into the stationary phase of growth (Stevens, 1974), therefore in this initial model, they were varied with batch culture time by decreasing the ribosome loading proportionally to the observed decrease in total cellular RNA content as a function of growth rate (Fig. 8). The rates of heavy and light chain consumption in the assembly process, R_H and R_L were determined as described in the following section.

Assembly of heavy and light chains

Covalent assembly of IgG_{2a} H and L chains in the endoplasmic reticulum (ER) has been shown to proceed primarily through a heavy chain dimer intermediate, in the following steps (Scharff and Laskov, 1970):

$$H + H \quad \rightarrow \quad H_2 \quad \text{(A)}$$

$$H_2 + L \quad \rightarrow \quad H_2L \quad \text{(B)}$$

$$H_2L + L \quad \rightarrow \quad H_2L_2 \quad \text{(C)}$$

Intracellular balances for the three assembly intermediates can be constructed:

$$d[H_2]/dt \quad = \quad R_A - R_B - \mu(t)*[H_2] \quad \text{(3)}$$

$$d[H_2L]/dt \quad = \quad R_B - R_C - \mu(t)*[H_2L] \quad \text{(4)}$$

$$d[H_2L_2]/dt \quad = \quad R_C - q - \mu(t)*[H_2L_2] \quad \text{(5)}$$

where q is the specific antibody secretion rate (molecules/cell/hr), and R_A, R_B and R_C are the rates of the reactions A, B and C (number of such reactions per unit time). The total rates of consumption of the heavy and light chains in the assembly process, R_H and R_L, involved in equations (1) and (2), can then be determined as:

$$R_H = 2 * R_A \quad \text{(6)}$$

$$R_L = R_B + R_C \quad \text{(7)}$$

The rates R_A, R_B and R_C were determined according to Percy et al., (1975) who developed a model describing the in vitro assembly of heavy and light chains. Second order kinetics were assumed for each assembly step:

$$R_A \quad = \quad 1/3*[H]*[H]*K_{HH} \quad \text{(8)}$$

$$R_B \quad = \quad 2*[H_2]*[L]*K_{HL} \quad \text{(9)}$$

$$R_C \quad = \quad [H_2L]*[L]*K_{HL} \quad \text{(10)}$$

K_{HH} and K_{HL} are the rate constants of formation of a disulfide bond between a heavy and a heavy or a heavy and a light chain respectively. The factor of 1/3 in equation (8) is included to account for the three disulfide bonds between the two heavy chains in murine IgG_{2a} (De Preval et al., 1970). Since there are two equivalent ways of linking a light chain to the heavy chain dimer, the rate constant of formation of a disulfide bond between the first light chain and the heavy chain dimer in this initial model was set equal to double the rate between the second light chain and the H_2L complex (equation 9). As an initial approximation, K_{HH} and K_{HL} were assumed to be equal ($K_{HH} = K_{HL} = K$). The time required for assembly of 95% of the antibody, starting with a given number of heavy and light chains available for assembly, was used as the criterion for essentially complete assembly (assembly time). This time has been reported to be between 10 and 30 min depending on the type of the immunoglobulin producing cell (Scharff and Laskov, 1970, Baumal and Scharff, 1973a).

Antibody secretion in the extracellular medium

The concentration of secreted antibody in the cell culture supernatant was calculated by the following equation:

$$[MAb]_t = [MAb]_{t=0} + \int_0^t x_v(t)*q(t)*dt \quad \text{(11)}$$

where $x_v(t)$ is the viable cell concentration in viable cells/ml at time t and $q(t)$ is the instantaneous specific antibody secretion rate at time t. The instantaneous specific secretion rate, $q(t)$, was determined at each time point by assuming that it is inversely proportional to the time required for immunoglobulin assembly, which in turn, varies during the batch cultivation period due to the varying intracellular concentrations of heavy and light chains. In equation (11) it has been assumed that at any given time all viable cells in the culture are actively secreting antibody.

Model simulations

Simulation of Immunoglobulin Assembly

By solving the set of equations (1) to (5) a radiotracer pulse chase experiment can be simulated as if a given amount of labeled heavy and light chains is synthesized in the cell during a short "pulse" with a radioactively labeled amino acid. The assembly of these chains can then be followed during the "chase" period starting at t=0. Since the characteristic times required for assembly are much shorter than both the cell's doubling time (approximately 17 hours) and the time elapsing before the assembled antibody is actually secreted (which could be as long as 150 min), as an initial approximation dilution by growth has been ignored in the above equations. Secretion of the assembled antibody during this short time period has been ignored in equation (5). The results of this simulation approach are (Figure 12a) in agreement with experimental observations of Scharff and Laskov (1970) and Baumal et al., (1971), in that the heavy chain dimer and the H_2L complex are only transiently present. Most of the antibody is assembled within a short period of time (about 3 min) with the rate of assembly decreasing due to the decreasing concentrations of intermediates, and the concentration of the assembled antibody increasing very slowly. The simulations were repeated for different sets of values of the assembly rate constant K, and the initial H and L chain concentrations. In these simulations the initial L/H chain concentration ratio was set equal to 1.0. As expected, the assembly time is inversely proportional to both the initial heavy

(and light) chain concentration and the value of the assembly rate constant, K (Fig. 12b). Fig. 12c shows the effect of the initial L/H chain concentration ratio on the assembly time. For all values of the assembly rate constant, K, and the initial heavy chain concentration, these simulations indicate that there may be a significant (45%) reduction in assembly time with an increase in the initial L/H chain concentration ratio from 1.0 to 1.5.

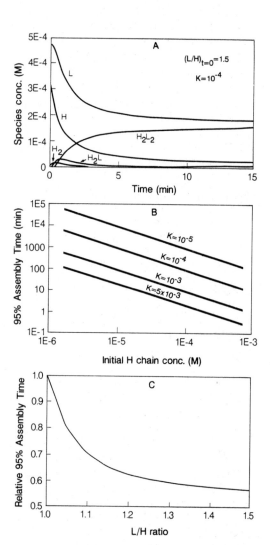

Fig. 12. Simulation of heavy (H) and light (L) chain covalent assembly; K = rate constant for disulfide bond formation. (Reproduced with permission from Bibila and Flickinger, 1991, Copyright 1991, John Wiley & Sons).

Simulations of Immunoglobulin Secretion

In order to simulate immunoglobulin secretion, equations (1) to (5) were solved numerically using a 4th order Runge-Kutta algorithm. The initial values of the intracellular H and L chain mRNA concentrations at the time of hybridoma inocculation in batch culture were set equal to 30,000 and 45,000 molecules/cell respectively, typical values for exponentially growing myeloma cells (Schibler et al., 1978). The initial values of the intracellular concentrations of H and L chains at inocculation were set equal to 4×10^8 and 6×10^8 molecules/cell respectively. An initial average assembly time of 20 minutes was used, an average of experimentally measured immunoglobulin assembly times in exponentially growing cells (Scharff and Laskov, 1970, Laskov et al., 1971, Baumal et al., 1971, Baumal and Scharff, 1973b). The initial secretion rate was 5 µg MAb/10^7 cells/hr, an experimentally determined average secretion rate for 9.2.27 cells in early exponential phase under the spinner growth conditions used. The initial values of the intracellular H_2 and H_2L concentrations were zero, and the initial intracellular concentration of assembled antibody (H_2L_2) was taken as 2×10^9 molecules/cell. This value was chosen based on experimental observations that the intracellular concentration of assembled antibody is approximately 3-5 times higher than the intracellular concentration of free light chains (Baumal and Scharff, 1973a).

At each time point during batch growth, the algorithm calculated the values of the intracellular H and L mRNA concentrations as well as the concentrations of the intracellular H and L chains and their ratio, L/H. At each time point the assembly time was adjusted to account for changes in the H chain concentration and the L/H ratio. Readjustment was done according to the results of simulations of assembly. Changes in the H chain concentration were accounted for by proportional changes in the assembly time, whereas changes in the L/H ratio affected the assembly time as shown in Fig. 12c. The results of those simulations on which readjustment of the assembly time was based were independent of the value of the assembly rate constant, K. The instantaneous specific antibody secretion rate, q(t), in this initial model was assumed to be inversely proportional to the assembly time. At each time point during batch culture, the secretion rate was corrected to account for the changing assembly time.

Figures 13 through 15 show the model simulation results. The intracellular content of H chain increases by 90% between the exponential and the stationary phase, indicating that although the synthesis rate of the H chain decreases, the intracellular H chain pool may increase due to a greater decrease in the dilution by growth term (Fig. 13a). The content of L chain increases by 40% within the same time (Fig. 13a). Their ratio decreases from 1.5 to less than 1.1 (Fig. 13b) in agreement with experimental observations that the L/H ratio is growth rate dependent (Stevens, 1974). Figure 14 shows the simulated variation in the intracellular levels of the H_2 and H_2L assembly intermediates and the assembled antibody (H_2L_2) with culture time. Results reported by other investigators indicate that intracellular antibody content may decrease as the growth rate decreases from the exponential into the stationary phase of growth (Ramirez and Mutharasan, 1989, Reddy and Miller, 1989). This behavior is predicted by this initial model. Assembly time was found to continuously decrease from 20 min to 13 min in the late stationary phase (Fig. 15a), as would be expected since the intracellular concentrations of all assembly intermediates continuously increase throughout the cultivation period (Fig. 15a, 16). The specific antibody secretion rate is predicted to increase by about 50% during the same time (Fig. 15b). Increases in the specific antibody secretion rate with decreasing growth rate or environmental stress have been experimentally observed by other investigators in batch hybridoma cultures (Ramirez and Mutharasan, 1989, Reddy and Miller, 1989, Suzuki and Ollis, 1989).

Figure 16 compares the experimentally measured and model predicted intracellular H and L mRNA levels and extracellular antibody concentration. Although in the case of intracellular H and L mRNA levels, the experimental points during the exponential phase are a little lower than the levels predicted by the model, this initial model can predict the constancy of the mRNA levels during the exponential phase and their decrease with time

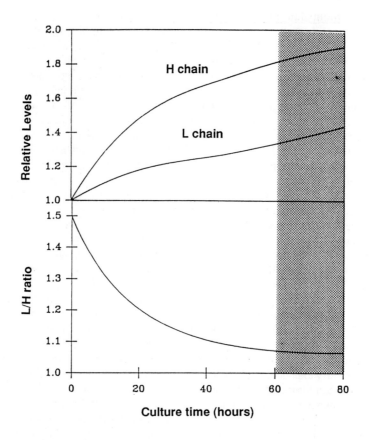

Fig. 13. Simulated relative intracellular levels of heavy (H) and light (L)chains and their ratio during batch cultivation. (Adapted from Bibila and Flickinger, 1991, Copyright 1991, John Wiley & Sons).

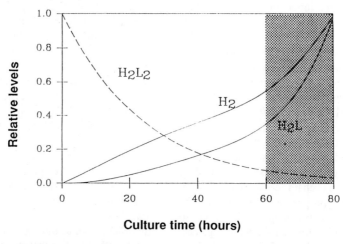

Fig. 14 . Simulated relative intracellular levels of assembly intermediates during batch cultivation. (Adapted from Bibila and Flickinger, 1991, Copyright 1991, John Wiley & Sons).

254

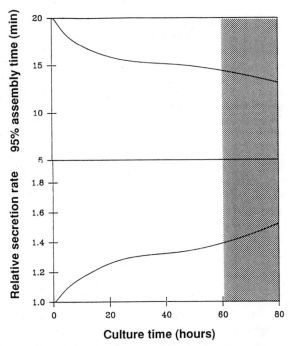

Fig. 15. Simulated variation of antibody assembly time and relative specific antibody secretion rate during batch cultivation. (Adapted from Bibila and Flickinger, 1991, Copyright 1991, John Wiley & Sons).

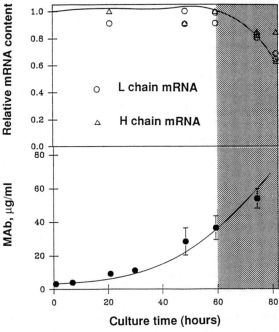

Fig. 16. Comparison of model simulations with experimentally determined relative intracellular heavy (H) and light (L) chain mRNA levels and MAb titer in the cell culture supernatant, during batch cultivation. (Adapted from Bibila and Flickinger, 1991, Copyright 1991, John Wiley & Sons).

255

with transition of cells into the stationary phase of growth (Fig. 16a). Although light and heavy chain mRNA levels decrease with decreasing growth rate, their ratio remains constant and equal to 1.5 throughout the exponential and stationary phases. There is good agreement between the experimentally measured and model predicted antibody concentrations in the cell culture supernatant (Fig. 16b). This good agreement between model and experimental data, despite the simplifying assumptions used in this initial formulation of the model, shows the potential usefulness of this structured unsegregated modeling approach.

Summary and Discussion

An initial structured kinetic model describing monoclonal antibody synthesis and secretion was developed based on the intracellular balances for heavy and light chain mRNA and heavy and light protein chains and the kinetics of covalent antibody chain assembly. The model was found to predict the experimentally observed changes in the intracellular levels of heavy and light chain mRNAs during entrance of 9.2.27 hybridoma cells from the exponential into the stationary phase of batch growth, as well as the accumulation of secreted antibody in the cell culture supernatant during batch cultivation.

We are currently in the process of extending this initial model in order to describe antibody synthesis and secretion in systems other than batch culture since specific antibody secretion rates under slow growth and/or stress conditions are reportedly higher than secretion rates in rapidly growing cells (Boraston et al., 1984, Tolbert et al., 1985b, Emery, 1986, Reuveny et al., 1986, Feder, 1988, Takazawa et al., 1988, Reddy and Miller, 1989, Suzuki and Ollis, 1989). Using a total cell recycle perfusion reactor it is probable however, that slow growth in the presence of continuous nutrient perfusion may be a significantly different physiological condition from slow growth during the stationary phase of a batch culture and hence the regulation of MAb synthesis may differ.

Extensions of this initial model may include the effect and action of ER resident proteins which have been shown to play an active role in immunoglobulin assembly, such as the heavy

chain binding protein (BiP, grp78) and protein disulfide isomerase (PDI) (Roth and Koshland, 1981, Freedman et al., 1984, Bole et al., 1986, Rothman, 1989). Interestingly enough, the intracellular levels of these assembly mediating protein factors have been shown to be growth phase dependent, to increase under stress conditions, (similar to the conditions which have been reported to result in specific antibody secretion rate increases), and correlate well with antibody secretion rates (Roth and Koshland, 1981, Myllyla et al., 1983, Freedman et al., 1984). Variations in the intracellular levels of these factors or their activity may affect the rate constant for disulfide bond formation and thereby H and L chain assembly rates in the ER. In addition to assembly mediating proteins, the modeling of steps in the immunoglobulin secretion process following assembly, such as the transport of the assembled antibody from the ER to the Golgi, intra-Golgi processing and final secretion to the extracellular medium are being investigated using compartmental modeling principles. The complete model should be a tool for identifying possible regulatory steps in the antibody synthesis and secretion process, and it may be useful in predicting and optimizing the performance of immunoglobulin secreting cells.

Acknowledgements

The authors would like to ackowledge the partial support of this work by Lilly Research Laboratories (Indianapolis, Indiana) and the Institute for Advanced Studies in Biological Process Technology. The authors are grateful to Drs. George Boder and Jon Schmidtke (Lilly) for use of the 9.2.27 hybridoma.

References

Altschuler, G.L.; Dziewalski, D.M.; Sowek, J.A.; Belfort, G.; Biotechnol. Bioeng., 1986, 27, 64.

Auffray, C.; Rougeon, F.; Eur. J. Biochem., 1980, 107, 303.

Baumal, R.; Potter, M.; Scharff, M.D.; J. Exp. Med., 1971, 134, 1316.

Baumal, R.; Scharff, M.D.; J. Immunol., 1973a, 111, 448.

Baumal, R.; Scharff, M.D.; Transplant. Rev., 1973b, 14, 163.

Bibila, T.A.; Flickinger, M.C.; *Biotechnol. Bioeng.*, **1991**, (in press).

Bole, D.G.; Hendershot, L.M.; Kearney, J.F.; *J. Cell Biol.*, **1986**, *102*, 1558.

Boraston, R.; Thompson, P.W.; Garland, S.; B Birch, J.R.; *Dev. Biol. Stand.*, **1984**, *55*, 103.

Castillo, F.J.; Thrift, J.; Mullen, L.; Bostic, G.; McGregor, W.C.; Chemical Society Meeting, Los Angeles, California, **1988**.

Chomczynski , P.; Sacchi, N.; *Anal. Biochem.*, **1987**, *162*, 156.

Craig, N.; Kelley, D.E.; Kelley, R.P.; *Biochim. Biophys. Acta*, **1971**, *246*, 493.

Dean, R.C.; Karkare, S.B.; Venkatasubramanian, K.; *Ann. N.Y. Acad. Sci.*, **1987**, *506*, 129.

De Préval, C.; Pink, J.R.L.; Milstein, C.; *Nature*, **1970**, *228*, 930.

Emery, N.; Chemical Industry Symposium on "Large scale production of monoclonal antibodies", London, **1986**.

Feder, J. ; *Adv. Biotechnol. Proc.*, **1988**, *7*, 125.

Flickinger, M.C.; Goebel, N.K.; Bohn, M.A.; *Bioproc. Eng.*, **1990**, *5*, 155.

Freedman, R.B.; Brockway, B.E.; Lambert, N.; *Biochem. Soc. Trans.*, **1984**, *12*, 929.

Kruse, P.F.; Patterson, M.K., Eds.; *Tissue culture: methods and applications*; Academic Press:**1973**; pp.406-408.

Laskov, R.; Lanzerotti, R.; Scharff, M.D.; *J. Mol. Biol.*, **1971**, *56*, 327.

Lee, S.M.; Gustafson, M.E.; Pickle, D.J.; Flickinger, M.C.; Muschik, G.M.; Morgan, A.C.; *J. Biotechnol.*, **1986**, *4*, 189.

Lee, G.M.; Huard, T.K.; Palsson, B.O.; *Hybridoma*, **1989**, *8*, 369.

Long, W.J.; Palombo, A.; Schofield, T.L.; Emini, E.A.; *Hybridoma*, **1988**, *7*, 69.

Morgan, A.C.; Galloway, D.R.; Reisfeld, R.A., *Hybridoma*, **1981**, *1*, 27.

Murakami, H.; *Cytotechnol.*, **1990**, *3*, 3.

Myllyla, R., Koivu, J.; Pihlajaniemi, T.; Kivirikko, K., *Eur. J. Biochem.*, **1983**, *134*, 7.

Percy, J.R.; Percy, M.E.; Dorrington, K.J.; *J. Biol. Chem.*, **1975**, *250*, 2398.

Potter, M.; *Physiol. Rev.*, **1972**, *52*, 631.

Ramirez, O.; Mutharasan, R.; AIChE Annual Meeting, San Francisco, California, **1989**.

Reddy,S.; Miller, W.S.; AIChE Annual Meeting, San Francisco, California, **1989**.

Reuveny, S.; Velez, D.; Miller, L.; Macmillan, J.D.; *J. Immunol. Method.*, **1986**, *86*, 61.

Revel, M.; Hiatt , H.H.; Revel, J.P.; *Science*, **1964**, *146*, 1311.

Roth, R.A.; Koshland, M.E.; *Biochemistry*, **1981**, *20*, 6594.

Rothman, J.E., *Cell*, **1989**, *59*, 591.

Scharff, M.D.; Laskov, R.; *Progr. Allergy*, **1970**, *14*, 37.

Schibler, U.; Marcu , K.B.; Perry, R.P., *Cell*, **1978**, *15*, 1495.

Singer, R.H.; Penman, S.; *Nature*, **1972**, *240*, 100.

Soeiro, R.; Amos, H.; *Biochim. Biophys. Acta*, **1966**, *129*, 406.

Stevens, R.H.; *Eur. J. Biochem.*, **1974**, *42*, 553.

Suzuki , E.; Ollis, D.F.; AIChE Annual Meeting, San Francisco, California, **1989**.

Takazawa, Y.; Tokashiki, M.; Hamamoto, K.; Murakami, H.; *Cytotechnol.*, **1988**, *1*, 171.

Tolbert, W.R.; Feder, J.; Lewis, C.; US Patent 4,537,860, **1985a**.

Tolbert, W.R.; Lewis, C.; White, P.J.; Feder, J. *In Large Scale Mammalian Cell Culture*; Feder, J.; Tolbert, W.R., Eds.; Academic Press:**1985b**; pp.97-123.

Velez, D.; Reuveny, S.; Miller, L.; MacMillan, J.D.; *J. Immunol. Meth.*, **1986**, *86*, 45.

Part IV
Bioreactor Engineering and Control

TRANSPORT OF NUTRIENTS AND OXYGEN in bioreactors is a problem that refuses to disappear. In pursuit of ever more concentrated biomass for more efficient catalysis, researchers have built models of immobilized microbial biocatalysts, have explored oxygen transport via solvents in which it is highly soluble, and have studied the turbulent transport of solutes at the micro scale.

Modeling of Immobilized Microbial Cells: Influence of Substrate or Product Inhibition on Activity and Biomass Spatial Profiles

David F. Ollis and Gregory D. Sayles, Department of Chemical Engineering, North Carolina State University, Raleigh, NC 27695-7905

The influence of substrate inhibition and of product inhibition on the kinetics and performance of immobilized cells is explored with simple cell growth kinetic models commonly found in the batch and continuous culture literature. Metabolic boundary conditions are used to allow prediction of the spatial location and distribution of viable cells at steady state; thus the biomass distribution is an outcome of the models rather than an initial assumption. Both product inhibition and substrate inhibition are predicted to have dramatic influences on performance of individual catalyst particles of immobilized cells. Convenient application to reactor design is indicated.

Immobilized microbial populations have long been of interest in waste water polishing (trickle filter), vinegar manufacturing, microbial corrosion of metallic surfaces, undesirable biofilm formation in cooling towers, heat exchangers and other process equipment, and in physiology (of the mouth and teeth, etc.). Over the last decade, interest in immobilization of non-film forming species via gel entrapment, microencapsulation, and hollow fiber containment has increased strongly. With such increased interest, the engineering modelling of cell kinetics in immobilized systems arises naturally. While an appreciable number of papers exist which describe models of immobilized cells, most cannot predict the biofilm or cell layer thickness. For example, Waner and Gujer (1986) reviewed biofilm mixed population models prior to 1986 and noted that only two of ten references cited "did not assume predefined microbial distributions".

An early attempt to address this model deficiency was that of Rittman and McCarty (1980) who used a conceptual model to predict biofilm thickness. They established a steady state model for which the total substrate flux J entering the biofilm was just sufficient to provide energy for an endogenous cell maintenance requirement. Thus

$$J \cdot Y_{X/S} = X \cdot L \cdot k_{maintenance} \qquad (1)$$

$$\text{where } J = (dS/dx)_{\text{film surface}} \qquad (2)$$

This formulation boundary condition is metabolism based; however it does not explicitly include any cell layer movement or net biomass synthesis.

A variant of this use of metabolism to establish a problem boundary condition was provided by Monbouquette and Ollis (1986), who used a (Monod + maintenance) specific growth rate,

$$\mu = \mu_m \frac{S}{K + S} - k_e \qquad (3)$$

to calculate the minimum substrate level (S_{maint}) in the steady state viable biofilm: this value was defined by applying a metabolic condition, $\mu_{net} = 0$, at the inner boundary:

$$S_{maint} = K[(\mu_m / k_e) - 1]^{-1} \qquad (4)$$

This boundary condition fixes the biofilm thickness L as a predicted variable of the model through the following algorithm:

Set S_{maint}, $dS/dx = 0$ at R^*

Integrate reaction diffusion equation up to

biofilm outer surface $S = S_S$ $(R = R^{**})$

Calculate biofilm thickness as $L = R^{**} - R^*$

For actual situations, R^{**} and S_S are characteristically known in advance (gel beads, microcapsules, hollow fibers). A modified algorithm is then used which involves a first guess of R^*, followed by integration to see if $S = S_S$ at $R = R^{**}$, and subsequent iterations on the guessed R^* to obtain closure. These authors used a simple two compartment structured model which predicted variations of viable cell layer thickness, cell growth rate, and cell composition with spatial position. Use of such a metabolic boundary condition to provide the unknown location of a steady state biofilm boundary has been usefully applied subsequently: Skowland and Kirmse (1989) applied this condition to obtain simplified particle and packed bed models for zero and first order microbial rate equations.

The notion of steady state moving biomass has also been considered. Monbouquette and Ollis (1986) assumed that at steady state in an immobilized gel bead, excess cell production was all exported via cell "leakage" from the support; this analysis predicts a maximal cell leakage rate, q, at the outer gel bead surface to be calculated as

$$q = C_x \int_{R^*}^{R^{**}} \mu \ r dr \qquad (5a)$$

where C_x = maximum achievable cell concentration in porous support. A more detailed model, applicable to either pure culture or mixed culture biofilms, was developed by Wanner and Gujer (1986), who calculated a surface boundary biofilm medium velocity, V_L as

$$V_L(t) = \int_0^L \mu_o dz + \sigma(t) \qquad (5b)$$

where the biomass production integral over the local averaged specific growth rate and the (negative) exchange of biofilm biomass with the outer nutrient fluid (sloughing, grazing by predators., etc.) were represented by the first and second right hand terms, respectively.

The present paper summarizes recent work by Sayles (1989) on two steady state kinetic cases of frequent occurrence: inhibition by either excess substrate or by product. The basic approach again uses Monbouquette's 1986 notion of a metabolic boundary condition to set the innermost location of viable cells, and to calculate the resultant steady state spatial profiles of substrate, product (second case only), specific growth rate and biomass film or layer thickness for each given external condition on substrate S and product P.

Product Inhibition: Model

This example provides the simpler of the two cases. The assumed form of the dimensionless specific growth rate (Sayles and Ollis, 1990a) is

$$\mu(S, P) = \left(\frac{S}{K + S}\right)(1 - K_i P), \qquad (6)$$

which is a convenient form, and common for which K_i values were calculated (Table 1c) from the maximum ethanol levels, P_{max}, which are reported to be tolerated by live cells: $P_{max}(g/L) = 25(C.thermocellum)$, $41(C.acetobutylicum)$, $60 (S.uvarum)$, $86(Z.mobilis)$, and $90(S.cerevisiae)$.

Inclusion of a specific endogenous maintenance and death rate, $\alpha \equiv k_e/\mu_{max}$, leads to a net specific growth rate of

$$\mu_{net} = \mu(S, P) - \alpha \qquad (7)$$

The model equations for this reaction and diffusion problem are summarized in Tables 1a-d. The particular growth inhibition form chosen, (Eq. 6) is popular but not unique; it has often been used to describe ethanol inhibition (Pamment, 1988) and is a particular example of general inhibition forms proposed previously, such as

$[1 - P/P_{max}]^n$, (Levenspiel, 1980), and

$[1 - (P/P_{max})^n]$, (Luong, 1985a).

261

Table 1a.   Steady state model dimensionless equations

$$\frac{1}{R^2}\frac{d}{dR}\left[R^2\frac{dS}{dR}\right] \;=\; \phi_S\,\mu(S,P) \tag{T1}$$

$$\frac{1}{R^2}\frac{d}{dR}\left[R^2\frac{dP}{dR}\right] \;=\; \phi_P\,\mu(S,P) \tag{T2}$$

$$[\mu_{net}(S,P)]_{R=R^*} \;=\; 0 \tag{T3}$$

where:

$$\frac{dS}{dR} \;=\; \frac{dP}{dR} = 0 \qquad\qquad at\ \ R = R^* \tag{T4}$$

$$S \;=\; S_S \ \ and \ \ P = P_S \qquad\qquad at\ \ R = 1 \tag{T5}$$

$$\mu_{net}(S,P) \;\equiv\; \mu(S,P) - \alpha \tag{T6}$$

$$\mu(S,P) \;=\; \left(\frac{S}{K+S}\right)(1 - K_iP) \tag{T7}$$

Table 1b.   Model parameter values

$$
\begin{aligned}
\phi_S &= 88.7 \\
\phi_P &= 27.4 \\
K &= 0.0023 \\
S_S &= 0.565 \\
\alpha &= 0.12
\end{aligned}
$$

Table 1c.   Ethanol inhibition constants

Organism	Maximum Ethanol Conc., P_{max} (g/l)	Corresponding $K_i = \rho_b/P_{max}$
C. thermocellum	25	7.1
C. acetobutylicum	41	4.3
S. uvarum	60	3.0
Z. mobilis	86	2.1
S. cerevisiae	90	2.0

Table 1d. Calculation of dimensionless quantities

$$\phi_S \equiv \frac{\mu_{max} \, r_p^2 \, C_{X,max}}{Y_S \, D_S^{eff} \, \rho_b}$$

$$\phi_P \equiv \frac{\mu_{max} \, r_p^2 \, C_{X,max}}{Y_P \, D_P^{eff} \, \rho_b}$$

$$PR \equiv \left(\frac{dP}{dR}\right)_{r=1} = \int_{R^*}^{1} \phi_P \, \mu \, R^2 \, dR$$

$$SA \equiv \frac{PR}{(C_{X,max}/\rho_b)\frac{4}{3}\pi(1 - R^{*3})}$$

$$E \equiv \frac{3 \, PR}{\phi_P \mu (R = 1)(1 - R^{*3})}$$

Product Inhibition: Results and Discussion

The model was used to calculate nutrient and product profiles, specific growth rate profiles, thickness of the steady state viable cell layer and the resultant particle productivity and average biomass specific activity for product formation (latter terms defined in Table 1d).

The influence of product inhibition was explored through variation of the intrinsic microbial sensitivity to product, K_i (higher K_i = increased sensitivity) and of external, bulk fluid product concentration, P_s (dimensionless, = P_{actual} / ρ_b (biomass intrinsic density)).

Figure 1 (top) presents the calculated profiles of net specific growth rate vs. radial position in an immobilized cell bead or particle, for the specific case of a high surface concentration of nutrient (S_s = .565 >> K = .0023, Table 1b). Curve A represents the uninhibited case ($K_i = 0$, no sensitivity to inhibition). Increasing K_i values result in net specific growth rate profiles which exhibit progressively lower values of μ_{net} at any given radial position. However, the resulting depression of local nutrient consumption allows deeper diffusional penetration of nutrient, leading to formation of progressively <u>thicker</u> viable biomass layers (curves B, C, D). Eventually,

however, the increasing sensitivity to product also results in achievement of toxic environments ($P = P_{max}$) within the internal bead volume, and the viable cell layer exhibits a <u>reduction</u> of thickness, leading to complete viable biomass extinction when μ_{net} is depressed to 0 at the surface. Since $P_s = 0.1$ and S_s >> K for this illustration, $\mu_{net} = 0$ at the bead surface when

$$\mu_{net} \sim (1 - K_i P_s) - \alpha = 0, \quad \text{or}$$

$$K_i = \frac{1 - 0.12}{0.1} = 8.8$$

The influence of substrate S and product P on the specific growth rate profiles is conveniently illustrated by considering the form of $\mu(S, P)$ as

$$\mu(S, P) \equiv \mu_s(S) \, \mu_p(P) \qquad (8)$$

where $\quad \mu_s(S) = S / (K + S) \qquad (9a)$

and $\quad \mu_p(P) = (1 - K_i P) \qquad (9b)$

The profiles of $\mu_s(S)$ and $\mu_p(P)$ appear in Fig.1 (center, bottom). These profiles show that $\mu_s(S)$ drops off sharply for modest values of K_i (curve B), but that $\mu_s(S)$ hardly varies for higher K_i (curves D, E). In other words at low K_i, μ_s

263

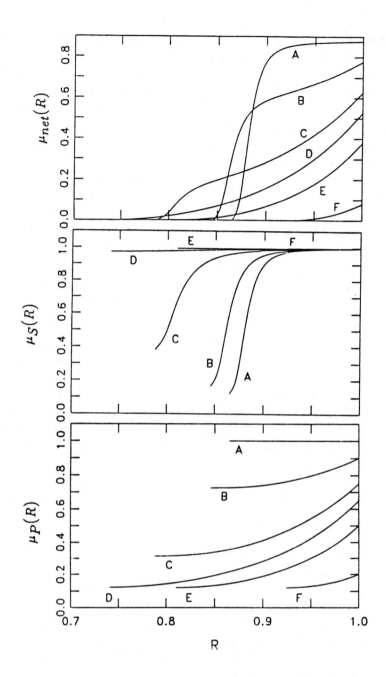

Fig. 1. Specific rate functions μ_{net}, μ_S, and μ_P as functions of radial position (R), for $K_i =$ 0(A), 1.0(B), 2.5(C), 3.5(D), 5.0(E), and 8.0(F). $P_S = 0.1$ for all cases. Reproduced with permission from Sayles and Ollis (1990).

(substrate limitation) dominates the behavior of $\mu(S,P)$, and at higher K_i, the function $\mu_p(P)$ (product inhibition) is primarily responsible for the spatial variation in $\mu(S, P)$ and therefore in $\mu_{net} \equiv \mu(S, P) - \alpha$.

The calculated variation of viable cell layer thickness L with K_i is shown in Fig. 2 for various particle surface concentrations of product, P_s. In each instance, increasing K_i results first in an increase of L, with further increases leading to a decrease of L and eventually to biomass extinction (L=0) at that value of K_i corresponding to $\mu_{net} = 0$ at the bead surface, i.e.,

$$\mu_{net} = 0 \cong (1 - K_i P_s) - \alpha$$

The calculated particle production rate, PR (Table 1d), is presented as a function of inhibition sensitivity, K_i, in Fig. 3 (top) for various P_s values. Increasing K_i results in a monotonic decline of PR, and increasing P_s sharply increases the average slope of this decline, signalling that in a high conversion plug flow reactor, later particles will be marginal contributors to overall reactor performance. Indeed, for an example K_i of 4.4, the vertical

dotted line represents the reactor axial variation of PR from inlet ($P_s = 0$) to maximal product concentration at the outlet ($P_s = 0.2$) corresponding to achievement of biomass extinction in the last bead before the reactor exit. Note here that the end of an active reactor is fixed by achievement of the condition $\mu_{net} = 0$ at the bead surface. A reactor of additional length would be pointless, since the product atmosphere is lethal at all points downstream of this zero net growth rate surface condition.

The calculated biomass volume-averaged cell specific activity SA (Table 1d) (Fig. 3, bottom) decreases sharply with increasing K_i in the domain where increasing K_i causes an increase in L; further increases of K_i eventually cause a decrease in L, and the resultant specific activity, SA, decreases much more slowly, since now both particle PR and biomass loading decrease with K_i where biomass loading

$$= (4/3) \pi (R^3 - (R - L)^3)\rho_b(1 - \varepsilon).$$

Product Inhibition: Reactor Design

Reactor design from these individual particle results is considered in two steps. First, the

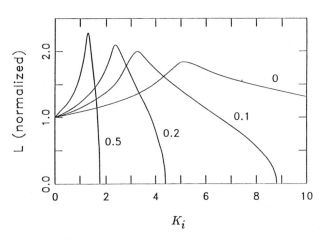

Fig. 2. Viable cell layer thickness (L) as a function of K_i. Parameter P_S. Reproduced with permission, from Sayles and Ollis (1990).

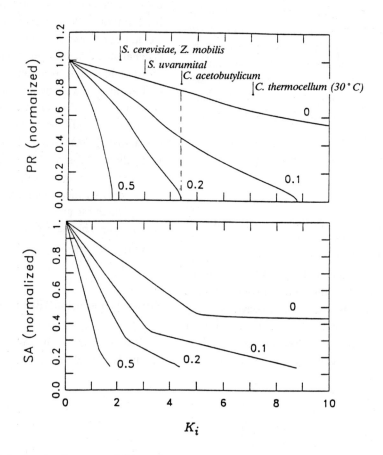

Fig. 3. Particle production rate (PR) and particle specific activity (SA), normalized to the same quantities calculated for uninhibited cultures ($K_i = 0$), as a function of K_i. Parameter: P_S. The arrows corresponding to known K_i values for several ethanol producing organisms. Dotted line: Example progression from entrance to (top, $P_S = 0$) to the end of a (viable cell) packed bed reactor ($P_S = 0.2$) for $K_i = 4.3$ (*C. acetobutylicum* example). Reproduced with permission, from Sayles and Ollis (1990).

calculated activities of each immobilized cell particle or bead are reduced to the conventionally calculated particle effectiveness E (Table 1); these results are presented in Fig. 4. Next, the effectiveness E (S_s, P_s) is incorporated in a reactor design equation of appropriate form. For reactor design equation of appropriate form. For example, a packed bed reactor of sufficient length to allow neglect of axial dispersion could be described by a single pair of equations (10a, b):

S profile:

$$\varepsilon V \frac{ds}{dz} + (1 - \varepsilon)\phi_s \mu(S, P)\left[1 - (1 - L(S, P))^3\right]$$

$$E(S, P)/3 = 0 \qquad (10a)$$

and P profile:

$$P(Z) - P_o(Z = 0) = \frac{\phi_p}{\phi_s}(S_o - S(Z)) \qquad (10b)$$

Substrate Inhibition: Model

Examples of substrate inhibition include *Z. mobilis* (glucose, ammonium ion), *N. winogradsky* (nitrite), *K. aerogenes* (sodium phenylacetate, benzoate or p-hydroxy benzoate), *H. polymorpha* (methanol), *C.utilis* (n-butanol), *C.lipolytica* (ethanol), and *S.*

cerevisiae (glucose). While specific inclusion of substrate inhibition in immobilized cells has been modelled previously by Atkinson and Mavituna (1983), Luong(1985b), and Monbouquette et al (1990), the more complete parametric study of Sayles (1989; Sayles and Ollis, 1990b) is summarized here as an example complementary to the product inhibition model discussed above.

The model approach is identical to that of Table 1, except that the net specific growth rate is now a function only of S,

$$\mu_{net} \equiv \mu(S) - \alpha \qquad (11a)$$

and the particular substrate inhibition form chosen is the most commonly cited version,

$$\mu(S) = \frac{\mu_{max}S}{(K_s + S)} \frac{K_i}{(K_i + S)}$$

or in dimensionless form,

$$\mu(S) = \frac{S}{(1 + S)(1 + \beta S)} \qquad (11b)$$

where μ and S are dimensionless. The dimensionless parameter β is defined as $\beta \equiv K_s / K_i$. From literature values for K_s and K_i, β values are calculated (Table 2); the pertinent

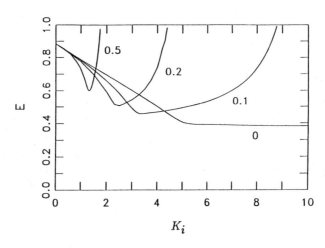

Fig. 4. Particle effectiveness (E) as a function of K_i. Parameter: P_S. $S_S = 0.565$. Reproduced with permission, from Sayles and Ollis (1990a).

Table 2a. Substrate inhibition constants

Organism	Substrate	K_S	K_i	β ($\equiv K_S/K_i$)
Z. mobilis CP4	glucose	0.50 g/l	2000 g/l	2.5×10^{-4}
S. cerevisiae	glucose	0.20 g/l	216 g/l	9.3×10^{-4}
K. aerogenes	hydrobenzoate	0.02175 mM	20.07 mM	0.0011
	phenylacetate	0.1746 mM	16.07 mM	0.011
	benzoate	0.3445 mM	4.50 mM	0.077
Nitrobacter sp.	nitrite	2.43 mM	149.2 mM	0.016
Candida utilis	sodium acetate	0.1481 %w/v	1.905 %w/v	0.078
Nitrosomomas sp.	ammonium	2.158 mM	11.07 mM	0.195
Pentane utilizing bacterium (JTP-271)	pentane	47.36 mm-Hg	47.36 mm-Hg	1.0

Table 2b. Calculation of dimensionless quantities

$$\phi_S \equiv \frac{\mu_{max}\, w^2\, C_{X,max}}{Y_S\, D^{eff}_{S,VCL}\, K_S}$$

$$SU \equiv \left(\frac{dS}{dX}\right)_{X=1} = \phi_S \int_{X_i}^{X_o} \mu[S(X)]\, dX$$

$$E \equiv \frac{SU}{\phi_S \mu(X=1)(1-X_i)}$$

$$\delta \equiv \frac{D^{eff}_{S,VCL}}{D^{eff}_{S,D}}$$

Table 2c. Various quantities as a function of β

β	S_M	S_L	$(X_i)_{max}$, MV layer	$(R_{tot})_{max}$ for MV layer, R_{tot} for MLV layer	$S_{S,max}$ $\delta = 1$	$S_{S,max}$ $\delta = 0.14$
10^{-3}	0.136	7332.	0.878	1.2×10^4	8796.	7537.
10^{-2}	0.137	732.	0.292	3.9×10^3	3493.	1119.
10^{-1}	0.139	72.2	0.105	1.1×10^3	1057.	210.
1	0.162	6.17	0.050	2.8×10^2	275.	43.8

Table 2d.   Steady state model dimensionless equations

The MV Model

$$\text{for } X_i \leq R \leq 1, \quad \frac{d^2S}{dX^2} - \phi_S\, \mu(S) = 0$$

$$S(X_i) = S_M$$

$$\left(\frac{dS}{dX}\right)_{X=X_i} = 0$$

$$S(X = 1) = S_S$$

The MLV Model

$$\text{for } X_i \leq R \leq X_o, \quad \frac{d^2S}{dX^2} - \phi_S\, \mu(S) = 0$$

$$\text{for } X_o \leq R \leq 1, \quad \frac{d^2\hat{S}}{dX^2} = 0$$

$$S(X_i) = S_M$$

$$\left(\frac{dS}{dX}\right)_{X=X_i} = 0$$

$$S(X_o) = \hat{S}(X_o) = S_L$$

$$\delta\left(\frac{dS}{dX}\right)_{X=X_o} = \left(\frac{d\hat{S}}{dX}\right)_{X=X_o}$$

$$\hat{S}(X = 1) = S_S$$

The LV Model

$$\text{for } 0 \leq R \leq X_o, \quad \frac{d^2S}{dX^2} - \phi_S\, \mu(S) = 0$$

$$\text{for } X_o \leq R \leq 1, \quad \frac{d^2\hat{S}}{dX^2} = 0$$

$$\left(\frac{dS}{dX}\right)_{X=0} = 0$$

$$S(X_o) = \hat{S}(X_o) = S_L$$

$$\delta\left(\frac{dS}{dX}\right)_{X=X_o} = \left(\frac{d\hat{S}}{dX}\right)_{X=X_o}$$

$$\hat{S}(X = 1) = S_S$$

269

range of β values for model simulation is seen to be $0 \leq \beta \leq 1.0$.

The form of the net specific growth rate, μ_{net}, provides two positive roots for the metabolic boundary condition $\mu_{net} = 0$. These roots lie on each side of the maximum positive value of m_{net} which occurs at $S = \beta^{-1/2}$. These roots are defined (Sayles and Ollis, 1990b) as the maintenance and lethal substrate values, respectively, where S_M = maintenance level = lowest value of S which provides cell growth to offset death and endogenous maintenance, and S_L = lethal level = highest value of S for which μ_{net} is non-negative (highest S which can still support cell viability).

Substrate Inhibition: Model Results and Discussion

In the previous case of product inhibition, the single metabolic boundary condition $\mu_{net} = 0$ (one root for P,S constrained by a mass balance) was used to determine the location of the inner viable cell layer boundary, and the outer boundary was that of the particle surface.

With the substrate inhibited case, two circumstances arise: when $S_s < S_L$, the boundary conditions on the viable cell layer are again $S=S_s$ at $R=1$ and $S=S_m$ at $R=1-L(S)$. When $S_s > S_L$, viable cells can no longer exist at the bead surface, and now both boundaries (L_1, L_2) of the viable cell layer are located by the dual metabolic boundary conditions of $S=S_L$ at $R=1-L_1$, and $S=S_m$ at $R=1-L_2$ with a viable cell layer thickness of $L=L_1-L_2 > 0$.

The full range of viable cell layer existence predicted in this model is shown in Fig. 5. Three different non-trivial biomass profiles are predicted at steady state: we define these distinct solutions as follow:

(i)   a maintenance viable cell layer (MV) when $S_m < S_S < S_L$.

(ii) a maintenance-lethal viable cell layer (MLV) when $S_S > S_L$ ($> S_M$ always) and

(iii) a lethal viable cell layer (LV) when $S_S > S_L$ and the particle center substrate value exceeds S_M so that the maintenance condition $\mu_{net} = 0$ cannot be used for the most deeply lying cells ($R = 0$).

The qualitative rationale for Fig. 5 is easily seen:

(i)   for $S_S < S_M$, no viable cells can be supported at any depth in the immobilization matrix, and the only steady state solution is the trivial one of zero biomass concentration everywhere.

(ii) for $S_M < S_S < S_L$, a viable cell layer exists between $R=1$ and $R=1 - L(S,\beta)$. Since the maximum nutrient level, S_S, is less than the smallest lethal concentration S_L, the viable cells exist from the outer surface inward to the point where $S_S = S_M$ which is established as the smaller positive root of the metabolic boundary condition $\mu_{net} = 0$.

(iii) for $S_S > S_L$, viable cells cannot exist at $R = 1$. However, if a viable cell population internal to the support can establish a substrate consumption rate, the consequent gradient of nutrient from $R = 1$ to the first cells allows for non-trivial solutions wherein the outer boundary of the viable cell layer occurs at $S = S_L$, the higher positive root of $\mu_{net} = 0$, and the inner boundary is at $S = S_M$, as before.

(iv) for $S_S > S_{s,max}$, no distribution of viable biomass exists that can reduce the internal nutrient level to S_L anywhere, hence the only steady state solution is the trivial one where complete biomass extinction occurs.

(v)   a final lethal viable (LV) solution is also predicted theoretically over the nutrient range identical to that for the MLV case [(iii) above]. Here a strongly

270

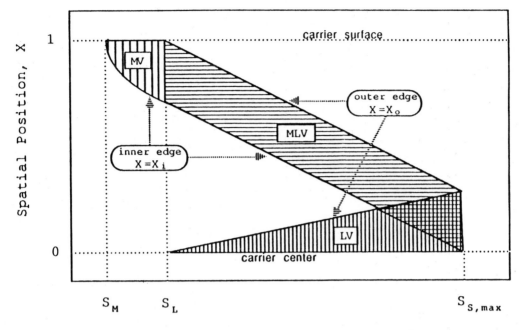

Fig. 5. Qualitative location and thickness of the viable cell layer as a function of surface substrate concentration S_S. The shaded regions contain viable cells; non-shaded regions lack any viable cells. See text for discussion of MV, MLV, and LV designations for viable cell layers. Reproduced by permission, from Sayles and Ollis (1990b).

inhibited culture finds a deeply lying niche wherein the culture's modest substrate consumption rate just matches a mild diffusion gradient.

Since the present model has boundary conditions determined by a restrictive condition between two variables (R^* and S_m), a stability analysis on the LV (or any other present) solution is not possible since the model does not allow independent perturbations of R^* and $S(R^*)$. Qualitative arguments suggest that both the MV and MLV layers are stable to perturbations, and hence experimental observation of these solutions is predicted. Similarly, the LV layer appears to be unstable, moving to either the MLV solution or the extinction solution when the outer edge boundary is perturbed toward or away from the outer surface, $R = 1$, respectively,

The spatial variation of net specific growth rate within the MV viable cell layer is shown in Fig 6 for three circumstances: (A) $S_S < \beta^{-1/2}$, so the cells nowhere exhibit the maximum value of μ_{net}. (B) $S_L > S_S > \beta^{-1/2}$, so the maximum in μ_{net} exists and viable cells can still grow at the surface, and (C) $S_L = S_S > \beta^{-1/2}$, so that $\mu_{net} = 0$ at the surface, $R = 1$. This last case (C) provides for existence of the thickest possible biomass layer in the MV regime.

For $S_S > S_L$, the MLV solution now exists (Fig.5), wherein an outer zone containing no viable cells provides a nutrient diffusion resistance that allows $S(R)$ to fall to S_L, which now defines the outer boundary of the MLV cell layer. This result is shown in Fig. 7 for two values of δ, the ratio of diffusivities in the viable cell zone to that in the outer region. The

271

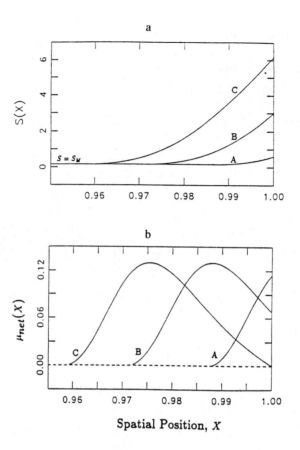

Fig. 6. (a) Substrate concentration (S) and (b) net specific growth rate (μ_{net}) within the immobilized cell particle as a function of radial position (R), for S_S =0.6(A), 3.0(B), and 6.17(C), and for β=1.0. Reproduced by permission, from Sayles and Ollis (1990b).

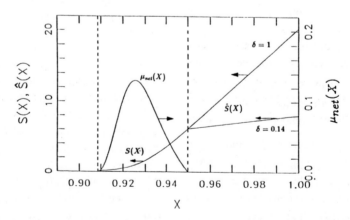

Fig. 7. The net specific growth rate (μ_{net}) and substrate concentration profiles for a typical MLV steady state. Vertical dashed line represents the inner and outer boundaries of the VCL. Substrate profiles in the "lethal" zone are shown for diffusivity ratios δ=1 and δ=0.14, as indicated. Reproduced by permission, from Sayles and Ollis (1990b).

value δ=1 corresponds conceptually to an outer region containing dead, non-lysed cells for which the effective nutrient diffusivity is the same as for the viable cell layer.   (Rigorously, dye permeation techniques such as trypan blue exclusion have been used to distinguish between live (dye impermeable) and dead (dye permeable) cells, so that our use of δ = 1 is a conceptual upper limit to d even with an outer layer full of dead cells).   The value δ = 0.14 represents the diffusivity ratio estimated for completely empty pores in the outer region.   The resultant viable cell layer location moves toward or away from the outer surface as S_S is decreased or increased, respectively.   (In planar geometry, the μ_{net} (R) profile would be invariant, and be simply shifted as suggested from Figs. 6 and 7.)

The thickness L of the viable cell layer regimes are, as expected, strong functions of the key parameter β ($\equiv K/K_i$). Figs. 8a, b, c indicate solutions for viable cell layer locations for MV and MLV cases when β values are 0 (no inhibition), 10^{-2} and 1.0 (maximum literature value).

Substrate Utilization: Reactor Design

As with the previous product inhibited case, these results are easily applied to reactor design. The corresponding particle substrate utilization rates (SU) and particle effectiveness factors E(S,β) are presented in Figs. 9 and 10. Only the presumably stable MV and MLV solutions are shown.

From the latter effectiveness factor, a plug flow reactor equation may again be written, now as eqns. (12a, b):

$$\varepsilon v \frac{ds}{dz} + (1 - \varepsilon) \phi_s \mu(S_s, \beta) \left[1 - (1 - L(S, \beta))^3 \right]$$

$$E(S, \beta) / 3 = 0 \qquad (12a)$$

and

$$P(Z) - P_o = \frac{\phi_p}{\phi_s}(S_o - S(Z)) \qquad (12b)$$

Summary

Cells grow or die according to the quality of the local nutrient and product atmosphere in an immobilization support.   Assuming that the support provides no direct influence on cell metabolism, the local atmosphere is completely described by the values of the local rate-determining concentrations (e.g., S, P).   With this assumption, the adaptive distribution of biomass and biomass specific activity may be predicted by solution of the appropriate reaction and diffusion rate equations. The key conceptual feature which provides for the biomass distribution to be a predicted rather than an assumed quantity is the use of one (ML) or two (MLV) metabolic boundary conditions corresponding to the single or dual roots of $\mu_{net} =0$.

This new approach awaits many experimental challenges which may well force modification and recasting of the steady state model. Additional features will have to be included, not the least of which is bulk displacement of cells due to growth and the corresponding influence of biomass displacement velocity. Wanner and Gujer (1986) have already provided an analytical framework for this velocity influence, and Stewart and Robertson (1988) have shown, via autoradiography, a convenient experimental technique for measurement of the predicted maximum biomass displacement velocity.

Unsteady state problems cannot conveniently make use of the metabolic boundary condition approach used here, and recourse to a full bead calculation at each point in time is necessary. Example references for this more complicated problem are Nakasaki, Murai and Akiyama (1989) (ethanol production, no inhibition, exogenous maintenance) and Monbouquette, Sayles, and Ollis (1990) (product inhibition, endogenous maintenance and cell death included).

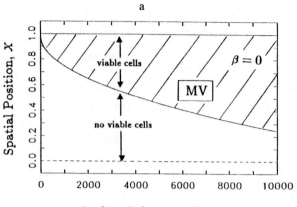

a

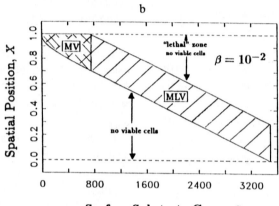

b

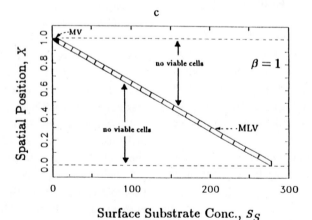

c

Fig. 8. Calculated location of the viable cell layers vs. the surface substrate concentration S_S for various degrees of substrate inhibition: $\beta=0$ (a, uninhibited reference state), $\beta=10^{-2}$ (b, weak inhibition), and $\beta=1.0$ (c, strong inhibition). The shaded regions represent all possible MV, MLV and LV steady states. Reproduced by permission, from Sayles and Ollis (1990b).

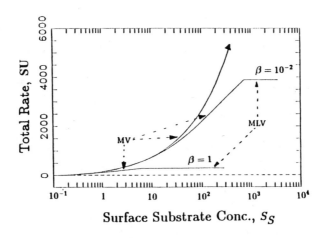

Fig. 9. Particle substrate utilization (SU) rate (Table 2b) vs. S_S for various values of β. Portions of each curve corresponding to the MV and MLV states are indicated. Reproduced by permission, from Sayles and Ollis (1990b).

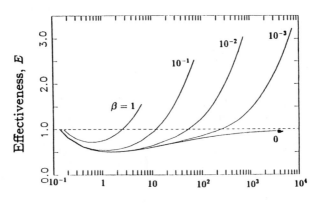

Surface Substrate Conc., S_S

Fig. 10. Viable cell layer effectiveness (E) defined in Table 2b vs. S_S for various values of β. Calculation of E only possible for MV state using Table 2b definition. For the MLV state, the quantity $m(S_S,\beta)$ $[1-(1-L)^3]E(S_S,\beta)$ in Eq. (12a) would be replaced by the directly calculated rate (since the reference surface rate $\mu(S_S, \beta)$ is negative!). Reproduced by permission, from Sayles and Ollis (1990b).

Nomenclature

C_x = maximum possible local cell concentration in support

K = Monod constant (M)

K_i = product inhibition constant (M^{-1})

k = maintenance constant (hr^{-1})

L = biomass layer thickness

P = product concentration (M)

q = cell leakage rate from bead

R = radial position (L)

R^* = inner boundary of viable cell layer

R^{**} = outer boundary of viable cell layer

S = substrate concentration (M)

S_{maint} = maintenance substrate level (M)

V_L = biofilm surface velocity

Greek Symbols

α = cell death and maintenance rate constant (dimensionless)

β = K_s/K_i

ε = support porosity

μ = specific growth rate (dimensionless)

μ_m = maximum specific growth rate

ρ_b = intrinsic biomass density

σ = biomass exchange (loss due to sloughing or grazing)

ϕ_p = product Thiele modulus

ϕ_s = substrate Thiele modulus

Acknowledgement

Portions of this work have been supported by Celgene, Inc., by the North Carolina Biotechnology Center, and by NCSU. The authors are pleased also to thank Professor Steven Peretti for his advice on numerical solutions and for provision of computer time

References

Atkinson, B.; Mavituna, F., *Biochemical Engineering and Biotechnology Handbook*, Nature Press, New York, **1983**, Ch. 10.

Levenspiel, O., *Biotechnol. Bioeng.*, **1980**, 22, 1671.

Luong, J.H.T., *Biotechnol. Bioeng.*, **1985(a)**, 27, 280.

Luong, J.H.T., *Biotechnol. Bioeng.*, **1985(b)**, 27, 1652.

Monbouquette, H.G.; and Ollis, D.F., *Ann. N.Y. Acad. Sciences*, **1986**, 469, 230.

Monbouquette, H.G.; Sayles, G.D.; and Ollis, D.F., *Biotechnol. Bioeng.*, **1990**, 35, 609.

Nakasaki, K.; Murai, T. and Akiyama, T., *Biotech. Bioeng.*, **1989**, 33, 1317.

Pamment, N.B., in *Alcohol Toxicity in Yeasts and Bacteria*, van Uden, N. (ed), CRC Press, Boca Raton, FL, **1988**, Ch.1.

Rittman, B.E.; and McCarty, P.E., *Biotechnol. Bioeng'g.*, **1980**, XXII, 2343.

Sayles, G.D., PhD thesis, NCSU, Raleigh, NC **(1989)**.

Sayles, G.D.; and Ollis, D.F., *Biotechnol. Prog.*, **1990(a)**, 6, 153.

Sayles, G.D.; and Ollis, D.F., *Chem. Eng. Science*, (submitted, **1990(b)**).

Stewart, P.S.; and Robertson, C.R., *Appl. Environ. Micro.*, **1988**, 54, 2464.

Skowland, C.T.; and Kirmse, D.W., *Biotechnol. Bioeng.*, **1989**, 33, 164.

Wanner, O.; and Gujer, W., *Biotech. Bioeng.*, **1986**, XXVIII, 314.

MICROMIXING IN FERMENTATION- The Response of Microorganisms to Hydrodynamic Forces

Eric H. Dunlop, Department of Chemical Engineering, Colorado State University,
Fort Collins CO 80523

When microorganisms are grown in fermentors, there exists a mismatch between the microscale of the turbulence (where the smallest eddy is typically 50 to 300 μm in diameter) and the cellular dimensions (1 to 5 μm). The cell can thus spend a significant fraction of its time in an eddy depleted of nutrients. The local fluid microscales were measured in a laboratory fermentor to confirm this. With *S. cerevisiae* in continuous culture, we found that the local microscales influence cell metabolism dramatically. The ultimate objective is to advance engineering science to the stage where rational design of fermentation and cell culture systems, with substantial reduction in empiricism, becomes possible. In particular to permit the design of fermentors using micromixing as a design parameter.

THE BASIC PROBLEM

It is intuitively obvious that fluid mechanical forces could influence how microorganisms, and other cells, grow. It is less obvious however which forces actually do what. The classical dilemma of fluid mechanics is that fluids in laminar flow can be well defined but are of limited practical significance in the typical fermentation environment. Turbulence on the other hand is difficult to define in adequate ways yet it is in this environment that most cells actually grow.

Theories of turbulence usually encounter the same problem of there being more unknowns than equations (the "closure" problem). In spite of the difficulties, a number of theoretical descriptions of restricted areas of turbulence have proved helpful and have withstood the test of time. The statistical theories of turbulence have proved useful, particularly those which treat turbulence as consisting of a spectrum of eddy sizes through which energy is cascaded and ultimately degraded to heat. This is best stated by Richardson's famous verse: "Big whorls have little whorls / Which feed on their velocities; / And little whorls have lesser whorls / And so on to viscosity", Richardson (1922) and Kolmogoroff (1961). The large eddies contain most of the kinetic energy, some of which is lost during the cascade process but most of which ends up being degraded to heat by much smaller eddies of almost negligible total energy, the energy being transferred to them through a cascade of instabilities. The smaller eddies are believed to adjust their motion to pass on the imposed energy flow to the smallest eddies that dissipate it as heat, Townsend (1976). The classical papers of Kolmogoroff (1961) describe this as a theory of local isotropy in which the motion of the smaller eddies depends only on the energy flow from the energy-containing eddies and on the fluid properties, otherwise being the same for all kinds of turbulent flow. Once the energy loss from the large eddies is known, the structure of the smaller ones is determined.

The issues to be examined are mixing on the scale of the biological cell, micromixing, and the micro-segregation of reactants. In particular whether they produce a biological response in a growing cell.

We begin with the realization that in many fermentors in which microorganisms are grown the diameter of the smallest turbulent eddy is about 50-300 μm. This compares with the diameter of the microorganism which is typically 1-5 μm. Consequently, regardless of how well the reactor may be mixed on a superficial basis, the bacterial cell may be sitting in a stagnant pool whose nutrients have been rapidly depleted. At first sight, increasing the power input to the fermentor stirrer should solve the problem. This is no real solution however as the size of the smallest eddy is proportional to the one-quarter power of the energy input, Tennekes (1972). Thus it would take 16 times as much energy to halve the eddy size -assuming the energy could actually be dissipated in the fermentor. On scale-up, when minimizing power costs are a concern, the problem is exacerbated.

Despite many major advances, there is also a lack of fundamental understanding of the

2181–2/92/0277$06.00/0 © 1992 American Chemical Society

nature of turbulent flows. This results in a deficiency in our theoretical and experimental tools for its effective study. Where turbulence theories have been most successful is in dealing with homogeneous isotropic flows, something that does not represent the fermentation environment well. It is hardly surprising that problems arise when microorganisms, whose biochemistry is complex, are grown in hard-to-define turbulent flows.

The problem is thus two-fold. First, can the fluid dynamics in the complex environment of a fermentor be defined in a useable way; and, second, does it really make a difference to the microorganism?

Ways in which the microscales of turbulence are known to be important occur in complex chemical reactions such as copolymerizations, which display a sensitivity to reaction order, Nauman (1983). First-order reactions are insensitive to micromixing while zero-order reactions exhibit moderate sensitivity. Liekus and Hanley (1984) calculate that Michaelis-Menten kinetics, typical of biological systems, can show major sensitivities to micromixing. Experimental verification has not until now been found in biological reactions. However some possible biological manifestations of the problem have been reported, Hansford and Humphrey (1966), Einsele et al. (1978), Tsai et al. (1969), Gerson (1980), Sawada et al. (1972), Faust et al. (1972), Ruklisha et al., (1989), and Rikmanis et al. (1987). The relationship between fluid turbulence and chemical reactions has been extensively studied, Brodkey (1981) and Patterson (1981) and some studies in biotechnology, Ho and Oldshue (1987), are appearing.

Microscales

The phenomenon of micromixing should be contrasted with the more conventional (bulk) macromixing. Macromixing typically deals with the distribution of nutrients on a large scale and how these nutrients are distributed through the fermentor. The phenomenon of micromixing on the other hand refers to the transport and the concentration gradients of nutrients at the microscopic scale of microorganisms. Micromixing requires totally different techniques for its study. Mixing, particularly micromixing, is expected to be of great importance in biochemical reactors, which are complex processes involving three phases. Individual biochemical reactions are usually very fast, but overall reactions, e.g., cell growth, are comparatively slow. Few studies have been concerned with the influence of fluid parameters on overall biological reactions.

Typically three scales are involved: a length scale (η), a time scale (τ) and a velocity scale (υ) given by Tennekes and Lumley (1972) and Hinze (1975):

$$\eta = \left(\frac{\nu^3}{\epsilon} \right)^{0.25} \quad 9\tau = \left(\frac{\nu}{\epsilon} \right)^{0.5} \quad \upsilon = (\nu \epsilon)^{0.25} \quad (1)$$

These are known as the Kolmogoroff scales and refer to the size, rotational time and rotational velocity of the smallest eddies in a turbulent flow. A Reynolds Number (the ratio of inertial to viscous forces) made from these scales has a value of one or less showing that the flow at these scales is quite viscous. These scales arise from the broadly accepted theories of turbulence advanced by Kolmogoroff (1961).

The important point is that only two parameters are involved: the kinematic viscosity (ν) and the energy input per unit mass (ϵ) to the fluid. Thus one can calculate, not the complete shape of the turbulence spectrum, but the convenient cut-off point of the Kolmogoroff eddies, i.e. the point beyond which little kinetic energy is contained in the turbulence.

There are two main ways of measuring micromixing reported in the literature: physical and chemical.

Physical methods are typified by hot film anemometry and laser doppler scattering from small particles. These methods have proved valuable in elucidating what is known of the structure of turbulence, but they have had, until recently, difficulties due to the presence of gas bubbles and cells and they require complex and expensive equipment.

Chemical methods are of two main types:
A. Extent of reaction systems: nitromethane and sodium hydroxide, Klein et al (1980), sodium thiosulfate and hydrogen peroxide, Keairns (1969), photochemical decomposition of cobalt salts, Treleaven (1972), ethyl acetate hydrolysis, Zoullian and Villermaux (1974), iodination of acetone, Plasari et al (1978), p-nitrophenyl acetate hydrolysis, Spencer and Lund (1980).
B. Product of reaction systems: 1-naphthol with diazotized sulfanilic acid (18-23), glycol diacetate hydrolysis, Zoullian and Villermaux (1974), iodination of para-cresol, Zoullian and Villermaux (1974), iodination of l-tyrosine, Bourne and Rohani (1983), bromination of resorcin, Bourne and Rys (1977).

Of these systems, the one that has been most extensively studied and best meet our criteria for use in fermentation modelling is the reaction between 1-naphthol (A) and diazotized sulfanilic acid (B). We therefore chose to use this reaction for measuring what happens to the microscales of turbulence in fermentors. This system has been extensively developed by Bourne et al (1977), Bourne, Crivelli et al. (1981), Bourne, Kozicki et al. (1981), Angst et al. (1982), Buldyga and Bourne (1984), and Bourne (1984), for measuring local microscales in flowing liquids. It is ideal for our purposes of characterizing the microscales in a fermentor in which yeast can be grown. Two reactions occur:

$$A + B \xrightarrow{k_1} AB$$

under diffusion control and

$$AB + B \xrightarrow{k_2} AB_2$$

under kinetic control.

AB and AB_2 are measured spectrophotometrically as they have distinct spectra in the visible region in the range 400 to 600 nanometers. By measuring the spectrum carefully at 2 nanometer steps with a high resolution (Carey 3) spectrophotometer it is possible to deconvolute the spectra using reference spectra of previously prepared pure A and pure B. Using the percentage of AB_2 measured and referring to the model developed by Bourne (1984), it can be shown that

$$\frac{[AB_2]}{[AB]} = fn\left(\frac{k_1}{k_2}\right), \left(\frac{[A]_0}{[B]_0}\right), \left(\frac{k_2 R^2 [B]_0}{D}\right) \quad (2)$$

where R is the radius of the effective diffusion sphere. Bourne and Kozicki (1981) and Angst (1982) showed that this sphere diameter is a close approximation to the Kolmogoroff eddy size. The appropriate rate constants and diffusivities have been accurately determined by Bourne et al. The precise numerical value for the microscale thus depends on the accuracy of Bourne's mathematical model. This is a potential weakness only as far as the precise values are concerned. An alternative that can be used is to define a micromixing index

$$X_s = 2[AB_2]/([AB] + 2[AB_2]) \quad (3)$$

in terms of observable parameters. While this is technically less ambiguous it yields no relationship to a readily useable concept such as the microscale diameter. In spite of its numerical imprecision the microscale is preferred. No information is lost and a useful physical picture is gained.

Time Scales

After the length scales, it is instructive to turn attention to the time scales associated with the turbulence. Unfortunately these are more difficult to quantify and interpret, and many different time scales exist.

a. The Fick's Law diffusion time, given by l^2/D, gives a measure of the time taken to deplete the eddy of substrate by diffusion. This is around 10 seconds for a 100 µm eddy.

b. The true micromixing times are more difficult to predict or measure. Attempts at measuring them have been made by Plasari et al. (1978). They predict micromixing times to be approximately 1-4 seconds. Thus the microscale eddies could persist for times comparable to those it takes for diffusion to deplete the eddy of substrate.

As the information is so difficult to obtain and interpret, it becomes necessary to find a way to measure the microscales directly, in the system of interest, and to look for relationships with cellular metabolism.

MICROSCALES IN FERMENTATION

Biological response to microscales

The most relevant literature is by Hansford (1966). They set up an experiment in a fermentor with four different injection substrate points using baker's yeast with glucose as a substrate and measured the yield as a function of dilution rate in a continuous-flow experiment. They found that when the substrate was injected behind the baffles the yield was reduced. When substrate was injected into the eye of the impeller increased yields resulted. When all four injection points were used simultaneously they obtained the highest yield.

In Figure 1 the data of Nagata (1975) show the distribution of microscales in a fermentor-type vessel to be highly non-uniform. The Hansford and Humphrey yeast experiment thus suggested to us that injection of substrate into different regions of the fermentor corresponded to injection into different zones of micromixing and that it produces important changes. To build on their observation and our interpretation, a 3 liter fermentor was set up with five different 0.5 mm diameter injection ports. The ports were positioned according to Nagata's indications of where extremes of micromixing could be anticipated. The ports were welded into place to ensure reproducibility.

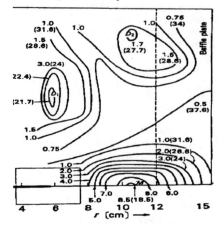

Figure 1. The microscale heterogeneity of a fermentor-like vessel is seen from Nagata's hot film studies. Dissipation energy (watts/kg) is shown with the length scale of the smallest eddies in parentheses (μm). Reproduced with permission from Nagata (1975). Copyright 1975 Halsted Press.

The questions addressed are straight forward:

1. Is there really a substantial mismatch between the size of the microorganism and the turbulent microscales? Are the microscales really as out of synchronization in a fermentor as is calculated?

2. Can we measure microscales in real fermentors?

3. Do they make a difference? In particular, do the varying levels of microscale induce a biological response such as yield changes?

4. Can we do anything about them in real fermentors either to minimize the problem or exploit the effect to advantage?

Development Of The Diazo System and Fermentor Characterization

To develop the diazo system for our use we carried out the following: a. prepared pure AB and AB_2. b. identified conditions, subsequently avoided, in which competing side reactions occurred. c. closed the mass balance consistently within 96% d. obtained regression

coefficients for the spectral deconvolution with $p < 0.005$. At this point it was felt that the diazo system, while non trivial to use, was capable of yielding the necessary information on fermentor performance.

Reactor characterization, using the diazo method, proceeds in three stages:

a. effect of stirrer speed at different injection points, b. effect of gas rate (VVM) at these injection points, c. comparison of these results with Bourne's stretching model, Bourne (1984) using bulk estimates of the power input, and hence microscale, obtained from Power Number relationships. The results obtained show that, as expected, microscales are reduced by increasing stirrer speed. Complex behavior is observed as a function of gas rate, with microscales either increasing or decreasing with increasing gas rate depending on the position of the injection port. For details see Dunlop and Ye (1990).

Preliminary Fermentation Experiments.

Yeast Flip-Flop Experiment

Using the characterized fermentor two extreme substrate injection ports were used; one with a microscale of turbulence roughly corresponding to approximately 50 μm, and the other with a microscale corresponding to approximately 160 μm. After an initial period of batch growth, a continuous fermentation was carried out, switching the substrate sequentially between these two ports. The results are shown in Figure 2.

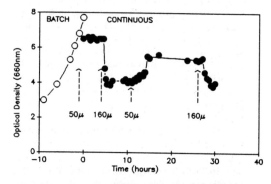

Figure 2. Substantial changes can be induced in yeast by switching the substrate injection from regions of poor micromixing (160 μm) to regions of much better micromixing (50 μm). Reproduced with permission from Dunlop and Ye (1990). Copyright John Wiley & Sons, 1990.

The most interesting point to emerge from this is the large optical density change observed by only switching from one port to the other. Overall, this was the first indication we had that changing the substrate injection port made a profound difference in the yield. It was an important difference and led us to believe that some interesting information was emerging.

Some attempts were made to examine changes in the biochemistry to see whether they supported the optical density changes. Significant changes were observed in mean cell volume, cell dry weight, yield on glucose, protein synthesis rate, aldolase specific activity and respiratory quotient. Non-significant changes were found in isocitrate dehydrogenase activity Dunlop and Ye (1990).

Carbon Conversion-Dilution Rate-Microscale Experiments

Setting the fermentor up as before but this time using all five glucose injection ports, six dilution rates were examined. The injection ports were found to correspond to Kolmogoroff microscales of 25, 50, 130, 160, 260 μm. Dilution rates were adjusted to 0.07, 0.16, 0.28, 0.30, 0.32, 0.36 hr^{-1}. Steady state was assured by multiple sampling and consistent off-gas composition on a mass spectrometer Dunlop and Ye (1990).

The results are shown in Figure 4 where carbon conversion (% of the carbon in glucose converted to cells) as a function of the microscale at the injection point is plotted with dilution rate as a parameter. The results are noteworthy. At the 0.16 hr^{-1} dilution rate, changing the microscale from 25 μm to 250 μm results in a major loss in carbon conversion from 56% down to 37%. A similar trend is seen at 0.07 and 0.28 hr^{-1} although the magnitude is not so large. This is a large change in the metabolic status of the cell and is of great scientific (and potential economic) significance. It is especially dramatic considering the effect was induced only by changing the injection point site and with it the local microscale.

The results are consistent with the idea that, with better micromixing (lower microscales), glucose is transported more effectively to the cell surface. Eventually, as the dilution rate becomes sufficiently large, excess glucose is present causing glucose repression (the Crabtree effect).

This is shown more clearly when the data are replotted in Figure 3 as cell yield (g cells/ g glucose) as a function of dilution rate. This allows us to superimpose the classic data of Von Meyenburg (1969), whose study involved substantial but unquantified power input into a small portion of a fermentor. Approximate calculation puts this overall (non-local) microscale at around 20 μm.

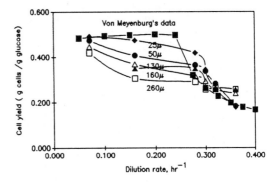

Figure 3. The glucose repression and its microscale effects are more clearly seen in this view of the results. Reproduced with permission from Dunlop and Ye (1990). Copyright John Wiley & Sons, 1990.

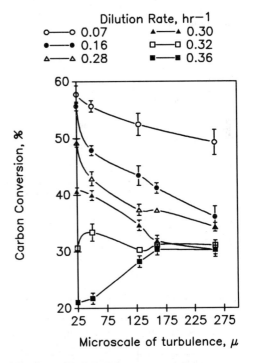

Figure 4. The metabolic efficiency is found to be quite dependent on local microscales of turbulence. Reproduced with permission from Dunlop and Ye (1990). Copyright John Wiley & Sons, 1990.

281

Three portions of the graph can be seen.

1. $D < 0.07$ hr^{-1}. All the cell yields are high and converge toward a single point. Adequate glucose is available for optimum metabolism.

2. $0.1 < D < D_c$. The necessary quantity of glucose can only be transferred at low microscales i.e., intense micromixing. As the microscale goes to the worst case of 260 µm failure to transfer adequate glucose results in progressively lower yields. In this range, carbon conversion is inversely proportional to microscale.

3. $D > D_c$. The onset of glucose repression occurs at D_c. $D_c = 0.24$ hr^{-1} for Von Meyenburg's data. For this study, onset was at 0.28 hr^{-1} for 25 µm eddies, 0.30 hr^{-1} at 50 µm. The exact onset is not accurately obtainable for the other microscales but exhibits the general trend consistent with the observation that D_c decreases with increasing bulk glucose concentration and, by implication, surface glucose concentration.

Our results to date are thus consistent with the picture of intense micromixing transferring more substrate to the cell surface, which is initially advantageous to metabolism but eventually overpowers it, causing onset of catabolite repression at lower dilution rates.

Injection port design

The kinetic energy of the substrate being injected, for a given flowrate, is inversely proportional to the 4th power of the orifice diameter. (Kinetic energy is proportional to the square of the liquid velocity which in turn is proportional to the square of the orifice diameter). Eventually the kinetic energy will become comparable to the energy in the local microscales. The ability of the microorganisms to respond to this becomes an issue. The problem being addressed can be seen from Table 1 below.

It is not obvious over what volume the energy is being dissipated but it is likely to be of the order of a milliliter. Some direct measurement of this will be obtained from laser doppler measurements to be made during this program. Thus at some point, probably around a 1 mm diameter port, the kinetic energy of the entering jet will be significant compared to the local energy dissipation rate and will overshadow its effect. This is important for two reasons i) such that experiments can be conducted where the dominant energy source is known and ii) to establish at what point in fermentor design the micromixing can be induced solely by appropriate design of the substrate injection ports.

Table 1. Somewhere around a 1 mm diameter feed port for the substrate the kinetic energy of the liquid becomes comparable with the local energy dissipation of the eddy. The addition of substrate then significantly alters the local environment

Feed Port Dia (mm)	Kinetic Energy Ratio	Watts Input by Jet (W)
3.0	1	0.00004
1.5	16	0.00074
1.0	81	0.0037
0.5	1296	0.06

Microscale (µm)	Point Energy Dissipation (W/ml)
38	0.0004
25	0.0026
18	0.0085

External Micromixing- Grid Induced Turbulence.

In Fowler and Dunlop (1989) we extended the concept to turbulence generation behind grids. The early onset of the Crabtree effect (catabolite repression) could be induced by the generation of turbulence behind a grid placed in an external loop fermentor (Figure 5). The response observed was that glucose repression took place earlier with grids in place, indicating that we successfully transferred more substrate to the yeast cell membrane resulting in repression. Some real changes were also seen in the ethanol concentration and the substrate level with and without the presence of screens. This then leads us to conclude that we can control the metabolic behavior of the yeast cell by changing the external concentration that the cell experiences and by changing the microscale of turbulence. This leads us strongly to conclude that yeast (and by implication other microorganisms) is in fact sensitive to changes in microscale.

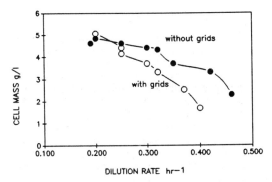

Figure 5. The effect of the grids is to induce such intense local turbulence that early onset of catabolite repression occurs. This is consistent with the effect that increased glucose concentrations have on the yeast (31).

Modelling Of Yeast Metabolism, Chaos Model

In examining other reasons for the results obtained to date, one other speculative explanation emerges. It derives from the concept of the yeast being intensively sensitive to the initial starting conditions. That is the behavior that one would expect from a system observing nonlinear dynamics in its metabolism and particularly one where the nonlinear dynamics shows chaotic behavior. It is possible to say that the real effect is "intense dependence on initial conditions". At a purely speculative level, this would suggest the presence of a fractal in the phase plane of the equations describing the metabolism in the yeast (Kopelman, 1988), where the reactants are spatially constrained.

Steinmeyer and Shuler (1989) proposed a metabolically structured model of baker's yeast growth which provides a convenient paradigm with which to explore non-linearities in the equations and, by implication, in yeast growth. Our very preliminary results exploring this possibility suggest that a bifurcation indicative of instability may exist. It is also possible that some way may be found whereby these effects can be related to the finding by Dudukovic (1977) that in chemical systems multiple steady states can sometimes arise as a result of micromixing phenomena. A separate but limited investigation into this area is being undertaken.

Laser Doppler Velocimetry Measurements.

The next logical step is to use Laser Doppler Velocimetry (LDV) to measure the **time** scales

of turbulence in fermentors and examine ways these can be coupled to or replace the length scales so far considered. The laser doppler velocimeter will be used to obtain the characteristic turbulent parameters- velocity macroscales and microscales; wavenumber cut offs; kinetic energy dissipation rate. We will obtain:

1. Autocorrelation functions with time delays to determine the integral scales of turbulence. i.e. the distance, L_s, and the time, over which the turbulent velocity fluctuations correlate with each other.

2. the Fourier transform of the data such that a power spectrum can be obtained as a function of wavenumber: E(k) vs. k.

3. using

$$\int_0^\infty k^2 E(k)\, dk = \epsilon_{turb} / \nu \qquad (4)$$

permits the turbulent energy to be obtained from the power spectrum.

4. the Corrsin (Brodkey, 1981) equation for mixing time will be used, i.e.

$$\tau = \frac{1}{2}\left[3\left(\frac{5}{\pi}\right)^{\frac{2}{3}}\left(\frac{L_s^2}{\epsilon}\right)^{\frac{1}{3}} + \left(\frac{\nu}{\epsilon}\right)^{\frac{1}{2}} \ln(Sc)\right] \qquad (5)$$

5. with the two-dimensional laser doppler we will obtain the Reynold's stresses, -u'v', and attempt to relate this to microbial yield. For turbulence enhanced kinetics for the reaction of species A, Brodkey (35) showed that

$$\frac{\partial \bar{C}_A}{\partial t} + (\bar{U} \cdot \nabla)\bar{C}_A = \bar{R}_A + D\nabla^2 \bar{C}_A - (\nabla \cdot \overline{(U'C'_A)}) \qquad (6)$$

i.e., the rate of reaction is changed by the Reynolds' stresses. Our objective is thus to experimentally obtain relationships between Reynolds' stresses (as a replacement for, or complement to, the Bourne diazo microscales) and the biological responses already described. Eventually it is anticipated that this will provide the data to distinguish between a time scale based hypothesis for micromixing vs a length scale based hypothesis.

CONCLUSIONS

We have shown that a powerful **correlation** exists between yield, cellular intermediates and the microscale of turbulence. We need to expand this to a **causal relationship**. Although extensive work has been done so far, it is still of a preliminary nature, needing substantial further elaboration to reach this desired state. This may be impossible to achieve but by approaching the problem from multiple angles the usefulness of the result is substantially enhanced.

SYMBOLS

A   Reactant, α-naphthol
B   Reactant, diazobenzenesulfonate
D'   Diffusion coefficient (m/sec)
D   Dilution rate (hr^{-1})
D_c Critical dilution rate (hr^{-1})
k   wavenumber of the eddy (m^{-1})
k_2 Reaction rate constant $(m^3/mole\ sec)$
L_s Integral scale of turbulence
V   Reaction volume, m^3
X   Micromixing index $= 2[AB_2]/([AB] + 2[AB_2])$
ν Kinematic viscosity $(m^2 sec^{-1})$
ϵ Energy dissipation $m^2 sec^3$
δ_0 Kolmogoroff turbulent length scale (m).
η length scale
τ time scale
υ velocity scale

REFERENCES

Angst, W.; Bourne J.R.; Sharma, R.N. *Chem. Eng. Sci.*; **1982**, *37*, 585.

Baldyga, J.; Bourne, J.R. *Chem. Eng. Comm.*; **1984**, *28*, 231, 243, 259.

Biotechnology Processes, Scale-up and Mixing; Ho, C.S.; Oldshue J.Y., Eds.; AIChE Press: **1987**.

BourNe J.R. *Inst. Chem. Eng. Symp. Ser.;* **1984**, *87*, 797.

BourNe, J.R.; Rohani, S. *Chem. Eng. Res. Des.;* **1983**, *61*, 297.

BourNe, J.R.; Rys, P.; Suter, K. *Chem. Eng. Sci.;* **1977**, *32*, 711.

BourNe, J.R.;. Crivelli, E.; Rys, P. *Helv. Chim. Acta*; **1977**, *60*, 2944.

BourNe, J.R.; Crivelli, E.; Kozicki, F.; Rys,P. *Chem. Eng. Sci.*; **1981**, *36*, (10), 1643.

BourNe, J.R.; Kozicki, F.; Moergeli U.; Rys, P. *Chem. Eng. Sci.*; **1981**, *36*, (10), 1655.

Brodkey, R.S. *Chem. Eng. Commun.*; **1981**, *8*, 1.

Dudukovic, M.P. *Chem. Eng. Sci.*; **1977**, *32*, 985.

Dunlop, E.H.; Ye S.J. *Biotech. Bioeng.*; **1990**, *36*, 854.

Einsele A.; Ristroph D.L.; Humphrey A.E. *Biotech. Bioeng.*; **1978**, *20*, 1487.

Faust, U.; Prave, P.; Sukatsch, D.A. *J. Ferment. Technol.*; **1972**, *55*, 609.

Fowler J.; Dunlop, E.H. *Biotech Bioeng.*; **1989**, *33*, 1039.

Gerson, D.F. *European J. App. Microbiol. Biotech.*; **1980**, *10*, 59.

Hansford G.S.; Humphrey A.E. *Biotech. Bioeng.*; **1966**, *8*, 85.

Hinze, J.O. *Turbulence*; 2nd Edition, McGraw Hill: 1975.

Keairns, D.L. *Can. J. Chem. Eng.*; **1969**, *47*, 395.

Klein, J.P.; David, R.; Villermaux, J. *J. Ind. Eng. Chem. Fundam.*; **1980**, *19*, 373.

Kolmogoroff, A.N. In *Turbulence: Classical Papers on Statistical Theory*" ed. Friedlander, S.K.; Topper, L. Interscience Publishers Inc, John Wiley & Sons, New York, **1961**.

Kolmogoroff, A.N., In "*Tubulence. Classical Papers on Statistical Theory*; Friedlander, S.K.; Topper, L., Ed.; Interscience Publishers Inc.: New York, 1961.

Kopelman R. *Science*; **1988**, *241*, 1620.

Liekus K.J.; Hanley T.R. *Biomedical Engineering III: Recent Developments*; Pergamon Press: **1984**, *107*.

Nagata S. *Mixing, Principles and Applications*; Halsted Press: **1975**, *341*.

Nauman, E.B.; Buffham, B.A. *Mixing in Continuous Flow Systems*; 172, John Wiley: New York; **1983**, *172*.

Patterson, G.K. *Chem. Eng. Commun.*; **1981**, *8*, 25.

Plasari, E.; David, R.; Villermaux, J., *ACS Symp. Chem. Reaction Eng.*; Houston, **1978**, No. 65, 125.

Richardson, L.F. *Weather Prediction by Numerical Processes*; Cambridge University Prss: Cambridge, 1922.

Rikmanis, M.A.; Vanags, J.J.; Viesturs, U.E. *Bioteknologiya;* **1987**, *1*, 72.

Ruklisha, M.P.; Vanags, J.J.; Rikmanis, M.A.; Toma, M.K.; Viesturs, U.E. *Acta Biotechnol,*; **1989**, *9*, 565.

Sadeh, W.Z. In *Refined Flow Modelling and Turbulence Measurements*; Iwasa, Y.; Tamai, N.; Wada, A., Eds.; Universal Academy Press: Tokyo, 1988.

Sawada, T.; Kojima, Y.; Takamatsu, T. *Proc. Pachec 72*, Tokyo, **1972**, *81*.

Spencer, J.L.; Lund, R.R. *Ind. Eng. Chem. Fund.;* **1980**, *19*, 142.

Steinmeyer, D.E.; Shuler, M.L. *Chem. Eng. Sci.*; **1989**, *44*, 2017.

Tennekes, H.; Lumley, J.L. *A First Course in*

Turbulence; The MIT Press: Cambridge, **1972**.

Townsend, A.A. *The Structure of Turbulent Shear Flow*; 2nd Edition, Cambridge University Press: 1976.

Treleaven, C.R.; Togby, A.H. *Chem. Eng. Sci.*; **1972**, *27*, 1653.

Tsai, B.I.; Erickson, L.E.; Fan, L.T. *Biotech Bioeng.*; **1969**, *11*, 181.

Von Meyenburg, K. *Catabolite Repression and the Germination Cycle of Saccharomyces Cerevisiae*; University Institute of Microbiology, Copenhagen, 1969.

Zoullian, A.; Villermaux, J. *J. Adv. Chem.*; (Ser. 133) **1974**, 348.

Gas-Liquid Oxygen Transfer in Perfluorochemical-in-Water Dispersions

James D. McMillan[*] and **Daniel I. C. Wang**, Department of Chemical Engineering, Massachusetts Institute of Technology, Cambridge, Massachusetts 02139

The maximum volumetric oxygen transfer rate, OTR_{max}, during 11.5-L fed-batch *E. coli* cultivation increases by 300% on a per liter aqueous phase volume basis in the presence of 36% (v/v) perfluorochemical (PFC). OTR_{max} increases by 180% on a total liquid volume basis. Enhancement in oxygen transfer in the presence of dispersed PFC is caused by higher liquid film oxygen permeability rather than by increased gas-liquid interfacial area or increased liquid film hydrodynamics. Unsteady state modeling is used to illustrate the mechanism of permeability enhancement.

It is difficult to adequately supply oxygen to actively respiring submerged cultures because of the low solubility of oxygen in aqueous solutions, typically less than 10 mg/L or 10 ppm at ambient conditions (Smith, 1928; Schumpe, 1985). As a result, the productivities of many aerobic cultivation processes are limited by oxygen transport rates. The many approaches for increasing oxygen transfer that are reported or proposed in the literature are reviewed by MacLean (1977) and Rols and Goma (1989). Despite the wealth of previous research aimed at improving oxygen supply, however, oxygenation remains a limiting factor in many processes and novel approaches for improving oxygen transfer continue to be explored. As we have reported previously, the use of perfluorochemicals (PFCs) is one method of improving oxygen supply (McMillan and Wang, 1987; McMillan and Wang, 1990).

History of Perfluorochemical Oxygenation

PFCs are attractive oxygen transport fluids for use with living systems because they are extremely biocompatible and solubilize oxygen to a high degree relative to most fluids. The earliest research on PFC oxygenation examined using PFCs to carry oxygen in artificial blood formulations (Riess and Le Blanc, 1982; Lowe, 1984). More recently, however, the potential of PFC oxygenation has become more widely appreciated and its use has expanded. For example, it is now being used in tumor therapy to increase the sensitivity of solid tumors to radiative and photodynamic treatments and to the action of cytotoxic drugs (King *et al.*, 1989). PFC oxygenation is also being used to supply oxygen to premature infants because it eliminates the risk of lung collapse that exists when conventional air or oxygen ventilation is used (Saltus, 1989).

The use of PFCs to improve oxygen supply to submerged cultures was patented in 1974 (Chibata *et al.*, 1974). Subsequently, various approaches for using PFCs have been considered. PFC has been dispersed in stirred tank bioreactors with aeration provided by conventional air sparging (McMillan and Wang, 1987; Junker, 1988; McMillan and Wang, 1990). PFC presaturated with oxygen has been sprayed into column reactors to provide droplet aeration of both microbial (Damiano and Wang, 1985) and mammalian (Biochem Technology, 1984; Cho and Wang, 1988) cell cultures. Oxygenation of immobilized cells has also been accomplished using presaturated emulsions of PFC (Adlercreutz and Mattiasson, 1982; Mattiasson and Adlercreutz, 1983).

We have chosen to use the cultivation of *Escherichia coli* in a conventional air-sparged

---

[*] present address: Biotechnology Research Branch, Solar Energy Research Institute, 1617 Cole Boulevard, Golden, Colorado 80401-3393

2181–2/92/0286$06.00/0 © 1992 American Chemical Society

stirred tank bioreactor as a model system for investigating the potential of improving bioreactor productivity using PFC dispersed by vigorous agitation. We refer to this technique of PFC-assisted oxygenation as cultivation in PFC-in-water dispersions. This report summarizes experiments we have performed to characterize oxygen transfer capabilities in this system, and describes unsteady state modeling we have used to elucidate the mechanism of oxygen transfer enhancement.

Background

Diffusion of oxygen into the liquid film adjacent to the gas-liquid interface controls the rate of oxygen absorption in sparged aqueous systems undergoing slow consumption of oxygen in the liquid phase (Danckwerts, 1970). Diffusive resistance within the liquid film is widely recognized to control the rate of gas-liquid oxygen transfer during submerged cultivation of unicellular aerobic microorganisms in which the cultivation broth remains Newtonian (Wang $et\ al.$, 1979; Bailey and Ollis, 1986).

The volumetric oxygen transfer rate, OTR, depends upon both the overall volumetric mass transfer coefficient, $K_L a$, and the driving force for mass transfer. $K_L a$ is usually based on the overall concentration driving force for mass transfer between the gas and liquid phases.

$$OTR = K_L a \left(C^* - C_L \right) \qquad (1)$$

Because of liquid film control, the overall mass transfer coefficient, K_L, is essentially equal to the individual liquid film mass transfer coefficient, k_L.

$$OTR = k_L a \left(C^* - C_L \right) \qquad (2)$$

Equation (2) shows that the oxygen transfer rate is equal to the product of the flux across the gas liquid interface, $k_L (C^* - C_L)$, and the specific gas-liquid interfacial area, a.

Improved oxygen transfer capacity in the presence of dispersed PFC therefore must be due to an increase in either interfacial area or flux. There are three mechanisms by which PFC could potentially influence oxygen transfer by affecting interfacial area or flux (McMillan and

Wang, 1987). First, PFC might contact the gas-liquid interface directly, thereby increasing the driving force for oxygen transfer into the liquid phase owing to the higher solubility of oxygen in PFC. The low surface tension of PFC may also cause the gas-liquid interfacial area to increase if PFC contacts the gas-liquid interface directly. It is widely recognized that the presence of an immiscible liquid phase increases specific gas-liquid interfacial area when the immiscible phase is surface active and decreases interfacial tension (Eckenfelder and Barnhart, 1961; Davies, 1963; Yoshida $et\ al.$, 1970; Linek and Benes, 1976; Hassan and Robinson, 1977; Hiemenz, 1986). Second, settling of dispersed PFC drops near or within the liquid film might cause convective or hydrodynamic enhancement. Hydrodynamic effects are proposed by Wise $et\ al.$ (1969), Hassan and Robinson (1977) and Andrews $et\ al.$ (1984a; 1984b). Third, PFC might improve solute flux through the film by increasing the overall oxygen permeability in the film. As shown in Equation (3), we define the permeability of solute in phase i, P_i, to be proportional to the maximum solute flux achievable in that phase.

$$P_i = D_i^{\frac{1}{2}} C_i^* \qquad (3)$$

Because the permeability of oxygen in PFCs is more than fifteenfold greater than in water, permeability enhancement is anticipated in PFC-in-water systems when the liquid film contains dispersed PFC (McMillan and Wang, 1990). The relative significance of the hydrodynamic and permeability mass transfer enhancement mechanisms on overall flux-associated enhancement depends upon the liquid film geometry and the velocity with which PFC drops move relative to the liquid film.

Approach

Experimentally, enhancement is quantified as a function of PFC loading by conducting a number of cultivation experiments at different PFC levels. Experiments are carried out at both the 1.25 L and 11.5 L scales and over a range in power input levels in order to determine the importance of vessel geometry and operating conditions on enhancement. The operative mechanism of enhancement is identified by distinguishing between increases in oxygen transfer caused by changes in specific interfacial area and increases caused by changes in the rate of oxygen flux across the gas-liquid interface. These two factors are separated by measuring the specific gas-liquid interfacial area in cell-free PFC-in-water dispersions, and then extrapolating these results to estimate the affect of PFC on gas-liquid interfacial area during cultivation (McMillan and Wang, 1990). This approach is used because high optical and acoustical opacity make *in situ* measurement of interfacial area during cultivation in the presence of PFC problematic. Time scale analysis is used to quantify the relative importance of the permeability and hydrodynamic mechanisms on flux-associated enhancement.

Materials and Methods

Perfluorochemical

Fluorinert Fluid FC-40 (3M, St. Paul, Minn.) is used in all experiments. FC-40 contains a mixture of PFCs, primarily trialkylamines with alkyl chain lengths between 3 and 5; the exact composition is proprietary. Physico-chemical properties of FC-40 are available in the product literature (3M, 1985; 3M, 1987).

Maximum Oxygen Transfer Rate Determination

Maximum oxygen transfer rate determinations are made during submerged cultivation of wild-type *Escherichia coli* K-12. Fed-batch PFC-in-water cultivations are conducted at 37°C and pH 6.9. Experiments are carried out in a 2-L (total volume) Setric 2M bioreactor and in a 14-L (total volume) Chemap LF bioreactor; vessel geometries are within the standard ranges cited by Wang *et al.* (1979). Using a working volume of 1.25 L, fermentations are conducted at an agitation rate of 500 rpm and an aeration rate of 0.8 vvm ($0.37 \ m^2/s^3$ or 0.5 hp/1000 L) in the 2-L bioreactor. These fermentations are initiated with no PFC, and PFC is added step-wise. Cultivation broth is removed prior to PFC addition to maintain constant total liquid volume. The 11.5-L (working volume) fermen-

tations are conducted at 800 rpm and 1.0 vvm ($0.37 \ m^2/s^3$ or 5 hp/1000 L) in the 14-L bioreactor. These fermentations are begun and maintained at constant PFC volume fraction throughout the fermentation.

The oxygen uptake rate is calculated directly from an on-line mass balance based on inlet and exit gas compositions determined by a Perkin-Elmer MGA 1200 mass spectrometer (Wang *et al.*, 1979). The pseudo steady state approximation is made that the oxygen transfer rate is equal to the oxygen uptake rate. Determinations of the maximum oxygen transfer rate, OTR_{max}, are made by averaging a minimum of 10 individual OTR values; individual measurements are made every 2 to 10 min, depending upon the experiment.

The PFC and aqueous phase volumes in the bioreactor are carefully monitored during all experiments. In addition to intermittent sampling for off-line analysis of cultivation broth, broth is withdrawn as required to maintain total liquid volume in the range 1.1-1.25 L or 11-11.5 L. Broth that is removed is transferred into a graduated cylinder so that both PFC phase and aqueous-cell phase volumes can be measured. One to five minutes of gravity sedimentation is required for phase separation, with settling time increasing as sample volume and cell concentration increase; settling times are longer when cell concentration is above 10 g dry cell weight (dcw)/L. Losses of PFC and aqueous phase because of sampling are determined by measuring the PFC phase (lower) and aqueous-cell phase (upper) volumes after phase separation. Losses caused by evaporation and stripping are estimated from the volume of condensate collected in the secondary exhaust gas condenser. Carryover of PFC and water in the gas phase following the secondary condenser is neglected.

Cell mass is determined using a Klett-Summerson Colorimeter equipped with a red filter and calibrated to actual dry cell weight (Wang *et al.*, 1979). A minimum of four dry cell weight determinations per cultivation are used for calibration.

Gas Holdup

Gas holdup is estimated from the difference in the dispersion height at the vessel wall. Each value reported represents the average

288

of three independent measurements. Holdup measurements are performed at 37°C in the 14-L Chemap bioreactor (11.5-L working volume) using phosphate buffered saline solutions of the same concentration as used in cultivation experiments, 6.0 g/L Na_2HPO_4 and 3.0 g/L KH_2PO_4. Gas holdup is measured in the cell-free PFC-in-water dispersions at identical operating conditions as those used during the 11.5-L cultivation experiments, i.e., agitation is at 800 rpm and aeration is at 1 vvm. In the gas holdup data reported herein, a floating baffle was used to improve the accuracy of the gas holdup measurement. This floating baffle is constructed out of two sheets of ¼-in.-thick polypropylene wired together; the overall baffle is ½-in. thick. One-inch holes are drilled throughout the baffle to permit the exhaust gas to exit the dispersion. The presence of the baffle nonetheless reduces the cross sectional area at the surface of the dispersion by 41%. As a consequence, absolute gas holdup values with the baffle present are somewhat higher than reported previously (McMillan and Wang, 1990). However, the presence of the baffle reduces fluctuations in height at the surface of the dispersion, particularly at high agitation rates, and therefore significantly improves the precision of the gas holdup measurement.

Further details on materials and methods and experimental protocol are available in McMillan and Wang (1987; 1990).

Experimental Results

Fig. 1 depicts typical oxygen transfer rate (OTR) (in mmol O_2 transferred per liter aqueous phase per hour) and dissolved oxygen (DO) (in percent air saturation) profiles during glycerol fed-batch cultivation of *E. coli* at the 1.25-L scale. This experiment was initiated with no PFC present in order to determine the maximum oxygen transfer rate in the absence of PFC, $OTR_{max}|_{\phi=0}$. Fig. 1 illustrates that during the early stages of cultivation, OTR increases exponentially with time, paralleling the increase in cell concentration during exponential growth phase (not shown). However, as the oxygen uptake rate of the culture increases (mmol oxygen consumed per liter aqueous phase per hour), the DO concentration progressively decreases. Fig. 1 shows that DO reaches zero at about 6 h, whereupon cultivation becomes oxygen-limited with the driving force for oxygen transfer at a maximum. At this point OTR plateaus at an $OTR_{max}|_{\phi=0}$ of 49 mmol/L_{aq}-h.

Upon subsequent addition of 14% (v/v) PFC (maintaining constant total liquid volume),

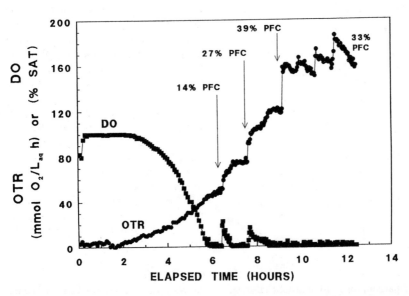

Fig. 1. Oxygen transfer rate and dissolved oxygen profiles during glycerol fed-batch cultivation of *E. coli* carried out at the 1.25-L scale. Arrows indicate the points at which perfluorochemical was added.

the DO level and the oxygen transfer rate immediately rise, demonstrating that the presence of PFC phase improves the oxygen transfer capacity of the bioreactor. After about 15 min, the oxygen uptake rate of the culture has increased sufficiently that the DO concentration again approaches zero. In the presence of 14% (v/v) PFC, the oxygen transfer rate plateaus at an $OTR_{max}|_{\phi=0.14}$ of 75 mmol/L_{aq}-h. When the PFC loading is further increased to 27% (v/v), similar behavior is observed. In this case an $OTR_{max}|_{\phi=0.27}$ of 120 mmol/L_{aq}-h is reached.

In this experiment several OTR_{max} values were measured after increasing the PFC loading to 39% (v/v). OTR_{max} values were measured at PFC loadings of 38%, 34%, and 33% (v/v) as the dispersed phase (PFC) loading steadily decreased from 9 to 12 h because of glycerol

addition. The fluctuations in the oxygen transfer rate data at the later stages of cultivation are the result of changes in the aqueous phase volume caused by carbon source addition and sample removal (for dry cell weight analysis and to maintain total liquid volume constant at 1.25 L). The effect of small fluctuations in the aqueous phase volume was reduced in the 11.5-L scale experiments.

The profiles from two experiments conducted at constant PFC loading at the 11.5-L scale are shown in Fig. 2. One experiment is a control in which no PFC is present and the other is an experiment carried out at a constant PFC loading of 36% (v/v). The top panel shows DO and cell concentration profiles, and the lower panel shows oxygen uptake rate profiles. The top panel shows that the presence of PFC extends the aerobic phase of cultivation

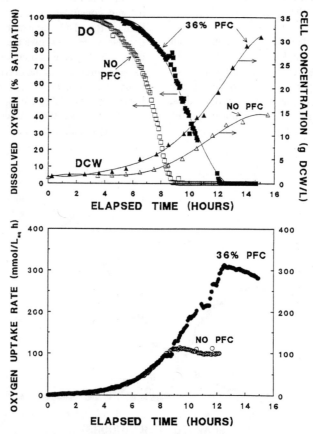

Fig. 2. Comparison of growth kinetics during 11.5-L glycerol fed-batch cultivation in the presence of 0% (open symbols) and 36% (v/v) (solid symbols) perfluorochemical. Upper figure shows dissolved oxygen (□,■) and biomass (△,▲) profiles. Lower figure shows oxygen uptake rate (○,●) profiles.

appreciably. Whereas the control fermentation becomes oxygen-limited at about 9 h at a cell concentration of less than 5 g dcw/L, in the presence of 36% (v/v) PFC, oxygen limitation is not reached until 12 to 13 h at a cell concentration of approximately 20 g dcw/L. After short periods of oxygen-limited growth, characterized by linear cell concentration versus time relationships, the cultures begin to lose their oxidative capabilities as a result of acetate accumulation (not shown). Because acetate accumulates more rapidly at higher cell concentration, the period of oxygen-limited growth is considerably shorter in the experiment conducted in the presence of PFC; when oxygen limitation begins the cell concentration in this experiment is roughly fourfold higher than in the control. Nonetheless, a more than twofold higher final cell concentration is obtained in the presence of PFC. The lower panel illustrates that improved bioreactor performance is a direct result of increased oxygen transfer capability.

A summary of OTR_{max} data obtained at the 1.25-L and 11.5-L scales is presented in Fig. 3. When calculated on a per liter aqueous volume basis, the trend in OTR_{max} as a function of PFC loading is linear at both scales. However, the slope of the OTR_{max} versus PFC volume fraction relationship is slightly higher at the 11.5-L scale than at the 1.25-L scale.

Fig. 4 shows a summary of 11.5-L scale OTR_{max} data plotted on both aqueous and total liquid volume bases. Whereas the increase in OTR_{max} is linearly related to PFC volume fraction when OTR_{max} is calculated on an aqueous volume basis, nonlinear saturation behavior is observed when OTR_{max} is calculated on a total liquid volume basis. Earlier analysis of OTR_{max} data obtained at the 1.25-L scale showed a similar trend (McMillan and Wang, 1987). At the 11.5-L scale, OTR_{max} increases by about 300% at 36% (v/v) PFC on an aqueous volume basis, and by about 180% on a total liquid volume basis.

Maximum volumetric productivities were calculated from the 11.5-L OTR_{max} cultivation data as the product of the maximum exponential growth rate of the culture (measured during each experiment) and the cell concentration at the point that the DO concentration decreases to zero and OTR_{max} is reached.

$$Q_p = \mu_{max} \, X \, |_{DO \rightarrow 0} \qquad (4)$$

Fig. 5 shows volumetric productivity and OTR_{max} data presented on total liquid volume bases. Productivity and OTR_{max} both exhibit saturation behavior on a total liquid volume basis, with productivity plateauing at about 5 g dcw/L-h above 0.2 volume fraction PFC. Total bioreactor productivity increases by roughly 150% in the presence of 20% (v/v) PFC. These data indicate that it is disadvantageous to operate at PFC loadings of greater than 20% (v/v) if the goal is to maximize bioreactor productivity.

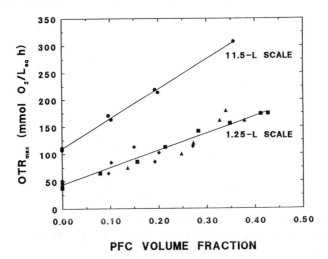

Fig. 3. Summary of maximum oxygen transfer rate data obtained at both the 1.25-L and 11.5-L scales. Values are on an aqueous liquid volume basis.

291

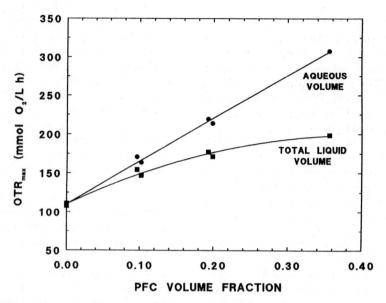

Fig. 4. Summary of maximum oxygen transfer rate data obtained at the 11.5-L scale. Values are on both aqueous (●) and total (■) liquid volume bases.

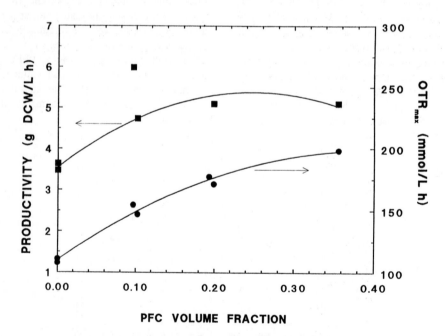

Fig. 5. Maximum volumetric productivity (■) as a function of perfluorochemical volume fraction. Maximum oxygen transfer rate data (●) is shown for comparison. Values are on a total liquid volume basis.

Mechanism of Enhancement

In a previous paper, we reported on experiments carried out to determine the effect of PFC loading on gas-liquid interfacial area in cell-free PFC-in-water dispersions (McMillan and Wang, 1990). Interfacial area was calculated from bubble Sauter mean diameters determined photographically and from gas holdup estimates made by measuring the change in dispersion height at the vessel wall. Our experiments showed that gas-liquid interfacial area was constant between PFC loadings of 0.5% and 15% (v/v). An independent confirmation of these results was desired, however, because the gas holdup measurement technique is reported to exhibit a ± 20% absolute error, and this introduces some uncertainty into the conclusions regarding constancy of interfacial area. The floating baffle is therefore used to improve the precision of the gas holdup measurement.

Fig. 6 shows the effect of PFC on gas holdup with the floating baffle present. Gas holdup increases rapidly from about 0.11 (m³ gas/m³ total liquid) in the absence of PFC to about 0.15 at 2.5% (v/v) PFC. Gas holdup remains relatively constant at 0.15 above 2.5% (v/v) PFC. The sharp increase in gas holdup between 0% and 0.5% (v/v) PFC indicates that

PFC is surface active at very low loadings. Although not shown, similar behavior is observed at 1200 rpm. Surface active behavior of PFC is anticipated because of the spreading behavior of PFC (McMillan and Wang, 1990). Apparently, the improved precision of the gas holdup measurement with the floating baffle present enables us to observe this behavior. We did not observe surface activity in measurements made without the baffle present. However, it is also possible that a small amount of residual PFC or traces of other surface active materials in the vessel may have prevented this behavior from being observed in the earlier gas holdup experiments. In any event, our new gas holdup data indicate that PFC is not affecting interfacial area at loadings above 2.5% (v/v). Because enhancement in oxygen transfer increases steadily with loading up to 20% (v/v) PFC, changes in oxygen flux rather than in interfacial area must be responsible for much of the improved oxygen transfer characteristics of PFC-in-water dispersions.

As discussed above, when PFC is present, improvements in the rate of solute flux across the gas-liquid interface can occur as a result of increases in liquid film hydrodynamics or increases in the permeability of oxygen in the liquid film. Time scale analysis is used to determine the relative importance of these two mechanisms.

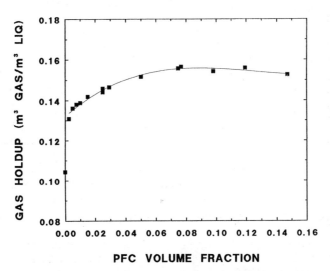

Fig. 6. Gas holdup as a function of perfluorochemical volume fraction in cell-free perfluorochemical-in-water dispersions.

Length and Time Scales

The characteristic time scale for an object (PFC drop or gas bubble) is defined as the characteristic length of the object divided by its characteristic velocity. The length scale of the momentum boundary layer, of individual liquid elements and of PFC drops depends upon the intensity of turbulence, and therefore on power input. Using the correlation of Michel and Miller (1962), power input into the 14-L bioreactor is calculated to be 3.7 m^2/s^3 (5 hp/1000 L) at 800 rpm and 1.0 vvm (11.5-L working volume) in an aqueous system. In a 50% (v/v) PFC-in-water dispersion at similar conditions, power input is also calculated to be 3.7 m^2/s^3 (7 hp/1000 L). Estimated power input remains constant on a per unit mass basis because of increasing dispersion density. The Kolmogorov turbulence microscale eddy length at a power input of 3.7 m^2/s^3 is about 25 µm (Hinze, 1959).

PFC drop Sauter mean diameter at a power input of 3.7 m^2/s^3 is estimated to be between 50 and 250 µm (Calabrese *et al.*, 1986; Bailey and Ollis, 1986). The drop size increases with PFC loading. Drop size is greater than the microscale eddy size primarily because of the high viscosity of the PFC phase (roughly four-fold higher than that of the aqueous phase).

Measurements of the bubble size distributions in cell-free PFC-in-water dispersions show that bubble size ranges from 0.5 to 2.5 mm, with a constant bubble Sauter mean diameter of about 1.9 mm (McMillan and Wang, 1990).

A drag coefficient plot is used to iteratively determine the terminal rise and terminal settling velocities of bubbles and large PFC drops ($d_d > 100$ µm) as a function of Reynolds number (Perry and Chilton, 1973). This approach is required because sedimentation and rise of large drops and bubbles does not occur in the Stokes' flow regime. The terminal settling velocities of smaller PFC drops, however, can be calculated using Stokes' Law.

Table 1 lists characteristic time scales for the movement of PFC drops and gas bubbles assuming that the characteristic velocities are equal to terminal settling (or rise) velocities. Bubbles of sizes 0.5 to 2.5 mm in free rise through both water and a 50% PFC-in-water dispersion exhibit a fairly constant characteristic time of about 10 msec. Characteristic bubble rise times are much shorter than characteristic sedimentation times for drops (30 - 200 msec).

Table 1

Time Scales During Cultivation in Perfluorochemical-in-Water Dispersions

Component of Interest	Diameter (µm)	Terminal Settling Velocity --- (µm/msec) --- $\times 10^{-5}$		Reynold's Number		Characteristic Time ----- (msec) -----	
		H_2O	50% PFC	H_2O	50% PFC	H_2O	50% PFC
PFC drops	25	2.9	1.5	0.0073	0.0037	110.0	210.0
	50	12.0	5.9	0.0590	0.0290	71.0	140.0
	100	37.0	16.0	0.3700	0.1600	27.0	64.0
Gas bubbles	500	570.0	600.0	29.0000	30.0000	8.7	8.3
	1500	1700.0	1700.0	260.0000	260.0000	8.6	8.6
	2500	2600.0	2600.0	640.0000	640.0000	9.8	9.8

Small drops sediment primarily in the Stokes' regime and move slowly relative to bubbles. This indicates that hydrodynamic effects caused by sedimentation of small ($d_d < 100$ µm) PFC drops are minor.

Although the characteristic settling time of large PFC drops ($d_d > 100$ µm) approaches that of bubble rise (not shown), it is anticipated that drop sedimentation within the liquid film remains in the Stokes' regime. There are two reasons for this. First, the fact that it takes several minutes for PFC drops to settle out of culture broth samples indicates that on a volume fraction basis most drops are between 50 and 100 µm in diameter, and thus move slowly relative to bubble rise. Second, drops that are large relative to the dimensions of the liquid film will be sterically hindered from interacting with the liquid film (discussed below).

Time scale analysis thus supports the hypothesis that convective transport caused by drop sedimentation is negligible, meaning gas-liquid oxygen transfer remains diffusion controlled in the presence of PFC. Mechanistically, this suggests that flux-related enhancement is caused by increased liquid film permeability rather than increased liquid film hydrodynamics, a conclusion that is in agreement with our previous findings (McMillan and Wang, 1987; McMillan and Wang, 1990).

Unsteady State Modeling

Because time scale analysis indicates that hydrodynamic enhancement is insignificant, flux modeling focuses on characterizing the mechanism of increased liquid film permeability. Modeling assumes that resistance within the liquid film controls the rate of gas-liquid mass transfer. Modeling also assumes that the gas-liquid interface is stabilized during cultivation in PFC-in-water dispersions so that the approximation of a stagnant liquid film is valid. Thus, Higbie's penetration theory (Higbie, 1935) is used to locally describe the process of diffusion into the liquid film.

The two key parameters required to establish a geometric representation of the liquid film geometry are the depth of the concentration boundary layer (the liquid film thickness) and the spacial distribution of PFC droplets within the liquid film (the liquid film geometry).

Film Thickness

The depth of the oxygen concentration boundary layer is a function of the time that surface elements are exposed to the gas-liquid interface. The characteristic exposure time for gas-liquid element contact estimated from bubble rise velocities is 10 msec.

Crank (1979) gives the solution for unsteady diffusion into a semi-infinite liquid that is initially at zero concentration and then at time zero subjected to a gas of constant composition, such that the interfacial concentration of the liquid is maintained at C_i^*.

$$C_i(x,t) = C_i^* \; erfc \; \frac{x}{\sqrt{4D_i t}} \qquad (5)$$

This equation can be evaluated to determine the penetration depth, l_p, defined as the distance into the liquid at which the concentration falls to 5% of C_i^*.

$$l_p = 2.8 \sqrt{D_i t} \qquad (6)$$

Values of l_p for both pure aqueous and pure PFC systems fall between 10 and 20 µm for characteristic exposure times of 10 msec.

Film Geometry

Modeling uses a geometric representation of the liquid film in which the dimensions of the drops are much larger than the film thickness. Modeling assumes that PFC does not contact the gas phase directly, an assumption supported by interfacial area results and a driving force analysis presented in McMillan and Wang (1990). Furthermore, gas holdup data show no surface activity at PFC loadings above 2.5% (v/v). If more than traces of PFC were contacting the interface directly, we would expect to see gas holdup and interfacial area continue to increase at higher PFC loadings.

Because calculations indicate that drop size is greater than the dimensions of the liquid film, fundamentally there can only be two types of surface elements: 1) elements in which oxygen diffuses into the aqueous phase alone (aqueous elements); and 2) elements into which oxygen diffuses into the PFC phase indirectly following the path gas → aqueous → PFC (layered medium elements). This is because PFC drops must be within a distance l_p of the gas-liquid interface for permeability enhancement to occur. Because the penetration depth ($l_p < 20$ μm) is considerably less than the PFC drop size (50-100 μm), only a single region of PFC phase is encountered during diffusion into a layered medium element. In other words, because the penetration distance is small relative to the drop size, solute diffusing into a layered liquid element only partially penetrates into a PFC droplet near the interface before a surface renewal event replaces the layered surface element with a fresh element from the bulk (either an aqueous or layered aqueous-PFC element).

Unsteady state modeling focuses on the problem of oxygen diffusion into a layered, semi-infinite medium, because the solution for the problem of diffusion into a semi-infinite homogeneous medium is well known [refer to Equation (5) or to Crank (1979)]. The mechanism of permeability enhancement is illustrated by examining the dynamic behavior of both types of liquid film elements.

The geometry used for unsteady state modeling is shown in Fig. 7. A layer of aqueous phase occupies the region $-l_{aq} < x < 0$, while a semi-infinite layer of PFC fluid occupies the region $x > 0$. The liquid-liquid interface is placed at the origin to simplify the solution.

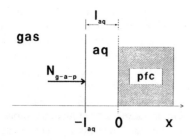

Fig. 7. Layered liquid film geometry used for transient modeling. Solute diffuses through an aqueous layer of depth l_{aq} into a perfluorochemical layer of infinite depth.

Governing Equations

The following parameters are defined in order to develop dimensionless equations for mass transfer into a layered medium.

$$C_a = \frac{C_{aq}}{C_{aq}^*} \qquad (7)$$

$$C_p = \frac{C_{pfc}}{C_{aq}^*} \qquad (8)$$

$$\tau = \frac{t \, D_{aq}}{l_{aq}^2} \qquad (9)$$

$$\xi = \frac{x}{l_{aq}} \qquad (10)$$

$$D = \frac{D_{pfc}}{D_{aq}} \qquad (11)$$

Using the variables defined by Equations (7) through (11), the governing equations for diffusion into a layered, semi-infinite medium become:

Conservation of mass in the aqueous phase:

$$\frac{\delta C_a}{\delta \tau} = \frac{\delta^2 C_a}{\delta \xi^2} \qquad (12)$$

Conservation of mass in the perfluorochemical phase:

$$\frac{\delta C_p}{\delta \tau} = D \frac{\delta^2 C_p}{\delta \xi^2} \qquad (13)$$

The initial and boundary conditions for the dimensionless equations are:

$$C_a = C_p = 0 \quad -1 < \xi < \infty \quad \tau = 0 \qquad (14)$$

$$C_a = 1 \qquad \xi = -1 \qquad \tau > 0 \qquad (15)$$

$$C_a = m \, C_p \qquad \xi = 0 \qquad \tau > 0 \qquad (16)$$

$$\frac{\delta C_a}{\delta \xi} = D \frac{\delta C_p}{\delta \xi} \qquad \xi = 0 \qquad \tau > 0 \qquad (17)$$

$$C_p \to 0 \qquad \xi \to \infty \qquad \tau > 0 \qquad (18)$$

Solution

The solution for this set of equations is obtained by the method of Laplace transforms. Carslaw and Jaeger (1959) provide details on the solution to the analogous heat transfer problem, which only differs in that the partition coefficient used to describe local equilibrium at the liquid-liquid interface is unity.

Concentration profile in the aqueous phase:

$$C_a(\xi,\tau) = \sum_{n=0}^{\infty} \alpha^n \left(\operatorname{erfc} \frac{(2n+1) + \xi}{2\sqrt{\tau}} \right.$$
$$\left. - \alpha \operatorname{erfc} \frac{(2n+1) - \xi}{2\sqrt{\tau}} \right) \qquad (19)$$

Concentration profile in the perfluorochemical phase:

$$C_p(\xi,\tau) = \frac{2}{m(1+\sigma)} \sum_{n=0}^{\infty} \alpha^n \operatorname{erfc} \frac{(2n+1)+\kappa\xi}{2\sqrt{\tau}} \qquad (20)$$

where:

$$\kappa = D^{-\frac{1}{2}} \qquad (21)$$

$$\sigma = \frac{1}{m} D^{-\frac{1}{2}} \qquad (22)$$

$$\alpha = \frac{\sigma - 1}{\sigma + 1} \qquad (23)$$

Application of Leibniz' Rule (Weast, 1979) for differentiation of an integral to Equations (19) and (20) yields expressions for the rate of solute flux through a layered semi-infinite medium.

Flux profile in the aqueous phase:

$$N_a(\xi,\tau) = -\frac{\delta C_a}{\delta \xi} \qquad -1 \le \xi \le 0 \qquad (24)$$

$$N_a(\xi,\tau) = \frac{1}{\sqrt{\pi\tau}} \sum_{n=0}^{\infty} \alpha^n \left(e^{-\left(\frac{(2n+1)+\xi}{2\sqrt{\tau}}\right)^2} \right.$$
$$\left. + \alpha\, e^{-\left(\frac{(2n+1)-\xi}{2\sqrt{\tau}}\right)^2} \right) \qquad (25)$$

Flux profile in the perfluorochemical phase:

$$N_p(\xi,\tau) = -D \frac{\delta C_p}{\delta \xi} \qquad \xi \ge 0 \qquad (26)$$

$$N_p(\xi,\tau) = \frac{1}{\sqrt{\pi\tau}} \frac{2\sigma}{1+\sigma} \sum_{n=0}^{\infty} \alpha^n e^{-\left(\frac{(2n+1)+\kappa\xi}{2\sqrt{\tau}}\right)^2} \qquad (27)$$

The flux across the gas-aqueous interface, found by evaluating Equation (25) at $\xi = -1$, is given by Equation (29).

$$N_a(\tau)\big|_{\xi = -1} = -\frac{\delta C_a}{\delta \xi}\bigg|_{\xi = -1} \qquad (28)$$

$$N_a(\tau)\big|_{\xi = -1} = \frac{1}{\sqrt{\pi\tau}} \sum_{n=0}^{\infty} \alpha^n \left(e^{-\frac{n^2}{\tau}} \right.$$
$$\left. + \alpha\, e^{-\frac{(n+1)^2}{\tau}} \right) \qquad (29)$$

The flux across the aqueous-perfluorochemical interface, determined by evaluating Equations (25) or (27) at $\xi = 0$, is given by Equations (30) or (31), respectively.

$$N_a(\tau)\big|_{\xi = 0} = \frac{1}{\sqrt{\pi\tau}} (1+\alpha) \sum_{n=0}^{\infty} \alpha^n e^{-\frac{(2n+1)^2}{4\tau}} \qquad (30)$$

$$N_p(\tau)\big|_{\xi = 0} = \frac{1}{\sqrt{\pi\tau}} \left(\frac{2\sigma}{1+\sigma}\right) \sum_{n=0}^{\infty} \alpha^n e^{-\frac{(2n+1)^2}{4\tau}} \qquad (31)$$

Using Equation (23), it can be shown that Equations (30) and (31) are equivalent due to the equality given by Equation (32).

$$1 + \alpha \;=\; \frac{2\sigma}{1 + \sigma} \qquad (32)$$

The total amount of solute transferred in time τ at any position ξ is found by integration of the appropriate flux expression.

$$F_\xi(\tau) \;=\; \int_0^\tau N_i(\eta)\Big|_\xi \, d\eta \qquad (33)$$

FORTRAN computer programs were written to evaluate the analytical expressions for solute concentration, flux, and total transfer as a function of distance and time.

Modeling Results

The top panels of Fig. 8 and Fig. 9 show the concentration and flux profiles, respectively, for transient diffusion into a layered medium using a gas phase partial pressure of 0.24 atm O_2, an aqueous layer depth of 1 µm, and a permeability ratio, σ, of 15 [see Equation (22)]. For comparison, the lower panels of these figures show the corresponding profiles for transfer into a single aqueous phase. Profiles are plotted in dimensional terms to facilitate direct interpretation in terms of length and time scales.

Fig. 8 shows that the presence of the PFC phase within the liquid film, even though it does not contact the gas-liquid interface directly,

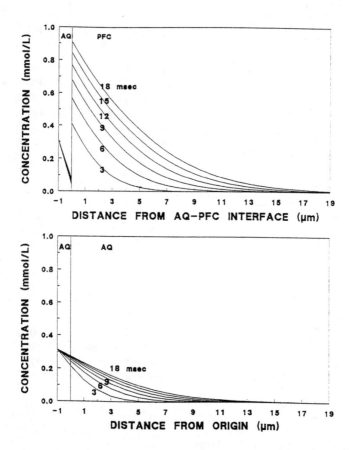

Fig. 8. Concentration profiles for oxygen diffusion into a layered aqueous-perfluorochemical medium. Upper figure shows profiles based on an aqueous layer depth of 1 µm, a permeability ratio of 15, and a gas phase oxygen partial pressure of 0.24 atm. Lower figure shows the corresponding profiles in an aqueous system.

significantly improves oxygen diffusion into the film. Compared to the control case of diffusion into an aqueous medium (shown in lower panel of Fig. 8), the oxygen concentration at the liquid-liquid interface remains much lower in the layered medium. Thus, the gradient in oxygen concentration that drives oxygen diffusion into the film stays higher when PFC is present. Moreover, because of higher oxygen permeability in the PFC phase, the depth of penetration is greater in the layered medium.

Fig. 9 illustrates the dramatic improvement in oxygen flux across the gas-liquid interface that results from the layered aqueous-PFC film geometry. Because of the high relative permeability of the PFC phase (characterized by a high σ value), the PFC phase acts as a rapid sink for oxygen diffusing through the aqueous film. Thus, as shown by Fig. 8, the driving force for oxygen transfer is maintained at a much higher level in the presence of PFC. Fig. 9 clearly demonstrates that from 3 to 18 msec the oxygen flux into the liquid film is much greater in the layered medium than in the aqueous control.

Aside from the permeability ratio, σ, the most critical parameter affecting the enhancement of oxygen transfer in the layered medium is the aqueous layer depth, l_{aq}. Fig. 10 shows total solute transfer across the gas-liquid interface as a function of exposure time for different values of l_{aq}. It is clear from this figure that potential enhancement in flux caused by a layered medium geometry diminishes rapidly as the aqueous layer depth is increased. For exposure times below 20 msec, very little enhancement in oxygen transfer occurs when l_{aq} exceeds 10 μm.

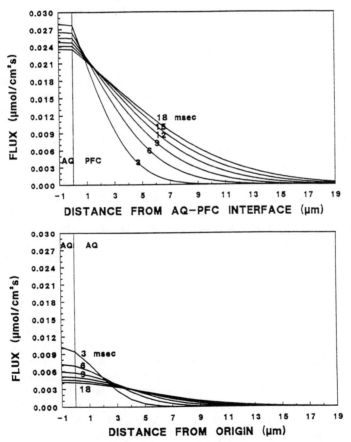

Fig. 9. Flux profiles for oxygen diffusion into a layered aqueous-perfluorochemical medium. Upper figure shows profiles based on an aqueous layer depth of 1 μm, a permeability ratio of 15, and a gas phase oxygen partial pressure of 0.24 atm. Lower figure shows the corresponding profiles in an aqueous system.

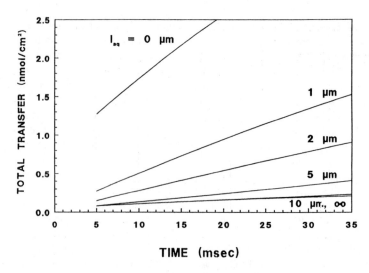

Fig. 10. Influence of aqueous layer depth on total oxygen transfer into a layered aqueous-perfluorochemical medium. The curve for $l_{aq} = 0$ μm depicts total solute transfer in a system with direct perfluorochemical contact, whereas the $l_{aq} = \infty$ curve illustrates the behavior in an aqueous system.

Conclusions

Cultivation in PFC-in-water dispersions increases oxygen transfer capabilities in sparged, agitated bioreactors and improves bioreactor productivity. Fed-batch cultivation experiments at both the 1.25-L and 11.5-L scales show that maximum oxygen transfer rates increase markedly during cultivation in PFC-in-water dispersions. The maximum volumetric oxygen transfer rate, OTR_{max}, during 11.5-L fed-batch *E. coli* cultivation increases by 300% on a per liter aqueous phase volume basis in the presence of 36% (v/v) PFC, and by 180% on a total liquid volume basis. The increase in OTR_{max} with PFC loading is linear on an aqueous volume basis; on a total volume basis OTR_{max} plateaus above 30% (v/v) PFC. A similar dependence of OTR_{max} on PFC loading is observed at the 1.25-L scale at a tenfold lower power input.

Cultivation experiments at the 11.5-L scale demonstrate that increased oxygen transfer capability can be translated into improved bioreactor productivity. Volumetric productivity increases up to a loading of 20% (v/v) PFC, at which point volumetric productivity has increased by 150%.

Interfacial area and gas holdup measurements indicate that enhancement in oxygen transfer is primarily flux-related, because increases in oxygen transfer capacity caused by PFC addition occur at much higher loadings than those at which PFC acts in a surface active manner. Time scale analysis demonstrates that hydrodynamic contributions to flux enhancement are negligible. Therefore, oxygen transfer enhancement in the presence of PFC is hypothesized to result from increased oxygen permeability in the liquid film. Dynamic modeling of permeability enhancement based on a layered medium geometry in which oxygen diffuses into the film by the pathway gas → aqueous → PFC demonstrates the importance of film geometry. Significant permeability enhancement occurs when the intervening aqueous layer depth in the liquid film is much smaller than the characteristic depth of penetration of the diffusing solute. When this condition is met the layered film geometry permits substantially higher fluxes into the aqueous phase to be maintained as a result of the high relative permeability of oxygen in the PFC phase.

We have recently developed a time-averaged surface renewal-based model of oxygen transfer enhancement in the PFC-in-water system. This model permits us to compare experimental measurements with theoretical predictions based on the mechanism of permeability enhancement and thereby gain further insight into the phenomenon of enhancement. The description of the time-averaged model development and of its comparison with our experimental data will be the subject of a future communication.

Nomenclature

a	specific gas-liquid interfacial area $[m^{-1}]$
C^*	hypothetical concentration of solute in equilibrium with bulk gas phase [mmol/L]
C	concentration of solute [mmol/L]
d_d	drop diameter [m]
D	diffusivity of solute $[m^2/h]$
dcw	dry cell weight $[g/L_{aq}]$
DO	dissolved oxygen concentration [% saturation]
erfc	complimentary error function,

$$erfc(\omega) = 1 - \frac{2}{\sqrt{\pi}} \int_0^\omega \exp(\eta^{-2})d\eta$$

k_L	liquid film mass transfer coefficient based on overall concentration driving force $[h^{-1}]$
K_L	overall mass transfer coefficient based on overall concentration driving force $[h^{-1}]$
l_{aq}	depth of aqueous layer in layered medium geometry (see Fig. 7) [m]
l_p	penetration depth, as defined by Equation (6) [m]
m	oxygen solubility ratio, $m = C_{aq}^{eq}/C_{pfc}^{eq}$
N	interfacial solute flux [mmol/m²-h]
OTR	volumetric oxygen transfer rate [mmol/L-h]
OTR_{max}	maximum volumetric oxygen transfer rate [mmol/L-h]
P	permeability of solute, as defined by Equation (3) $[mmol/cm^2-s^{\frac{1}{2}}]$
Q_p	volumetric productivity [g dcw/L-h]
rpm	revolutions per minute $[min^{-1}]$
t	time [h]
vvm	volume of gas per total volume of liquid per minute $[min^{-1}]$

x	distance into liquid normal to the gas-liquid interface [m]
X	cell concentration [g dcw/L]
ϕ	perfluorochemical volume fraction
μ_{max}	maximum exponential growth rate $[h^{-1}]$

Dimensionless variables

C	concentration, $C = C_i/C_{aq}^*$
D	diffusivity ratio, $D = D_{pfc}/D_{aq}$
F	total solute transferred at position ξ in time τ
N	flux normal to the gas-liquid interface
α	permeability parameter defined by Equation (23)
κ	diffusivity parameter defined by Equation (21)
ξ	distance normal to the gas-liquid interface, $\xi = x/l_{aq}$
σ	permeability ratio defined by Equation (22)
τ	time, $\tau = (tD_{aq})/l_{aq}^2$

Subscripts

a	aqueous phase (dimensionless)
i	liquid phase i
L	bulk liquid phase
p	perfluorochemical phase (dimensionless)

Acknowledgments

The authors gratefully acknowledge the considerable financial support provided for this research. During the initial period over which this research was conducted, J. D. McMillan was partially supported by a United States National Science Foundation Graduate Fellowship. Financial support for this research has also been provided by the Sun Company of Radnor, Penn., and by the Massachusetts Institute of Technology Biotechnology Process Engineering Center through the NSF-ERC Initiative under cooperative agreement CDR-88-03014.

References

Adlercreutz, P.; Mattiasson, B. *Appl. Microb. Biotechnol.* **1982**, *16*, 165-170.

Andrews, G. F.; Fonta, J. P.; Marrotta, E.; Stroeve, P. *Chem. Eng. J.* **1984a**, *29*, B39-B46.

Andrews, G. F.; Fonta, J. P.; Marrotta, E.; Stroeve, P. *Chem. Eng. J.* **1984b**, *29*, B47-B55.

Bailey, J. E.; Ollis, D. F. *Biochemical Engineering Fundamentals* (second edition); McGraw-Hill: New York, **1986**; p 464, pp 474-476.

Biochem Technology UPDATE. **1984**, *1*(1), November.

Calabrese, R. V.; Chang, T. P. K.; Dang, P. T. *AIChE. J.* **1986**, *32*, 657-681.

Carslaw, H. S.; Jaeger, J. C. *Conduction of Heat in Solids* (second edition); Oxford University Press: Oxford, UK, **1959**, pp 319-323.

Chibata, I. S.; Yamada, S. T.; Wada, M. N.; Izuo, N. Y.; Yamaguchi, T. Y. U.S. patent no. 3,850,753. **1974**.

Cho, M. H.; Wang, S. S. *Biotech. Lett.* **1988**, *10*, 855-860.

Crank, J. *The Mathematics of Diffusion* (second edition); Clarendon Press: Oxford, UK, **1979**.

Damiano, D.; Wang, S. S. *Biotech. Lett.* **1985**, *7*, 81-86.

Danckwerts, P. V. *Gas-Liquid Reactions*; McGraw-Hill: New York, **1970**.

Davies, J. T. *Adv. Chem. Eng.* **1963**, *4*, 1-50.

Eckenfelder, W. W., Jr.; Barnhart, E. L. *AIChE J.* **1961**, *7*, 631-634.

Hassan, I. T. M.; Robinson, C. W. *Biotech. Bioeng.* **1977**, *19*, 661-682.

Hiemenz, P. C. *Principles of Colloid and Surface Chemistry* (second edition); Dekker: New York, **1986**; pp 314-317.

Higbie, R. *Trans. AIChE.* **1935**, *31*, 365-389.

Hinze, J. O. *Turbulence: An Introduction to its Mechanism and Theory*, McGraw-Hill: New York, **1959**; pp 142-248.

Junker, B. H. Ph. D. Thesis, Massachusetts Institute of Technology, Department of Chemical Engineering, **1988**. p 33 and p 324.

King, A. T.; Mulligan, B. J.; Lowe, K. C. *Bio/Technology.* **1989**, *7*, 1037-1042.

Linek, V.; Benes, P. *Chem. Eng. Sci.*, **1976**, *31*, 1037-1046.

Lowe, K. C. *Comp. Biochem. Physiol.* **1987**, *87A*, 825-838.

MacLean, G. T. *Proc. Biochem.* **1977** November, 22-24.

Mattiasson, B.; Adlercreutz, P. *Ann. N. Y. Acad. Sci.* **1983**, *413*, 545-547.

McMillan, J. D.; Wang, D. I. C. Biochemical Engineering V. *Ann. N. Y. Acad. Sci.*, **1987**, *506*, 569-582.

McMillan, J. D.; Wang, D. I. C. Biochemical Engineering VI. *Ann. N. Y. Acad. Sci.* **1990**, *589*, 283-300.

Michel, B. J.; Miller, S. A. *AIChE J.* **1962**, *8*, 262-266.

Chemical Engineers Handbook (fifth edition). Perry, R. H.; Chilton, C. H., Eds.; McGraw-Hill: New York, **1979**.

Riess, J. G.; Le Blanc, M. *Pure and Appl. Chem.* **1982**, *54*, 2383-2406.

Rols, J. L.; Goma, G. *Biotech. Adv.* **1989**, *7*, 1-14.

Saltus, R. *Boston Globe*, **1989**, August 2.

Schumpe, A. In *Biotechnology*; Rehm, H.-J.; Reed, G., Eds.; VCH: Weinheim, FRG, **1985**, Vol. 2, pp 159-170.

Smith, D. F. In *International Critical Tables*; McGraw-Hill: New York, **1928**; Vol. 3, pp 254-272.

3M. Product manual on fluorinert electronic liquids. **1985**, p 9 and p 66.

3M. Fluorinert electronic liquids brochure #98-0211-2588-9(27.5)NPI. **1987**.

Wang, D. I. C.; Cooney, C. L; Demain, A. L.; Dunnill, P.; Humphrey, A. E.; Lilly, M. D. *Fermentation and Enzyme Technology*, Wiley: New York, **1979**, pp 157-193.

CRC Handbook of Chemistry and Physics (60th edition); Weast, R. C., ed., CRC Press: Boca Raton, FL, **1979**.

Wise, D. L.; Wang, D. I. C.; Matelles, R. I. *Biotech. Bioeng.* **1969**, *11*, 647-681.

Yoshida, F.; Yamane, T.; Miyamoto, Y. *Ind. Eng. Chem. Process Des. Develop.* **1970**, *9*, 570-577.

Part V
Bioseparations Scaleup

ADSORPTION TECHNIQUES CONTINUE TO BE ASCENDENT, and the search for ideal adsorbents through which unimpeded flow of eluent can occur continues. In the area of adsorption, this is the era of the membrane—the affinity membrane in particular. At least two firms have staked claims in this field, and both describe their products in scientific terms in this and the next section. Extraction methods using two immiscible aqueous phases have been under-utilized, but their versatility is becoming apparent through investigations of a variety of potential applications.

Strategies to Minimise Fouling in the Membrane Processing of Biofluids

Anthony G. Fane, K.J. Kim, P.H. Hodgson, G. Leslie, C.J.D. Fell, A.C.M. Franken, V. Chen and K.H. Liew, Centre for Membrane and Separation Technology, University of New South Wales, Kensington, Australia, 2033

Efficient separation and purification techniques are essential to the successful application of bioprocessing. The pressure-driven membrane processes of ultrafiltration (UF) and crossflow microfiltration (MF) are obvious candidates for processing the products of fermentations. Table 1 shows that these two techniques cover a broad range of components found in fermentation liquors. In principle the techniques can be used for the following important separations:

(i)   concentration of macromolecules;
(ii)   harvesting of biomass;
(iii)   recovery of macromolecules from biomass suspensions;
(iv)   fractionation of macromolecular mixtures.

Furthermore using membrane technology the processing should be achievable at modest temperature, without undue stress on the bio-products, in sterile conditions, and in batch or continuous mode. However, whilst membrane technology has had many successful applications in bioprocessing it is not a universal panacea.

One of the major obstacles to the use of membranes and the translation from lab scale to practical scale is the problem of fouling (Fane and Fell, 1987). Solutions to this problem are coming from a better understanding of the mechanisms of interactions between solutes or particles and membranes. This paper reviews our recent work on membrane fouling in UF and MF and suggests strategies to overcome the problem. The discussions focus on the concentration of macromolecules (retentive UF), the harvesting of biomass (retentive MF) and the passage of macromolecules through solute-permeable UF and MF membranes.

The Various Forms of Fouling

The process of separation and filtration which occurs in UF and MF results in an increased concentration of retained species at the membrane surface. The degree of concentration polarisation (sometimes referred to as gel-polarisation in UF, or particle polarisation in MF) is determined by the balance of convection towards the membrane and shear-induced 'back-transport' from the membrane. Thus concentration polarisation increases with pressure and feed concentration and decreases with high

Table 1 Typical Components of Fermentation Broths

Components	Molec. Wt (Daltons)	Size (nm)	Components Retained (useful range)		
			RO	UF	MF
Yeasts, fungi		10^3 - 10^4			
Bacteria		300 - 10^4			
Colloids		100 - 10^3			
Virus		30 - 300			
Proteins	10^4 - 10^6	2 - 10			
Polysaccharide	10^4 - 10^6	2 - 10			
Enzymes	10^4 - 10^6	2 - 5			
Antibiotics	300 - 400	0.6 - 1.2			
Simple Sugars	200 - 10^3	0.8 - 1.0			
Organic acids	100 - 500	0.4 - 0.8			
Inorganic ions	10 - 100	0.2 - 0.4			

crossflow or stirring. Ideally polarisation and flux achieve a steady state, and the release of pressure allows all the polarised solutes or solids to return to the bulk solution. However, in practice membrane fouling usually occurs. By definition, fouling results in one or more of the following;

(i)   a gradual decrease in flux with time,
(ii)   a gradual increase in retention of intitially permeable species,
(iii)   a used membrane with a reduced pure water flux.

Fouling may be reversed by cleaning, but not by simply altering operating conditions, such as pressure or crossflow. Fouling is not the same as concentration polarisation, but is a consequence of it. The foulants become attached to the membrane by processes such as adsorption, precipitation or convectively-driven plugging, and these processes may continue as more foulants are brought to the fouled membrane. In this paper the term deposition is used to describe the irreversible accumulation of foulant. Figure 1 depicts various forms of fouling; the pictures are idealised, particularly since most membranes have tortuous, non-cylindrical pores, and possibly irregular or mesh-like surfaces. The figure represents certain scenarious, as follows;

(a)   surface deposition, with pore plugging - represents a likely model for fouling of a retentive UF membrane by a macrosolute,
(b)   internal deposition - represents fouling of a large pore UF or MF membrane which is initially permeable to a macrosolute,

(c)   surface 'cake' deposition, with particle-particle binding and pore obstruction - represents a possible model for biomass above an MF membrane,
(d)   as for (c) except that the biomass is surrounded by biopolymer.

Other scenarios are possible, for example model (b) combined with (c) [or (d)] could depict an MF membrane which retains biomass and is partially permeable to macrosolutes. Our definition of fouling could also be extended to include changes to the original cake layer which renders this layer gradually less permeable, for example by growth of a biofilm or by excretion or retention of macromolecules. This situation could be pictured as a transition from model (c) to (d).

Factors which appear to influence the various forms of fouling are listed in Table 2 along with the preferred strategies to minimise the problem. Examples which illustrate the importance of the various factors are described below.

Retentive Ultrafilters

Retentive ultrafilters, used for the concentration of macromolecular solutions, typically have molecular weight cut-offs of 30K to 50K Daltons. It has been known for some time (Fane et al., 1981) that such membranes usually have a rather low surface porosity and a wide distribution of pore sizes. Recent observations (Kim et al., 1990) by Field

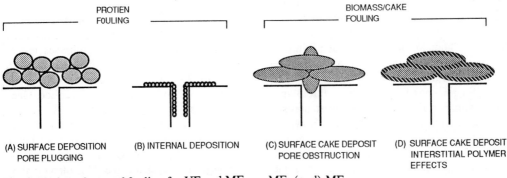

Fig. 1. Various forms of fouling for UF and MF (a) Low MWCO UF, (b) High MWCO UF or MF, (c, d) MF.

Table 2 Factors Which Influence Fouling

Factor	Preferred Strategy
Membrane: Porosity	> 50%
Pore Size Distribution	Isoporous
Rugosity	Smooth
Hydrophilic/hydrophobic	Hydrophilic
Charge	Neutral or same as solute*
Solute/Particle:	
Aggregation	Discrete
Hydrophilic/hydrophobic	Hydrophilic
Charge	High rather than low*
Operating Conditions:	
Pressure	Low
Shear Rate	Modest
Start-up	Gradual
Ionic Environment	Low Strength*

Note* Strategy may differ for solute passage through 'permeable' membranes

Emission Scanning Electron Microscopy (Figure 2) have confirmed that surface porosities are typically less than 5%. These surface porosity characteristics are far from ideal and our earlier conclusions (Fane and Fell, 1987, Fane et al. 1981) remain valid, namely that such surfaces will be sparsely and irregularly permeable and susceptible to local polarisation, pore plugging and fouling.

Various membranes, detailed in Table 3, have been compared in a standard 3 hour test in a stirred cell using 0.1% BSA as a model foulant at its isoelectric point (pH 4.8). Performance is expressed as UF flux loss = (Δ flux over 3 hours)/(initial UF flux at 5 minutes). Similar results were obtained in a thin-channel crossflow cell (Suki et al., 1984).

Figure 3 provides evidence that fouling is linked with porosity, since the lowest porosity membranes are more readily fouled. However, other characteristics of the selected membranes, such as rugosity and contact angle, provide similar trends and it is not possible to isolate clearly the magnitude of the individual effects. For example, UF membrane surfaces viewed by electron microscopy show qualitative differences in roughness which coincide with performance. We have attempted to quantify the 'nano-scale' roughness by use of photogrammetry (analysis of topography from stereo micrographs) details of which are given elsewhere (Kim, 1987). Preliminary results, given in Figure 4, show an apparent correlation between flux loss and surface roughness (standard deviation). The most effective membrane tested was an Amicon PM30 polysulphone membrane pretreated with a non-ionic surfactant, which was significantly smoother than the untreated membrane. Further discussion of surfactant treatment is given below.

Membrane hydrophilicity was assessed by contact angle, and for the membranes tested it was in the range 45 to 91°. Figure 5 shows that UF flux loss tends to increase with contact angle, or hydrophobicity, of the surface. One of the most effective membranes was a polysulphone pretreated with ethyl cellulose (EC). However, contact angle does not provide a general correlation as can be seen from the 'anomalous' ST membrane. This membrane was a PM30 membrane coated by a Langmuir-Blodgett monolayer of stearic acid (Kim et al. 1989), and although it was hydrophobic the UF

Fig. 2. FESEM micrograph of Millipore PTTK membrane.

Table 3 Membranes Used for Retentive Ultrafiltration of BSA [see Figures 3-5]

Type		Material	Nominal Mol. Wt. Cut Off
Amicon	PM30	Polysulphone	30,000
	XM100A	Polyco(acrylonitrile/vinyl chloride	100,000
	YM5	{Regenerated	5,000
	YM30	{Cellulose	30,000
Millipore	PTGC	Polysulphone	10,000
Rhone-Poulenc	IRIS3038	Polyacrylonitrile	25,000

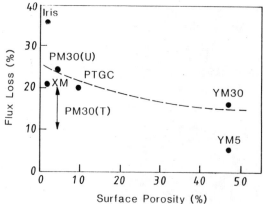

Fig. 3. Effect of surface porosity on UF flux loss.

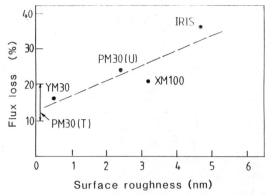

Fig. 4. Effect of surface roughness on UF flux loss.

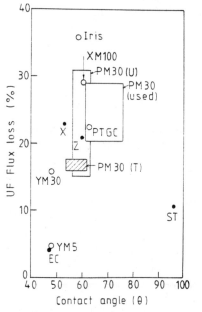

Fig. 5. Effect of contact angle on UF flux loss.

flux loss was relatively low. It is possible that the detrimental effects of hydrophobicity were counter balanced by the reduced surface roughness (Figure 4). Surface charge would not have been a factor in this case because the protein was at its isoelectric point.

The operating conditions which affect fouling are ionic and pH environment, which influences surface charge interactions, and ΔP and shear rate (or crossflow velocity), which influence concentration polarisation. Figure 6 (from Fane and Fell, 1987, and Suki et al. 1984) shows the results for the UF of 0.1% BSA, at pH 5, in a thin-channel crossflow cell. Protein deposition, which is directly related to fouling resistance, is strongly influenced by the following operating conditions;

(i)   pH at the isoelectric point - protein is more likely to aggregate and deposit;

(ii)   raised ionic content - suppresses the 'double layer' interactions and allows aggregation;

(iii)   low crossflow - low surface shear increases concentration polarisation and local concentrations.

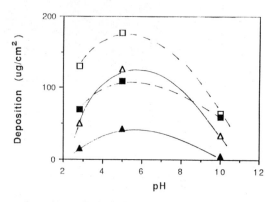

Fig. 6. Protein deposition vs pH, [5 hr, 0.1% BSA],
□ crossflow 0.3 m/s, 0.2 M NaCl
△ crossflow 1.1 m/s, 0.2 M NaCl
■ 0.3 m/s no salt, ▲ 1.1 m/s no salt.

Concentration polarisation will also be increased by an increase in transmembrane pressure (ΔP). Table 4 (Kim et al. 1988) shows that UF flux loss varies significantly as pressure is raised from 50 to 200 kPa. These results provide further evidence of the benefits of surface treatment.

Table 4 Transmembrane pressure vs. UF flux loss[a] for untreated and treated (0.1% Methyl Cellulose, 5 min. contact) PM30 membranes (0.1% BSA, 3 hrs. UF, 300 rpm, pH 5) [b]

Pressure (kPa)	50	100	200
Untreated	21.1	23.1	26.5
Treated	2.8	13.0	18.6

(a) Flux loss (%) = 100 $(J_{5min} - J_{3hrs})/(J_{5min})$.

(b) 3 Hr fluxes were 75 and 80 1/m^2 hr, for 50 and 200 kPa (Untreated)
80 and 95 1/m^2 hr, for 50 and 200 kPa (Treated)

Retentive Microfilters - Biomass

Crossflow microfiltration provides a means for cell harvesting and under these conditions a cake of bacteria forms on the membrane (Figures 1(c) and (d)). The nature of the membrane has a significant effect on performance.

For example, we have compared (Gatenholm et al. 1988) a range of MF and large pore UF membranes, as detailed in Table 5, for the processing of an Escherichia coli fermentation broth (3 g/litre). Figure 7 shows that the 'steady state' (90 minute) flux was inversely related to the initial water flux of the membrane. The MF membranes exhibited a rapid flux decline in the first few minutes of operation to a steady-state lower than for the UF membranes. Electron microscope examination (Gatenholm et al., 1988a) of the cakes on the surfaces of a typical MF and a UF membrane showed radically different deposits.

The layer on the MF membrane was composed of densely packed cells, with evidence of pore blockage due to cells trapped vertically in the pores. The layer on the UF membrane appeared to be composed of fine debris, possibly protein, and a few cells. These results suggest that MF membranes may be susceptible to fouling due to their large 'pluggable' pores as well as their high initial flux which provides rapid convection of cells to the surface. For this reason low operating pressures are recommended. When conditions permit retention of some macromolecules the use of smooth UF membranes may give higher performance than MF membranes. However, in many applications the use of MF membranes is dictated by the need to pass dissolved solids with the filtrate.

In order to compare MF membranes of different type and surface morphology a series of tests have been performed using an E.coli suspension on membranes detailed in Table 6. These membranes all have a similar nominal pore size, but provide a variety of porosities, from highly porous/isoporous (Anopore), low porosity/isoporous (Nuclepore) and highly porous/pore size range (Millipore).

Tests involved batch filtration of a fixed inventory of bacteria to produce cake layers of about 5 μm thickness. Relative fluxes were compared for applied pressures of 30, 60 and 100 kPa and for conditions of low and high ionic strength (10^{-4} and 10^{-1} M NaCl respectively). In addition, after a filtration run the 'cleanability' was assessed by measuring water flux after a period of unpressurised stirring; an estimate of 'cake lift off' was obtained by analysis of the bulk suspension. Details are provided elsewhere (Fane et al. 1990).

Under all conditions of ΔP and ionic strength the high porosity, isoporous Anopore membrane showed a higher flux than the other membranes. Figure 8 plots ratios of Anopore flux (J_A) to Nuclepore flux (J_N) for the range of pressures; improvement ratios range from 1.2 to 2.0 x. Similar improvements were observed relative to the Millipore, although in this case the ratios decreased with pressure.

From the flux values and the known cake inventory it is possible to estimate the average 'specific cake resistance' (α) (Fane et al, 1990).

Table 5 Membranes used for Retentive Microfiltration of E.coli broth
    [See Figure 7]

Membrane Type	Material	Pore Size (μm) or Cut-off (D)	Initial Water[a] Flux ($1/m^2$ hr)
Durapore[c]	(Modified) PVDF (b)	0.22μm	7,900
Accurel[d]	Polypropylene	0.2 μm	6,700
PKMK[c]	Polysulphone	300,000	1,100
PTMK[c]	PVDF	300,000	2,800
YM100[e]	Regen. Cellulose	100,000	840
UF-PS-100[f]	(Modified) Polysulphone	100,000	100

(a)   At 100 kPa and 25°C   (b)   PVDF = polyvinylidene fluoride
(c)   Millipore   (d)   Enka
(e)   Amicon   (f)   Kalle

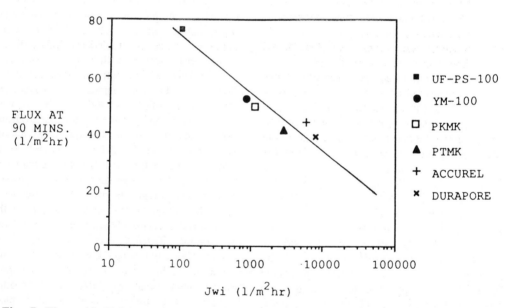

Fig. 7. Flux with E.Coli broth versus initial water flux for various UF and MF membranes.

Table 6 Membranes used for Comparison of MF Membrane Surface Morphology - E.coli Filtration (Figures 8 - 9)

Membrane Type	Structure	Thick-ness (μm)	Porosity	Pore Range (μm)	Rough-ness	Water Flux[c] (1/m² hr)
Anopore (0.2) (a) (Anodised Alumina)	Isoporous Cylindrical Pore	60	50	0.15 - 0.22	Smooth	2880
Nuclepore (0.2) (a) (Polycarbonate)	Isoporous Cylindrical Pore	10	7 - 10	0.21 - 0.24	Smooth	2880
Millipore (0.22) (PVDF) (a) (b))	Mesh-tortuous Interconnected Pores	100	60	0.21 - 0.61	Rough	4160

(a) hydrophilic   (b) hydrophobic   (c) flux at 50 kPa.

The following trends are evident;

(i) α values for Nuclepore cakes are greater than for Anopore cakes; Millipore cakes have the highest α values at low pressure (30 kPa) but were less than Nuclepore at high pressure (100 kPa);

(ii) α values were higher for cakes formed from higher ionic strength feeds;

(iii) the cakes were compressible with compressibility factors n (from $\alpha = \alpha_o \Delta P^n$) in the range 0.5 to 1.7.

The results for flux and specific resistance show that the high porosity/isoporous surface gives the best performance. This can be

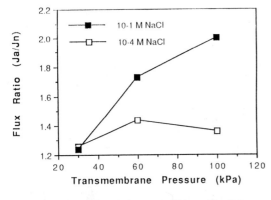

Fig. 8. Flux ratio of Anopore (J_a) to Nuclepore (J_n) as a function of pressure at 10^{-1} M (closed symbol) and 10^{-4} (open symbol) NaCl.

explained in terms of the effect of surface porosity on the apparent resistance of the deposit layer. Sparse porosity leads to lateral flow through the deposit with an increase in the effective flow path through this layer. We have reported a similar effect in the UF of protein solutions (Fane et al. 1981). Although the Millipore membrane has high surface porosity it has a relatively wide pore size distribution, so that flow through the surface and the deposit may be unevenly distributed.

'Cleanability' test results are summarised in Figure 9, where cake lift off (after unpressurised stirring) is plotted versus pure water flux recovery. There is a clear relationship between the amount of cake removed by shear and the water flux recovery. The results show only a marginal difference between the membranes except for the hydrophobic Millipore (MB) which failed to respond well even under the most favourable operating conditions.

Both applied pressure and ionic environment had a major effect in these series of experiments, and the trends mirror those found for UF of macrosolutes. Raised ionic strength reduces flux (by 20 to 80%) and produces a higher specific resistance, probably due to the suppression of double layer interactions which promote cake voidage (Fane et al. 1990). Fouling, which is manifest through poor cake lift off and low water flux recovery, was also more serious for the higher ionic strength suspensions (closed symbols in Figure 9). This is probably because the cakes formed are more cohesive and possibly because electrostatic double layer interaction between the membrane and the cake are suppressed at high ionic strength. Higher operating pressure appears to compound the effect of high ionic strength (large closed symbols in Figure 9). Since the cakes are compressible it is likely that the high pressure promotes aggregation and binding to the membrane. Estimated cake voidages (Fane et al. 1990) are low (< 0.2), which suggests that interstitial space between individual E.coli is partially filled with exopolymer (model (d) in Figure 1). This polymer may explain the compressibilities observed, and the binding at high pressure. In practical terms the results

suggest a strategy to use low to modest applied pressures and low ionic strength feeds. The latter may not be an option if the feed is a fermentation liquor with significant amounts of nutrient electrolytes present.

A series of tests, in which the E.coli was suspended in a solution of 0.1 wt% BSA was also performed to simulate a broth containing both suspended and dissolved solids. At 100 kPa, with and without electrolyte the Anopore achieved a higher flux than the Nuclepore (improvement ratios up to 1.6). Surprisingly the 'cake lift off' and water recovery with BSA present were significantly better than in the absence of BSA (see Table 7). The presence of the protein appears to have protected the membranes from irreversible adhesion or fouling

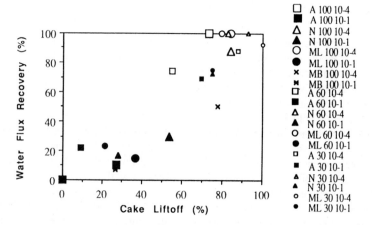

Fig. 9. Water flux recovery and cake lift off. Membranes are Anopore (A), Nuclepore (N), Millipore hydrophilic (ML), Millipore hydrophobic (MB). Pressures 100, 60, 30 kPa. Ionic environment is 10^{-1} M and 10^{-4} M NaCl.

Table 7 Cake Lift-Off and Water Recovery for E.Coli Feed With and Without 0.1 wt% BSA

Pressure (kPa)	Membrane	NaCl (M)	Cake Lift-Off (%)		Water Recovery (%)	
			No BSA	0.1% BSA	No BSA	0.1% BSA
100	A	10^{-4}	73	89	100	100
100	A	10^{-1}	27	76	10	74
100	N	10^{-4}	84	93	88	100
100	N	10^{-1}	54	80	30	100

A = Anopore (0.2 μm); N = Nuclepore (0.2 μm)

by the bacteria. However, as we discuss below, there is considerable potential for internal fouling of solute-permeable membranes by macrosolutes.

Solute-Permeable Membranes - Protein Transport

The recovery of a macromolecular product from a broth requires membranes with low or negligible retention of the desired solute. Membranes which should achieve this are high molecular weight cut-off (MWCO) ultrafilters and the smaller pore size microfilters. However if these membranes foul they not only lose water flux but also acquire rejection properties as the pores close up (Figure 1b). The 'composite' membrane produced by solute deposition may also develop a charge character similar to the solute and this can also influence rejection.

Figure 10 shows the flux and rejection profiles of a range of partially permeable membranes subject to 0.1% BSA solutions of different pH. The membranes respond in different ways depending on their structure and properties and the solution environment. At least four types of response are possible, as described below.

Type I: High cut-off UF - 'Adsorptive' - this membrane rapidly adsorbs the solute and the pore size closes to such an extent that almost complete rejection is achieved. Figure 10 (a) shows the Amicon XM100A membrane with a nominal cut-off of 100 KD rejecting BSA with a molecular weight of 67 KD.

Type II: High cut-off UF - 'Non-adsorptive' - this is exemplified by the Amicon YM100 Figure 10 (b) which is an alternative to the more adsorptive XM100. Rejections are lower than for the XM100 and at pH9 the rejection is particularly low. Explanations for the trends are that at pH5 (close to the isolectric point) and pH3 (BSA is positive) sufficient adsorption occurs to close up the pores and allow rejection of protein aggregates (pH5) or positively charged (pH3) protein molecules. At pH9 (BSA is negative) the adsorption is very low, possibly because the membrane itself carries a residual negative charge. The surface charge on the membrane would have to be small enough and the pores large enough to obviate a charge repulsion rejection.

Type III: Microfilter - 'Adsorptive' - Figure 10 (c) is a polyamide microfilter (Lefebvre et al. 1980) with pores of about 0.2 to 0.5 µm size, and Figure 10 (d) is a Goretex Teflon membrane (0.02 µm nominal pore size). The two membranes show similar trends. At the isoelectric point, pH5, the rejection falls to a low value, at pH3 rejection rises to greater than 80%, and at elevated pH the rejection is moderate to high (40% (pH7) to 70% (pH9)). This behaviour can be explained at pH values either side of the isoelectric point by a charge repulsion mechanism between similarly charged proteins in solution and adsorbed to the membrane. The fact

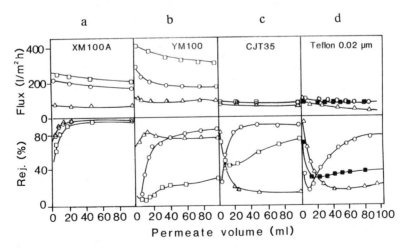

Fig. 10. UF flux and rejection profiles for various membranes at different pH values [0.1% BSA, 100 kPa, 300 rpm, pH3 (o); pH5 (Δ); pH7 (■); pH9 (□).

that proteins expand at high and low pH may also be important. At pH5 there will be no charge interaction and if the pores remain sufficiently open the solute will pass with a small rejection.

Type IV: - Microfilter - 'Non-Adsorptive' - This membrane shows no rejection of the macromolecule and represents an ideal case for recovery of solute product in the filtrate.

Figure 11 compares two isoporous microfilters, the highly porous Anopore (0.2 µm) and the low porosity Nuclepore (0.2 µm); membrane characteristics are given in Table 6. The results show for the same pore size, but for different porosity.

(i)   the low porosity Nuclepore suffers more significant flux decline (its final flux is less than half that of the Anopore),

(ii) the low porosity membrane gradually develops greater rejection (a 'steady-state' rejection of 20% for the Nuclepore and only 2 to 3% for the Anopore).

In terms of the types of response, the low porosity membrane behaves as Type III and the high porosity as Type IV. This effect could be due to differences in the membrane materials, but both are relatively hydrophilic and possess a residual negative charge at pH7 (the conditions used). A more likely explanation is based on the differences in the local shear forces at the

entrance to the pores and within the pores. For cylindrical pores, radius R, and a membrane porosity ε, the shear force at the pore wall τ_R can be obtained from (Franken et al. 1990),

$$\tau_R = 4 \, \mu \, J_v/\varepsilon \, R = R \, \Delta P/2L \quad (1)$$

where J_v is the flux, μ the solution viscosity and L is pore length.

For an initial flux of about 10^{-3} m/s (3600 $1/m^2$ hr) the shear forces for the Anopore and the Nuclepore are,

$$\tau_R \text{ (Anopore)} \approx 80 \text{ Pa}$$
$$\tau_R \text{ (Nuclepore)} \approx 570 \text{ Pa}$$

These values can be compared with those typical of a concentric viscometer, which reaches values up to 25 Pa. Clearly the Nuclepore membrane can impose very high shear forces which could distort the proteins and facilitate their adsorption onto the pore walls or increase the tendency to protein aggregation at the mouths of pores. However irreversible changes in protein conformation are less likely due to the very short time of exposure in the membrane. Shear-time product (τ t) values can be estimated from (Franken et al. 1990),

$$\tau \, t = 8 \, \mu \, L/3 \, R \quad (2)$$

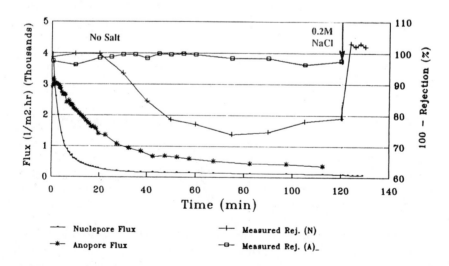

Fig. 11. Flux and Protein passage (100 - Rejection) for 0.2 µm Anopore and 0.2 µm Nuclepore (0.1 wt% BSA, pH7) without salt and with salt injection.

where L is membrane thickness, or pore length. For the membranes considered this gives 0.27 Pa.s for the Nuclepore and 1.6 for the Anopore; both values are much less than the 50 Pa.s suggested as being detrimental to enzymes (Charm and Wong, 1970). In support of this we have found only marginal changes to protein (BSA) conformation on passage through a Nuclepore filter by analysis of spectra from FTIR and Circular Dichroism (Franken et al. 1990).

If shear forces within the pores, or the mouths of pores, are important in determining the internal adsorption and fouling of microfilters it should be possible to see a detrimental effect of applied pressure; equation (1) shows that for the same membrane τ_R increases linearly with ΔP. Figure 12 (Fane and Hodgson, 1989) shows that filtration of BSA through an Anopore 0.2 μm membrane at 200 kPa achieves a lower final flux than filtration at 100 kPa or even 60 kPa.

The effect of stirring speed on the filtration of BSA through a Nuclepore 0.2 μm membrane is shown in Figure 13; the ordinate is 100-Rejection (%), i.e. percent solute passage. Unstirred, the rejection remains at 0% for the duration of the experiment, for 120 rpm stirring speed the onset of rejection occurs at about 28

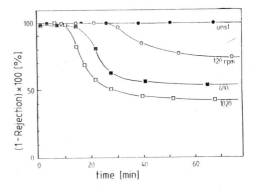

Fig. 13. Rejection profiles for Nuclepore 0.2 μm versus stirring speed (0.1% BSA)

minutes, for 400 rpm at about 15 minutes and for 1020 rpm at about 10 minutes. We also see a greater 'steady-state' rejection at higher stirring speed. The effect of stirring speed could be two-fold. Firstly increased stirring improves back-diffusive mass transfer, which decreases polarisation and lowers the permeate concentration, thereby increasing the observed rejection. Secondly the effect of stirring the bulk solution can be to increase protein flocculation and aggregation (Franken et al. 1990). For the conditions of our experiment the estimated shear force τ imparted on the bulk solution at 400 rpm is about 0.38 Pa. For 10 minutes of stirring this gives a shear-time product of 230 Pa.s and for 1 hour it gives 1380 Pa.s. Shear-time products of this magnitude may have induced protein aggregation, leading to greater retention with higher stirring speed.

At first sight the results suggest operation at low stirring speed if low rejection of solute is required. However this conflicts with the need to minimise 'cake' build up and fouling when biomass is present or the need to minimise local concentrations of macromolecules to facilitate fractionation or reduce surface fouling. Although a strategy of low to zero stirring could not be recommended it may be possible to use

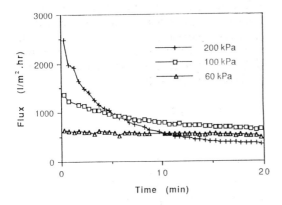

Fig. 12. Flux histories for 0.2 μm Anopore versus applied pressure (0.1 wt% BSA, 400 rpm).

To summarize at this point, the optimal strategies for minimisation of fouling when using solute-permeable membranes are to use membranes with high porosity (and by implication a narrow pore size distribution) and with low adsorptivity (hydrophilic) and to use them at low pressure rather than high. These strategies are the same as required for retentive ultrafilters and retentive microfilters. The strategies regarding crossflow/stirring and ionic environment are less clear cut.

'parametric pumping' by flow variation or pulsations, which is an approach reported to enhance flux (Jaffrin, 1987).

The effect of ionic environment also raises a dilemma. On the one hand a greater tendency to aggregation, adsorption and fouling is observed with solutions of raised ionic concentration (Figures 6 and 9), on the other hand the presence of ions suppresses charge interactions between proteins and adsorbed layers. This can lead to subtle effects. Table 8 gives BSA rejections obtained using a polyamide microfilter (see also Figure 10 (c)), with and without 0.2 M NaCl (Waters, 1982). At pH5 (close to the isoelectric point) the addition of salts raises rejection, presumably because of the formation of protein aggregates that can be retained. At the pH extremes the rejections are substantially lower, and this is believed to be because the charge-repulsion mechanism is less effective as the ions shield the charges on the proteins in solution and adsorbed onto the membrane.

The influence of ions is also shown dramatically in Figure 11. Towards the end of this BSA filtration using a Nuclepore 0.2 μm membrane at pH7 the rejection was about 20%. Following the addition of salt to give the BSA solution a concentration of 0.2 M NaCl the BSA rejection dropped rapidly to zero. Suppression of the effective negative charges on the proteins by the NaCl is the most likely explanation. The use of ionic environment to improve fractionation of protein mixtures by ultrafiltration has been described by Ingham et al. (1980) who used KCl to suppress complex formation between BSA and lysozyme and achieved better separations.

In summary, ionic environment is clearly a factor which controls the rejection of passage of proteins through microfilters or ultrafilters as well as influencing fouling and flux. Manipulation of ionic environment to improve the performance of solute-permeable membranes will depend on the application. The significant point is that in this case the presence of ions is not necessarily a disadvantage.

Membrane Pretreatment

In addition to the selection of the best available membrane (preferably isoporous, highly porous, smooth and hydrophilic) and the optimum operating conditions the membrane user may be able to achieve improved performance by membrane pretreatment.

We have examined the use of surfactants and polymers for surface treatment by means of passive adsorption (Kim 1987, Kim et al. 1988, Fane et al. 1985), convective (ultrafiltration) adsorption (Kim, 1987) and by deposition of Langmuir Blodgett films (Kim et al. 1989). Passive adsorption appears to be the most effective and simple approach. Enhanced performance for retentive membranes could be due to the more homogeneously permeable nature of the coated surface, its smoothness and its hydrophilicity; some data for treated membranes are included in Figures 3, 4 and 5 and Table 4. Performance improvement for some of the treatments is summarised in Figure 14, which plots the UF flux loss versus the UF flux at 3 hours for treated and untreated membranes. This shows that the relatively easily fouled polysulphone, PM30, is significantly improved by treatment with methyl cellulose (5 minutes contact with 0.1% MC). The advantage of this type of pretreatment is that it can be regenerated after a cleaning cycle.

Partially permeable microfilters and ultrafilters also respond to pretreatment. Figure 15 compares the performance of an untreated

Table 8 Effect of Ionic Environment on BSA Rejection (Waters, 1982)

Salt Content	BSA Rejection (%)		
	pH3	pH5	pH9
Nil	93	5	85
0.2 M NaCl	65	75	65

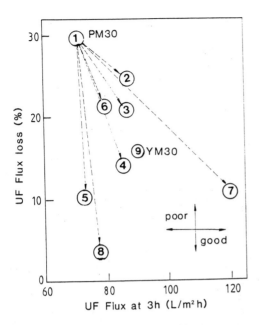

Anopore (0.02 µm) and one treated with a non-ionic surfactant, N100 polyethylene oxide. The treated membrane achieved a 'steady-state' flux about 1.6 x that of the untreated membrane and a rejection of less than 2% compared with more than 25% for the untreated membrane. Apparently the surfactant, which is small enough to enter the pores, has reduced fouling and the development of solute rejection. The use of surfactant in this role is believed to be favoured by some membrane manufacturers (Brock, 1983). However the recommended strategy is to use pretreatment agents which are compatible with the products and which can be simply regenerated by the user.

Antifoams

It is common to use additives during fermentations to control foaming problems. These antifoams are surface active and during membrane processing could accumulate at the membrane surface. Even at very low concentrations some antifoams can have a dramatic effect on flux. For example, Figure 16

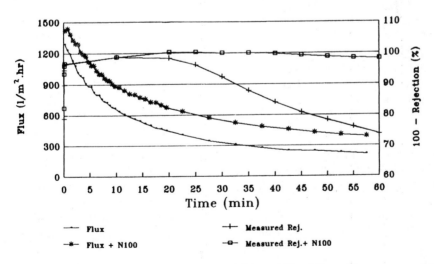

Fig. 15. Flux and rejection profiles for untreated and treated 0.02 µm Anopore membranes (0.1% BSA, 100 kPa. Treatment with Teric polyethylene oxide surfactant).

317

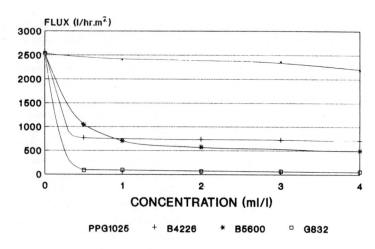

Figure 16. Effect of antifoams on 30 minute flux for hydrophobic 0.22 µm Millipore membrane.

Table 9 Flux at 30 Minutes with Various Antifoams (2 ml/l)

Membrane	Flux ($1/m^2$ hr)				
	PPG(a)	B422(b)	B5600(c)	G832(d)	Pure Water
Anopore (0.2)	1264	516	438	62	1801
Millipore (0.2) Hydrophilic	2240	41	413	66	2250
Millipore (0.2) Hydrophobic	2400	735	574	62	2530

(a)   PPG, polypropylene glycol, Molecular Wt. 1025
(b)   B4226,glycol esters/polyglycols
(c)   B5600,glycerides, polyglycols
(d)   G832, silicones

shows the 30 minute ('steady state') flux for a hydrophilic 0.2 µm Millipore Durapore membrane processing solutions of 4 different antifoams (details in footnote to Table 9). In some cases flux dropped below 100 $1/(m^2.hr)$, from the water flux of about 2550 $1/(m^2.hr)$.

Only the PPG (polypropylene glycol) provided negligible effect.

The responses of 3 different membranes are summarised in Table 9, which compares the 30 minute fluxes for the 0.2 µm Anopore and 0.22 µm Millipore (hydrophilic and hydrophobic). These data show that the silicone-based antifoam was particularly detrimental to flux for all 3 membranes. The hydrophobic Millipore was the least sensitive to 3 out of 4 antifoam types.

Caution is clearly required in the use of antifoams if membrane processing follows. Silicones should be avoided, if possible. The most prudent approach would be to perform pilot tests before selection of the antifoam or the membrane.

Conclusions

The application of UF and MF to the processing of biofluids requires strategies to overcome the fouling which reduces flux and separation capabilities. Fouling may be due to surface deposition of proteins on a retentive ultrafilter, or a biomass 'cake' on a retentive microfilter, or internal deposition for a solute permeable membrane.

For retentive UF and MF membranes the strategies for minimisation of fouling are similar, namely use of highly porous and isoporous membranes which are smooth and hydrophilic. Also fouling is less and membrane recovery is easier for operation at low to modest pressures and with low ionic strength solutions.

For solute-permeable membranes the passage of solute is also favoured by high porosity, isoporosity and hydrophilicity. Similarly low transmembrane pressure gives less flux decline, possibly because it produces lower shear conditions within or near the membrane.

The effect of stirring is to increase rejection through reduction in concentration polarisation and possibly due to induced aggregation of proteins via rather high shear-time effects in the bulk solution. Selection of optimal stirring or crossflow may be a compromise between the needs to achieve solute transport and the need to control fouling, including 'cake' deposition.

Ionic environment may similarly require a compromise since the use of high ionic strength encourages fouling but under some conditions it also allows better protein passage.

Pretreatment of membranes with surfactants and polymers provides another strategy for the reduction of fouling in retentive UF and solute-permeable MF applications.

Guidelines for this type of pretreatment still need to be developed. Antifoaming agents used in fermentation liquors can drastically decrease flux. Selection of these agents needs careful consideration.

Acknowledgements

Much of the work reported here was supported by the Australian Government through funding for the Commonwealth Special Research Centre for Membranes and Separation Technology. The support for one of us (P.H.) from Alcan International is also acknowledged. This review includes the results of previous co-workers; the input of Drs P. Gatenholm, A. Suki and A.G. Waters is gratefully acknowledged. We are also grateful to Professor Hans Coster and Professor Kevin Marshall for stimulating discussions.

References

Brock, T.D., *Membrane Filtration*, (**1983**), Springer-Verlag.

Charm, S.E., Wong, B.L., *Biotech. Bioeng.* (**1970**), 1103-1109.

Fane, A.G., Fell, C.J.D., Waters, A.G., (**1981**), *J.Memb.Sci.*, 9, 245-262.

Fane, A.G., Fell, C.J.D. and Kim, K.J., (**1985**), *Desalination*, 53, 37-56.

Fane, A.G., Fell, C.J.D., (**1987**), *Desalination*, 62, 117-136.

Fane, A.G., Hodgson, P.H., (**1989**), *Proc. 1st Int. Conf. on Inorg. Membranes*, Montpellier, 501-506.

Fane, A.G., Fell, C.J.D., Hodgson, P.H., Leslie, G., Marshall, K.S., (**1990**), *Proc. Vth World Filt. Congress*, Nice, 320-329.

Franken, A.C.M., Sluys, J.T.M., Chen, V., Fane, A.G., Fell, C.J.D., (**1990**), *Proc.Vth World Filt. Congress*, Nice, 207-213.

Gatenholm, P., Paterson, S., Fane, A.G., Fell, C.J.D., (**1988**), *Proc. Biochem.* 79-81.

Gatenholm, P., Fell, C.J.D., Fane, A.G., (**1988a**), *Desalination*, 70, 363-378.

Ingham, K.C., Busby, T.F., Sahlestrom, Y., Castino, F., (**1980**), *Ultrafiltration Membranes and Applications* (Cooper, A.R. [Ed], Polym. Sci. and Tech. 13, 141-158.

Jaffrin, M.Y. (**1987**), *Life Support Systems*, 267-271.

Kim, K.J. **(1987)** *Ph.D. Thesis*, University of New South Wales.

Kim, K.J., Fane, A.G., Fell, C.J.D., **(1988)** *Desalination*, 70, 229-249.

Kim, K.J., Fane, A.G., Fell, C.J.D., **(1989)**, *J. Membr. Sci.*, 9, 245-262.

Kim, K.J., Fane, A.G., Fell, C.J.D., Suzuki, T. and Dickson, M., **(1990)**, *J.Membr.Sci.*, 54(1-2), 89-102.

Lefebvre, M.S., Fell, C.J.D., Fane, A.G., Waters, A.G., **(1980)**, *Ultrafiltration Membranes and Applications*, (Cooper, A.R. [Ed.]), Polym. Sci. and Tech., 13, 79-98.

Suki, A.B., Fane, A.G., Fell, C.J.D., **(1984)**, *J.Memb.Sci.*, 21, 269-283.

Waters, A.G., **(1982)**, *Ph.D. Thesis*, University of New South Wales.

Role of Functionalized Membrane Separation Technology in Downstream Processing of Biotechnology-Derived Proteins

Vipin K. Garg, Stephen E. Zale, Abdul R.M. Azad and O. Dile Holton, Sepracor Inc., 33 Locke Drive, Marlboro, MA 01752

Purified proteins are being increasingly used for therapeutic and diagnostic purposes as a result of the current biotechnological revolution. As the commercialization of these proteins progresses, the need for more efficient bioseparation technologies is becoming a critical issue. The major technical challenge is to develop efficient and economical processes for downstream recovery of proteins that meet regulatory requirements.

Recently, there has been a growing interest in the use of membrane-based separation techniques to facilitate rapid, efficient and scalable purification of biomolecules (Brandt, et al, 1988; Nachman, et at, 1990; Pietronigro, 1990; Zale, et al, 1990). Brandt and co-workers were first to report the use of functionalized affinity membranes for protein purification in 1988. Since then, several examples of membrane-based protein purification have been reported in the literature. Both flatsheet (Pietronigro, 1990) and hollow-fiber (Zale, 1990) membrane devices have been developed and are commercially available. Applications of these membranes have been shown for large-scale purification of monoclonal antibodies using Protein A affinity (Pietronigro, 1990; Zale, et al, 1990), and for several recombinant proteins using immunoaffinity on membrane-based systems (Garg, et al, 1990; Nachman, et al, 1990).

The data presented in this article will discuss the utility of functionalized hollow fiber membranes in downstream processing of proteins, and describe the benefits of membrane-based adsorptive separations. The role of adsorptive membranes in overall purification strategy will be discussed. A particular emphasis will be given to the membrane-based affinity purification systems for therapeutic grade monoclonal antibodies.

Functionalized Membrane Technology

As shown in Table 1, the use of membranes in downstream processing is widespread. Traditionally, the most frequent use of membranes has been in the areas of clarification and concentration. A newer application of membranes in downstream processing is their use as purification matrices. Affinity membrane matrices have been shown to have distinct advantages over conventional column matrices for processing large volumes of dilute feedstreams by providing high productivity per unit matrix volume, and ease of scalability and automation (Nachman, et al, 1990; Zale, 1990). Affinity membrane modules described in this paper are based on microporous hollow-fiber membranes (Sepracor Inc.) activated by covalent attachment of affinity ligands to their interior pore-wall surface. These membranes can be used to process dilute protein feedstreams at high volumetric throughputs due to their superior mass transfer characteristics, high permeability and uniquely small aspect ratio as compared to chromatographic columns.

Hollow fiber membranes provide extremely short fluid-flow path lengths in comparison to the superficial area provided for flow (Fig. 1). More importantly, convection of protein-containing solution through ligand-activated membrane pores minimizes diffusional resistances associated with the porous or gel-type particles typically used in column separations. In a packed bed column, the characteristic diffusion pathlength for a target protein is on the order of the bead diameter (typically 50-100 µm), the corresponding diffusion distance within the membrane matrix is on the order of the pore diameter (0.5-1 µm). The decrease in diffusion distance of this orders of magnitude leads to a vast reduction in characteristic diffusion time and hence, a vast reduction in the fluid residence

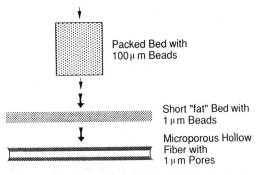

Packed Bed with 100 µm Beads

Short "fat" Bed with 1 µm Beads

Microporous Hollow Fiber with 1 µm Pores

Fig. 1. Diagrammatic illustration of an ideal short fat bed affinity column as compared to hollow fiber affinity membrane.

Table 1. Conventional versus membrane technologies for large-scale processing of biotechnology-derived proteins

Unit Operation	Conventional Process	Membrane Process
1. Cell harvesting	- Centrifugation	- Microfiltration (MF) - Ultrafiltration (UF)
2. Clarification & debris removal	- Centrifugation	- MF - UF
3. Protein concentration	- Precipitation (salt, solvent) - Extraction	- UF - Diafiltration - Membrane-modulated protein precipitation
4. Protein purification	- Fractionation (heat, salt, solvent) - Column chromatography (affinity, IEX, HIC, etc.)	- UF - Membrane chromatography (affinity, IEX, HIC, etc.)
5. Desalting	- Size exclusion chromatography (gel filtration)	- UF - Dialysis - Electrodialysis
6. Final product sterilization	- Heat treatment - Radiation treatment	- Sterile filtration

time required for mass transfer. As a consequence, the speed of affinity-based bioseparations can be dramatically improved, and ligand turnover frequency can be increased manyfold. As depicted in Fig.1, an affinity membrane can be thought of as an extremely short, fat, microporous bed.

A simplified illustration of a hollow fiber affinity membrane separation is shown in Fig. 2. The feed solution containing the target protein is loaded onto the membrane module in the crossflow filtration mode. Fluid that permeates the membrane wall (filtrate or flow-thru) is depleted of target protein as the membrane becomes saturated. After the loading step, a wash buffer is used to remove entrained feed solution and loosely bound protein from the membrane. The pure product is dissociated from the membrane and collected during the elution step. During reequilibration, the membrane is brought back to conditions

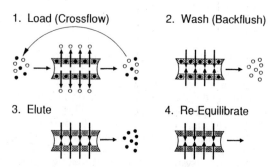

Fig. 2. Phases of membrane-based affinity adsorption. [●] desired protein; [O] impurities.

favorable for target protein capture. The membrane is then ready to begin another cycle.

Direct convective contact of process fluids with the affinity membrane sorbent during all steps of the purification cycle makes cycle times as short as a few minutes feasible when an affinity membrane module is operated in conjunction with automated fluid management systems.

Scale-up of membrane affinity separation is accomplished by increasing membrane matrix volume (i.e., number and length of fibers) while maintaining constant bed height (fiber wall thickness). Flow rates are scaled linearly with respect to matrix volume and frontal area. Fluid residence times are kept constant; breakthrough behavior, transmembrane pressure, purification cycle time and volumetric productivity remain essentially unchanged as module size is increased.

MAb Purification

In the recent years, generic ligand affinity separation has become an increasingly preferred method of purifying MAbs. In fact, after ion exchange, affinity chromatography is the most frequently used method for the purification of MAbs (Garg, 1990). We have immobilized several generic ligands (e.g., protein A, protein

G, and others) on hollow-fiber membranes for the purification of different mouse and human IgGs. In particular, we have developed alternative approaches to purify large quantities of therapeutic grade MAbs by membrane affinity chromatography (Garg, et al, 1990; Zale, 1990).

An example of the purification of a mouse IgG from serum containing media using a protein A membrane module is shown in Fig. 3. Using a 10 ml matrix volume module, in just 15 min, 1.2 L of crude MAb was purified to near homogeneity with a yield of 97%. Under these conditions, the fluid residence time of the feed solution within the membrane matrix is approximately 2 seconds. Analysis of the filtrate (flow through) material indicated that essentially all of the target protein was captured during a single pass through the membrane. At this scale, the projected productivity of the system would be approximately 8 g of purified antibody per 24 h day. As further scale-up is required, several fold higher quantities of MAb can be processed on larger size modules. As shown in Table 2, depending upon the antibody titer of the feed solution up to 2 g/h (or 40-45 g/day) of MAb can be purified using a Model 1500 (50 ml matrix volume) protein A or protein G module.

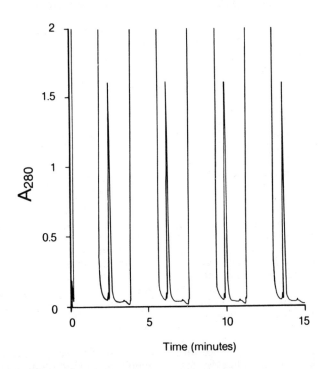

Fig. 3. Multicycle purification of a mouse IgG from cell culture supernatant. Module: Model 300 Protein A (Sepracor Inc.); volume processed: 1.2 L; feed titer: 72 mg/l; IgG yield: 97%.

Table 2. Membrane affinity performance of model 1500 (50 ml matrix vol) protein A and protein G modules

Feed titer	Protein A/human IgG1		Protein G/mouse IgG1	
(mg/L)	Process rate (L/h)	(g/h)	Process rate (L/h)	(g/h)
30	42.0	1.1	39.0	1.0
100	21.0	1.8	19.0	1.6
200	12.4	2.1	11.0	1.9

Protein A membrane-purified MAb was tested for purity by SDS PAGE, and GPC HPLC. The presence of contaminating bovine serum albumin (the principal protein component of calf serum added to the cell culture media), and protein A originating from the protein A membrane module, were each determined by ELISA. Reduction in nucleic acids as a result of purification was assessed using fluorimetry.

SDS PAGE (Fig. 4) and GPC HPLC analyses indicated that the product was essentially pure. No aggregates were detected by GPC. Analysis for bovine albumin by ELISA indicated the presence of approximately 3% albumin in the purified product. Comparison of feed and product albumin levels indicated that protein A purification removes >99.5% of the albumin present in the starting material. More complete removal of contaminating proteins can be attained by employing more extensive wash steps.

Because protein A is immobilized to affinity membrane matrix via stable, covalent secondary amine linkages, leaching of ligand from the support is minimal. The presence of leached protein A in affinity membrane purified MAb was assessed using a commercial ELISA-based kit. The sensitivity of the assay was approximately 10 ppm. Assay of protein A-purified MAb yielded a value of 11 ppm at the limit of detection of the assay.

Fluorimetric analysis of cell culture supernatant and purified MAb indicated that 95% of contaminating nucleic acids had been removed in the rapid membrane affinity purification process.

Scale-up and Process Development Considerations

As indicated earlier, it was expected that scale-up of affinity membrane separations could be performed on the basis of maintaining constant fluid residence time during loading. This expectation is based on the fact that in a membrane based affinity separation, the effective bed height is equivalent to the fiber wall thickness. As affinity membrane modules are scaled up, the number and length of the fibers are increased, but the bed height remains constant. A second consequence of the constant bed height approach is that transmembrane pressure is independent of scale.

These expectations were verified in experiments where the dynamic capacity of protein A modules was determined as a function of the fluid residence time at two different scales: 0.5 ml (Model 15) and 10 ml (Model 300) matrix volumes, respectively (Fig. 5). Even at residence times as short as 2 sec, performance was comparable at both Model 15 and 300 scales, and the dynamic capacity scaled linearly with membrane matrix volume. As the MAb loading approaches the saturation capacity of the membrane, the efficiency of capture gradually falls off (Fig. 5). The gradual nature of this decrease is thought to stem from axial dispersion of target protein together with a heterogeneous pore size distribution.

Fast cycle times characteristic of membrane based separations make it possible to optimize a purification protocol in a relatively short time. For example, flow rates and times of various steps in a purification protocol, and

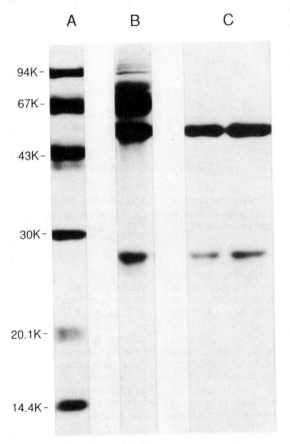

Fig. 4. SDS-PAGE (reduced) analysis of a mouse IgG purified by a one-step protein-A membrane process. A: MW Std.; B: starting material after concentration; C: purified antibody after the membrane step.

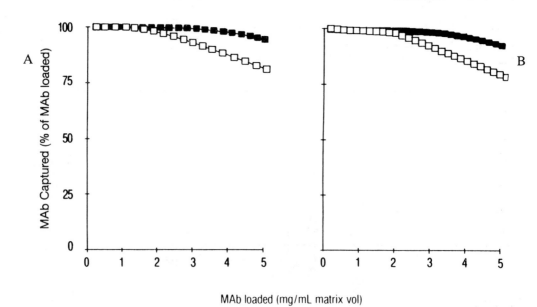

Fig. 5. MAb dynamic capacity determination of protein-A membrane at two different scales. A: Model 15 (0.5 ml matrix vol); B: Model 300 (10 ml matrix vol). Feed: clarified CCS (100 g/ml MAb). Fluid residence time: [□] 2 sec; [■] 12 sec.

composition of feed, wash, elution, and re-equilibration buffers can all be optimized in as little as a few hours. Furthermore, these optimized parameters can then be scaled up easily to a pilot or production scale run.

In the example shown in Table 3, the effect of the wash protocol on MAb purity and recovery and on cycle time was examined. The highly permeable membrane matrix can be flushed with many matrix volumes in a short period of time. Short cycle times (< 5 min) are maintained while passing > 20 matrix volumes during the wash step. With a wash time as short as 90 seconds, essentially homogeneous MAb was obtained in good yield.

Once the target protein is bound to the membrane matrix, wash solution composition can be varied to enhance product purity. For instance, in addition to changes in pH and ionic strength, small percentages of detergents and organic solvents (i.e., 0.1% Tween or 10% alcohol) can be introduced in certain applications. Our preliminary data indicates that such wash regimes can be used with many proteins, including MAbs, to improve virus removal and DNA clearance during purification (Garg, et al 1990).

With regards to the scale-up considerations, the high productivity of membrane-based purification will impact substantially on the way commercial protein purification schemes are laid out. Ideally, affinity processes should be deployed as far upstream as possible, in order to best utilize their inherent specificity. Low volumetric productivity of conventional affinity columns generally leads to positioning of the affinity step relatively late in the purification scheme, after the product has been concentrated and partially purified. The high productivity associated with membrane affinity separations permits the use of affinity immediately after the bioreactor harvest step. This eliminates the need for a separate concentration step, such as ultrafiltration, as well as other chromatographic purification steps prior to the affinity step, thereby improving overall process yield.

Other Applications

In addition to MAb purification, we have developed several other applications using various immobilization chemistries on hollow fiber affinity membranes. In order to successfully immobilize different biological

Table 3. Effect of wash parameters on MAb recovery and purity

Wash time (sec)	Wash volume (ml of wash buffer)	Cycle time (min)	Percent recovery	Percent purity
10	24	3.5	88	83
30	71	3.8	88	94
60	140	4.3	88	97
90	210	4.8	85	>99
180	420	6.3	84	>99

Module:	Model 300 protein A module
Feed titer:	86 mg/L
Estimated feed purity:	4%
MAb loaded per cycle:	36 mg

ligands to the hollow fiber membranes, a variety of base linking chemistries, aldehyde (Wright and Hunter, 1982), hydrazide (O'Shannessy and Hoffmann, 1987), and FMP (2-fluoro-1-methylpyridinium Ptoluene-sulfonate) (Ngo, 1986) have been developed. This provides a significant advantage for process development, as various chemistries can be evaluated by the protein purification chemists to best optimize the immobilization and the functionality of each ligand before scale-up.

The base linking chemistries range from a relatively fast chemistry, aldehyde, with a reaction time in the order of 10-60 min, to the FMP chemistry, which typically requires 12-20 h coupling incubation time. The hydrazide chemistry requires a 1-5 h immobilization incubation time. Although some chemical manipulation of the ligand to be immobilized is needed, the site directed immobilization provided by hydrazide chemistry has shown significant advantages when antibodies are attached to the affinity fiber. The three chemistries together also offer a wide range of immobilization pH, from pH 5 in the case of hydrazide to pH 10 in the case of FMP chemistry.

Fig. 6 illustrates the membrane-based immunoaffinity purification of a growth factor produced by a genetically engineered microorganism as monitored by GPC-HPLC. During the loading phase, the target protein binds to the membrane together with a high mol wt impurity (Fig. 6B). By washing the membrane with 0.5 M NaCl, the impurity is removed (Fig. 6C), and elution at low pH yielded a product that is essentially pure (Fig. 6D). In another example, affinity membrane was used to capture Factor VIII, the blood clotting protein. Factor VIII complex is a labile protein, and in view of its large size one might have expected to encounter sluggish antigen/antibody binding kinetics. However, this did not prevent successful Factor VIII capture (81% at 12 sec residence time) and 115-fold purification factor (Zale, et al, 1990). Subsequently, we have demonstrated Factor VIII purification directly from blood plasma.

Immobilization of other ligands such as plant lectins, heparin and fetuin to the base chemistries produces affinity matrices with binding activities comparable to that of particulate supports. Table 4 shows the amount of ligand immobilized and the measured activity

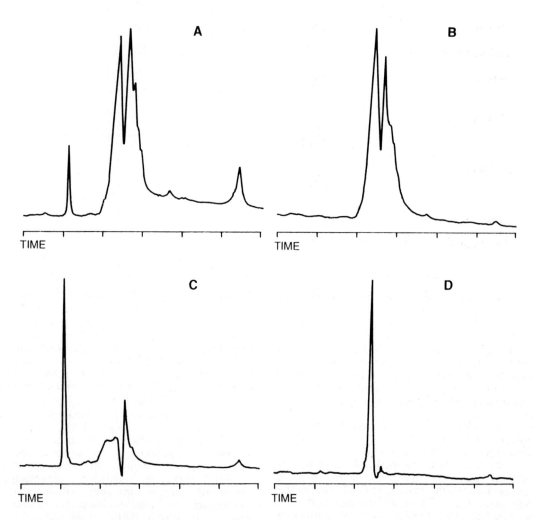

Fig. 6. Size exclusion HPLC analysis of a membrane-based immunoaffinity purification of a microbially produced recombinant growth factor. A: feed; B: flow-thru; C: wash; D: eluted growth factor.

Table 4. Amounts of various ligands immobilized on activated hollow fiber membranes

Ligand	Immobilization chemistry	Ligand density (mg/ml matrix vol)
Lentil lectin	FMP	1.5 - 2.5
	ALD	1.5 - 2.0
Con-A	FMP	0.8 - 1.0
Jacklin lectin	FMP	0.8 - 1.5
Fetuin	FMP	1.0 - 2.0
	ALD	1.5 - 2.5
Murine IgG	FMP	1.0 - 3.0
	ALD	1.0 - 3.5
	HYD	1.5 - 2.5
Rabbit IgG	FMP	1.5 - 2.5
	ALD	1.5 - 2.5
	HYD	1.0 - 2.0
Goat IgG	HYD	1.0 - 3.0
Ovalbumin	FMP	0.5 - 1.0

FMP: 2-Fluoro-1-methylpyridinium *P*-toluenesulfonate; ALD: Aldehyde; HYD: Hydrazide.

on the membrane matrix. The plant lectins, lentil lectin and con A, were both effective in capturing genetically engineered, cell culture produced growth factor glycoproteins and hybridoma produced glycoproteins such as IgM.

In addition to direct immobilization of biologically active ligands, the ability to insert bifunctional crosslinkers to extend the distance from the membrane matrix has been shown to increase the likelihood that space sensitive ligands will retain their activity on the matrix (Scouten, 1981). These additional spacers can be derivative to give a range of functional immobilization chemistries, such as epoxide and sulfhydryl moieties, which have been shown to be good functional groups on hollow fiber membranes.

Conclusions

Existing column-based affinity separation systems are constrained by pressure drop and mass transfer limitations. Membrane-based systems minimize these constraints and allow operation at rates limited only by the inherent kinetics of ligand-ligate interactions. The use of membrane-based purification has been demonstrated in the recovery of MAbs and other biotechnology-derived proteins.

Membrane-based purification systems will find increasing use in downstream processing of proteins. The principal advantages are: fast process development; high volumetric throughput; high capture efficiency of target protein from dilute feedstreams; concentration and purification in one step; efficient ligand utilization; and easy scale-up from pilot to production scale.

References

Brandt, S.; Goffe, R.A.; Kessler, S.B.; O'Conner, J.L.; Zale, S. E. *Bio/Technology*, **1988**, *6*, 779-782.

Garg, V.K. In *Targeted Therapeutic Systems*;

Tyle, P.; Ram, B.P., Eds.; Marcel Dekker, New York, NY, 1990, pp 45-73.

Garg, V.K.; Zale, S.E.; Holton, O.D.; Khazaeli, M.B. *Influence of Separation Parameters on Structural Integrity and Functional Efficacy of Biologic Products*, 200th ACS National Meeting, Washington, D.C., August 26-31, 1990.

Nachman, M.; Azad, A.R.M.; Bailon, P. *Membrane-Affinity Chromatography: A High-Efficiency Affinity Purification Method*, 200th ACS National Meeting, Washington, D.C., August 26-31, 1990.

Ngo, T.T. *Bio/Technology*, **1986**, *4*, 134-137.

O'Shannessy, D.J.; Hoffman, W.L. *Biotech. Appl. Biochem.*, **1987**, *9*, 488-496.

Pietronigro, D. *Membrane Affinity Separation for Biomolecule Purification and Diagnostics, Protein Purification and Biochemical Engineering*, UCLA Symposium, Lake Tahoe, CA, March 19-25, 1990.

Scouten, W.H. *Affinity Chromatography*, Wiley-Interscience, 1981, pp 102-105.

Wright, J.F.; Hunter, W.M. *J. Immunol. Methods*, **1982**, *12*, 311.

Zale, S.E. *Membrane-Based Affinity Separations*, UCLA Symposium, Lake Tahoe, CA, March 19-25, 1990.

Zale, S.E.; Holton, O.D.; Garg, V.K. *Affinity Membrane Mediated Protein Purification, Symposium on Membrane Bioseparation*, 199th ACS National Meeting, Boston, MA, April 22-27, 1990.

Extraction of Lipases in Aqueous Two-Phase Systems: The Strategy for an Optimization

U. Menge, Gesellschaft für Biotechnologische Forschung mbH, Department of Enzymetechnology, Mascheroder Weg 1, D-3300 Braunschweig, Germany

Abstract: A factorial design of experiments and a graphical presentation of results facilitated the optimization of aqueous two-phase systems as used for the extraction of proteins significantly. Parameters varied during phase partitioning experiments are interdependent and their influences were easily analyzed by this method.

The first partitioning experiments made evident that phase systems featuring a single efficient step for purifying lipase from *Mucor miehei* were not likely to be developed. Thus, different strategies for optimizing were applied: variation of the phase composition and the volume ratio, as well as the combination of different phase systems. The latter procedure resulted in a purification of the lipase to 69% at a yield of >80%.

A comparison of partition coefficients of lipase from *Mucor miehei, Staphylococcus carnosus* and *Pseudomonas spec.* demonstrated a high species specifity of their partition coefficients.

Introduction

During the last decade aqueous two-phase systems have found a broad interest for the separation of proteins (Hustedt et al, 1985). However, relatively few industrial processes have been announced to use this method. One of the successful applications is the extraction of β-interferon from crude production medium. This process was developed in the GBF and is used now for some years in production scale by an industrial company. During the phase partitioning β-interferon is concentrated into the top-phase of an aqueous two-phase system and purified several hundred-fold without any significant loss of activity (Menge, 1983).

Purification of proteins by aqueous two-phase systems has unique advantages: gentleness, minimal time consumption, simple scale up to any volume, high volume capacity, as well as high preservation of sample hygiene and protein stabilization. The last two advantages of aqueous two-phase systems are of high benefit for the purification of labile proteins. A main factor for the successful application of aqueous two-phase systems for the purification of β-interferon was an extensive optimization of the phase extraction. A similar optimization procedure was now per-

formed with lipase from *Mucor miehei*. The corresponding methodology has so far been neglected and there are only few publications in this field (Menge et al, 1983; Backman and Shanbhag, 1984).

Lipases are of increasing interest in biotechnology, as they enable stereospecific reactions with many organic compounds and they are active even in the absence of water. There is an increasing number of publications on these enzymes and very recently the 3D-structures of two lipases have been published (Brady et al, 1990; Winkler et al, 1990).

The purification of lipases is hampered in some cases by their hydrophobicity, which may cause unspecific adsorption of the enzyme to surfaces of column supports and membranes. In aqueous two-phase systems solid surfaces are minimized. Moreover, we investigated liquid-liquid extractions for the purification of lipase from *Staphylococcus carnosus* and *Pseudomonas spec.* ATCC21808 in view of the increasing demand on different lipases.

Material and Methods

All lipases used were concentrated, crude fermentation broth (*Mucor miehei* gift from

NOVO Industri A/S, Bagsvaerd, DK, *Pseudomonas spec.* ATCC21808 from M. Kordel, GBF, *S. carnosus* from H. Erdmann, GBF).

Lipase from *Mucor miehei* was assayed photometrically at 571nm and 37°C with 136 μM 1,2-o-dilauryl-rac-glycero-3-glutaric acid-resorufin ester (Boehringer-Mannheim). The substrate was dissolved in 100 μl of dioxane/thesit (1/1) and added to 850 μl of 0.1 M potassium phosphate, pH 6.6, and 1% thesit .

The activity of lipase from *S. carnosus* and *Pseudomonas spec.* was analyzed with 0.7 mM p-nitrophenyl palmitate in presence of 0.4 mg Triton X100/ml and 50 mM Tris-HCl, pH 8.0, at 37°C.

SDS-gel electrophoreses were performed in 8-25% gradient gels from Pharmacia and silver-stained.

Protein was analyzed by coomassie brilliant blue G (Bradford, 1976) with human serum albumin as standard. 3% PEG_{6000} and 1% NaCl were added to the protein assays to compensate interfering effects of these phase components. All assays were performed at least in triplicate.

Aqueous two-phase systems (total mass 5 g) were prepared by weighing in concentrated solutions of the components and last of all crude lipase was added. The mixtures were slowly turned over for 10 min at room temperature and separated by short low speed centrifugation.

Results

Optimization Methodology for Phase Partitioning Experiments

Partitioning in aqueous two-phase systems during protein purification is not widely accepted. One reason n.ay be that there are only few rules for the selection of phase systems (Ålbertsson, 1985; Walter et al, 1985). The possible variations in the composition of aqueous two-phase systems are numerous (Table 1; Ålbertsson, 1985). Thus, the key for a successful optimization of a phase extraction is an efficient strategy for planning and evaluating partitioning experiments.

During our search for suitable phase compositions more than a hundred different phase systems were tested. Phase compounds were not varied in a linear fashion or following the simplex method (Backman and Shanbhag, 1984), but by choosing experiments by factorial design. During this method two, three, or more factors, e.g. the concentration of

Table 1. Parameters controlling partitioning in aqueous two-phase systems, theoretical number of variations planning an optimization and obligatory evaluation parameters

factors	variations (in theory)	evaluation parameters
top phase components	3	partition coefficients of:
their concentration	5	- product
bottom phase components and	3	- contaminants
their concentration	5	
pH	5	yield
additives (salt, organic compounds)	3	purification factor
their concentration	5	
volume ratio	-	volume reduction
sample content	-	
total number: >9	16875	5

332

phase compounds, the pH or a salt additive were varied on distinct levels simultaneously.

A second complication in the application of phase systems arises from the necessity to evaluate different interdependent data, e.g. partition coefficients, volume ratio, specific activity and yield (Table 1). Most essential were the partition coefficients of the product and that of the total protein (Ålbertsson, 1985), as these indicate influences of phase components on the partition very evidently. But, partition coefficients give only an indirect hint on the resulting yield and the purification factor of a product, which are decisive for an efficient application of phase extraction. Both yield and purification factor observed, depended on the volume ratio of the phases, which could be controlled by the concentration of the phase constituents (Ålbertsson, 1985).

It takes some experience not to be confused by this multiplicity of variables, resulting data and their interdependence. A mathematical evaluation of factorial-designed experiments by multilinear regression is inapplicable as a large number of data would be necessary. Conventional tables are too inflexible for the presentation of multivariate experiments, as the sequence of results can be changed only by editing a completely new table.

In contrast, a 3D-graphical presentation (Fig. 1, 2) proved to be very helpful for an analysis of results in alternating arrangements. Parallel planes in the graphic represent distinct levels of one parameter and its influence can be analyzed simply by comparison of different sets of parallel planes.

Moreover, a graphical representation allows the documentation of different data (Table 1) within boxes, and these can be easily surveyed on the same graph.

A further advantage of a graphical presentation of data arises from an inherent statistical proof of effects. E.g., similar tendencies in the change of K_a were found on different parallel axes (Fig. 1). Now, the conclusion was founded not only on one but on up to six pairs of data and error sensitive partitioning data were confirmed. Such a confirmation proved to be very useful, especially in the case of biological assays, during which high variations of results are unavoidable. The resulting

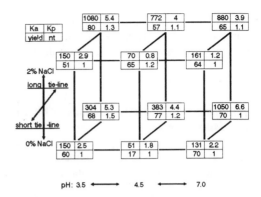

Fig. 1. Results of initial experiments in the partitioning of crude lipase from *Mucor miehei* in PEG$_{1550}$/phosphate systems (4/25% w/w, long tie-line; 9/16% w/w, short tie-line) in presence of 0 and 2% NaCl at pH 3.4, 4.5 and 7.0. Data listed in box: K_a (partition coefficient of lipase), K_p (partition coefficient of total protein), yield in top phases (%) and n_t (purification factor of lipase in the top phase).

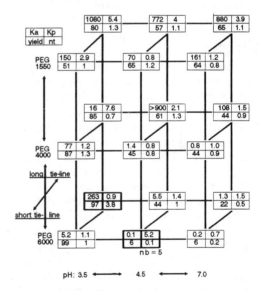

Fig. 2. Results of experiments in partitioning of crude lipase from *Mucor miehei* obtained in systems of different molecular weight of PEG. PEG and phosphate concentration (% w/w) at short and long tie-line: PEG$_{1550}$ (9/16 and 4/25), PEG$_{4000}$ (8/14 and 4/20), PEG$_{6000}$ (4/18 and 7/12). nb: purification factor of lipase in bottom phase. See Fig. 1.

error in partition coefficients can be even more severe as these are quotients of such more or less imperfect experimental data.

Optimization of Phase Extractions of Lipase from *Mucor miehei*

For the first trials in partitioning of lipase from *Mucor miehei* PEG/phosphate systems were selected from an economical point of view. Moreover, it was assumed that lipases may interact with interfaces, the normal location of their emulsified substrates. To avoid such an interfering adsorption, systems with a relatively low interfacial tension, containing PEG_{1550}, were chosen. To study influences of different interfacial potentials between top and bottom phase, systems with a long and a short tie-line were selected from the corresponding phase diagram (Ålbertsson, 1985). As a second factor the pH was varied from below (pH 3.5), at (pH 4.5) and above the isoelectric point of lipase (pH 7.0). Besides, NaCl was varied between 0 and 2%.

Fig. 1 shows the results of the first 18 factorial-designed experiments. The most significant data found were extremely high partition coefficients for lipase ranging from 51 to up to 1080, to find on the back plane of the graph. Second, in systems with a higher PEG concentration in the top-phase the extraction of lipase was increased by a factor of up to eleven. Thus, the most influencing factors arose from the length of the tie-line in the corresponding phase diagram.

Thirdly, recoveries of up to 80% of lipase were observed in the top phases. Unexpected for a lipase, the yield was higher in many systems with the relatively higher interfacial tension. This phenomenon may also indicate that lipase from *Mucor miehei* did not interact with the interface and remained freely solubilized in the phases.

The influence of NaCl was not pronounced, but the salt favored the extraction of lipase into the top phase of at least some systems. Unfortunately, also most of the accompanying proteins were extracted into the top phase and, consequently, lipase was enriched only very little.

In the second stage during the optimization we searched for phase systems with an increased differentiation in the partition of lipase and accompanying proteins. In many cases partition coefficients could be lowered by the use of higher molecular weight PEG (Ålbertsson et al, 1987). Thus, we tested phase systems containing PEG_{4000} and PEG_{6000} and, again, systems with a long and a short tie-line. As the third factor the pH was varied as above, whereas NaCl was kept constant at 2% to limit the number of experiments.

Fig. 2 shows the results of these experiments. The partition coefficient of lipase decreased with increasing molecular weight of PEG in nearly all systems studied, maximum decrease being 700-fold. The maximum decrease found for partition coefficients of the accompanying proteins was 8-fold only. The observed extreme decrease of the partition coefficient of lipase was characteristic for this protein and exceeded 100-fold those observed in PEG/dextran systems for ovalbumin, a protein of similar size (Ålbertsson et al, 1987).

However, among 18 phase systems tested only two resulted in a purification of lipase (Fig. 2, bold boxes). These systems contained PEG_{6000}/phosphate (4/18% w/w, pH 3.5 and 7/12% w/w, pH 4.5). In the latter system the partition of lipase was even reversed to the bottom phase whereas the mass of protein remained in the corresponding top phase. These systems were later combined to a two-step procedure (see below).

The volume capacity of phase systems is a further essential parameter to be considered for an industrial purification process. During the first approach to the optimization of phase extraction systems contained only 1% crude lipase to study phase partitioning under more or less ideal conditions. Such a small volume capacity is not acceptable during a technical process. However, at increased sample contents of up to a maximum of 48% the partition coefficients of lipase dropped from several hundreds to 1 (data not shown) and clouded interphases formed in most of the systems.

To enable an increased sample input the effect of different salts was studied. These influence the solubility of proteins in both phases, as well as electrostatic and hydrophobic interactions. Thus, they influence the partition coefficients of proteins, even in systems containing 20% phosphate (Menge et al, 1983; Ålbertsson, 1985). In fact, 8% NaCl, K_2SO_4 or $(NH_4)_2SO_4$ increased the partition coefficient of lipase from 26 to 241, 125, and 87 in systems containing PEG_{4000}/phosphate (4/18% w/w, pH 3.4). Also in systems containing PEG_{6000}/phosphate (7/12% w/w, pH 4.5) the partition coefficient of lipase increased by additional salts from 0.03 to 20.

The partition coefficient of the total protein increased by salt addition too, but at a significantly lower extent from 2 to 4 and 0.5 to 2.

The most promising salt effect was found with $(NH_4)_2SO_4$ and this was analyzed in more detail by linear variation of its concentration between 0 and 12%. At 8% a pronounced maximum in the partition coefficient of lipase was found (Fig. 3). However, purification factors for lipase did not exceed 1.5. Also the use of ammonium sulfate as phase forming compound instead of phosphate did not improve the purification of lipase.

Despite many variations in phase composition we so far found no aqueous two-phase system useful for an effective single step purification of lipase from *Mucor miehei*. Thus, significant problems in the optimization of phase systems for the extraction of lipase became evident and the strategy was changed.

Influence of the Phase Volume on the Purification Factor

As an alternative approach to the optimization of a phase system the volume ratio can be reduced if the partition coefficient of the total protein is in the neighborhood of 1 or smaller. The purification factor of a compound extracted into the top phase (n_t) depends on the yield in the top phase (Y_t), the partition coefficient of the total protein (K_p), and the volume ratio (Q_{Vol}) as described by the following equation:

$$n_t = Y_t \cdot (1 + 1/(K_p \cdot Q_{Vol})).$$

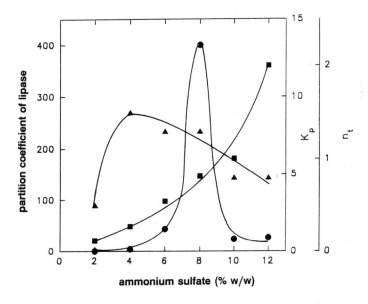

Fig. 3. Influence of ammonium sulfate on the partition coefficient of lipase from *Mucor miehei*, the partition coefficient of total protein (K_p) and the purification factor in the top phase (n_t) in a phase system containing PEG_{6000}/phosphate (7/12% w/w, pH 4.5, 20% crude lipase).

Fig. 4 shows a graphical presentation of this equation for several relevant values of Y_t and Q_{Vol}. This equation is exactly valid only if K_p is similar for all proteins in a sample and is not influenced by the product constituting only a minor part of the total protein. Nevertheless, this equation can be used for an estimation of a hypothetical increase of the purification factor, that can be achieved by reduction of the top phase volume.

Usually top and bottom phase are of similar size, $Q_{Vol} \approx 1$. At such a volume ratio a partition coefficient of the total protein of 0.25 is necessary to purify a product 5-fold by

quantitative extraction into a top phase. However, at a reduced top phase ratio, e.g. $Q_{Vol} = 0.2$, the same degree of purification can be realized at a significant higher partition coefficient of the total protein. The loss of product in the relatively large bottom phase is minimal if the partition coefficient of the product is $>> 10$ (Menge et al, 1983).

These considerations were verified by experiments, e.g. we found an increase in the purification factor from 3.8 to 6.5 after reducing the volume ratio from 0.4 to 0.07. The yield decreased from 83 to 66% only. Thus, by a 6-fold reduction of the volume ratio the

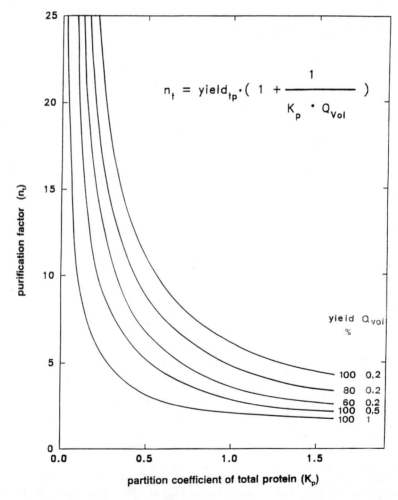

$$n_t = yield_{tp} \cdot \left(1 + \frac{1}{K_p \cdot Q_{Vol}} \right)$$

Fig. 4. Purification factor (n_t) of enzymes extracted into top phases as a function of the partition coefficient of the total protein (K_p), the yield in the top phase and the volume ratio.

purification factor was nearly doubled. Moreover, by the small top phase the product was concentrated in a significantly smaller volume.

Combination of Two Phase Extractions

In a last trial of improving the purification of lipase two phase extractions were combined. As shown in Fig. 2, there were two systems of PEG_{6000} by which lipase could be purified, one with a top phase and one with a bottom phase extraction. To realize the second extraction step the isolated top phase of the first system was adjusted to the composition of the second system by addition of PEG_{6000}, phosphate, NaOH and water. The progress of this purification procedure was analyzed by gel electrophoreses (Fig. 5). The final bottom phase contained >80% of the lipase and its purity was estimated to be 69%. The overall purification factor was 12.

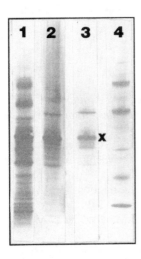

Fig. 5. Analysis of the two-step extraction of lipase by SDS-gel electrophoresis. Lanes 1: crude lipase; 2: top phase of the first system (PEG_{6000}/sodium phosphate/NaCl/lipase (3.3/15/1.9/20% w/w, pH 3.4); 3: bottom phase of the second system (first top phase 67%, additional: 5.4% (w/w) $NaH_2PO_4 \cdot H_2O$, 0.67% NaCl and water, adjusted to pH 6.9); 4: molecular weight marker proteins. (x): lipase (adapted from Menge, 1989).

Comparison of the Phase Partitioning of Three Lipases

Apart from lipase from *Mucor miehei* the phase partitioning of crude lipase from *Staphylococcus carnosus* and *Pseudomonas spec.* ATCC21808 was studied (Fig. 6). Lipase from *Mucor miehei* is distinguished by high partition coefficients of up to 1000 whereas those of the enzyme from *Pseudomonas spec.* reached a maximum of 87. In the case of lipase from *Staphylococcus carnosus* very small partition coefficients (0.02 - 3.5) were found and an extraction into the bottom phase dominated.

The general order of decreasing partition coefficients with increasing molecular weight of PEG was even reversed for lipase from *Pseudomonas spec.* in systems with the longer tie-line at pH 4.5 and 7.

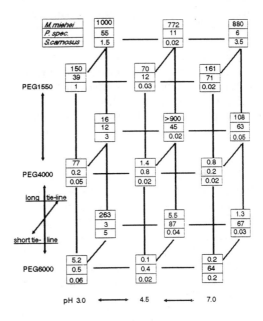

Fig. 6. Comparison of the partition coefficient of lipase from *Mucor miehei*, *Pseudomonas spec.* ATCC21808 and *Staphylococcus carnosus* (see Fig. 1).

Discussion

Factorially designed experiments were found to be a very useful methodology for optimizing phase extractions. Different factors affecting phase partitioning could be analyzed simultaneously. In contrast to experiments with linearly varied parameters this methodology could easily reveal an interdependence of these factors.

An alternative method to find an optimal phase composition in the maze of multidimensional variability is the simplex method (Backman and Shanbhag, 1984). This method utilizes a two-dimensional graphical presentation of phase compositions as points in the field of two variables. Three points, representing different phase compositions, form a triangle and new phase compositions are selected by geometrical manipulation of this triangle. Depending on the increase or decrease of partition coefficients the triangle is expanded, reflected, or contracted.

There are three essential drawbacks of the simplex method. It is directed to a minimized number of experiments. But, a significant delay can be encountered by the serial performance of analyses as these are time-consuming, in particular biological assays. Moreover, in contrast to factorial experiments there is no statistical element in the simplex method.

Finally, this method does not reveal the interdependence of different factors on phase partitioning. All these disadvantages could be eliminated by application of factorial design of experiments.

The graphical presentation of data from factorial experiments represents a three-dimensional table. It enabled a quick analysis of experimental factors considering different sets of parallel axes. This methodology proved to be very helpful in the case of our partitioning experiments during which three experimental factors and four evaluation parameters were optimized simultaneously. It allowed a quick identification of systems offering the best chance of a further optimization.

On the other side, unpromising results of a set of experiments became more evident, e.g. the low efficiency of all single step extractions, as discussed above. This was discouraging and in contrast to the optimization of phase extractions realized, e.g. for β-interferon (Menge et al, 1983).

However, the liquid-liquid extraction of lipase from *Mucor miehei* could be optimized by combination of two systems of a lower efficiency and the choice of a reduced volume ratio during the second extraction. The complete development of this purification procedure took a few weeks only.

The partition coefficient of lipases from different sources, such as fungal and microbial, were very different indicating the general hydrophobic character of proteins like lipases was not controlling their extraction into the hydrophobic, PEG containing top phase. Lipase from *Mucor miehei* and *Pseudomonas spec.* are very similar in respect to their isoelectric point (IP = 4.3 and 4.5) and molecular weight (32,000 and 33,000). Still, their partition behavior differed significantly. This may correspond to the low homology of their primary structure (30%).

The much lower partition coefficients of lipase from *S. carnosus* may be attributed to two factors: 1. its higher positive charge (IP = 5.6), that may favor the extraction into the negatively charged bottom phase (Zaslavsky et al, 1982). 2. its molecular weight is significantly higher (85,000) and, thus, it may be exposed to size exclusion effects of the highly concentrated polymeric PEG in the top phase.

References

Albertsson, P. Å. *Partition of Cell Particles and Macromolecules*; John Wiley & Sons: New York, **1985**.

Albertsson, P.Å., Cajarville, A., Brooks, D.E., Terneld, F. *Biochim. Biophys. Acta*, **1987**, 926, 87-93.

Backman, L., Shanbhag, V.P. *Anal. Biochem.*, **1984**, 138, 372-379.

Bradford, M.M. *Anal. Biochemistry*, **1976**, 72, 248-254.

Brady, L., Brzozowki, A.M., Derewenda, Z., Dodson, E., Toiley, S., Christiansen, L., Norskov, L., Menge, U. *Nature*, **1990**, 343, 767-770.

Hustedt, H., Kroner, K.H., Menge, U., Kula, M.R. *Trends in Biotechnology*, **1985**, 3, 139-144

Menge, U., Morr, M., Mayr, U., Kula, M.-R. *J. Appl. Biochem.* **1983**, 5, 75-90.

Menge, U., Schmid, R.D., in *Proceedings of the 15th Scandinavian Symposium on Lipids*,

Shukla, V.K.S., Hølmer, Rebild Bakker, DK, **1989**, 306-316.

Winkler, F.K., D'Arcy, A.D., Hunziker, W., Nature, **1990**, *343*, 771-774.

Walter, H., Brooks, D., Fisher, D., Eds.; *Partitioning in Aqueous Two-phase systems*; Academic Press: Orlando, USA, **1985**.

Zaslavsky, B.U., Miheeva, L.M., Mestechkina, N.M., Rogozhin, S.V. *J. Chromatography*, **1982**, *253*, 149-158.

Salt-PEG Two-Phase Aqueous Systems to Purify Proteins and Nucleic Acid Mixtures

Kenneth D. Cole
National Institute for Standards and Technology
Biotechnology Division
Boulder, CO 80303

This paper examines the effect of variables on protein and nucleic acid partitioning in aqueous two-phase extraction systems composed of salt and polyethylene glycol (PEG). The variables that can be changed in these systems include: pH, PEG concentration, PEG molecular mass, salt type, salt concentration and addition of protein denaturants such as chaotropic salts and detergents. Methods for the preparation of high molecular mass DNA from crude mixtures are described. These methods offer the advantages of simplicity, speed and use of nontoxic materials.

There is a great need for nucleic acid separations that are rapid, efficient, economic and suitable for automation. Partition of biomolecules in aqueous two-phase extraction systems (ATPS) satisfies many of these criteria. Although there are many methods for purifying DNA, the most commonly used method is extraction of aqueous solutions containing the nucleic acids with organic solvents, typically phenol and chloroform (Blin and Stafford, 1967). This paper describes a method using ATPS that rapidly, efficiently and inexpensively purifies DNA. This paper also examines the variables in ATPS that affect the efficiency of the extraction.

The partitioning of nucleic acids in aqueous two-phase systems has been studied in systems composed of the polymers dextran and polyethylene glycol (PEG)(reviewed in Albertsson, 1986 and Müller, 1985). The partitioning of nucleic acids in the dextran-PEG systems are extremely sensitive to the phase conditions. Small changes in the ionic strength or of salt type added can change the partition coefficient of the nucleic acid several orders of magnitude (Albertsson,1965). Under the appropriate conditions in the dextran-PEG system it is possible to completely separate double-stranded DNA from single-stranded (denatured) DNA (Albertsson, 1965; Alberts, 1967).

Several methods have been developed for the purification of nucleic acids from crude sources by using the dextran-PEG ATPS. The procedure of Rudin and Albertsson (1967) uses high concentrations of sodium chloride (4 M) to partition proteins to the top (PEG-rich) phase and the DNA to the bottom (dextran-rich) phase. Successive extractions with new top phase without salt transfers the DNA to the top phase. The method of Favre and Pettijohn (1967) depends on careful removal of ionic components of the growth media by washing of the bacteria and organic solvent extraction to remove most of the proteins. The DNA initially separates to the top phase. The conditions are changed and the DNA then separates to the bottom phase in subsequent extractions. Both of these methods require careful attention to the ionic strength and pH of the phase system. It has been reported that mammalian DNA prepared using the method of Rudin and Albertsson (1967) is contaminated by mucopolysaccharides (Pritchard et al. 1971). Viede et al. (1983) used ATPS composed of potassium phosphate and PEG 6000 to separate β-galactosidase from *E. coli*. The cellular debris, the majority of the proteins and nucleic acids (DNA and RNA) were present in the denser salt phase.

The partitioning of nucleic acids in the dextran-PEG ATPS has been found to depend on size. The size dependent partitioning of DNA in the dextran-PEG system has been exploited of to purify DNA fragments produced by cutting the DNA with restriction enzymes. Müller et al. (1979) used a liquid-

liquid partition column to separate DNA fragments from 150 to 22000 base pairs (bp). Cellulose or celite was used to immobilize the dextran-rich phase in the column and a PEG solution was used as the mobile phase.

A technique for the isolation of circular plasmid DNA using the dextran-PEG ATPS has been described (Ohlsson et al. 1978). This method takes advantage of the rapid reassociation rate of the covalently closed circular DNA compared to that of the _E. coli_ genomic DNA. In this method a plasmid preparation was prepared by lysing the bacteria with detergent, removing proteins by organic solvent extraction and precipitating the DNA with ethanol. The DNA preparation was denatured by heat and added to a dextran-PEG ATPS. The renatured circular DNA was then recovered in the PEG phase. The partitioning of DNA from the polyoma virus has been investigated (Pettijohn, 1967). The partition coefficient of supercoiled polyoma DNA was approximately one-tenth that of the partition coefficient of the open circular or linear forms of the polyoma DNA in the dextran-PEG ATPS used.

The ability of dextran-PEG ATPS to separate single-stranded nucleic acids from double-stranded nucleic acids has been used to purify DNA-DNA hybrids from unhybridized RNA molecules (Mak et al. 1976). Mak et al. hybridized poly (A) containing RNA from HeLa cells infected with adenovirus type 2 to denatured (single-stranded) DNA from adenovirus type 2. The mixture was added to a dextran-PEG system. Under the appropriate conditions used the DNA-RNA double-stranded hybrids and double-stranded DNA molecules separated to the top (PEG-rich) phase while unhybridized RNA and single-stranded DNA were in the bottom (dextran-rich) phase. The RNA-DNA hybrid could be melted apart by heating and then applied to an oligo dT cellulose column. The poly (A) containing RNA bound to the column and the DNA did not bind. The specific mRNA molecules were then recovered from the oligo dT column. They judged that their recovery of the specific RNA molecules was nearly quantitative with a purity of 70% to 90%.

A technique for the purification of supercoiled plasmid and eukaryotic high molecular mass DNA using salt-PEG ATPS has also

been developed (Cole 1991). To better understand and use salt-PEG ATPS for bio-separations we studied the variables that affect the partitioning of proteins and nucleic acids in these systems. The study described in this paper examines the variables that affect the partitioning of a mixture of standard proteins in two salt-PEG ATPS, one composed of ammonium sulfate as the phase-forming salt (group A) and sodium or potassium phosphate as the phase-forming salt (group P). It also examines the use of proteinase K to treat liver homogenates in order to prepare DNA of high molecular mass, thereby avoiding the use of the denaturant, guanidine isothiocyanate used in the previous study (Cole 1991).

Materials and Methods

Preparation of Mouse Liver DNA

Mouse livers were dissected and rinsed with phosphate-buffered saline (0.9% NaCl and 0.05 M sodium phosphate pH 7.5) and chopped into small pieces with a razor blade. The liver (1 g) was added to 20 ml of a solution containing sodium dodecyl sulfate (SDS) (1% m/v), tris(hydroxymethyl)amino methane (TRIS)(0.05 M), ethylenediaminetetraacetic acid (EDTA)(1 mM) and proteinase K (0.1 mg/ml). The final pH of the mixture was 8.0. The mixture was homogenized at 10000 rpm (1 cm generator) for 1 min. The homogenate was incubated at 42°C for 2 h. The homogenate (10 g) was added to a phase system (final mass of 50 g) of the composition indicated in Table 1. The mixture was rocked gently for 10 min at room temperature. The phase system was centrifuged (5000g for 10 min at 25°C). The bottom of the tube was punctured, and the bottom phase was drained into a fresh tube. The layer of precipitated material present at the interphase region was carefully avoided. Fresh top phase (prepared from a 50 g system with water substituted for liver homogenate) was added and the resulting phase system mixed at room temperature for 10 min. This phase system was centrifuged

Table 1. Composition of the Phase Systems

Phase System	PEG M.M.	PEG Conc.	Salt Type	Salt Conc.	Addit.	Addit. Conc.	pH	Phase Ratio (T/B)
A1	1000	15	AS	14			7.5	0.80
A1S	1000	15	AS	14	SDS	1	7.4	0.77
A1G	1000	15	AS	14	GuSCN	24	7.4	0.89
A2	1000	18	AS	16			7.5	0.80
A3	1000	21	AS	20			7.5	0.84
A4	1000	25	AS	24			7.4	1.13
A5	1000	15	AS	14			9.3	0.80
A6	1000	21	AS	20			9.3	0.84
A7	8000	10	AS	14			7.5	0.42
A8	8000	14	AS	20			7.5	0.59
P1	1000	15	KP	12			7.5	1.20
P2	1000	21	KP	18			7.5	1.24
P3	8000	15	KP	10			7.5	0.67
P4	8000	21	KP	18			7.5	0.5
D1	1000	15	NP	8.5	EDTA SDS	0.15 1	8.0	1.35
D2	1000	14.4	KP*	11.5	EDTA	0.15	8.0	1.17
D3	1000	21	KP*	22.5	EDTA	0.15	8.0	0.88

All compositions are given in mass %. PEG: polyethylene glycol. M.M.: average molecular mass. AS: anhydrous ammonium sulfate.KP: anhydrous potassium phosphate (KH_2PO_4/K_2HPO_4 ratio of 1/5.25).KP*: anhydrous KH_2PO_4. NP: anhydrous NaH_2PO_4. SDS: sodium dodecyl sulfate. EDTA: ethylenediaminetetraacetic acid. GuSCN: guanidine isothiocyanate. pH: the final pH of the phase system after addition of the sample. Ratio: ratio of the volume of the top phase to the volume of the bottom phase.

(5000g for 10 min at 25°C), and the bottom phase was collected in a fresh tube. Extraction with fresh top phase was repeated two more times for a total of four extractions. The final bottom phase was dialyzed against 10 mM TRIS and 1 mM EDTA pH 7.5 for 4 h at 25°C.

Extraction of Protein Standards

A mixture of 8 proteins (described in the legend of Fig.1) was dissolved in 30 mM sodium phosphate buffer, pH 7.5 containing 6.5 mM dithiothreitol. The mixture of proteins (0.67 mg of each protein in 1.6 ml buffer) was added to a phase forming mixture with a final mass of 10 g. The pH of the phase system was adjusted with concentrated HCl or NaOH. The mixture was gently rocked at 25°C for 10 min and then centrifuged (5000g for 10 min at 25°C). The top phase was removed and the bottom phase collected by puncturing the tube. The samples were dialyzed against 1 % (w/v) ammonium bicarbonate for 3 h at 25°C and then lyophilized. The dried samples were redissolved in sample buffer (Laemmli, 1970) and subjected to electrophoresis in the presence of SDS in a 12.5% polyacrylamide slab gel according to Laemmli (1970). The gels were then stained using Coomassie Brilliant Blue G-250 (Neuhoff et al., 1988).

Image Analysis of Stained Gels

The gels were then scanned and analyzed using a computerized image analysis system. The band boundaries and integrated optical intensity were measured using a commercial program. The integrated optical intensity was used to calculate protein concentration by comparison to a standard curve generated by using different amounts of each individual protein.

Reverse Phase HPLC

A mixture of RNase A (bovine pancreas), lysozyme (chicken egg) and bovine serum albumin (fraction V) was dissolved in 30 mM sodium phosphate buffer of pH 7.5 containing 6.5 mM dithiothreitol to give a concentration of 10 mg/ml of each protein. One ml of protein solution was added to a phase system (composition in Table 1) with a total mass of 10 g. The pH was adjusted using concentrated NaOH or HCl. The mixture was equilibrated at 25°C for 30 min and then rocked gently for 10 min at 25°C. The mixture was centrifuged (5000g at 25°C for 10 min) to separate the phases. The protein mixtures were separated on a C18 reverse phase column using a mobile phase (1 ml/min) of 0.1 % trifluoroacetic acid and a gradient of acetonitrile. The proteins were detected by absorbance at 280 nm. A standard curve for each protein was generated by injecting varying amounts of each protein and measuring the integrated area under each peak. The top and bottom phase were diluted with water so that the protein concentration fell into the linear range. The partition coefficient was calculated by dividing the protein concentration in the top phase by the protein concentration in the bottom phase. Recovery was calculated by multiplying the protein concentration by the phase volume and dividing by the total amount of protein added to the phase system.

Results

Two methods of analysis were used to examine the partitioning of protein mixtures in this study. SDS polyacrylamide gel electrophoresis (PAGE) can be used to examine the partitioning of a wide spectrum of proteins simultaneously. The SDS PAGE analysis is limited to analysis of systems with low molecular mass (1000) PEG because PEG 8000 interferes with electrophoresis and is not easily removed from the protein mixtures. The presence of even small amounts of PEG 8000 in the samples caused extreme smearing. PEG 8000 could be removed by column chromatography or precipitation of the proteins with acetone. Using these techniques it is experimentally difficult to ensure complete protein recovery in each of the phase systems. SDS PAGE is also limited in the amount of sample that can be loaded on the gel and analyzed. Separation of proteins on reverse

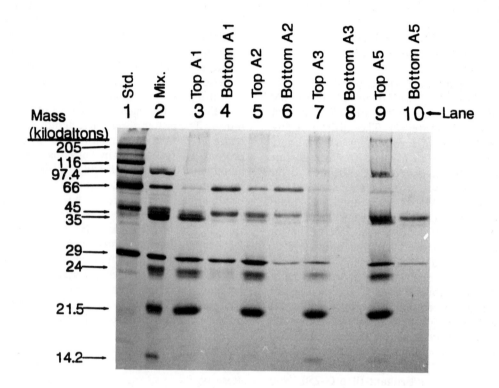

Fig.1 SDS PAGE of top and bottom phase samples of protein mixtures. Lane 1 contains a mixture of high molecular mass standards. Lane 2 is a sample of the starting protein mixture used in the partitioning experiment. The identity and mass of the 8 proteins in the mixture used in the partitioning experiments are (starting from the top): rabbit muscle phosphorylase B (97400),bovine serum albumin (66000), ovalbumin (45000), yeast alcohol dehydrogenase (35000), bovine carbonic anhydrase (29000), bovine pancreas PMSF treated trypsinogen (24000), soybean trypsin inhibitor (21500) and bovine alpha-lactalbumin (14200). Equal amounts of top or bottom phase system were loaded in the lanes. Lane 3 (top phase) and lane 4 (bottom phase) are the samples from extraction system A1. Lane 5 (top phase) and lane 6 (bottom phase) are the samples from extraction system A2. Lane 7(top phase) and lane 8 (bottom phase) are the samples from phase system A3. Lane 9 (top phase) and lane 10 (bottom phase) are the samples obtained from extraction system A5.

phase HPLC is limited to proteins of sufficient purity to give one or a few peaks resolved from the other components. RNase A, egg white lysozyme and bovine serum albumin (BSA) gave single peaks well resolved from each other and were chosen for the study.

Fig. 1 and Fig. 2 shows the SDS PAGE gels of the mixture of proteins separated in ammonium sulfate-PEG 1000 systems. As can be seen from this figure precipitation plays an important role in these systems. Increasing the tie-line length (increasing concentrations of salt and PEG), going from system A1 through A2 through A3 up to A4 increased partition coefficients and decreased the recovery of the proteins. The low recovery of proteins is dramatically seen in the recovery of the bottom phase of system A3 (Fig 1, lane 8) or the bottom phase of system A4 (Fig. 2, lane 10). Phosphorylase B was not very soluble in these systems and was only detected in small amounts in the top phase of system A1 (Fig. 1, lane 3) and slightly greater amounts in phase A5 (Fig. 1 lane 9). Bovine serum albumin (BSA) had a partition coefficient of 0.12 in phase system A1 (Fig. 1, lanes 3 and 4) which increased to 0.43 in phase system A2 (Fig. 1 lanes 5 and 6). The partition coefficient of ovalbumin in system A1 was 0.35 and increased to 1.44 in system A2. Carbonic anhydrase had a partition coefficient of 1.27 in system A1, and this increased to 5.11 in system A2. Trypsinogen, soybean trypsin inhibitor and alpha-lactalbumin were present in such high amounts in the top phase and low amounts in the bottom phase that it is difficult to measure their partition coefficient accurately on gels.

The effect of increased pH is shown in phase systems A5 and A6. The bottom phase of system A5 (Fig. 1, lane 10) contains only a small amount of ovalbumin and carbonic anhydrase ,and the top phase contains most of the proteins (Fig. 1, lane 9). The bottom phase of the high tie-line length and high pH system A6 (Fig. 2, lane 4) contains no detectable proteins and the top phase of system A6 (Fig. 2, lane 3) contains only low amounts of the proteins.

Phase system A1S had 1% SDS substituted for the equivalent amount of water in phase system A1. The addition of SDS increased the partition coefficient of most proteins. The partition coefficient of BSA and ovalbumin in changing from system A1 to system A1S, increased from 0.11 to 1.25, 0.35 to 1.08, respectively. Phase system A1G was made by the substitution of 24 % guanidine isothiocyanate for the equivalent amount of water in phase system A1. The incorporation of guanidine isothiocyanate into the phase system A1G resulted in undetectable amounts of protein in the bottom phase with the exception of soybean trypsin inhibitor. The partition coefficient of soybean trypsin inhibitor actually decreased.

· The HPLC was used to measure the protein concentration and total protein in the top and bottom phase. We measured protein recovery in the top phase and bottom phase to determine the amount of precipitation in these systems. The results of protein partitioning measured by HPLC are shown in Table 2. These data also illustrate the effect of increasing tie-line length on the protein partitioning. Going from phase system A1 to A3, the partition coefficient of RNase, lysozyme, and BSA increased at least 10 fold for RNase and BSA and approximately 75-fold for lysozyme. Total yield of the proteins decreases dramatically with increasing tie-line length. For the systems with low total yield of proteins there is significant precipitated protein at the interphase region. These systems could be regarded as having a third phase, and the partition coefficient values for these systems may not be typical of the case of systems containing lower concentrations of proteins. This effect is under further investigation.

The effect of using higher molecular mass PEG in the ammonium sulfate-PEG system is also shown in Table 2. Phase systems A7 and A8 utilize PEG 8000. The RNase, lysozyme and BSA all had much lower partition coefficients in the PEG 8000 systems compared to the corresponding PEG 1000 systems (A1 and A3). Increasing the tie-line length (from A7 to A8) doubled the partition coefficients for lysozyme and BSA but these coefficients are still less than one. The increased tie-line length in the PEG 8000 systems also results in lower yields of the protein. A significant amount of protein is still present in the bottom phase at high tie-line lengths of the ammonium sulfate-PEG 8000 system (system A8). This result is compared to the case of

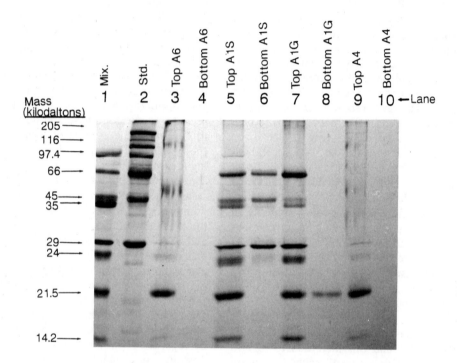

Fig. 2 Photograph of a 12.5% SDS protein electrophoresis gel. Lane 1 contains a sample of the protein mixture used in this partitioning experiment. Lane 2 contains a mixture of high molecular mass protein standards. The identity of these proteins are described in Fig 1. Lane 3 (top phase) and lane 4 (bottom phase) are the samples obtained from extraction system A6. Lane 5(top phase) and lane 6 (bottom phase) are the samples from extraction system A1S. Lane 7 (top phase) and lane 8 (bottom phase) are the samples from extraction system A1G. Lane 9 (top phase) and lane 10 (bottom phase) are the samples from extraction system A4.

Table 2. Protein Partitioning Data determined by HPLC

Phase System/ Protein	Partition Coefficient	% Total Recovery	% Recovery Top Phase	% Recovery Bottom Phase
A1 RNase	0.796 ± 0.001	99.44 ± 0.31	38.84 ± 0.11	60.60 ± 0.40
Lysozyme	2.67 ± 0.06	85.36 ± 0.68	58.41 ± 0.77	26.90 ± 0.16
BSA	0.126 ± 0.006	96.67 ± 1.31	8.94 ± 0.28	89.87 ± 1.56
A3 RNase	7.08 ± 0.89	35.23 ± 0.59	29.92 ± 0.89	5.13 ± 0.53
Lysozyme	199.94 ± 51.1	16.05 ± 0.83	15.95 ± 0.80	0.11 ± 0.04
BSA	23.93 ± 1.08	3.39 ± 0.34	3.22 ± 0.34	0.16 ± 0.01
A7 RNase	0.017 ± 0.006	83.2 ± 10.6	0.60 ± 0.25	82.6 ± 10.4
Lysozyme	0.169 ± 0.022	83.94 ± 5.68	5.64 ± 0.65	78.30 ± 5.67
BSA	0.015 ± 0.007	80.14 ± 8.26	0.41 ± 0.25	79.73 ± 8.24
A8 RNase	0.014 ± 0.004	60.23 ± 6.73	0.49 ± 0.20	59.73 ± 6.52
Lysozyme	0.379 ± 0.019	18.42 ± 1.76	3.24 ± 0.52	15.81 ± 1.37
BSA	0.030 ± 0.005	36.03 ± 7.17	0.66 ± 0.24	35.36 ± 6.93
P1 RNase	0.270 ± 0.005	71.55 ± 1.85	17.45 ± 0.71	54.11 ± 1.15
Lysozyme	1.723 ± 0.308	74.77 ± 7.26	50.58 ± 7.27	24.19 ± 0.12
BSA	32.65 ± 2.69	69.49 ± 0.64	67.75 ± 0.55	1.74 ± 0.14
P2 RNase	28.16 ± 4.44	17.21 ± 2.11	16.66 ± 2.11	0.46 ± 0.03
Lysozyme	37.95 ± 4.65	12.37 ± 3.10	12.11 ± 3.03	0.26 ± 0.08
BSA	10.48 ± 1.07	2.04 ± 0.18	1.89 ± 0.15	0.15 ± 0.23
P3 RNase	0.016 ± 0.001	83.87 ± 0.77	0.88 ± 0.06	82.99 ± 0.72
Lysozyme	0.134 ± 0.006	92.45 ± 0.74	7.58 ± 0.28	84.87 ± 0.91
BSA	0.003 ± 0.001	95.20 ± 2.18	0.18 ± 0.09	95.02 ± 2.11
P4 RNase	0.194 ± 0.002	57.88 ± 0.44	0.66 ± 0.06	57.22 ± 0.48
Lysozyme	1.995 ± 0.098	7.05 ± 0.70	3.33 ± 1.13	3.72 ± 0.50
BSA	0.076 ± 0.022	18.62 + 2.20	0.82 ± 0.35	17.80 ± 1.92

All means are the values obtained from three phase systems analyzed by the HPLC. The standard deviation values are shown next to the mean values.

the ammonium sulfate-PEG 1000 system where there is very little protein in the bottom phase at high tie-line length.

The partition and recovery of the same three proteins in phase systems using potassium phosphate (group P) are also shown in Table 2. Trends in protein partitioning are similar compared to the ammonium sulfate systems. There are, however, important individual protein differences, such as the partitioning of BSA in system A1 (partition coefficient of 0.12) compared to its behavior in system P1 (partition coefficient of 32). The phase system for the most efficient removal of proteins from the bottom phase is the high tie-line length potassium phosphate-PEG 1000 (system P2). This systems leaves less than 0.5% of any of the three proteins in the bottom phase.

The utility of the salt-PEG system in preparing DNA of high molecular mass is shown in Fig. 3 (lanes 3 to 8), this figure includes the result of an experiment using three salt-PEG 1000 phase systems to prepare DNA of high molecular mass from mouse liver. The compositions of the phase systems (D1, D2 and D3) are shown in Table 1. The final yield of high molecular mass DNA was estimated by comparison of the fluorescence intensity of the stained high molecular mass DNA band to that of known standards. Phase system D1 gave approximately 2 mg DNA per g of mouse liver (wet mass). Phase system D2 (potassium phosphate, low tie-line length) gave approximately 2.6 mg DNA per g of liver. Phase system D3 (potassium phosphate high tie-line length) gave approximately 1.5 mg DNA per g of liver. Variable amounts of RNA are purified along with the DNA. This is indicated by fluorescence with low molecular mass observed on the gels that is removed by treatment of the samples with DNase-free RNase (results not shown). The DNA preparation is judged to be free of protein by protein gel electrophoresis and absorbance readings. The optical density ratios at 260 nm and 280 nm were 1.9 to 2.0.

Phase system D1 is a sodium phosphate-PEG 1000 system containing 1% SDS. SDS is soluble in the sodium phosphate systems (at least up to 5%) but only sparingly soluble in the potassium phosphate systems (although other ionic detergents are soluble). The

solubility of sodium phosphate limits its utility in two-phase systems. The concentration of sodium phosphate in this system D1 is near its limit of solubility and if the phase system is allowed to cool a few degrees the salt will crystalize and there will no longer be two liquid phases. The potassium phosphate systems (D2 and D3) did not contain SDS due to the low solubility of SDS in these systems. These systems gave satisfactory yields, and they allow phase systems with high tie-line lengths to be used.

Discussion and Conclusions

Aqueous two-phase extraction offers many advantages such as speed, simplicity and economy for efficient DNA extractions. Very large or very small samples are easily accommodated. The PEG-salt systems are easily emulsified, resulting in rapid mass transfer. This is an important consideration in bioprocessing. The resulting rapid removal of nucleases or their inactivation is necessary to avoid degradation of the nucleic acid preparations. The same argument could be used in protein separations. In this case removal of proteases is important to avoid modification of the protein product.

Nucleic acids are polyanions and in the salt PEG two-phase system remain in the salt-rich phase under most conditions. To achieve a separation it would be most convenient to partition the proteins, cellular debris and other components to the top phase. The extraction conditions used must be balanced against denaturation of the DNA. Denatured DNA and single stranded nucleic acids are not soluble in high salt solutions. It has been shown that denatured DNA was not soluble in these salt-PEG systems (Cole, 1990).

The variables available to manipulate in the salt-PEG two phase system are: salt type, salt concentration, pH, temperature, PEG molecular weight, PEG concentration and additives. Temperature is best held constant at 25°C as a practical point to simplify possible future automation. The salt type is, of course, limited to those that form two phases with PEG, the sulfates and phosphates. Salt concentration is limited by the solubility and necessary concentration to form a two phase system.

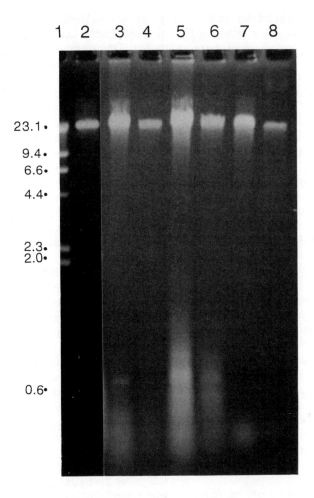

Fig. 3 Agarose gel of mouse liver DNA. The samples were electrophoresed in a 1% gel at 100 V for 4 hr. The gel was stained with ethidium bromide and photographed under ultraviolet illumination. Lane 1 contains the DNA fragments from lambda DNA cut by the restriction enzyme Hind III. The sizes of the fragments (in kilobases) are indicated on the side. Lane 2 contains intact lambda DNA (48.5 kilobase). The DNA prepared from the final bottom phase after extraction of mouse liver with system D1 is in lanes 3 and 4 (lane 4 has one-tenth the amount of DNA in lane 3).DNA prepared using phase system D2 is in lanes 5 and 6 (lane 6 has one-tenth the amount of DNA in lane 5). DNA prepared using phase system D3 is in lanes 7 and 8 (lane 8 has one-tenth the amount of DNA in lane 7).

The PEG used in this study had a molecular mass of either 1000 or 8000. Additives to the system would include detergents or chaotropic (water order disrupting) salts.

Detergents are commonly used in biotechnology to stabilize or solubilize membrane or other insoluble proteins. They are often used to disrupt protein structure and inhibit enzymatic activity. The ionic detergent SDS has proven to be very useful in fully denaturing enzymes that might modify nucleic acids during their purification. The chaotropic salts such as guanidine isothiocyanate have also proven useful for solubilizing proteins and disrupting protein structure to inhibit enzymatic activity. SDS and guanidine isothiocyanate act by different mechanisms to cause protein denaturation. SDS strongly interacts with proteins to block hydrophobic interactions and is believed to disrupt protein structure by fairly uniformly imparting a negative charge per mass. The chaotropic agents in high concentrations are thought to compete with hydrogen bonding in proteins and cause an unfolding of the protein so that the proteins assume a structure more like a random coil. Adding additional components to a two-phase system is complicated by the possibility that the additive may partition unequally between the two phases. The unequal partitioning of a detergent or chaotropic agent between the two phases may significantly alter the phase properties. However it is of interest to determine the effect of SDS or guanidine isothiocyanate in two phase systems because of their utility in purifying nucleic acids.

Svensson et al. (1985) studied the partitioning of the detergents Triton X-100 and octyl glucoside in ATPS composed of dextran and PEG. Triton X-100 separated strongly to the upper phase and octyl glucoside separated approximately equally to both phases. It has been reported that SDS separated approximately equally to both phases in dextran PEG two-phase systems (Albertsson, 1986). We are interested in determining the partitioning of detergents and chaotropic agents in these salt PEG systems.

The two different salt systems gave significantly different individual protein partitioning results. The PEG 1000 systems gave significantly higher partition coefficients than the PEG 8000 systems. Increasing the pH of the phase system results in higher protein partition coefficients and increased protein precipitation. Detergent and chaotropic salts that result in protein denaturation cause increased protein concentration in the top phase. A combination of high tie-line length, low molecular-mass PEG, high pH and addition of detergents or chaotropic salt is very efficient at removing proteins from the bottom phase through protein precipitation and high partitioning. This combination of conditions promotes protein denaturation and probably hydrophobic interactions between proteins and the PEG phase. To achieve a satisfactory yield of nucleic acids the extraction conditions must be balanced against effective removal of proteins and other constituents and denaturation of nucleic acids.

The recommended new DNA isolation procedure is illustrated diagrammatically in Fig. 4. It consists of a limited number of simple steps. The procedure should lend itself to automation. It is also rapid and flexible to meet the demands of particular separations. The DNA prepared is of high molecular mass and has not been subjected to organic solvents. In this method the liver homogenate is treated with proteinase K in the presence of SDS to inactivate nucleases before addition to the phase systems.

The DNA purification method described in this paper have significant advantages over existing purification methods. Proteins are rapidly denatured and removed from the DNA phase. This prevents modification of the nucleic acids by nucleases. The method is simple and requires no special equipment. It uses inexpensive chemicals and avoids exposure to toxic organic solvents. This process is also suitable for many samples or for large scale separations. We are in the process of further characterizing this method for rapid, gentle separation of nucleic acids and proteins from complex mixtures.

Purification of High Molecular Mass Eucaryotic DNA

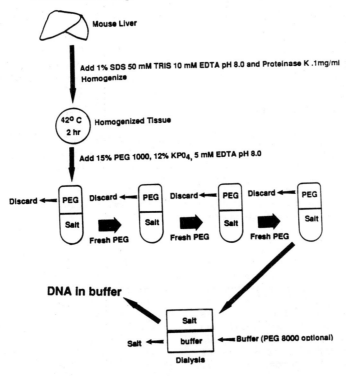

Fig. 4 Diagram of the PEG-Salt extraction of mouse liver DNA.

References

Alberts, B.M. *Biochem.* **1967**, *6*, 2527-2532.

Albertsson, P.-Å. *Biochim. Biophys. Acta* **1965**, *103*, 1-12.

Albertsson, P-Å. *Partition of Cell Particles and Macromolecules.* John Wiley & Sons, NY, 1986.

Blin, N.; Stafford, D.W. *Nuc.Acid Res.* **1976**, *3*, 2303-2308.

Cole, K.D. *Biotechniques* (in press) **1991**.

Favre, J.; Pettijohn, D.E. *Eur. J. Biochem.* **1967**, *3*, 33-41.

Laemmli, U.K. *Nature* **1970**, *227*, 680-685.

Mak, S.; Öberg, B.; Johansson, K.; Philipson, L. *Biochem.* **1976**, *15*, 5754-5761.

Müller, W.; Schuetz, H.J.; Guerrier-Takada, C.G.; Cole, P.E.; Potts, R. *Nuc. Acid Res.* **1979**, *7*, 2483-2499.

Müller, W. In *Partitioning in Aqueous Two-Phase Systems*; Walter, H.;Brooks, D.E.; Fisher,D., Eds. Acad. Press, Orlando, FL 1985; pp 227-266.

Neuhoff, V.; Arnold, N.; Taube, D.; Ehrhardt, W. *Electro.* **1988**, *9*, 255-262.

Ohlsson, R.; Hentschel, C.C.; Williams, J.G. *Nuc. Acids Res.* **1978**, *5*, 583-590.

Pettijohn, D.E. *Eur. J. Biochem.* **1967**, *3*, 25-32.

Pritchard, D.G.; Halpern, R.M.; Smith, R.A. *Biochim. Biophys. Acta* **1971**, *228*, 127-134.

Svensson, P.; Schroder, W.; Akerlund, H.-E.; Albertson, P.-Å. *J. Chrom.* **1985**, *323*, 363-372.

Rudin, L.; Albertson, P.-Å. *Biochim. Biophys. Acta* **1967**, *134*, 37-44.

Veide, A.; Smeds, A.-L.; Enfors, S.-O. *Biotech. Bioeng.* **1983**, *25*, 1789-1800.

A Polyethylene glycol-Sodium Chloride Multiphase System for Extraction of Acid Hydrolysates

Robin M. Stewart and Paul Todd, National Institute of Standards and Technology, Boulder, Colorado 80303

Polyethylene glycol (PEG) and sodium chloride (NaCl) form a multiphasic system at temperatures above 40°C. This system was used for the extraction of peptide flavorings from yeast acid hydrolysates. The phase diagram of the PEG-NaCl system was characterized at 60°C and was found to have a closed envelope containing a liquid/liquid region and a liquid/liquid/solid region. In the extraction of flavorings from yeast acid hydrolysates the cell wall debris partitioned exclusively into the bottom, salt-rich phase, while a significant fraction of the amino acids partitioned into the upper, PEG-rich phase.

The use of aqueous two-phase polymer-polymer or polymer-salt systems in bioprocessing research has been well documented (Albertsson, 1986; Walter, 1985). The major uses have been in protein, organelle and cell separation at the bench scale. The main advantages of the aqueous systems are their biocompatibility, ease of scale-up, and high purification yields (Albertsson, 1986; Walter, 1985; Diamond, 1989). Polymer-polymer systems such as polyethylene glycol (PEG)-dextran have been extensively studied (Albertsson, 1986; Walter, 1985; Raghava Rao, 1990) and while their properties are well understood, they have certain disadvantages such as high cost and slow separation rates. PEG-salt systems have better properties from a process point of view, such as low cost and fast separation rates. To date all PEG-salt systems used in bioseparations have contained multivalent anions such as phosphate and sulfate, under the assumption that monovalent anions, such as chloride would not form two phases with PEG.

It was discovered that at supersaturated salt conditions two liquid phases could be formed with PEG and NaCl. This system should be suitable for extraction processes in the food flavorings industry, where the harsh effects of the high salt concentrations and high temperature would have no effect on an acid hydrolyzed material, such as yeast. The PEG-NaCl system also takes advantage of the existing high salt concentration in the neutralized hydrolysate to form the phases. PEG is an approved food additive (Sax, 1975), so its presence in the product does not pose any health problems for human consumption.

Acid hydrolyzed yeast is a source of food flavorings and enhancers. Whole yeast, such as spent brewers yeast, is treated with approximately 6M HCl at elevated temperatures for a specific period of time then neutralized using NaOH. This treatment breaks down nucleic acids and proteins into nucleotides, amino acids and dipeptides, respectively. This process also results in a product that contains up to 38% NaCl in the final spray-dried product. While some salt is preferred for flavor, the excess salt and the portions of the yeast which do not contribute to flavor, such as the cell wall, can be removed to provide a low salt, concentrated flavoring product.

A study of the applicability of the multiphase PEG-NaCl system in flavorings extraction was therefore undertaken. Specifically, a phase diagram at 60°C was characterized, and the partitioning of yeast hydrolysate constituents was determined qualitatively.

Materials and Methods

Phase System. The phase systems were formed by weighing PEG (MW=8000), NaCl and water into 50 ml polypropylene tubes and heating to 60°C. The system was kept in a 60°C water bath for 24 h, with intermittent shaking of individual tubes to dissolve the solids. The phases were allowed to separate in the polypropylene tubes at 60°C for approximately 4 h, then removed separately by pipetting.

Yeast Hydrolysate. Commercial bakers yeast was hydrolyzed by the addition of 6M HCl, heating to 45°C for 22 h, then increasing the temperature to 60°C for 3 h. The hydrolysate was then neutralized using 8M NaOH solution. The final concentrations were (w/w): 40% yeast cake, 30% salt, and 30% water (not taking into account the water in the yeast cake). When making the phase systems, the amounts of water, dry salt, and PEG were adjusted according to the amount of hydrolysate in the final phase system.

Phase Composition Determination. The phase compositions were determined gravimetrically using lyophilization (freeze-drying) and ashing of aliquot portions of the separated phases. Portions of each phase were weighed into tared Pyrex test tubes. This initial sample mass for the top and bottom phases was m_t and m_b, respectively. These samples were lyophilized and reweighed after all water had been removed. From this the percent water in each phase, $[H_2O]_t$ and $[H_2O]_b$, was calculated directly.

$$[H_2O]_t = 100 - [(mass\ solids)_t/m_t]$$
$$[H_2O]_b = 100 - [(mass\ solids)_b/m_b]$$

Ashing was then performed on the lyophilized samples, utilizing the fact that PEG vaporizes at high temperature. The sample was heated for several days until the residual solids were white or ash gray and no change in mass occurred with subsequent heating. This residue was the mass of salt in the sample. The concentration of salt in top and bottom phases, $[NaCl]_t$ and $[NaCl]_b$, was calculated directly from the mass of residue.

$$[NaCl]_t = (mass\ residue)_t/m_t$$
$$[NaCl]_b = (mass\ residue)_b/m_b$$

The concentrations of PEG, $[PEG]_t$ and $[PEG]_b$, were calculated from the differences between the masses of lyophilized solids and ashed residue.

$$[PEG]_t = [(mass\ solids)_t - (mass\ residue)_t]/m_t$$
$$[PEG]_b = [(mass\ solids)_b - (mass\ residue)_b)/m_b$$

When a solid phase was present in the phase system, it was assumed to be pure NaCl with no precipitated PEG. This assumption was based on the fact that PEG 8000 melts at 58-62°C (Sax, 1975). Variability in melting point depends on chain length, which is slightly different from lot to lot. The observed melting point for the PEG used in these experiments was slightly below 60°C.

Thin Layer Chromatography. Qualitative analysis of amino acid partitioning between the aqueous phases was performed by thin layer chromatography (TLC). Silica gel G plates, 20 X 20 cm were used as the stationary phase. The mobile phase was n-propanol/water (70:30) in a glass developing chamber. A 5 microliter sample of each phase was spotted 2.5 cm from one end of the plate and allowed to migrate until the solvent front had moved 10 cm. The resulting chromatogram was developed by spraying with a ninhydrin-propanol solution and subsequently heating with a heat gun until all spots were strongly colored. Glutamic acid was used as a reference material. The spots were not analyzed quantitatively.

Results and Discussion

The phase diagram for the PEG-NaCl aqueous phase system was characterized at 60°C as shown in Fig. 1. This figure is a synopsis of all system compositions investigated; each different plotting symbol represents total system composition and falls within a different region of the phase diagram. The boundary of the envelope containing the liquid/liquid and liquid/liquid/solid phases was determined by observation of numerous phase systems having varying total

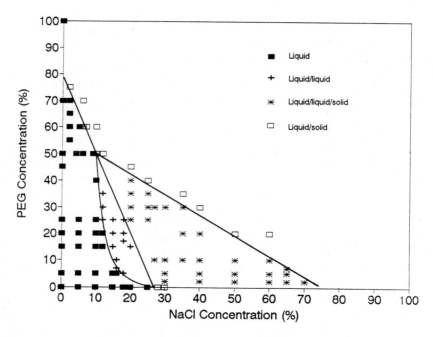

Fig. 1
Phase diagram of the PEG-NaCl system at 60°C showing all compositions investigated. Each different plotting symbol falls within a different region of the phase diagram.

compositions and recording, where appropriate, the compositions of the phases.

Fig. 2 shows the phase boundaries and tie lines for the same system. The symbols correspond to phases whose compositions were determined by gravimetric analysis after phase separation. The phase compositions in the liquid/liquid region had an average deviation of 2.00% PEG and 0.64% NaCl for the upper phase and 0.63% PEG and 1.01% NaCl for the lower phase. The average deviation for the liquid/liquid/solid region was 1.69% PEG and 1.01% NaCl for the upper phase and 0.01% PEG and 0.11% NaCl for the lower phase. The plait point for this system was determined by using the law of rectilinear diameters, and is shown in the figure as the solid point at approximately 16% PEG and 13% NaCl.

The composition of the top and bottom phases within the liquid/liquid region show that the lower, salt-rich phases contained only very small amounts of PEG. Phase compositions and phase volume ratios within this region were shown to be dependent upon starting composition, as in other polymer-salt systems (Albertsson, 1986). In contrast, phase compositions within the liquid/liquid/solid region were independent of starting composition. The upper phase consisted of approximately 50% PEG and 10% NaCl, while the lower phase consisted of approximately 0.1% PEG and 26.5% NaCl. The solid salt phase volume did vary considerably with starting composition, as did the liquid phase volumes.

The addition of 20 or 30% yeast hydrolysate (by weight) to phase systems consisting of (total, adjusted for hydrolysate) 25% PEG and either 15 or 25% NaCl at 60°C (Fig. 3) shows that the yeast cell wall debris partitions into the lower, salt-rich phase as predicted (Hustedt, 1985). Qualitative tests on these phases using thin layer chromatography indicate that amino acids partition approximately equally between the phases under these conditions.

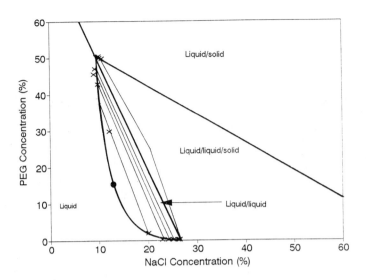

Fig. 2
Phase diagram of the PEG-NaCl system. Tie lines showing top and bottom phase compositions in liquid/liquid and liquid/liquid/solid regions. The solid circle represents the plait point.

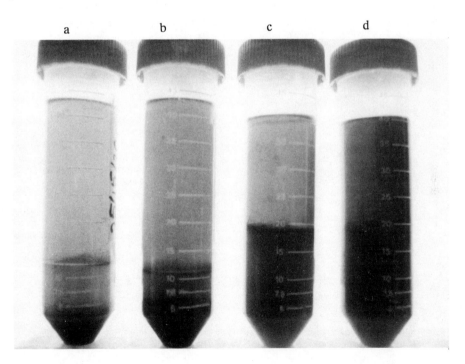

Fig. 3
Photograph of a phase system at 60°C, composition from left to right:
(a) 25% PEG, 15% NaCl, 20% hydrolysate
(b) 25% PEG, 15% NaCl, 30% hydrolysate
(c) 25% PEG, 25% NaCl, 20% hydrolysate
(d) 25% PEG, 25% NaCl, 30% hydrolysate
Phase systems in (a) and (b) are within the liquid/liquid region of the phase diagram while (c) and (d) are within the liquid/liquid/solid region. Solid NaCl can be seen at the bottom of the tubes (c) and (d). Cell wall debris partitions to the bottom, salt-rich phase.

355

Conclusions

The PEG-NaCl multiphase system forms two liquid phases under supersaturated salt conditions at temperatures above 40°C. A three phase system containing a solid phase develops at high salt concentrations due to the solubility limit of NaCl.

The cell wall debris from yeast acid hydrolysate partitions into the lower, salt-rich phase. Qualitative tests using thin layer chromatography indicate that the amino acids partition approximately equally between the phases at 60°C.

References

Albertsson, P.-Å. Partition of Cell Particles and Macromolecules; 3rd Ed.,John Wiley and Sons: New York, NY, 1986.

Diamond, A. D.; Hsu, J. T. Biotech. Bioeng. 1989, 34, 1000-1014.

Hustedt, H.; Kroner, K. H.; Kula, M.-R. In Partitioning in Aqueous Two-Phase Systems; Walter, H.; Brooks, D.; Fisher, D., Eds.; Academic Press: Orlando, FL, 1985, pp. 529-587.

Raghava Rao, K. S. M. S.; Stewart, R. M.; Todd, P. Sep. Sci. Technol. 1990, 25, 985-996.

Sax, N. I. Dangerous Properties of Industrial Materials, 4th Ed.; Van Nostrand Reinhold Co.: New York, NY, 1975.

Walter, H.; Brooks, D.; Fisher, D., Eds.; Partitioning in Aqueous Two-Phase Systems; Academic Press: Orlando, FL, 1985.

Part VI
Emerging Technologies
in Bioseparations

IF PRECIPITATION, FILTRATION, AND CHROMATOGRAPHY are considered to be conventional technologies, then emerging technologies include combined methods, biphasic aqueous extraction, affinity filtration, and electrokinetic techniques. Fundamental research on process integration and aqueous phase thermodynamics and two recently commercialized separation technologies are presented.

An Expert System for Selection and Synthesis of Protein Purification Processes

J. A. Asenjo and F. Maugeri, Biochemical Engineering Laboratory, University of Reading, Reading RG6 2AP, England

This chapter discusses recent developments in rational process design and their potential application in present and future biotechnology for large scale protein separation and purification processes. It includes the use of expert systems that, in addition to 'expert' empirical knowledge, have access to data bases and rigorous mathematical knowledge. The development of a prototype expert system and the interfacing and use of a database of properties of main protein contaminants is presented. An expert system for selection of optimal protein separation sequences (process synthesis) will give the user a number of alternatives that can be chosen on the basis of extensive data on proteins and unit operations. Such a system constitutes a clear case of 'expert amplification' and not of simple 'expert replacement'.

A protein purification process is usually composed of a sequence of separation and purification operations whose final aim is to obtain the required product at a prespecified level of purity. On a large scale it is necessary to obtain the highest possible yield while minimizing the resources utilized and hence the cost. For the manufacture of chemical and biochemical products one can consider two extreme cases: 1. The product is of high value in a virtually competition free market, or 2. The product is a high-volume chemical with many producers in a highly competitive market. In the first case one would probably choose the first successful purification procedure found in order to enter the market ahead of possible competitors (Prokopakis and Asenjo, 1990).

Modern biotechnology processes are still virtually all very close to case 1, but as products become more competitive and therefore their use more widespread (e.g. hepatitis B vaccine) they will move closer to the second case. As this happens the importance of rational design tools in biotechnology will grow and its economic importance will evidently increase. In order to achieve this we can recognize two different tasks: 1. Selection of Separation Sequence: giving maximum yield and using minimum resources (number of steps and cost of each step); and, 2. Design of Individual Unit Operations: using design equations and correlations as well as properties of operations, proteins and contaminants. It is important to bear in mind that a process designed for small quantities is usually not optimal for large scale.

Defining Final Product

At this stage it is necessary to define the final product and have information on its uses. How the product is going to be used (if therapeutic, what size doses, how many, how often?) Questions regarding the purity required are vital (e.g. 99%, 99.9% or 99.98%) as well as allowable ranges of impurity concentrations. With therapeutic proteins any impurities have to be minimized whereas for the production of bulk industrial enzymes this is not the case. For instance in vaccine production it is neccesary to remove all traces of unwanted immunogens to prevent potentially catastrophic immunological side reactions. Experience with foetal calf serum contamination of therapeutic proteins has given rise to a concensus that levels of contaminating protein of around 100 p.p.m. should be acceptable for product safety (Cartwright, 1987). The major perceived hazard appears to be the persistence of nucleic acids with oncogenic potential. State of the art assays (Cartwright, 1987) will allow detection of around 10 pg of nucleic acids, and it is generally agreed that final product contamination should not exceed 10 pg per dose. Pyrogens that mainly originate from bacterial cell walls will have side reactions even if present in very small amounts. For these in vivo testing using established methods is necessary, hence on-line control of such processes is very difficult to achieve.

Characterization of Starting Material

Large scale process design will be mainly dependent on the physical, chemical, and biochemical properties of the contaminating materials in the original broth and those of the protein that will constitute the final product. The properties of the starting material will be partially determined by its fermentation source viz: bacterial, yeast or mammalian cell, the type of cultivation medium used (e.g. presence of albumin, calf serum, proteases, solid bodies like whole cells or cell debris) and whether the product is intracellular or extracellular. To these we must add the actual physicochemical properties of the product (surface charge/titration curve, surface hydrophobicity, M.W. and shape (Stokes radius), biospecificity towards certain ligands (e.g. dyes), pI, stability) as compared to those of the contaminant components in the crude broth. The stability of the final product is also of utmost importance as this will affect the types of operations that can be used as well as the conditions and processing times that can be afforded.

SELECTION OF SEPARATION SEQUENCE

The next stage is to define realistic separation steps on the basis of all the information provided. The following five main heuristics or rules

of thumb provide a good basis for process selection:

Rule 1 : 'Choose separation processes based on different physical, chemical or biochemical properties'.

Rule 2 : 'Separate the most plentiful impurities first'.

Rule 3 : 'Choose those processes that will exploit the differences in the physicochemical properties of the product and impurities in the most efficient manner'.

Rule 4 : 'Use a high resolution step as soon as possible'.

Rule 5 : 'Do the most arduous step last'.

Although it may appear difficult to define the most arduous step a clear example is the use of gel filtration for protein separation which is a rather inefficient step from the point of view of resolution (poor), low flow rate and compressibility of the matrix but needs to be used when molecular weight is the only property difference between the product protein and the contaminant (e.g. dimer of the protein product).

Purification Process and Unit Operations

One must always consider that once the purification procedure is established and regulatory approval of the product is underway the purification procedure can only be changed at great expense. Only a particular product obtained by a specific procedure obtains regulatory approval. Hence, even in the very early stages of protein purification one should only use in the laboratory procedures that can be realisticaly used in large scale, ie. for which suitable large scale equipment either exists or might be developed in the foreseeable future. Otherwise more rationalized and efficient processes will have to wait for the second or third generation of process and plant design which can be a very wasteful exercise.

The number of necessary separation and purification steps in a large scale protein purification procedure should not be more than four or five and they can be divided into two main subprocesses of protein recovery (Figure 1) and protein purification (Figure 2). Most processes in use today

INTRACELLULAR PRODUCT

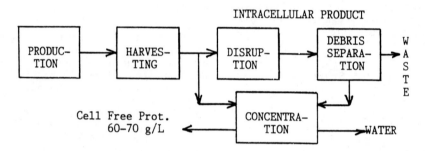

Figure 1. The sub operations involved in the RECOVERY block of a
separation process.

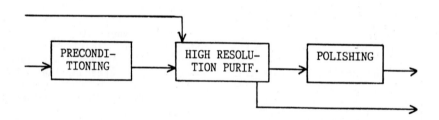

Figure 2. The sub operations involved in the PURIFICATION block of a
separation process.

have far more steps than are optimally required. In many cases the rationale for the inclusion of a particular step is not clear. Most processes have been developed using the philosophy: 'If the product is not pure enough then add another step', and the selection of steps has rarely been decided taking into account the molecular properties of the main contaminants, particularly in the early stages of the purification.

Obviously there will be interactions between the fermentation system and conditions and protein recovery, hence production methods used will affect the later purification steps. Such interactions have been discussed in some detail (e.g. Wang, 1983; Fish and Lilly, 1984; Cartwright, 1987). It is important then to consider the process of fermentation and downstream processing as a single system so that, for example, the effect of decisions about the fermentation conditions on subsequent purification stages are made clear. Product concentration will partly depend on the reactor system used (stirred tank or air- lift or hollow fibre). The presence of nucleic acids and proteases as well as bacterial contamination have to be minimised which results in a need for rapid processing. The presence of calf serum will usually increase the number of purification stages required. Recombinant proteins are in many cases present in particles that need to be solubilized and refolded. In conclusion, not only is it important to discuss the upstream processing in the light of all the protein purification stages but also to make the necessary decisions which will improve the recovery of the protein product early on in the process development stages.

Scale-up and Novel Large Scale Operations

In virtually all chromatographic procedures, particularly in those based on adsorption type interactions, scale up is achieved by increasing the radius of the column while maintaining the column height. The effect of process parameters on resolution and throughput in chromatographic procedures have been reviewed (Asenjo and Patrick, 1990). The resolution of proteins is mainly determined by the elution strategy (desorption) and not by the length of the column. Relatively short columns are favoured in the range of 15 - 30 cm in height. The largest available column for protein purification appears to be a 1700 to 2500 liter one, 2 m in diameter with an adjustable bed height between 55 and 80 cm (Amicon). In such systems the appropriate design of flow distributors is a crucial concern in obtaining reproducible scale-up. Radial flow chromatography, a radically new column design which has been developed by Sepragen, constitutes a more rational approach to the scale up of chromatography (Asenjo and Patrick, 1990). The mobile phase flows radially and hence scaling up is achieved by increasing column length. Flow rates seem high and hence processing times shorter than with traditional column design; however, scale-up is very hardware intensive (approximately linear increase of equipment with size).

When using gel filtration for protein fractionation the main problem is that long columns are necessary to obtain relatively good resolution. This dictates the quality of resolution as opposed to the elution strategy as in other chromatography operations based on adsorption. This leads to problems in large scale gel filtration since the porous matrices required are mechanically weak and compress easily. The way this has been partially solved is by the use of stack systems. For example, to obtain a 90 cm bed six 15 cm beds have been linked in series.

The development of protein separation and purification techniques for efficient use on a large scale (eg. separation and renaturation of proteins from inclusion bodies or separation of VLPs, virus like particles cloned in yeast) is in its infancy. Most operations in use today are just large versions of laboratory procedures and have not been developed specifically for large scale use. In the next 5 to 15 years we should see important new developments in the conception of procedures and apparatus specifically developed for large scale use. For instance an alternative to chromatographic affinity adsorption that can be used continuously at large scale is the use of two stages of adsorption and desorption in which the affinity support phase is recycled between the adsorption and desorption stages. The two phases can be either liquid or solid (eg. a liquid fluorocarbon (Eveleigh, 1987), or solid agarose (Gordon et al., 1990)). Such a continuous system can also be used for ion-exchange adsorption. Another important development for large scale processing is aqueous two-phase partitioning, in which it is now becoming possible to manipulate systems to separate proteins on the basis of hydrophobicity, charge, affinity and molecular mass, as is done in chromatography (Albertson et al., 1990; Asenjo et al., 1991a). Other examples of emerging technologies are reverse micelles, differential product release (Andrews et al., 1991), and fluidized bed adsorption (Draeger and Chase, 1990).

PROCESS DESIGN

Procedure

Process design and selection of operations is a complex procedure in which a design evolves from preliminary to final stage in a trial-and-error fashion by repeatedly revising and refining the initial assumptions and restrictions: 1. the flowsheet generation (qualitative/ semiquantitative), 2. quantitative design of units, 3. revisions of these (1. and 2.) until some objective is reached. An important aspect of process design involves the selection of operations and design of plant equipment. In the initial stages this process is more or less done using heuristics: using rules of thumb to arrive at a rapid (and reliable) specification of equipment type, size and possibly cost.

The design of a protein recovery and purification process shares many characteristics with other engineering design activities. To design a process or an operation requires the satisfaction of a number of constraints (purity, quality, process temperature, desired yield) using what is known about the materials (chemical and biochemical properties, thermodynamics and fluid dynamics of the process material) to end with a sequence of equipment interconnected in a particular order. It is important to stress that the type of reasoning behind the design process does not rely only upon strict mathematical

models. Equations could only provide the information necessary to conclude that a particular piece of equipment is appropriate, but the inclusion or not of the step is left for the designer in a job which is based mainly on judgement (Asenjo et al., 1989).

The optimization of a process design is carried out after one or more purification strategies have been chosen and performed and thus process conditions are known. The interest at this point is to find the optimal operating conditions for specific separation operations so that their performance can be compared with alternative operations. It would also be desirable to compare alternative chromatographic sequences giving a similar final product in terms of overall economics (Duffy et al., 1988; Kosti, 1989).

Quantitative information on the performance of individual operations including design correlations and data bases of properties of materials is vital. For instance, in the case of mechanical cell disruption the design information available has been described (Schutte et al., 1983). It includes flow rate, type of agitator or operating pressure, cell concentration and type, fraction of product released and size of the disrupter. For a chromatography operation the information necessary has been described in several sources on the theoretical and design analysis of adsorption type chromatography (Yamamoto et al., 1987; 1988; Wang, 1990); it concerns characteristics of columns (size, geometry) and properties of gels and other adsorbents (binding capacity, dissociation constants, flow rate, compressibility, half life and breakthrough curves).

Modelling and Simulation of Unit Operations

Mathematical models and mathematical correlations of the operations will allow simulation of performance and also may be used to scale up individual operations. Computer simulations are a useful tool with which to optimize individual separations (Hedman et al., 1989). Examples of useful downstream process simulations and investigation of process conditions are microbial cell breakage with

362

selective product release using enzymes (Hunter and Asenjo, 1988; Liu et al., 1988) and investigation of the affinity and ion–exchange chromatography of proteins (Chase, 1988; Arve, 1989). The first is an example of cell breakage and intracellular product release, and the second is an example of protein separation by chromatography.

Recently substantial attempts to model enzymatic cell lysis and product release have been carried out. Three models of yeast lysis developed by Hunter and Asenjo serve different purposes (Hunter and Asenjo, 1986; 1988; 1990; Liu et al., 1988). The simple model is a lumped, two–step model which follows the major features of the data and may prove useful for design of lysis reactors (Hunter and Asenjo, 1986; 1987). The structured model, which accounts for the source of protein within a cell, was developed to gain a mechanistic basis for predicting the effects of untested process conditions and to give insight into the physical processes at work during lysis and product release (Hunter and Asenjo, 1988). A schematic diagram of the

reactions of this model is shown in Figure 3. The structured mechanistic model of the kinetics uses a number of differential equations, one for each cellular component. This model was used for process simulations where the release of proteins from different cellular locations can be predicted, analysed and improved (Hunter and Asenjo, 1988). The concept of DPR (differential product release) was originally coined as a result of the development of this model. Process simulations were performed to predict the release of proteins from the wall, cytosol and intracellular particles such as organelles or recombinant protein inclusions.

A simulation of site–linked product recovery is presented in Figure 4 and Table I. In the first lysis step, using lytic enzyme and osmotic support, 93% of the wall protein was released from the cell wall. Some protein is hydrolyzed by the presence of some "destructive protease", but 74% of it survived to be recovered at the end of the first hour. Since only 3% of the protoplasts burst during this step,

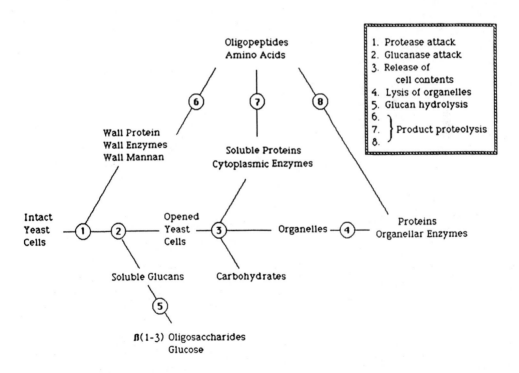

Figure 3. Mechanism of enzymatic cell lysis in structured model
(from Hunter and Asenjo, 1986)

363

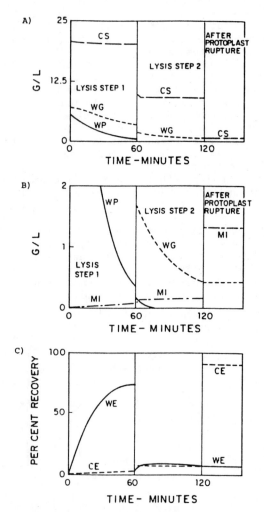

Figure 4. Enzyme recovery and breakdown products from subcellular structures
a. Cell structure breakdown products
——————— wall protein (WP)   – – – – wall glucan (WG)
—— —— cytosol (CS)   —— •—— mitochondria (MI)
b. Cell structure; expanded ordinate showing mitochondrial products released
c. Enzyme release, per cent of original enzyme in cell
——————— wall enzyme (WE)   – – – – cytoplasmic enzyme (CE)
(from Hunter and Asenjo, 1986)

Table 1. Process Conditions for Enzyme Release Simulation

First lysis step

Yeast	36.33g
Enzyme	40%
Buffer	0.3 Os/L
Total Volume	1L

Reaction mixture centrifuged; 5% of supernatant and 100% of pellet retained and resuspended in twice the initial volume of enzyme/buffer solution.

Second lysis step

Digested yeast	26.90g
Enzyme	20%
Buffer	0.3 Os/L
Total volume	2L

Breakage: by stirring or passage through a pump

Protoplast rupture	95%
Mitochondrial rupture	0%

from: Hunter and Asenjo (1986)

little cytoplasm was released. The second digestion was included to decrease the amount of structural glucan from about 50% to about 13% of its original mass. This also made the cells more fragile for rupture in the third stage. Only a small amount of cytoplasmic protein is released from the protoplasts during the second stage, a desirable result since protein located inside the protoplasts is not attacked by "destructive protease". At the end of the second hour the protoplasts are easily broken by stirring or centrifugation. Almost all of the mitochondria (96%) are released during protoplast rupture, but they remain whole since the buffer osmolarity is kept above 0.3 Os/L.

This performance was analysed, and means of improving and maximizing the recovery of wall, cytosol and mitochondrial proteins in three different stages were investigated. This analysis resulted in the conditions recently shown by Andrews et al. (1991) where the results previously obtained by computer simulations using the structured model (Figure 4) could be analysed and improved in support of the concept of DPR, differential product release. This example is an important application of the appropriate use of mathematical modelling for process simulation, process investigation and process development.

More recently a population balance model that takes into account the heterogeneity of microbial cell populations has been described (Hunter and Asenjo, 1990).

pH and temperature optima for the glucanase (that degrades the inside wall, Figure 3) and protease (that degrades the outside wall, Figure 3) can be very different (Liu et al., 1988). Hence a pH shift can be used to produce a high protease activity in the initial stages of reaction and high glucanase and low protease activities later to minimize protein product degradation (Figures 5 and 6). The use of protease inhibitors after the initial stages of reaction has also been considered (Liu et al., 1988).

Process Synthesis

The design of a large-scale process for purifying a protein to a high yield yet obtaining a high-purity product while also minimizing the cost requires three main considerations: (i) clearly defining the final product objective, (ii) characterizing the

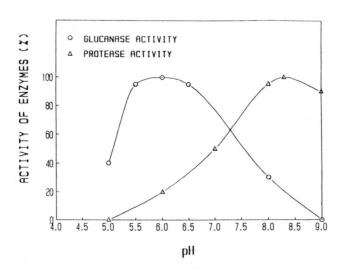

Figure 5. The activities of protease (Δ) and glucanase (O) of the <u>Oerskovia</u> lytic system. The solid lines are approximating polynomial functions (from Liu et al, 1988).

366

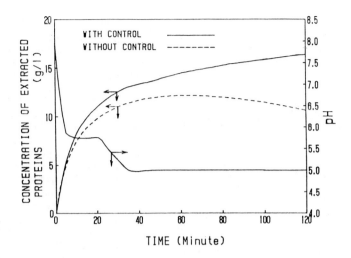

Figure 6. Concentration profiles of extracted proteins with and without pH control (from Liu et al, 1988).

starting material (product and impurities) in terms of its physico-chemical properties, and with these two pieces of information (iii) defining a minimum number of efficient separation steps by using some established heuristics such as those used in chemical process synthesis.

The problems that have to be solved in process synthesis and optimization of downstream protein separations, are of two types: (i) choosing between alternative operations (e.g. homogenizer vs. bead mill or centrifugation vs. cross flow microfiltration) and (ii) the design of an optimal chromatographic sequence with maximum yield and minimum number of steps, a problem that is combinatorial in nature. The first type of problem can be adequately solved provided that appropriate mathematical correlations and mathematical models that can be used as useful simulation tools are developed. The second type of problem has been partially tackled in classical chemical engineering by finding a rigorous solution using numerical methods such as mathematical programming techniques (eg. resolution of 'tree structure' (Prokopakis and Asenjo, 1990)) or more recently by using an Expert Systems (ES) approach. For the design of an optimal sequence the first solution has limited use in biotechnology due to a lack of useful design equations and data bases. The

second approach appears more attractive since it allows the use of empirical knowledge which is not rigorous in nature and is typical of that used by experts. Computer based expert systems are an important tool in the field of Artificial Intelligence (AI). Efforts have been made to develop AI systems for this purpose (Siletti and Stephanopoulos,1986; Wacks, 1987; Siletti, 1989) or to adapt existing systems (called 'shells') (Asenjo et al., 1989) both for the manipulation of heuristics (rules of thumb), databases and simple algebraic design equations.

DEVELOPMENT OF AN EXPERT SYSTEM

Prototype

Today there are well developed expert software systems or "shells" (e.g., Personal Consultant Plus (PC Plus, Texas Instruments) or Expert Systems Environment (ESE, IBM)) that help develop an organised knowledge base from the domain knowledge and also provide the inference engine. Asenjo et al. (1989) have found that for prototype systems shells can be adequate, particularly if they are capable of evaluating uncertainties associated with the inference process. They implemented a second generation of protein purification rules in two shells, ESE and PC Plus. Expert knowledge was obtained partially from the literature but mainly from industrial experts working on the large scale separation and purification of therapeutic, diagnostic and analytical proteins. Soon it became apparent that the true bottleneck in the development of expert systems for protein purification is not in its implementation but in the acquisition, clarification, formalization and structuring of the domain knowledge.

The knowledge was expressed in about 65 rules, some of which carry a degree of uncertainty (Asenjo et al., 1989). The downstream process was divided into two distinct subprocesses in order to structure the knowledge (Figure 1 and Figure 2). Processing of recombinant proteins present in intracellular inclusion bodies or other particles was not considered in this prototype but has now been included in an expanded version (next section and Asenjo et al., 1991b).

The Recovery subprocess comprises harvesting, cell disruption, separation of solid debris and precipitation of nucleic acids (these last three steps only if disruption is required) and concentration (also called dewatering). This subprocess is characterised by the objective of recovering the product from the production system. The Purification subprocess takes the 60 – 70 g/1 protein solution and purifies the individual protein product to a high purity with a high yield. It comprises preconditioning

or cleaning (to obtain a 'sparkling clear' solution), high resolution purification, which can be carried out in one or two steps, and polishing, when necessary (usually to remove traces of minor contaminants such as for therapeutical applications).

The various parameters used to characterize the broth and the culturing system are:

- MICROBIAL SOURCE (Bacteria, Yeast, Fungus, Mammal)
- PRODUCT OF INTEREST: Name
- Cellular Location (Intracellular, Extracellular, Unknown)
- Titration Curve (charge as a function of pH), Isoelectric point
- Dipole moment
- Surface Hydrophobicity
- Biospecificity data base
- Molecular Weight/shape (Stokes radius)
- Two Phase Aqueous Systems data base

The proposed process consists of a sequence of operations to obtain the stated design objective. There might be several different sequences of operations that will accomplish the same objective. In those cases, a quantitative degree of performance (given by the 'certainty factor') of each operation is assigned by the expert and carried by the system into the proposed design. A few of the rules used in this expert system are the following (Asenjo et al., 1989):

e.g. The rules that select the initial harvesting equipment (H-EQUIPMENT parameter) are:

		Certainty factor
IF	MICROORGANISM = FUNGI	
THEN	H-EQUIPMENT = Microporous Membrane System	0.4
	H-EQUIPMENT = Rotary Vacuum Filter	0.3
	H-EQUIPMENT = Filter Press	0.3
IF	MICROORGANISM = YEAST	
THEN	H-EQUIPMENT = Disc Centrifuge	0.6
	H-EQUIPMENT = Microporous Membrane System	0.4
IF	MICROORGANISM = BACTERIA	
THEN	H-EQUIPMENT = Microporous Membrane System	1.0
IF	MICROORGANISM = MAMMALIAN	
THEN	H-EQUIPMENT = Disc Centrifuge	0.7
	H-EQUIPMENT = Microporous Membrane System	0.3
IF	MICROORGANISM = UNDEFINED	
THEN	H-EQUIPMENT = Disc Centrifuge	0.5
	H-EQUIPMENT = Microporous Membrane System	0.5

We shall now describe a few rules from the Purification subprocess shown in Figure 2. If the solution is already 'sparkling clear' the first step could directly be a high resolution purification. Preconditioning is a stage which is necessary if the solution is not 'sparkling clear' due to fine suspended material or other contaminants which may produce fouling of the high resolution purification columns. In a few cases the high resolution purification can be achieved in one step; more commonly two or more steps are required.

High resolution purification is usually carried out by chromatography. Different techniques of chromatography exploit different physico-chemical properties, and some are much more efficient than others in exploiting each of these properties. The quantitative evaluation of 'efficiency' is closely related to the numerical evaluation of resolution in chromatography (e.g. Janson and Ryden, 1989). However, for the purposes of this paper it has been treated empirically in a rather generic form (see next section) as evaluated by an expert. Ion-exchange chromatography will separate the proteins based on their difference in charge. The charge on a protein changes with pH following the titration curve. Hence, if carried out at significantly different pH's at which the difference in charge of three or more proteins is very different, this technique can be used twice to purify a protein from different protein contaminants. Ion-exchange can use small differences in charge to give a very high-resolution separation and hence it is an extremely efficient operation for separating proteins.

Affinity chromatography can have a very high specificity for a particular protein or a small group of proteins; hence it can also separate at very high resolution. The matrix can be expensive but it can be reused for long periods. Ligand leakage into the product can be a problem. The expert system uses information supplied by the user about purification of the protein (usually generated at a laboratory scale) or a database containing known biospecific ligands and other physicochemical information. Regarding cost, affinity chromatography will usually be more expensive than ion exchange (Duffy et al., 1988; Kosti, 1989). Hydrophobic Interaction Chromatography (HIC) has been proposed as a pretreatment step or as a first high resolution purification (HRP) step. The resolution can be high, but as the distribution of surface hydrophobicity in a protein can be very 'localized' this will lower resolution. Gel filtration for protein fractionation is normally not used as a high resolution operation in the large scale due to the low efficiency in exploiting differences in molecular weight.

This prototype expert system (65 rules) did not have a data-base on physico chemical properties of main protein contaminants so the selection of high resolution purification operations was rather empirical and based on knowledge of the efficiency of the different techniques of separating a protein from its main contaminants. Efficiency was classified as high, medium or low. The following set of rules is for the first high-resolution purification step (HRP-EQUIPMENT STAGE 1):

		Certainty factor
IF	ION EXCHANGE EFFICIENCY = HIGH	
THEN	HRP-EQUIPMENT STAGE 1 = ION EXCHANGE	0.95
IF	ION EXCHANGE EFFICIENCY = MEDIUM	
THEN	HRP-EQUIPMENT STAGE 1 = ION EXCHANGE	0.65
IF	ION EXCHANGE EFFICIENCY = LOW	
THEN	HRP-EQUIPMENT STAGE 1 = ION EXCHANGE	0.35
IF	AFFINITY CHROMATOGRAPHY EFFICIENCY = HIGH	
THEN	HRP-EQUIPMENT STAGE 1 = AFFINITY CHROMATOGRAPHY	0.75
IF	AFFINITY CHROMATOGRAPHY EFFICIENCY = MEDIUM	
THEN	HRP-EQUIPMENT STAGE 1 = AFFINITY CHROMATOGRAPHY	0.45

369

```
IF      HYDROPHOBIC INTERACTION CHROM. (HIC) EFFICIENCY = HIGH
THEN    HRP-EQUIPMENT STAGE 1 = HIC                                    0.70

IF      HYDROPHOBIC INTERACTION CHROM. (HIC) EFFICIENCY = MEDIUM
THEN    HRP-EQUIPMENT STAGE 1 = HIC                                    0.40
```

If a second purification step is to follow (usually the case as a higher purity than that obtained after one stage of high resolution purification is normally required) the buffer necessary will usually be different from the one the sample is in after the first step. Gel filtration is commonly used for desalting as there is no need to concentrate the protein sample. Diafiltration can also be used. Finally, polishing is used to eliminate trace contaminants and obtain ultra high purity.

In the development of the prototype expert system it was found that the overall process was satisfactorily divided into two subprocesses with clear objectives: recovery and purification. Selection of operations in the recovery subprocess could be well structured. In the second subprocess (purification) the structuring of the knowledge was more difficult. The main deficiency of available information was found to be in that required for the selection of high resolution purification operations and for the separation of minor contaminants present that are removed in the final polishing stage. The selection of high resolution purification operations (which are usually one or more chromatographic steps) should be based on the physico-chemical properties of the proteins and those of the major contaminant proteins. This information is vital to the selection of the right operations and their best possible order according to their relative efficiencies (Asenjo, 1990; Asenjo et al., 1991b; see next section).

For selection of operations, information generated at a very small scale in terms of 'efficiency' of separation or alternatively information on physico-chemical properties (charge-titration curve, bioaffinity, surface hydrophobicity, pI, M.W., shape) will be used. In this case the deviation of the value for the product protein from those of the main contaminants should be used. A factor for efficiency of the operation in exploiting these

differences also must be included in the evaluation (Asenjo, 1990), as discussed in detail in the next two sections (Expansion and Implementation).

Expansion of Prototype Expert System

To predict the selection of many operations in the Purification subprocess it is necessary to characterize both the product and the main contaminant proteins in terms of their physicochemical properties. Main sources of proteins (for physicochemical characterization) in modern biotechnology industries are few: e.g. E. coli, mammalian cells and yeast. Good characterization of proteins from these sources will allow selection of purification operations on a much more rational basis.

Selection of actual operations is based on information generated at a small scale to determine performance and efficiency of particular separations. Alternatively, information on physical, chemical and biochemical properties of product and contaminants can be used (Table 2). In this case the deviation (DF = deviation factor) of the value of the protein product from those of the main contaminants should be found. A factor for efficiency (η) of the separation operation in exploiting this difference and deviation (DF) of physicochemical property must be included in the evaluation. It is possible then to define a 'separation coefficient' (SC) that can be used to characterize the ability of the separation operation to separate two or more proteins (Asenjo, 1990).

$$SC = f(DF, \eta)$$

In the cases of ion exchange or affinity chromatography there are differences in the cost of matrices used (e.g. protein A affinity chromatography uses a more expensive matrix than CM-Sepharose ion exchange) although most of the cost in such a process is associated with the hardware (columns,

Table 2. Properties to be Exploited for the Separation
and Purification of Different Proteins

1. Charge (Titration Curve)
2. Biospecificity
3. Surface Hydrophobicity
4. pI (Isoelectric Point)
5. M.W. (Molecular Weight)
6. Shape (Stokes Radius)

accessories, control system) as most matrices can be reused many times resulting in reduced matrix costs. Also different adsorption capacities and flow characteristics of the matrices will result in columns of different size. However, column hardware cost is only a fraction of the total cost hence the total hardware cost of a chromatographic step is relatively constant. Differences in the cost of a purification operation can be taken into account by the use of a cost factor (CF) giving an expression for the economic separation coefficient (ESC).

$$ESC = f(SC, CF)$$

The values of the parameters in these two expressions should range between $0 < \eta \leq 1$ and $0 \leq DF \leq 1$. As such values are relative the maximum value for DF for individual properties has to be defined within this range and the value for η given to a particular operation will also depend on the range (or maximum possible value for the deviation of a specific protein property) which must be standardized for different operations. The value of the CF (cost factor) will be < 1 or > 1 and a standard operation (such as Ion-exchange using CM-Sepharose) should be given a value of 1. We have made a first attempt to define both a separation coefficient, SC, and an economic separation coefficient, ESC. They are shown in Table 3 (efficiency = η; DF = deviation factor). We have recently suggested the inclusion of a term for concentration as this will

affect the selection criteria since the contaminants in higher concentrations have to be removed first (Rule 2). However, as concentration does not appear to intrinsically affect the actual separation coefficient, the suggestion of using the term 'Separation Selection Coefficient', SSC, when including the concentration term θ has been preferred (SSC = DF·η·θ).

It should be noted that the two parameters η and the cost factor, CF, are thus far empirical and rather subjective. A more rigorous estimation is presently under study in our group. The cost factor is not based on a rigorous economic evaluation, such as has recently been carried out in our group (Kosti, 1989), but on a very 'approximate' evaluation of the cost involved in using such an operation. There are many elements apart from the direct variable and capital costs that affect the choice of process and hence the 'approximate' evaluation of cost and thus CF (e.g. availability of matrix in the pilot plant, reliability, robustness with variation in feedstock, speed of process implementation or quality control). This role of other elements is partly related to the fact that cost of production of a therapeutic or diagnostic protein is still only a small fraction of the final price. Hence the cost differences found in a rigorous economic evaluation are much more marked than those shown in the expressions in Table 3. All values shown in Table 3 will be subjected to modifications as the rationale proposed is tested in real cases.

Table 3. List of Expressions, Variables and Numerical Values
for Separation Coefficients

$$SC = DF \cdot \eta$$

DF = Deviation factor for hydrophobicity,
molecular weight and pI

$$DF = \frac{\text{Protein Value} - \text{Contaminant Value}}{\text{Max.[Protein Value, Contaminant Value]}}$$

DF = 1.0 for Affinity Chromatography

$$\eta = \text{Efficiency} = \begin{cases} 1.00 & \text{for Affinity Chromatography} \\ 0.70 & \text{for Ion Exchange} \\ 0.35 & \text{for Hydrophobic Interaction Chromatography} \\ 0.20 & \text{for Gel Filtration} \end{cases}$$

$$SSC = DF \cdot \eta \cdot \theta$$

θ = concentration factor

$$\theta = \frac{\text{Concentration of the Contaminant Protein}}{\text{Total Concentration of Contaminant Proteins}}$$

$$ESC = \frac{SSC}{CF}$$

$$CF = \text{Cost Factor} = \begin{cases} 1.0 & \text{for affinity chromatography} \\ 0.6 & \text{for gel filtration} \\ 0.3 & \text{for ion exchange} \\ 0.3 & \text{for hydrophobic interaction} \\ & \quad \text{chromatography} \end{cases}$$

Implementation and Testing of Prototype Expert System

The rationale for selection of high-resolution purification operations has been implemented into our prototype expert system. This was done by interfacing a program in PASCAL in which the main molecular properties of a target product protein were compared with those of the main protein contaminants and then used to select the most appropriate high-resolution purification operations as shown in Figure 7. The rationale discussed in the previous section and Table 3 was used.

The main sources used for the production of recombinant and mammalian proteins today are few. For the purpose of our prototype only 3 main production systems were chosen for the characterization of the main protein contaminants present in these sources. These are E. coli (intracellular proteins), yeast (intra and extracellular) and mammalian cells (extracellular proteins). Initial results of our present work on the characterization of the main proteins in these sources is shown in Tables 4, 5 and 6 (Keeratipibul, 1989; Noble, 1990). This approach appears conceptually valid for molecular weight and for hydrophobic interaction chromatography, but care has to be taken in the selection of ion-exchange chromatography as a suitable method. Values of the isoelectric point, pI, of proteins are useful for the selection of operating conditions when using an anion or a cation exchange matrix (Scopes, 1987) but not for the selection of operations that will give better separation resolution between proteins. Data on the charge of the proteins as a function of pH (Asenjo, 1990) or on its adsorption properties on the different matrices is necessary, as has been discussed in much more detail previously (Asenjo et al., 1991b). pI is only directly relevant for the selection of chromatofocusing as a separation operation, but chromatofocusing is not a particularly feasible operation for large scale use, owing not only to the high cost of a polybuffer but mainly due to the fact that such buffers are unacceptable for use with therapeutic proteins. As more accurate and detailed information on the protein contaminants is made available, it will be appropriately implemented in the database.

The 'expanded' prototype expert system with access to the databases and the more rational selection of high resolution separation operations discussed in this paper resulted in an expert system with approximately 130 rules in addition to the Pascal interface. We are presently testing the use of this expert system for selection of purification operations.

Work on artificial intelligence and expert systems (ES) clearly shows that properly developed ES can be a vital tool to assist in solving the knowledge-intensive and heuristic-based problem of process synthesis in biotechnology. Rigorous methods will not be appropriate to solve the overall synthesis problem, as rigorous information and mathematical correlations are not as readily available as they are in chemical process engineering. The overall downstream process synthesis problem in biotechnology does not have a strict combinatorial nature, whereas the high-resolution purification stages within the purification subprocess (1, 2 or more purification stages where several alternatives in different sequences can be used) do. However, rigorous models have an important role in the simulation of individual operations (e.g., for process evaluation and comparison of performance and cost of individual operations). It appears that the limiting factor in the development of ES's for protein purification is not the implementation of new AI programs but the acquisition, clarification, formalization and structuring of the domain of expert knowledge.

Hybrid and Evolutionary Systems

It is clear that heuristic methods (such as those that have been implemented in expert systems) are one end of the spectrum of available process synthesis techniques; the other end consists of rigorous methodologies such as mathematical programming techniques. A second stage in the development of expert systems should consider the introduction of quantitative models (mathematical correlations, design equations and short-cut methods) for the

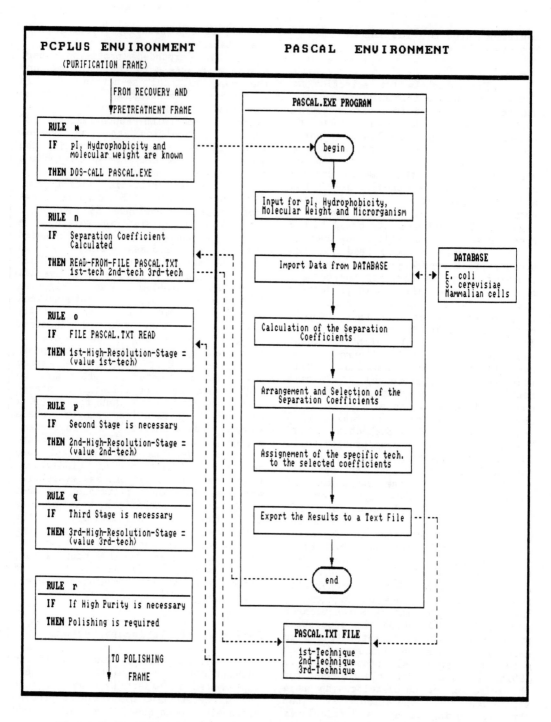

Figure 7. Interfacing of Expert System shell with PASCAL programme for selection of high resolution purification operations.

Table 4. Properties of Main Protein Contaminants in
         E. coli lysate (from Keeratipibul, 1989)

Band No.	Molecular Weight	Hydrophobicity	pI
1	120,200	0.02 M	4.98
2	145,000	not seen	5.20
3	82,000	0.13 M	5.20
4	200,000	not seen	4.56
5	13,800	0.64 M	5.20
6	27,200 – 22,900	0.26 M	4.56
7	39,500	0.13 M	5.73
8	39,500	0.64 M	4.56
9	39,500	0.13 M	4.56
10	120,200	0.02 M	5.46
11	145,000	not seen	4.64

Cell lysate was prepared by bead milling.
Molecular Weight was measured by gel filtration, Isoelectric point
(pI) was measured by isoelectric focusing using a Sephadex gel and
hydrophobicity was measured by hydrophobic interaction chromatography
(HIC) using a Phenyl–Superose gel in an FPLC and a gradient elution
from 2.0 M to 0.0 M $(NH_4)_2SO_4$ in 0.1 M KH_2PO_4. Units used are
the concentration of $(NH_4)_2SO_4$ at which the protein eluted.

Table 5. Properties of the 10 Main Protein Bands Present in
the S. cerevisiae lysate (from Noble, 1990)

Band No.	Molecular Weight	Hydrophobicity	pI
1	38,000	0.42 – 0.27 M	6.1 – 6.4
2	60,300	0.34 M	5.7
3	154,900	0.34 M	7.0
4	14,800	0.04 – 0.00 M	4.9
5	8,500	0.42 – 0.27 M	5.1 – 5.7
6	12,300	not seen	6.7 – 7.0
7	41,700	ppt.	6.4 – 6.7
8	41,700	0.64 – 0.57 M	5.4 – 7.0
9	55,000	etOH	7.6
10	3,400	ppt.	8.1

Cell lysate was prepared by bead milling
Molecular Weight, pI and hydrophobicity were measured as in Table 4.,
but using Octyl-Sepharose as the hydrophobic gel (more hydrophobic) and
a gradient elution from 1.5 M to 0.0 M $(NH_4)_2SO_4$ to avoid protein
precipitation. Some protein bands still precipitated (ppt. in table).
etOH means tightly bound band that needed to be eluted with 24% ethanol
in deionized water.

Table 6. Properties of the 10 Main Protein Bands Present in the Supernatant of CHO[*] cell culture (from Noble, 1990)

Band No.	Molecular Weight	Hydrophobicity	pI
1	66,100	0.97 - 0.00 M	4.8
2	204,200 - 141,300	0.97 - 0.56 M	5.4 - 8.7
3	295,000	0.97 - 0.71 M	6.0
4	72,400	ppt.	5.4
5	53,000	ppt.	5.2
6	72,400	ppt.	5.7
7	169,800	ppt.	5.4
8	2,800	1.25, 0.56 M	5.7
9	6,500	ppt.	5.2
10	169,800	0.71 M	4.8

[*]Culture supernatant of CHO (Chinese Hamster Ovary Cells) Molecular Weight, pI and hydrophobicity were measured as in Table 4., but using a gradient elution from 1.7 M to 0.0 M $(NH_4)_2SO_4$ to avoid protein precipitation. Some protein bands still precipitated (ppt. in table).

design and evaluation of individual operations and their alternatives, and for introducing basic cost calculations into the selection procedure of alternative processes. Such a hybrid system that will include heuristic rules, more rigorous information and design correlations in addition to databases is particularly attractive for process biotechnology, as rigorous correlations and detailed information are not readily available. It should also become possible to establish evolutionary methods to improve the design of an existing or initial process sequence. Such a system should also incorporate a database of previous designs and partial designs, so that new designs may be generated from concrete past experience. This capability is valuable for i) establishing a reasonable initial flowsheet to start the evolutionary procedure and, ii) incorporating well tested separation subtasks into new designs.

CONCLUSIONS

The design of an optimal protein separation sequence is an important problem, which is gaining in importance as biotechnology products become more competitive. Both modern chemical engineering and computer tools can be used for optimizing such design and selection processes. These include the use of numerical methods as well as Expert Systems (ESs) that can use both numerical, and heuristic information as well as databases.

The interfacing of the prototype expert system with a database of physico–chemical and thermodynamic properties of main protein contaminants in typical production streams is an important improvement that will allow the selection of high resolution purification operations on a much more rational basis. For the further development of this field there is an important need for generating more detailed databases for protein products, fermentation streams and contaminants.

For selection of optimal protein separation sequences the ES will give the user a number of process alternatives that can be chosen based on extensive data back–up on protein sources and unit operations as well as algebraic correlations.

Acknowledgements

Part of this work was supported by postdoctoral research grants from the European Economic Community and from the Conselho Nacional de Desenvolvimento Cientifico e Tecnologico of Brazil to whom thanks are due.

REFERENCES

Albertsson, P.-A.; Johansson, G.; Tjerneld, F. In Separation Processes in Biotechnology. Asenjo, J.A., Ed.; Marcel Dekker, New York, 1990, pp. 287-327.

Andrews, B.A.; Huang, R.-B.; Asenjo, J.A. In Biologicals from Recombinant Microorganisms and Animal Cells - New Strategies in Production and Recovery. White, M.; Reuveny, S.; Shaffermann, A., Eds.; VCH Publishers, Weinheim, Germany, 1991.

Arve, B. Simulation and Modelling of Chromatographic Processes. 32nd International IUPAC Congress, Stockholm, 2-7 August, 1989.

Asenjo, J.A. In Separation Processes in Biotechnology. Asenjo, J.A., Ed.; Marcel Dekker, New York, 1990, pp. 3-16.

Asenjo, J.A.; Franco, T.; Andrews, A.T.; Andrews, B.A. In Biologicals from Recombinant Microorganisms and Animal Cells - New Strategies in Production and Recovery. White, M.; Reuveny, S.; Shaffermann, A., Eds.; VCH Publishers, Weinheim, Germany, 1991a.

Asenjo, J.A.; Herrera, L.; Byrne, B. J. Biotechnol., 1989, 11, 275-298.

Asenjo, J.A.; Patrick, I. In Protein Purification Applications: A Practical Approach, Harris, E.L.V.; Angal, S., Eds; IRL Press, U.K., 1990, pp. 1-29.

Asenjo, J.A.; Parrado, J.; Andrews, B.A. Ann. N.Y. Acad. Sci. (1991b), in press, (Proc. Int. Conf. on Progress in Rec. DNA Technology and Applications, Prokop, A.; Bajpai, R., Eds.)

Cartwright, T. Trends Biotechnol., 1987, 5, 25-30.

Chase, H.A. Symp. on Antibodies for Purification, SCI, London, March 1988.

Draeger, N.M.; Chase, H.A. In *Separations for Biotechnology II*, Pyle, D.L., Ed.; Elsevier, 1990, pp. 325-334.

Duffy, S.A.; Moellering, B.J.; Prior, C.P. Optimal Large Scale Purification Strategies for the Production of Highly Purified Monoclonal Antibodies for Clinical Application. *196th ASC National Meeting*, MBTD division, Los Angeles, CA, 25-30 Sept., 1988.

Eveleigh, J.W. Fluorocarbon liquid and solid affinity support, *5th Int. Conf. on Partition in Aqueous Two-phase Systems*, Oxford, U.K., August, 1987.

Fish, N.M.; Lilly, M.D. *Bio/Technology*, 1984, 2, 623-627.

Gordon, N.F.; Tsujimure, H.; Cooney, C.L. *Bioseparation*, 1990, 1, 9-21.

Hedman, P.; Janson, J.C; Arve, B.; Gustafsson, J.G. *Proc. of 8th Int. Biotechnol. Symp.*, 1, Durand, G.; Bobichon, L.; Florent, J., Société Francaise de Microbiologie, 1989, pp. 612-622.

Hunter, J.B.; Asenjo, J.A. In *Separation, Recovery and Purification in Biotechnology: Recent Advances and Mathematical Modeling*, Asenjo, J.A.; Hong, J., Eds.; ACS Symposium Series, Amer. Chem. Soc., Washington, D.C., 1986, pp. 9-32.

Hunter, J.B.; Asenjo, J.A. *Biotechnol. Bioeng.*, 1987, 30, 481-490.

Hunter, J.B.; Asenjo, J.A. *Biotechnol. Bioeng.*, 1988, 31, 929-943.

Hunter, J.B.; Asenjo, J.A. *Biotechnol. Bioeng.*, 1990, 35, 31-42.

Janson, J.-C.; Ryden, L. *Protein Purification*, VCH Publishers, Weinheim, Germany, 1989.

Keeratipibul, S. *Characterization of Proteins from E. coli and S. cerevisiae for the Design of Protein Separation Operations*, M.Sc. thesis, University of Reading, 1989.

Kosti, R. *Economic Evaluation of Large Scale Protein Purification Operations*, M.Sc. thesis, University of Reading, 1989.

Liu, L.C.; Prokopakis, G.J.; Asenjo, J.A. *Biotechnol. Bioeng.*, 1988, 32, 1113-1127.

Noble, I. *Characterization of the Main Contaminant Proteins Present in a Mammalian Cell Culture Supernatant and a S. cerevisiae Lysate*, B.Sc. dissertation, University of Reading, 1990.

Prokopakis, G.J.; Asenjo, J.A. Synthesis of downstream Processes, In *Separation Processes in Biotechnology*. Asenjo, J.A., Ed.; Marcel Dekker, New York, 1990, pp. 571-601.

Scopes, R.K. *Protein Purification*, 2nd. ed., Springer Verlag, New York, 1987.

Schütte, H.; Kroner, K.S.; Husted, H.; Kula, M.R. *Enzyme Microb. Technol.*, 1983, 5, 143-148.

Siletti, C.A.; Stephanopoulos, G. Computer Aided Design of Protein Recovery Processes, *192nd ASC National Meeting*, Anaheim, CA, Sept. 1986.

Siletti, C.A. *Computer Aided Design of Protein Recovery Processes*, Ph.D. thesis, MIT, USA, 1989.

Wacks, S. *Design of Protein Separation Sequences and Downstream Processes in Biotechnology: Use of Artificial Intelligence*, M.Sc. thesis, Columbia University, New York, 1987.

Wang, H.Y. *Ann N.Y. Acad. Sci.*, 1983, 413, 313-321.

Wang, L. In *Separation Processes in Biotechnology*, Asenjo, J.A., Ed.; Marcel Dekker, New York, 1990, pp. 359-400.

Yamamoto, S.; Nomura, M.; Sano, Y. *AIChE Journal*, 1987, 33, 1426.

Yamamoto, S.; Nakanishi, K.; Matsuno, R. *Ion-Exchange Chromatography of Proteins*, Marcel Dekker, New York, 1988.

Thermodynamics of Aqueous Mixtures of Salts and Polymers

Heriberto Cabezas, Jr. and Mostafa Kabiri-Badr, Department of Chemical Engineering, University of Arizona, Tucson, Arizona 85721, U.S.A.
Steven M. Snyder and David C. Szlag, Center for Chemical Technology, National Institute of Standards and Technology, 325 Broadway, Boulder, Colorado 80304, U.S.A.

The reliable aqueous two-phase extraction of proteins is presently hindered by the lack of a basic understanding of phase formation and composition. We have developed a simple and general statistical thermodynamic model for the component chemical potentials in aqueous mixtures of salts and polymers including the dependence on polymer molecular weight. We illustrate the utility of the model by predicting the phase compositions at 25°C and ambient pressure for two polyethylene glycol-dextran and two polyethylene glycol-salt systems.

Aqueous two-phase extraction is an important emerging technique in bioseparations as evidenced by several recent monographs and review papers (Albertsson, 1986; Kula et al., 1982; Walter et al., 1985; Hustedt et al., 1990). The application of this technique is presently hindered by the apparent lack of predictability of the phase formation and protein partitioning processes. Although the subject has been extensively studied (Edmond and Ogston, 1968; Brooks et al., 1985; King et al., 1988; Kang and Sandler, 1988; Forciniti and Hall, 1990; Abbott et al., 1990; Cabezas et al., 1990a), there still are casual anecdotes as to how polymers obtained from different sources, but supposedly similar, do not always form two phases as expected. These observations are a reflection of the relatively poor level of understanding of the molecular effects at work in these systems. Some of the observed effects are due to changes in the actual molecular weight of the polymers or to the presence of various impurities including salts (Cabezas et al.; 1990b). These issues can be clarified by the application of classical thermodynamics together with models for the necessary physical properties of the

solution. Consequently, the objective of this work is to present a fundamental thermodynamic model, based on statistical mechanics, for the component chemical potentials and to illustrate the use of the model together with classical thermodynamics in the calculation of phase diagrams.

Chemical Potential Model

Classical Thermodynamics (Gibbs, 1961) tells us that the equilibrium phase compositions of a two-phase system can be calculated from the equality of the component chemical potentials in each of the phases. However, in order to perform a practical calculation one requires a model for the chemical potentials in terms of temperature, pressure, and, most importantly, composition. We address this topic first.

We present a general model for a solution that contains two polymers (d and p), a salt (s) and water (w). The expressions can later be specialized to systems that have no salt or only one polymer. The model represents an extension of our previous work on polymer-polymer (Cabezas et al., 1990a) and salt-polymer aqueous two-phase systems (Kabiri-Badr, 1990).

We assume that the polymers are neutral, monodisperse, and we do not account for the fractionation of the polymers between the phases. The salts are treated as strong electrolytes that are completely dissociated into ions. The ions are modelled as charged hard spheres with non-electrostatic attractive interactions at short ionic separation. The polymers are represented as random coils or blobs occupying a roughly spherical volume of diameter "R_i", separated from other polymers by a region of solvent. Strictly speaking, this limits the applicability of the model to solutions at polymer concentrations below the onset of chain overlap. The physical situation is illustrated in Fig.1.

With the aforementioned assumptions, applying the model to some of the more concentrated solutions represents a successful but empirical extension. Throughout this treatment, the solvent, water, is treated as a background field in which the solutes are imbedded. This approach simplifies the statistical mechanical problem enormously but at the expense of obscuring the role of solute-solvent interactions.

The model expressions are based on the isobaric-isothermal osmotic virial expansion (Hill, 1957) together with a separate contribution for electrostatic effects. The contribution from non-electrostatic forces is given by second (C_{ij}) and third (C_{ijk}) osmotic virial coefficients between the ions (C_{ss} and C_{sss}), the ions and the polymers (C_{ds} and C_{ps}), and the polymers (C_{dd}, C_{dp}, C_{pp}, C_{ddd}, and C_{ppp}). For salt-polymer systems, we found that neglecting the osmotic third virial coefficient between the ions (C_{sss}) yielded unsatisfactory phase diagrams. For the case of two-polymer systems, it is possible to neglect the osmotic third virial coefficients (C_{ddd} and C_{ppp}) and still obtain satisfactory phase diagrams (Edmond and Ogston, 1968; King et al., 1988) which indicates that their contribution is small. Although, it is

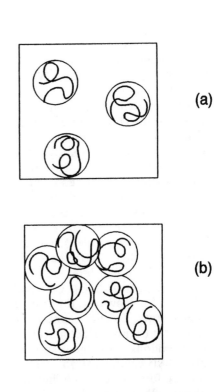

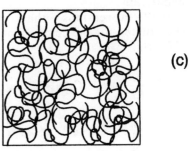

Fig. 1. Schematic diagrams of polymer solution regimes in (a) dilute solution, (b) onset of chain overlap, and (c) semidilute solution.

known that they are important at the higher polymer concentrations (Haynes et al., 1989) and including them significantly improves the accuracy of calculated phase diagrams without increasing the input of experimental data required by the model. Lastly, we have neglected third virial coefficients between unlike components (C_{ddp}, C_{dpp}, C_{dds}, C_{dss}, C_{pps}, and C_{pss}) because their contribution was found to be within the uncertainty ($\pm 0.02\%$) of the experimental data that would have been used in their evaluation. We concluded that they were not significant.

The contribution from long range electrostatic forces ($\ln \gamma_{\pm s}^{lr}$) is given by a new model for the chemical potentials in electrolyte solutions (Kabiri-Badr, 1990). This part appears as an additive contribution to the constant pressure osmotic virial expansion. It is possible to theoretically include the electrostatic interactions in screened or resumed McMillan-Mayer (1945) ion-ion osmotic virial (B_{ij}, B_{ijk}) coefficients (Friedman, 1962). These can later be converted to constant pressure (C_{ij}, C_{ijk}) coefficients (see Eq. 5). However, the resulting virial coefficients do not contain the very important non-electrostatic forces between ions nor is the result accurate at high salt concentration. These other interactions have to be included semi-empirically.

The expressions that we have constructed for the component chemical potentials are

$$\frac{\mu_d}{RT} = \frac{\mu_d^o}{RT} + \ln m_d + 2C_{dd}m_d + 2C_{dp}m_p$$
$$+ 2v_s C_{ds}m_s + \frac{3}{2}C_{ddd}m_d^2 \qquad (1a)$$

$$\frac{\mu_p}{RT} = \frac{\mu_p^o}{RT} + \ln m_p + 2C_{pp}m_p + 2C_{dp}m_d$$
$$+ 2v_s C_{ps}m_s + \frac{3}{2}C_{ppp}m_p^2 \qquad (1b)$$

$$\frac{\mu_s}{RT} = \frac{\mu_s^o}{RT} + v_s \ln m_{\pm s} + 2v_s^2 C_{ss}m_s$$
$$+ 2v_s C_{ds}m_d + 2v_s C_{ps}m_p \qquad (1c)$$
$$+ \frac{3}{2}v_s^3 C_{sss}m_s^2 + v_s \ln \gamma_{\pm s}^{lr}$$

$$\frac{-\mu_w}{RT} = \frac{-\mu_w^o}{RT} + m_d + m_p + v_s m_s + C_{dd}m_d^2$$
$$+ C_{pp}m_p^2 + v_s^2 C_{ss}m_s^2 + 2C_{dp}m_d m_p$$
$$+ 2v_s C_{ds}m_d m_s + 2v_s C_{ps}m_p m_s \qquad (1d)$$
$$+ C_{ddd}m_d^3 + C_{ppp}m_p^3 + v_s^3 C_{sss}m_s^3$$
$$+ v_s \int_0^{m_s} m_s \frac{\partial \ln \gamma_{\pm s}^{lr}}{\partial m_s}\Big|_{T,P} dm_s$$

where μ_i^o is the reference chemical potential of component i in pure water, m_i is the dimensionless molality of component i (moles-i/moles-w) which is related to the conventional molality (moles-i/Kg-water) by the factor $1000/Mw$ where Mw is the molecular mass of water, v_s is the number of ions in the salt, the C_{ij}'s are the second osmotic virial coefficients for components i and j, the C_{iii}'s are the third osmotic virial coefficients for component i, $m_\pm$ is the salt mean molality, and $\ln\gamma_{\pm s}^{lr}$ is the electrostatic contribution (Kabiri-Badr, 1990). The expression for $\ln\gamma_{\pm s}^{lr}$ is given by

$$\ln\gamma_{\pm s}^{lr} = \frac{-S_\gamma}{v_s\omega_s}\left[\frac{v_{+s}^2 Z_+^4 I^{1/2}}{1+a_{++}B_\gamma I^{1/2}}\right.$$
$$\left. + \frac{v_{-s}^2 Z_-^4 I^{1/2}}{1+a_{--}B_\gamma I^{1/2}} + 2v_{+s}v_{-s}Z_+^2 Z_-^2 F_s(I)\right] \qquad (2)$$

where $F_s(I)$ is defined by

$$F_s(I) = \frac{1}{q}\left[\frac{(2c-b^2)I^{1/2}-bcI}{1+bI^{1/2}+cI}\right.$$
$$\left. + \frac{4c}{q^{1/2}}\tan^{-1}\frac{(qI)^{1/2}}{2+bI^{1/2}}\right] \qquad (3)$$

and b, c, q, I, and ω_s are defined by

$$b = \frac{a_{++} + a_{--}}{2} B_\gamma$$

$$c = \frac{a_{++} + a_{--}}{2} \alpha_{+-} B_\gamma^2$$

$$q = 4c - b^2$$

$$I = \frac{m_s}{2}\left[v_{+s}Z_+^2 + v_{-s}Z_-^2\right]$$

$$\omega_s = v_{+s}Z_+^2 + v_{-s}Z_-^2$$

and where Z_+ and Z_- are ionic valences, S_γ is a solvent quantity from the Debye-Hückel theory with a value of 8.7629 (mole-w/mole-s)$^{1/2}$ at 25°C and 0.1 MPa; B_γ is another Debye-Hückel solvent quantity with a value of 2.4487 (mole-w/mole-s)$^{1/2}$/Å at 25°C and 0.1 MPa; v_+ and v_- are the number of cations and anions in the salt, I is the ionic strength, a_{++} and a_{--} are the ionic radii (Marcus, 1985), and α_{+-} is an empirical ionic interaction parameter.

The size or molecular mass of the polymers is known to affect the solute concentrations at which two phases are formed (Albertsson, 1986; Walter et al., 1985). Thus, we have developed scaling expressions for the osmotic virial coefficients from the Renormalization Group theory of polymer solutions (DesCloizeaux and Jannink, 1987; Freed, 1987) which account for this effect. Our expressions were developed by firstly considering each polymer as a blob (see Fig. 1). Then, for purposes of polymer-polymer interactions, each blob can be treated as a hard sphere of diameter R_i in a solvent at constant chemical potential or McMillan-Mayer (MM) conditions (McMillan and Mayer, 1945). Theory (DesCloizeaux and Jannink, 1987; Freed, 1987) tells us that for very long polymers under MM conditions, R_i is proportional to N_i^v where N_i is the degree of polymerization of polymer i and v is a universal scaling exponent valid for all polymers. Statistical mechanics (Croxton, 1975) gives expressions for the MM virial coefficient

(B_{ij}) between two hard spheres of diameters R_i and R_j. We illustrate the general case for any two polymers i and j. Note that for the self interaction of a polymer, $i = j$ (d and d, p and p), then the diameters R_i and R_j are equal and $N_i = N_j$.

$$B_{ij} = \tilde{b}_{ij}\left(\frac{R_i^v + R_j^v}{2}\right)^3 = b_{ij}\left(\frac{N_i^v + N_j^v}{2}\right)^3 \quad (4)$$

where $\tilde{b}_{ij}$ and b_{ij} are proportionality coefficients representing the interactions between monomers of polymers i and j.

These expressions are related to those in our previous work (Cabezas et al., 1990a). However, here we correct them from constant solvent chemical potential or MM conditions to constant pressure. This is important since all of our calculations and experiments are done at constant pressure. The final results, including a simple conversion that is valid for an incompressible solution, are given below. We do not give a detailed derivation of this result because it is available elsewhere (Hill, 1957). The result in its most general form for an incompressible solvent is

$$C_{ij} = \frac{\rho_w}{2}\left[2B_{ij} - \bar{V}_i^o - \bar{V}_j^o\right] \quad (5)$$

where C_{ij} is a constant pressure second osmotic virial coefficient, ρ_w is the molar density of pure water, B_{ij} is an MM second osmotic virial coefficient, and $\bar{V}_i^o$ is the partial molar volume of component i at infinite dilution.

The working expressions for the polymer-polymer constant pressure second virial coefficients (C_{ij}), including the assumption of monomer additivity for the polymer partial molar volumes at infinite dilution ($\bar{V}_i^o = \bar{V}_{mi}^o N_i$), are

$$C_{dd} = \rho_w(b_{dd}N_d^{3v} - \bar{V}_{md}^o N_d) \quad (6a)$$

$$C_{pp} = \rho_w(b_{pp}N_p^{3v} - \bar{V}_{mp}^o N_p) \quad (6b)$$

$$C_{dp} = \frac{\rho_w}{2}\left[2b_{dp}\left(\frac{N_d^v + N_p^v}{2}\right)^3 - \overline{V}_{md}^o N_d \right.$$

$$\left. - \overline{V}_{mp}^o N_p \right] \quad \text{(6c)}$$

where ρ_w is equal to 0.055345 mole/ml at 25°C and 0.1 MPa, b_{ij} is an interaction coefficient between monomers of polymers i and j, N_i is the degree of polymerization of polymer i, v is equal to 0.593, $\overline{V}_{mi}^o$ is the formal partial molar volume of a monomer of polymer i at infinite dilution in water.

The value of the exponent v was obtained from isopiestically determined experimental solvent activities (Ochs et al., 1990) in mixtures of water and polyethylene glycols of various molecular masses (Kabiri-Badr, 1990). Our value of 0.593 is essentially the same as the value of 0.60±0.2 obtained by Watts (1974) from a computer simulation of a self-avoiding walk on a lattice and the value of 0.588±0.001 obtained by LeGillou and Zinn-Justin (1977) from a perturbation theory. The agreement is remarkable.

The best theoretical value for the exponent v is that of LeGillou and Zinn-Justin (1977), 0.588, rather than the more aesthetically pleasing 0.6 or 3/5. This reflects the fact that v is not a rational number in a three dimensional space. Moreover, the calculation of phase diagrams is very sensitive to the value of v so that neither 0.588 nor 0.6 will yield satisfactory results. The difference between our value of v and that of the aforementioned workers reflects the approximate nature of the DeGennes analogy between the statistical mechanics of a polymer and an ensemble of magnetic spins (DeGennes, 1972).

Theory does not provide explicit guidance as to how salt-polymer interactions scale with polymer molecular mass. However, we postulate that a scaling law similar to that for polymer-polymer interactions exists, but with a different exponent, τ, since the polymer blobs will often shrink in the presence of salt. Using isopiestically determined experimental solvent activities in aqueous mixtures of salts and polymers (Ochs et al., 1990), a value of 1.036 was determined for τ. This value was adequate to represent the solvent activity data for six different aqueous solutions containing either MgSO$_4$, Na$_2$SO$_4$, or Na$_2$CO$_3$ with either polyethylene glycol 1000 or 8000. From this experience we concluded that our hypothesis seemed to be generally valid. The expression for the salt-polymer second osmotic virial coefficient is given by

$$C_{ps} = \frac{\rho_w}{2v_s}\left[2b_{ps}N_p^\tau - \overline{V}_{mp}^o N_p - \overline{V}_s^o\right] \quad \text{(7)}$$

where b_{si} is an effective interaction coefficient between a monomer of polymer i and the ions, τ is the scaling exponent for salt-polymer interactions, and $\overline{V}_s^o$ is the partial molar volume of the salt at infinite dilution in water, and the other quantities have been previously defined.

We have also developed scaling expressions for the third osmotic virial coefficients (C_{iii}) of polymers. Theory indicates that for very long linear polymers under conditions of constant solvent chemical potential or MM conditions, the ratio of the MM third osmotic virial coefficient (B_{iii}) to the square of the second (B_{ii}) is a universal number (α) which is valid for all polymers (DesCloizeaux and Jannink, 1987; Freed, 1987). This is expressed in general form by

$$\frac{B_{iii}}{B_{ii}^2} = \alpha \quad \text{(8)}$$

The above result is valid for a solution at MM conditions. We convert it to constant pressure using the results of Hill (1957) for an incompressible solution. The general result is

$$C_{iii} = \rho_w^2\left[\alpha B_{ii}^2 + (\overline{V}_i^o)^2 - 2B_{ii}\overline{V}_i^o\right] \quad \text{(9)}$$

We were not able to confirm our hypothesis regarding α possibly due to the experimental uncertainties involved in the determination of third virial coefficients or simply because dextran is a branched polymer. However, we did find that a single value (α_p) sufficed for the polyethylene glycols and another for the dextrans (α_d).

Our working expressions for the polymer constant pressure osmotic third virial coefficients are

$$C_{ddd} = \rho_w^2 \left[\alpha_d \left(b_{dd} N_d^{3v} \right)^2 + \left(\overline{V}_{md}^o N_d \right)^2 \right. \tag{10a}$$
$$\left. - 2 b_{dd} N_d^{3v} \left(\overline{V}_{md}^o N_d \right) \right]$$

$$C_{ppp} = \rho_w^2 \left[\alpha_p \left(b_{pp} N_p^{3v} \right)^2 + \left(\overline{V}_{mp}^o N_p \right)^2 \right. \tag{10b}$$
$$\left. - 2 b_{pp} N_p^{3v} \left(\overline{V}_{mp}^o N_p \right) \right]$$

Note that the only new polymer parameter in these equations is α_p and that it can be obtained from polymer-water data. The other parameters are already present in the scaling expression for B_{ii} and C_{ii}.

Model Parameters

The value of those model parameters or properties which are generally applicable to many systems and which are not specific to a particular solute have been given in the section describing the model. Parameters in this category are the Debye-Hückel quantities (S_γ and B_γ) which are solvent properties, the scaling exponents (v and τ) which are assumed to be applicable to all polymer solutions, and the pure water density (ρ_w). The values for all of the solute specific model parameters and properties are listed in Tables 1 and 2.

For a system consisting of water and two polymers (d and p), the model contains five solute specific parameters: b_{dd}, b_{pp}, b_{dp}, α_d, and α_p. In addition, the formal monomer partial molar volumes at infinite dilution ($\overline{V}_{md}^o$ and $\overline{V}_{mp}^o$) are required.

Our values for b_{pp} and α_p were determined from our own experimental water activity data on solutions of polyethylene glycol (Ochs et al. 1990). The values for b_{dd}, b_{dp}, and α_d were calculated from phase diagram data (King et al., 1988). The monomer partial molar volumes were obtained from our own density measurements on water-polymer solutions.

In contrast to the relative simplicity of polymer solutions, the presence of a salt creates a great deal of complexity. Thus, for a system consisting of water, a salt (s), and a polymer (p) the model contains eight solute specific parameters: b_{pp}, b_{ps}, C_{ss}, α_p, α_{+-}, a_{++}, a_{--}, and C_{sss}. Note that most of the new parameters refer to the salt. Again, we also require the partial molar volumes at infinite dilution for the monomers ($\overline{V}_{mp}^o$) and the salt ($\overline{V}_s^o$).

The value of the polymer parameters (b_{pp}, α_p, V_{mp}^o) is the same for salt-polymer and polymer-polymer systems. The value of b_{ps} was obtained from solvent activity data on aqueous mixtures of salt and polymer (Ochs et al., 1990). The ionic radii (a_{++} and a_{--}) are the independently determined values of Marcus (1985). The value of α_{+-}, C_{ss}, and C_{sss} was obtained from salt activity data on salt-water mixtures (Golberg, 1981; Rard and Miller, 1981). The value of the salt partial molar volume is that of Millero (1972).

The results obtained from the model seem most sensitive to the values of some of the polymer parameters (b_{ij}, v, and τ) and the salt virial coefficients (C_{ss} and C_{sss}).

Phase Behavior of Polymer-Polymer Systems

According to classical thermodynamics (Gibbs, 1961), the equilibrium state of the two phases formed by an aqueous (w)

Table 1. Salt and Polymer Interaction Coefficients
b_{ij} (ml/mol); C_{ss} (mol-w/mol-s)

	PEG	DEXTRAN	MgSO$_4$	Na$_2$CO$_3$
PEG	$b_{pp} = 20.8$	$b_{dp} = 56.3$	$b_{ps} = 103$	$b_{ps} = 151$
DEXTRAN	----	$b_{dd} = 135$	----	----
MgSO$_4$	----	----	$C_{ss} = -0.139$	----
Na$_2$CO$_3$	----	----	----	$C_{ss} = -3.368$

Table 2. Salt and Polymer Properties

	PEG	DEX	MgSO$_4$	Na$_2$CO$_3$
α_i	0.075	0.011	----	----
α_{+-} (Å)	----	----	2.74	2.00
a_{++} (Å) a_{--} (Å)	----	----	1.30 4.50	1.90 4.50
C_{iii}(mol-w/mol-i)^2	Eq. 10b	Eq. 10a	33.12	16.77
∇°_{mi} or ∇°_s (ml/mol)	37.0	111	-7.1	-6.5

mixture of polyethylene glycol (p) and dextran (d) is determined by

$$\mu_d^t(T,P,m_d^t,m_p^t)=\mu_d^b(T,P,m_d^b,m_p^b) \qquad (11a)$$

$$\mu_p^t(T,P,m_d^t,m_p^t)=\mu_p^b(T,P,m_d^b,m_p^b) \qquad (11b)$$

$$\mu_w^t(T,P,m_d^t,m_p^t)=\mu_w^b(T,P,m_d^b,m_p^b) \qquad (11c)$$

where μ_i^j is the chemical potential of component i in phase j, T is the temperature, P is the pressure, and m_i^j is the dimensionless molality of component i in phase j.

Equations (11) represent a system of three equations and four unknowns (m_d^t, m_p^t, m_d^b, m_p^b) at a given temperature and pressure. To calculate the phase composition (or molalities) we first close the system of equations by choosing a value for one of the molalities, then insert the model for the chemical potentials in terms of molalities, and finally solve the system of equations for the remaining three molalities.

We have calculated the entire phase envelope at 25°C and ambient pressure for the two different polyethylene glycol-dextran aqueous two-phase systems depicted in Figs. 2 and 3. In these calculations we have used the chemical potential model (Equations 1 to 3, 6,7 and 10) where $m_s=0$ since there is no salt, along with the aforementioned parameters. The experimental phase compositions and the polymer molecular masses are those of King et al. (1988). The number average polymer molecular masses were used in the calculations.

Phase Behavior of Salt-Polymer Systems

Similarly to the case of two-polymer systems, the equilibrium state of the two phases formed by an aqueous (w) mixture of polyethylene glycol (p) and a salt (s) is determined by

$$\mu_p^t(T,P,m_p^t,m_s^t)=\mu_p^b(T,P,m_p^b,m_s^b) \qquad (12a)$$

$$\mu_s^t(T,P,m_p^t,m_s^t)=\mu_s^b(T,P,m_p^b,m_s^b) \qquad (12b)$$

$$\mu_w^t(T,P,m_p^t,m_s^t)=\mu_w^b(T,P,m_p^b,m_s^b) \qquad (12c)$$

where, as before, μ_i^j is the chemical potential of component i in phase j, T is the temperature, P is the pressure, and m_i^j is the dimensionless molality of component i in phase j.

Again, equations (12) represent a system of three equations and four unknowns (m_p^t, m_s^t, m_p^b, m_s^b) at a given temperature and pressure. We close the system of equations by setting the value of one of the molalities and then solve for the remaining three molalities.

We have used the model to calculate the entire phase envelope at 25°C and ambient pressure for the two different polyethylene glycol-salt aqueous two-phase systems depicted in Figs. 4 and 5. The details of the calculation and the model are the same as for the polymer-polymer case, except that $m_d=0$ since there is no second polymer present. The salt and polymer parameters used in the calculations are given in Tables 1 and 2 and in the text. The experimental phase compositions are those of Snyder and Szlag (1990). The polymer molecular masses used in the calculation are the approximate nominal values, i.e., 1000 for PEG-1000 and 8000 for PEG-8000.

Summary and Conclusions

We have developed a fundamental statistical thermodynamic model for the component chemical potentials in aqueous mixtures of salts and polymers. We have successfully applied the model to the prediction of phase diagrams of polymer-polymer and salt-polymer aqueous two-phase systems. We have shown that the model predictions are sufficiently accurate for technological

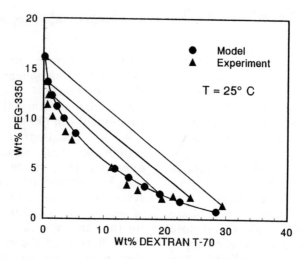

Fig. 2. Phase diagram for an aqueous ($Mn = 3690$) and dextran t-70 mixture of polyethylene glycol 3350 ($Mn = 37000$).

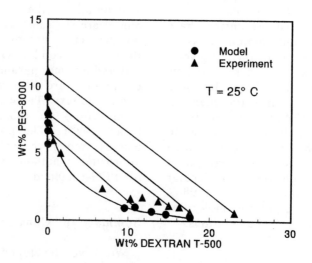

Fig. 3. Phase diagram for an aqueous ($Mn = 8920$) and dextran t-500 mixture of polyethylene glycol 8000 ($Mn = 167000$).

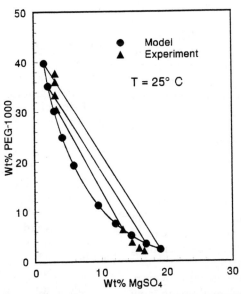

Fig. 4. Phase diagram for an aqueous (Mn ≈ 1000) and MgSO₄. mixture of polyethylene glycol 1000

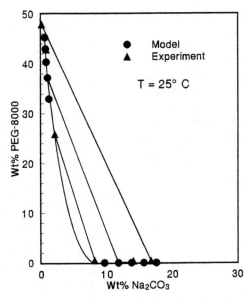

Fig. 5. Phase diagram for an aqueous (Mn ≈ 8000) and Na₂CO₃. mixture of polyethylene glycol 8000

applications. Previously, the same model had been applied to the prediction of salt partition coefficients and phase changes with the addition of small amounts of salt to polymer-polymer aqueous two-phase systems (Cabezas, et al., 1990b).

The model incorporates the dependence of the chemical potentials on polymer molecular weight. This has been accomplished by use of a scaling theory for polymer solutions and by the development of new scaling expressions for salt-polymer interactions. These results are particularly useful for predicting the phase formation and composition changes that might be expected with changes in the molecular mass of the phase forming polymers. The model also incorporates the known theory of electrolyte solutions including the Debye-Hückel limiting law.

Extending the model to other polymers or salts is a matter of evaluating new values for the parameters. For salts, the necessary data can generally be obtained from the literature (Golberg, 1981; Rard and Miller, 1981; Marcus, 1985). For polymer-polymer and polymer-salt interactions, the necessary data can in principle be obtained from any number of common physical measurements such as osmotic pressure, vapor pressure, or isopiestic experiments. We have found the latter most useful in this respect.

We are presently extending the model to include the partitioning of proteins in salt-polymer aqueous two-phase systems. We are also developing predictive models for other phase properties such as density and viscosity.

Acknowledgements

The authors are grateful for the financial and material support of the U.S. National Institute of Standards and Technology (NIST). H. Cabezas, Jr. is also particularly grateful to D.G. Friend and J.R. Fox of the Themophysics Division at NIST for some enlightening discussions regarding scaling in polymer solutions.

References

Albertsson, P.-Å. Partition of Cell Particles and Macromolecules; Wiley-Interscience: New York, NY, 1986.

Abbott, N.L.; Blankschtein, D.; Hatton, T.A. Bioseparation 1990, 1, 191.

Brooks, D.E.; Sharp, K.A.; Fisher, D. In Partitioning in Aqueous Two-Phase Systems: Theory, Methods, Uses, and Applications to Biotechnology; Walter, H.; Brooks, D.E.; Fisher, D., Eds.; Academic Press: New York, NY, 1985, p 11.

Cabezas, H. Jr.; Evans, J.D.; Szlag, D.C. In Downstream Processing and Bioseparation; Hamel, J.-F., P.; Hunter, J.B.; Sikdar S.K., Eds.; ACS Symposium Series No. 419; American Chemical Society: Washington, D.C., 1990a; p 38.

Cabezas, H. Jr.; Kabiri-Badr, M.; Szlag, D.C. Bioseparation 1990b, 1, 227.

Croxton, C. Introduction to Liquid State Physics; John Wiley &Sons: New York, NY, 1975.

DeGennes, P.G. Phys. Lett. 1972, 38A, 339.

DesCloizeaux, J.; Jannink, G. Les Polymeres en Solution: leur Modelisation et leur Structure; Les Editions de Physique: Paris, France, 1987.

Edmond E.; Ogston A. G. Biochem. J. 1968, 109, 569.

Forciniti, D.; Hall C.K. In Downstream Processing and Bioseparation; Hamel, J.-F., P.; Hunter, J.B.; Sikdar S.K., Eds.; ACS Symposium Series No. 419; American Chemical Society: Washington, D.C., 1990; pp 53.

Freed, K.F. Renormalization Group Theory of Macromolecules; John Wiley & Sons: New York, NY, 1987.

Friedman, H.L. Ionic Solution Theory; Interscience: New York, NY, 1962.

Gibbs, J.W. The Scientific Papers of J. Willard Gibbs, Thermodynamics; Dover, New York, NY, 1961; Vol. 1.

Golberg, R.N. J. Phys. Chem. Ref. Data 1981; 10: 3.

Haynes, C.A.; Beynon, R.A; King, R.S.; Blanch, H.W.; Prausnitz, J.M. J. Phys. Chem. 1989, 93, 5612.

Hill, T.L. J. Am. Chem. Soc. 1957, 79, 4885.

Hustedt, H.; Johansson, G.; Tjerneld, F., Eds.-Spec. Issue Bioseparation 1990, 1, 181.

Kabiri-Badr, M. PhD Dissertation; University of Arizona, Tucson, AZ, 1990.

Kang, C.H.; Sandler, S.I. Macromolec. 1988, 21, 3088.

King, R. S.; Blanch, H. W.; Prausnitz, J.M. AIChE J. 1988, 34, 1585.

Kula, M.-R.; Kroner, K.H.; Hustedt, H. (1982) In Advances in Biochemical Engineering; Fiechter, A., Ed.; Springer-Verlag: New York, NY, 1982, Vol. 24; pp. 73.

LeGillou, J.C.; Zinn-Justin, J. Phys. Rev. Lett. 1977, 39, 95.

Marcus, Y. Ionic Solvation; Wiley-Interscience, New York, NY, 1985.

McMillan, W.G.; Mayer, J.E. J. Chem. Phys. 1945, 13, 276.

Millero, F.J. (1972) In Water and Aqueous Solutions: Structure, Thermodynamics, and Transport Processes; Horne, R.A., Ed.; Wiley-Interscience: New York, NY, 1972; p 519.

Ochs, L.R.; Kabiri-Badr,M.; Cabezas, H. Jr. AIChE J. 1990, 36, 1908.

Rard, J.A.; Miller, D.G. (1981) J. Chem. Eng. Data 1981, 26, 33.

Snyder, S.M.; Szlag, D.C., personal communication, 1990.

Walter, H.; Brooks, D.E.; Fisher, D. Partitioning in Aqueous Two-Phase Systems: Theory, Methods, Uses, and Applications to Biotechnology; Academic Press: New York, NY, 1985.

Watts, M.G. J. Phys. A: Math. Nucl. Gen. 1974, 7, 489.

Membrane Affinity Technology For Biomolecule Purification and Diagnostics

Fook Hai Lee, Kent Murphy, Paul Lin, Sara Vasan, Darany An, Suzanne DeMarco, Vincent Forte, Dennis Pietronigro, NYGene Corp., One Odell Plaza, Yonkers, New York, 10701

Membrane Affinity Separation System (MASS®) is a technology that exploits a series of chemical modifications of microporous membranes to produce affinity supports. We have demonstrated that these affinity membranes can be utilized to rapidly isolate antibodies and other biomolecules with high purity and recovery from complex biological mixtures. It is expected that similar affinity substrates will also provide significant advantages in diagnostics and extracorporeal therapeutics.

Affinity based separation is the most powerful method available currently for biomolecule purification. It delivers the highest purity and highest recovery of a biological product from a complex sample (Pharmacia, 1983). The power of the affinity method is based upon its specificity. Affinity separation, when performed correctly, is a binary process; that is, only the desired product and nothing else is bound.

Conventional affinity separation has utilized gels as the substrate to which ligands are bound. Gels greatly limit the power of affinity separation because of the relatively low flow rates through them, compressibility at high flow rates and pressures, difficulty in scaling up, ligand leaching and low number of reuses. Membrane Affinity Separation System (MASS®) is a core technology that exploits a series of proprietary chemical modifications of microporous membranes to produce affinity supports. These modified microporous membranes can be used in diagnostics, drug screening, enzyme bioreactors and extracorporeal therapeutics (Fig. 1).

Technical Data and Product Performance

1. <u>High purity</u> of 98+% and <u>recoveries</u> of 85-95% are obtained in a single pass (Fig. 2). Due to the large pore size structure (0.45 - 5 um) and specificity of MASS, contaminating DNA and viruses should be largely eliminated. Recoveries are independent of product concentration from 10 µg/ml to 10 mg/ml (Fig. 3). Therefore, no concentration or dilution of feed stock is necessary prior to affinity separation using MASS.

2. <u>High capacity</u> binding, for instance, the Protein A based products bind 25-30 mg IgG/ml membrane (0.45 um) and 10-12 mg IgG/ml membrane (3.0 um).

3. <u>Reusability</u> greater than 100 times without compromising product performance using 50 mg and larger MASS devices (Fig. 4)

4. <u>Very fast process cycle times</u> of four to five minutes. These fast process cycle times are uniformly obtained by 1-mg to 1-gram capacity MASS systems. The product must only be in contact with the membrane for approximately a few seconds for efficient binding to occur. This is because the large pore size,

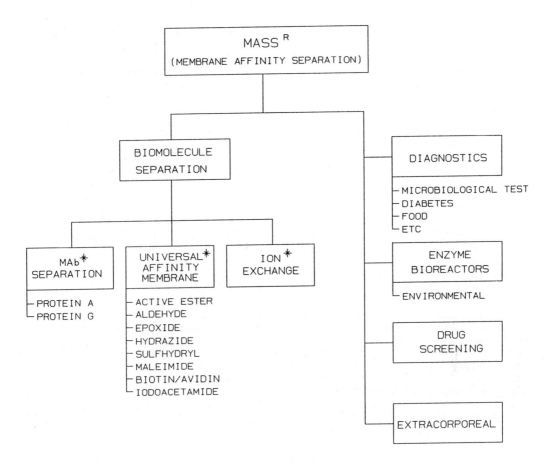

Fig. 1 MASS technology overview, indicating diagnostic and prep- parative applications and meth- ods.

0.45 um to 3.0 um, allows the sample flow to bring the product in contact with its ligand. Flow rates of up to 2 liters per min- ute are efficiently processed by NYGene's one-gram-capacity de- vice. The slow flow rate of standard gels and beads is due to the pore size structure of typically 300 Angstroms which impose an artificial diffusion limitation on the affinity bind- ing process (Stimpson, 1986; Klein, 1991). The resultant pro- ductivity enhancement of MASS compared to HPLC and Gel columns is shown in Fig. 5.

5. Cost effectiveness based upon speed which greatly decreas- es labor time (typically the largest cost associated with biomolecule purification) and high reusability.
6. Low ligand leaching. We have routinely determined li- gand leaching of MASS Protein A products by ELISA and obtained values between 5-10 parts per million which are below the FDA's limitation. These data correlate with the high reusability of MASS.
7. Easy sanitization using 0.2N NaOH as determined by endo-

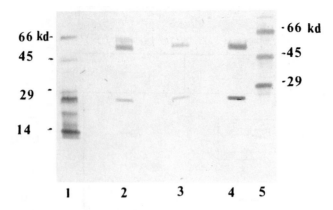

Fig. 2 SDS-PAGE of IgG purified from rabbit serum by MASS Protein A device. Rabbit serum (Pel-Freeze) containing approximately 50 mg IgG was passed through a MASS Protein A device (50 mg capacity) with a flow rate of 150 ml/min. After washing away the contaminants, IgG was eluted by 0.1M glycine-HCl, pH2.5 The IgG was dialysed in H_2O to remove salt content for electrophoresis. SDS-PAGE was performed in a gradient 10-15% gel in reducing buffer with Pharmacia's Phast system using the method of Laemmli (Laemmli, 1970) From left= Lane:1 - Low molecular weight standards (Sigma); 2 - IgG (H&L chains) purified by another manufacturer's Protein A device; 3 and 4 - IgG (H&L chains) purified by MASS Protein A device; 5 - High molecular weight standards.

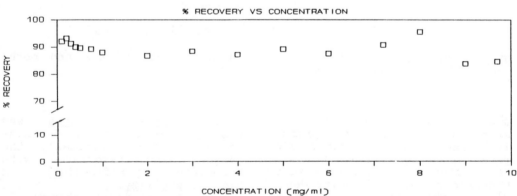

Fig. 3 Percentage of recovery vs. concentration. A MASS Protein A (50 mg capacity) device was challenged at capacity with varying concentrations of rabbit IgG. Neat rabbit sera with IgG concentrations of 7.2 and 9.7 mg/ml were diluted in a non-antibody producing cell line tissue culture fluid with 10% fetal calf serum. The flow rate throughout was 250 ml/min. The IgG was eluted by 0.1M glycine-HCl, pH2.5 and the concentration was measured with a spectrophotomer at 280 nm.

MASS Reusability

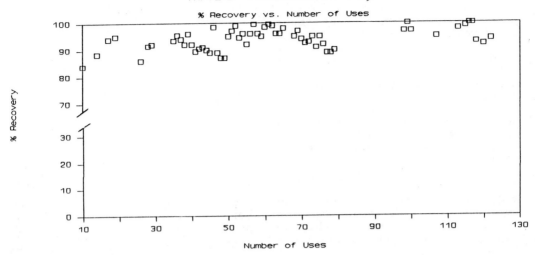

Fig. 4 Percentage of recovery vs. number of uses. A MASS Protein A (50 mg capacity) device was repeatedly challenged at capacity with rabbit IgG in serum diluted to a concentration of 100 ug/ml in PBS. The flow rate throughout was between 150 and 250 ml/min.

IgG Purification vs Time

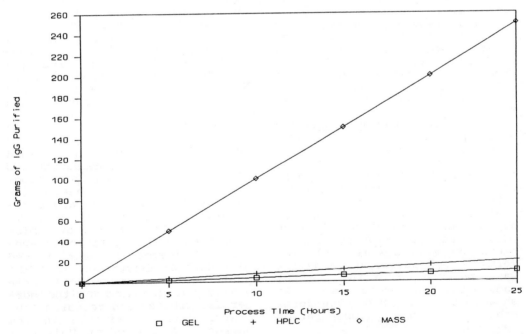

Fig. 5 Productivity enhancement of MASS compared to a gel column and HPLC each using a one-gram capacity device.

toxin challenge testing using the LAL assay. The membranes as supplied do not contribute endotoxins to the process and endotoxin in feed stocks are largely separated from the product during purification on MASS even in the absence of sanitization.

8. Easy storage in 50% ethanol at 4°C for therapeutic applications or in azide containing buffer for non-therapeutic applications.

9. No catastrophic dry out to ruin either the separation media and/or the products. The membranes are hydrophilic and retain water and even if they do dry out there is no loss of activity.

10. Direct scalability - a purification performed on a one milligram device will scale up to the one gram device linearly as the surface area of the membrane scales.

11. Simple and reproducible use - the devices are exceptionally simple to use requiring very little technical training and have been demonstrated to yield reproducible results time and time again.

12. Ligand flexibility - with the Universal Affinity Membranes essentially any ligand containing amino or hydroxyl groups can be coupled via stable covalent bonds to MASS. The chemical process is summarized in Fig. 6.

13. Low operating pressures of 1-3 psi result in less complicated process equipment and lower capital expenditures.

Table 1 shows direct comparisons of MASS Protein A devices with other methods and products. The feed stock was tissue culture supernatant containing IgG$_{2a}$ from 55.2 mouse hybridoma cells supplemented with 20% fetal bovine serum. Each purification device was operated according to the manufacturer's protocols. As shown, MASS Protein A outperforms other available bioseparation devices.

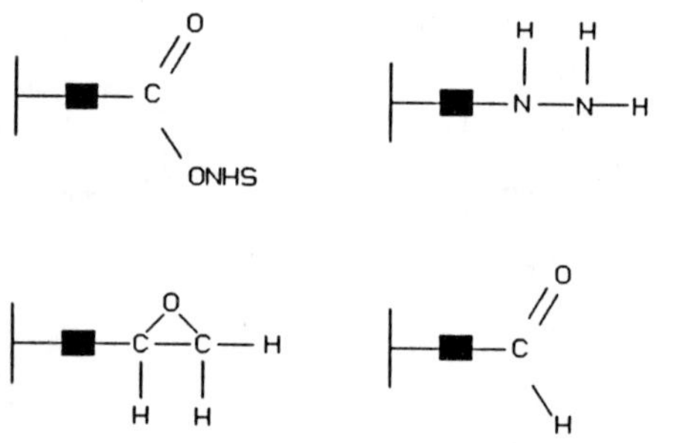

Fig. 6 Universal Affinity Membrane (UAM) for the specific affinity purification of drugs, receptors, blood factors, cytokines, hormones, etc. The active ligands on the membranes are clockwise from upper left= active ester, hydrazide, aldehyde and epoxide. Other chemistries are available as mentioned in Fig. 1.

MASS Protein A devices have also been used to purify mouse IgG$_1$ antibodies. Figure 7 shows SDS-PAGE analysis of a therapeutic grade IgG$_1$ antibody which is a very poor Protein A binder. However, by adjusting the binding conditions this antibody is recovered at 80% yield with 98+% purity, in a single pass.

Table 2 shows direct comparisons of MASS protein G devices with other products. The feed stock was IgG$_1$ in serum free tissue culture supernatant. Again, MASS Protein G devices show superiority over other products.

Discussion

The performance of MASS outlined above demonstrates the advantages of this emerging new technology over the gels or beads for biomolecule separation and purification.

In the field of biopharmaceutical separations, Protein A and Protein G derivatized microporous membranes for the purification of monoclonal antibodies for use in both the commercial process and research laboratory are available in the market. A series of Universal

Table 1. Comparison of Protein A Separation Technologies

Company	Technology	Purity	Recovery	Time*
NYGene	Membrane	98+%	85-90%*	1-5 min
Co. A	Membrane	20%	40-50%	45 min
Co. B	Membrane	98+%	30-40%	15 min
Co. C	Column/gel	95%	50-80%	1-4 hrs
Co. D	Column/gel	82%	65-75%	1+hrs

* To separate 1 mg IgG

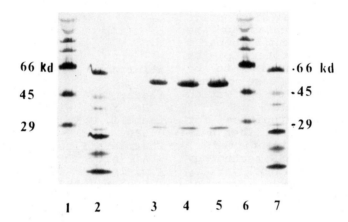

66 kd
45
29

1 2 3 4 5 6 7

Fig. 7 SDS-PAGE of a therapeutic mouse IgG$_1$ monoclonal antibody. The experimental conditions were as mentioned in Fig. 2, except that the binding buffer was 1.5M glycine and 3M NaCl, pH8.9. From left= Lane 1 and 6 - High molecular weight standards (Sigma); 2 and 7 - Low molecular weight standards; 3, 4 and 5 - Purified IgG$_1$ antibody (H&L chains).

Table 2. Comparison of Protein G Separation Technologies

Company	Technology	Purity	Recovery	Time
NYGene	Membrane	>98%	>98%	7 min
E	Membrane	>98%	>53%	30 min
F	Membrane	95%	20%	80 min
G	Gel	>98%	41%	30 min

Affinity Membranes (UAM®) which will allow a user to covalently attach his own ligand for specific biomolecule purifications and diagnostic tests (Figure 6) have recently been developed.

Currently under development is a series of microporous ion exchange membranes to complement the affinity product line thus giving MASS the capability to totally purify most biopharmaceuticals and other biomolecules.

All MASS purification products including Protein A, Protein G, UAM and ion exchange devices will be fitted in a computerized instrument AutoMASS®, allowing the simple, rapid and reproducible purification of most biomolecules. The output of this system with the 1 gram device is about 10 grams/hour which is about 10 times the productivity of similar purification instrument systems.

Membrane microtiter plates for specific clinical diagnostics and laboratory immunoassays are also under development. Compared to plastic microtiter plates these membranes promise to be superior in speed, ease of use and the wide range of activation chemistries available.

Some key opportunities for MASS in extracorporeal therapeutics has also been identified.

In summary, MASS technology has been leveraged into a variety of new applications. Each of these promise to provide substantial benefits in clinical medicine and biological research.

Conclusion

In conclusion:
. Affinity separation is a powerful method for the purification of biomolecules.
. Microporous membranes are ideal substrates for affinity separations.
. MASS is a core technology which has applications in a growing number of biomolecule separations, diagnostics, and extracorporeal therapeutics.

References

Klein, E., Affinity Membranes, New York, Wiley, 1991.

Laemmli, U.K., Nature, 1970, 277, 680.

Pharmacia Fine Chemicals AB., Affinity Chromatography, Principles and Practice, Uppsala, Sweden, 1983.

Stimpson, D.J., Polym. Prep. Am. Chem. Soc., Div. Polym. Chem., 1986, 27, 424.

Recycling Free Flow Focusing: Apparatus and Applications

James A. Ostrem, Glen Ward, Leon Barstow, and Rodolfo Marquez, Protein Technologies Incorporated, Tucson, Arizona 85719
Thomas R. Van Oosbree, Promega Corporation, Madison, Wisconsin, 53711
Buck A. Rhodes, RhoMed Incorporated, Albuquerque, New Mexico, 87106
Catherine C. Wasmann, University of Arizona, Tucson, Arizona, 85721

A new instrument for preparative-scale isoelectric focusing of proteins in solution is described. The RF3™ consists of a focusing cell, heat exchanger, multichannel peristaltic pump and power supply. The protein/ampholyte mixture is exposed to the electric field for 3-4 seconds per cycle while passing through a narrow (0.75 mm) gap in the focusing cell. We describe its use in purification of monoclonal IgGs, a single-strand nuclease from *Aspergillus oryzae*, and neomycin phosphotransferase II overexpressed in *Escherichia coli*.

Recycling isoelectric focusing has potential for large scale separation of proteins based on differences in isoelectric point (pI), a physical parameter rarely used in preparative purification schemes. One advantage is that separations are carried out in free solution. This eliminates the need to elute protein from gels or chromatographic media, and minimizes losses due to protein absorption to solid supports.

Early instruments designed by Bier and coworkers (Bier et al., 1979; Bier, 1986) relied on nylon screens to stabilize liquid flow in the focusing cell and counteract hydrodynamic forces generated by the electric field. An improved design that relies on shear-stabilized flow through the focusing cell was described recently (Bier et al., 1989). The first commercial prototype of this instrument (RF3™, Protein Technologies Inc., Tucson, Arizona) has now been developed. We describe the RF3™ in some detail and show examples of its application to the purification of proteins from several sources.

Instrument Design and Operation

The general layout of the RF3™ is illustrated in Fig. 1. Thirty teflon tubes convey solution through a closed loop consisting of the focusing cell, heat exchanger, peristaltic pump, and bubble trap. At the end of the focusing process, solution in the tubing bundle is diverted into a fraction collector for simultaneous collection of all thirty fractions.

Separation of proteins takes place during transit of solution through the flow cavity (0.75 mm X 4 cm X 20 cm) in the focusing cell (Fig. 2). Electrodes are located on either side of the flow cavity such that the electric field is applied perpendicular to the flow of solution through the cell. The front and back surfaces of the flow cavity are formed by two parallel blocks of machined plexiglass (Fig. 3). The electrode compartments, containing highly conductive electrolyte solutions (normally 0.1 M phosphoric acid and 0.1 M NaOH), are separated from the flow cavity by cation and anion selective membranes, as shown in Fig. 3. The electrolytes are recirculated through the electrode chambers and the electrolyte buffer reservoirs by a two channel peristaltic pump.

The display for the high voltage power supply, controls for the electrolyte buffer pump and the main peristaltic pump, and the temperature display are located on a panel to the right of the central compartment (Fig. 4). The ampholyte solution is recycled through the system by a thirty-two channel variable-speed peristaltic pump located below the focusing cell (Fig. 4).

During focusing, the temperature of the ampholyte solution increases as it passes through the focusing chamber. A temperature probe buried in the tubing bundle near the top of the focusing cell measures the maximum

2181–2/92/0399$06.00/0 © 1992 American Chemical Society

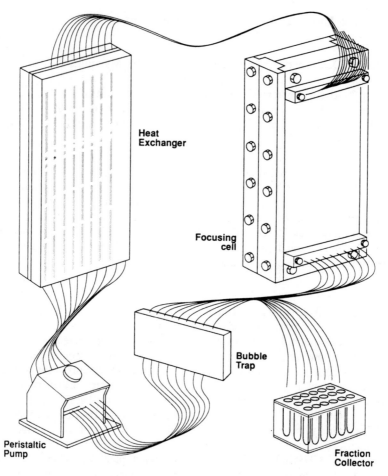

Heat
Exchanger

Focusing
cell

Bubble
Trap

Peristaltic
Pump

Fraction
Collector

RF3
Recycling Free Flow Focusing

Figure 1. Idealized drawing of the RF3™ emphasizing the solution flow path. A closed loop of thirty teflon tubes circulates the ampholyte/protein solution through the focusing cell where the electric field is applied. The solution passes into a water cooled heat exchanger where heat is dissipated through thin-walled teflon tubing immersed in coolant. The solution next passes over the multichannel peristaltic pump, through a bubble trap/pulse dampener, and finally back to the bottom of the focusing cell. At the end of the separation, the solution is diverted to a fraction collector for simultaneous collection of all thirty fractions.

temperature that the ampholyte solution reaches during each cycle, just before entering the heat exchanger. A thermal limiter built into the temperature sensor shuts off the high voltage power supply if a preset temperature maximum (adjustable from 0 - 100 C) is exceeded during the focusing procedure. The solution is cooled in the heat exchanger before crossing the main pump and entering the top of the bubble trap/pulse dampener (lower front, Fig 4). Solution exits the bubble trap and passes directly to the bottom of the focusing cell. This cycle is repeated two to three hundred times during the focusing procedure. After focusing is complete, as evidenced by stable current readings, fractions are collected by diverting the

400

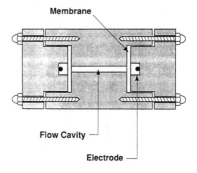

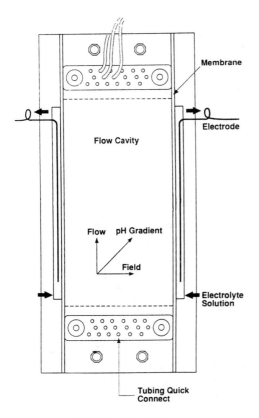

RF3 Focusing Cell

Figure 2. Close-up of the focusing cell in the RF3™. Solution passes through the lower connector into the flow cavity and flows rapidly up through and out the upper connector. The electric field is applied perpendicular to the solution flow, as indicated. Cation and anion selective membranes separate the flow cavity from the electrolyte buffers on either side.

RF3 Cross Section

Figure 3. A cross section of the focusing cell showing the thin (0.75 mm) gap between the front and back acrylic plates, and the electrode chambers separated from the flow cavity by membranes.

During the filling procedure, solution is pumped from the top of the focusing cell to the bottom, opposite to the direction of solution flow during focusing. This helps remove air trapped in the system. Small bubbles that remain after filling are flushed into the bubble trap by changing the direction of flow to the normal run direction (from the bottom to the top of the focusing cell). Protein load is typically from 5 mg to 500 mg per run in the RF3™. Total time per run is normally 2-3 hours. Setup time is approximately 15 minutes. The entire fluid path is cleaned without disassembly in 5 - 10 minutes by recirculating distilled water and 0.1 M NaOH or sodium hypochlorite (bleach) through the system.

An example of the stable laminar flow of proteins through the focusing cell is shown in Fig. 5. A mixture of two colored proteins, hemoglobin (50 mg) and bovine serum albumin (50 mg) stained with bromphenol blue were injected into a prefocused pH gradient consisting of 8 mM arginine, 10 mM cycloserine, 5 mM p-aminobenzoic acid and focused for 30 minutes at 200 V. Although the proteins are moving rapidly from the bottom to the top of the focusing cell, very little lateral motion can be detected (Fig. 5). The electric field drives each protein to its isoelectric point in the pH gradient, and maintains this position in the pH gradient during transit through the focusing cell.

solution to the fraction collector located on the left hand side of the main compartment (Fig. 4).

The RF3™ is normally filled with a 1% (w/v) ampholyte/10% (v/v) glycerol solution and prefocused for 30-60 minutes before protein is introduced. Alternatively, ampholyte can be added directly to a dilute protein solution. The RF3™ is then filled with the protein/ampholyte mixture using a 60 cc syringe to inject the solution through a port near the base of the focusing cell. Air escapes through a vent and reservoir located at the upper left corner of the central compartment (Fig. 4).

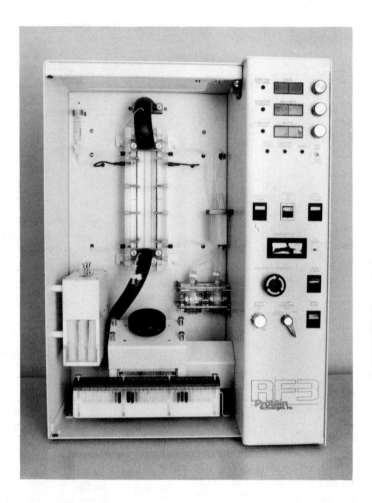

Figure 4. A front view of the RF3™. The main compartment contains the focusing cell and directly below it, the thirty-two channel peristaltic pump and bubble trap/pulse dampener. The two channel peristaltic pump that moves the electrolyte solutions through the electrode chambers is mounted on the right hand side, and the thirty tube fraction collector is mounted on the left hand side of the main compartment. The vent and reservoir are located in the upper left hand corner. The controls for the high voltage power supply, pumps, and temperature display are visible on the right panel.

More complex mixtures of proteins can be separated using wide or narrow-range carrier ampholytes available from several commercial sources.

Fig. 6 shows the results of a separation of a mixture of five protein standards. Twenty-five milligrams of each protein were mixed and injected into a prefocused pH 3-10 gradient. After focusing at 1500 V for 90 minutes, the fractions were collected and analyzed by IEF-PAGE. Proteins with pIs differing by as little as 0.5 pH units (bovine serum albumin, pI 4.7; glucose oxidase, pI 4.2) were separated using wide range (pH 3-10) ampholytes. The instrument is capable of resolving proteins with pIs differing by as little as 0.05 pH units in narrow range pH gradients.

Applications

Three examples of the use of the RF3™ in

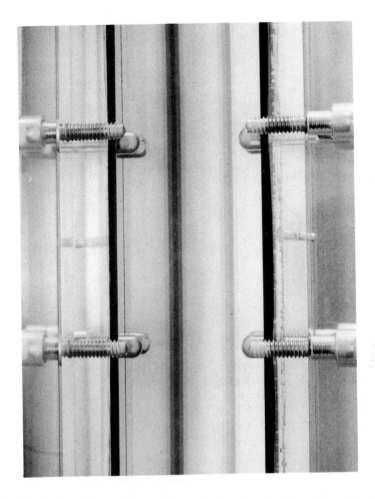

Figure 5. Stable laminar flow of proteins through the focusing cell. Bromphenol-blue stained bovine serum albumin (50 mg) and rabbit hemoglobin (50 mg) were mixed and injected into a prefocused pH gradient generated by focusing 8 mM arginine, 10 mM cycloserine, and 5 mM p-aminobenzoic acid. The proteins were focused for 30 minutes at 200 V and photographed when focusing was nearly complete.

purification of proteins from relatively crude protein mixtures will be presented: 1) monoclonal antibodies from ascites fluid, 2) S1 nuclease from *Aspergillus oryzae*, and 3) overexpressed neomycin phosphotransferase from *Escherichia coli*.

To prepare monoclonal IgGs, ascites fluid (cell line MOPC 173) was first defatted with lipid clearing solution (Beckman, Fullerton, CA). Ammonium sulfate was added to 50% saturation. The precipitate was collected by centrifugation and dialyzed against phosphate-buffered saline before lyophilization. Lyophilized protein was dissolved in, and

dialyzed against, 20 mM Tris-Cl pH 8.0 before separation in the RF3™.

Fig. 7 illustrates a typical time course for the separation of monoclonal IgGs from ascites. The pH gradient was prefocused at 1500 V limiting voltage, 150 W limiting power for 60 minutes. Protein (40 mg) in 5 ml of solution was injected into the RF3™ and focused for approximately 90 minutes. Thirty fractions were collected simultaneously at the end of the separation and analyzed for pH and protein composition. The pH profile (Fig. 8) shows that a nearly linear gradient from pH 2 to pH 10 was generated in the RF3™. Fractions were reduced and

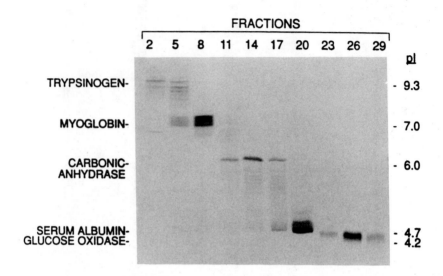

Figure 6. Separation of protein pI standards in a pH 3-10 gradient. A 1% pH 3-10 gradient was prefocused for 1 hour at 1500 V. The proteins (25 mg each) were dissolved in 50 mM Tris-Cl, pH 7.6 and passed through a 0.45 μ syringe filter before injecting into the RF3™. The proteins were focused for 90 minutes at 1500 V and collected. Fractions were analyzed by IEF-PAGE and stained with Coomassie Brilliant Blue R-250.

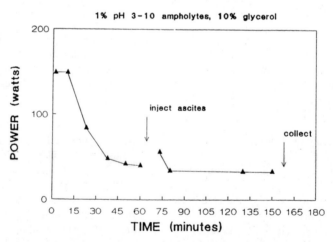

Figure 7. Time course for the separation of mouse monoclonal IgGs from ascites protein. The RF3™ was set to run at 150 W limiting power, 1500 V limiting voltage and prefocused for 60 minutes. Protein (40 mg) in 5 ml volume was injected into the RF3™ and focused for an additional 90 minutes before fractions were collected.

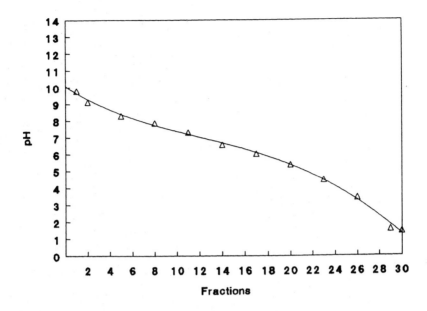

Figure 8. pH gradient generated in the RF3™ using 1% pH 3-10 carrier ampholytes in 10% glycerol. The pH of every third fraction was determined following the separation outlined in Fig. 7.

denatured with SDS and analyzed by SDS-PAGE to examine the distribution of proteins in the pH gradient. Heavy (50 kD) and light chain (25 kD) immunoglobulin polypeptides were clearly separated from contaminating acidic proteins in the ascites preparation (Fig. 9, fractions 5-14). The antibody in these fractions was estimated to be >90% pure. Approximately 80% of the antibody was recovered in fractions 5-14, as estimated by A_{280} using blanks containing 1% ampholyte/10% glycerol. A small amount of antibody was trapped in precipitate that recirculated through channels 21-22 (Fig. 9).

Purification of S1 nuclease from *Aspergillus oryzae* is an example where the RF3™ has been used to separate a group of very acidic proteins using a narrow-range pH gradient. Soluble protein prepared from *A. oryzae* focused near the anode wick in pH 5-6 analytical IEF gels (Fig. 10). Best resolution of the proteins was achieved in analytical IEF gels containing pH 3-4 ampholytes. To prepare protein for separation in the RF3™ *A. oryzae* powder (Enzyme Development Corporation, New York, N.Y.) was mixed with distilled water, filtered through Whatman #3 filter paper, and precipitated with 0.1% Bioprocessing Aids™ (BPA-1000, TosoHaas, Philadelphia, PA). The supernatant from this step was mixed with DEAE-Sepharose and proteins were eluted with a linear NaCl gradient. Fractions with single strand nuclease activity were pooled and dialyzed against 30 mM Na-acetate pH 4.6, 0.1 mM $ZnCl_2$ before isoelectric focusing in the RF3™.

Approximately 50 mg of protein was injected into a prefocused pH gradient consisting of 1% (w/v) pH 3-4 ampholytes, 20% (v/v) glycerol. Fig. 11 shows the pH gradient and nuclease activity in fractions from this separation. Proteins that elute as a homogeneous group from DEAE-Sepharose were clearly resolved into several distinct fractions in the RF3™ (Fig. 12). The peak of nuclease activity was associated with protein which focused in fractions 12-16, between pH 3.6 and pH 3.8. This corresponds to the pI reported for highly purified S1 nuclease (Olesun and Sasakuma, 1980). Previous separations with up to 250 mg protein using similar conditions in the RF3™

FRACTIONS

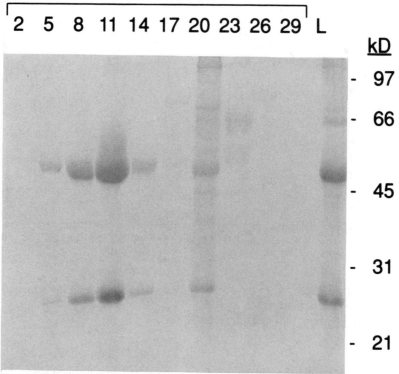

Figure 9. SDS-PAGE analysis of the fractions from the pH 3-10 separation of ascites proteins containing monoclonal IgGs. Aliquots of every third fraction were mixed with SDS sample buffer containing 2-mercaptoethanol and boiled for 5 minutes. Proteins were resolved by SDS-PAGE in a 12.5% acrylamide gel and stained with Coomassie Brilliant Blue R-250. The position of the molecular weight markers are indicated in the right margin. kD, kilodalton.

have yielded protein with a specific activity of 54,000 u/mg protein, with an overall recovery of 82% of the activity (Ostrem et al., 1990).

Narrow range pH gradients have also been used to purify protein overexpressed in *Escherichia coli*. In this case a gene coding for neomycin phosphotransferase II was fused with lacz in a pUC plasmid. Overexpression of the protein was induced with isopropyl thiogalactopyranoside. The cells were lysed by passing them twice through a French press, then through an 18 gauge needle to reduce the viscosity of the lysate. Debris was removed by centrifugation. A 5% (w/v) stock solution of polyethyleneimine (Polymin P, Bethesda Research Laboratories, Gaithersburg, MD) was adjusted to pH 8.0 with HCl and added to a final concentration of 0.4% (w/v) to the supernatant. The solution was mixed for 20 minutes at 4 C and the precipitate was removed by centrifugation. Solid ammonium sulfate was added to 50% saturation to the supernatant and stirred overnight at 4 C. The precipitate was collected by centrifugation. The pellet was dissolved in 50 mM Tris-Cl pH and dialyzed against 20 mM Tris-Cl pH 8.0 before isoelectric focusing.

Most of the proteins in the lysate focused between pH 4.0 and pH 6.5 in analytical gels. Fractions from separation of the *E. coli* proteins in 1% pH 5-6 ampholytes are shown in Fig. 13. Neomycin phosphotransferase II focused in fractions 14-17 (arrowhead, Fig. 14). This protein made up approximately 30% of the total

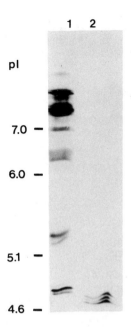

pl

7.0 —

6.0 —

5.1 —

4.6 —

1 2

Figure 10. Analytical isoelectric focusing of pooled DEAE column fractions with single-strand nuclease activity. An extract of *A. oryzae* was prepared by heat denaturation and ammonium sulfate precipitation as described by Vogt (1980). Soluble protein was mixed with DEAE Sepharose, and eluted with a linear NaCl gradient. Fractions with single-strand nuclease activity were pooled and analyzed by isoelectric focusing in a pH 5-6 gel. Lane 1, isoelectric protein standards (BioRad, Richmond CA). Lane 2, Pooled DEAE column fractions with nuclease activity. The gel was stained with Coomassie Brilliant Blue R-250.

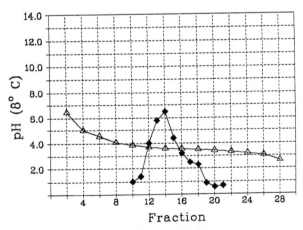

Figure 11. pH (△) and S1 nuclease activity (◆, units x 10^{-1}/ug) in fractions from a separation in 1% (w/v) pH 3-4 ampholytes/20% (v/v) glycerol. An extract of *A. oryzae* was prepared by precipitation with polycationic Bioprocessing Aids™ (BPA-1000, TosoHaas, Woburn, MA) followed by DEAE column chromatography. Approximately 60 mg of protein with S1 nuclease activity was injected into a prefocused pH gradient and focused for 2 hours at 1500 V, 90 minutes at 800 V, and 30 minutes at 500 V. Fractions were collected and checked for pH and S1 nuclease activity as described (Ostrem et al., 1990).

FRACTIONS

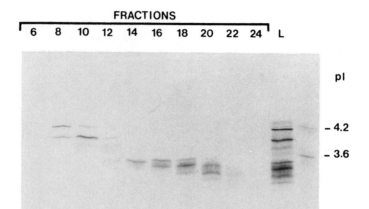

Figure 12. Analytical isoelectric focusing of fractions from the pH 3-4 separation shown in Fig. 11. Aliquots of every other fraction (#6 - #24) were applied to an IEF gel containing 2% (w/v) pH 3-4 ampholytes, 10% (w/v) sorbitol, 5.5% (w/v) acrylamide (29:1 acrylamide:bis) and focused for 60 minutes at 1500 V. L, protein sample loaded on the RF3™. pI, isoelectric points of protein standards: glucose oxidase, pI 4.2, and amyloglucosidase, pI 3.6, from *Aspergillus niger*.

FRACTION

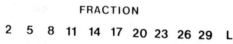

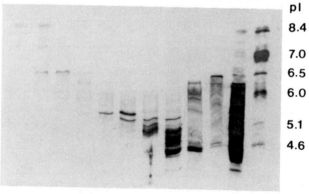

Figure 13. Fractions from separation of an *E. coli* extract containing overexpressed neomycin phosphotransferase II. Protein was prepared from *E. coli* as described in the text, injected into a prefocused pH gradient consisting of 1% (w/v) pH 5-6 ampholytes, and focused for an additional 2 hours at 1500 V. Aliquots of every third fraction across the pH gradient were applied to an IEF gel containing 2% (w/v) pH 5-6 ampholytes, 10% (w/v) sorbitol, 5.5% (w/v) acrylamide (29:1 acrylamide:bis). A silver stained gel is shown. L, protein loaded on the RF3™. pI, isoelectric point. Protein pI standards are in the right most lane, and isoelectric pH values are indicated in the right hand margin.

FRACTION

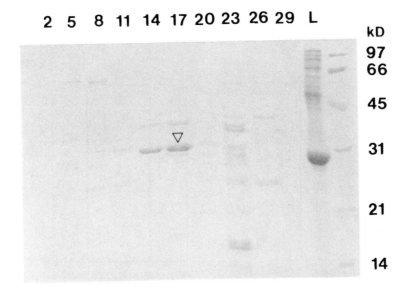

Figure 14. Analysis of the fractions shown in Fig. 13 by SDS-PAGE. Aliquots of every third fraction were mixed with SDS sample buffer and boiled for 5 minutes. Samples were applied to a 12.5% (w/v) acrylamide gel. A gel stained with Coomassie Brilliant Blue R-250 is shown. An open arrow head indicates the position of the recombinant neomycin phosphotransferase II. L, protein loaded on the RF3™; kD, kilodalton. Protein standards are in the right most lane, and molecular weights are indicated in the right hand margin.

protein present (L, Fig. 14) and was highly enriched by a single isoelectric focusing step in the RF3™.

Summary

The RF3™ is the first commercial instrument that combines laminar flow through an electric field with recycling, to achieve separation of proteins in free solution. Rapid flow through the narrow gap in the focusing cell counteracts convective and electrohydrodynamic forces that normally destroy the resolution of proteins in the absence of polyacrylamide or other solid matrices. Brief residence times in the flow cavity expose proteins to very short periods of increased temperature. For example, although the temperature of the ampholyte solution may rise 8-10 C during transit through the focusing cell, the total length of time any given protein molecule resides in the cell is approximately 6 minutes during a 90 minute separation.

Separation by isoelectric point complements chromatographic techniques based on protein size, charge, or hydrophobicity. The RF3™ overcomes problems associated with earlier preparative IEF methods by eliminating density gradients, screens, gels, or chromatographic media used to stabilize the pH gradient. These materials often interact with proteins, or disrupt focusing if precipitation occurs during the focusing procedure. Although it is useful to know the isoelectric point of a protein prior to purification in the RF3™, wide range carrier ampholytes spanning pH 2 - pH 11 can be used for an initial separation. The location of the protein in the pH gradient can then be determined by SDS-PAGE, ELISA, or activity assays.

Isoelectric focusing offers an attractive

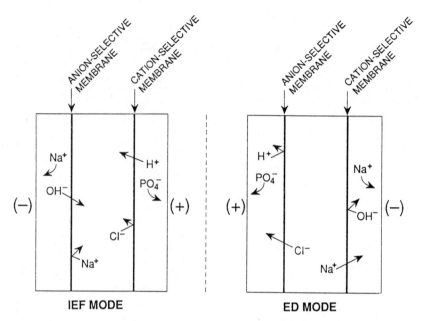

Figure 15. Isoelectric focusing (IEF MODE) and electrodialysis (ED MODE) configurations of the RF3™. Sodium and chloride ions in a protein buffer injected into the RF3™ are driven next to the cathode and anode, respectively, by the electric field. The ion selective membranes lining the focusing cell cavity trap the ions in fractions next to the electrode chambers in the IEF mode. Reversing the polarity of the electric field and the electrolyte buffers (ED mode) allows the sodium and chloride ions to pass through the membranes and into the electrolyte chambers.

alternative to conventional purification of antibodies from complex protein mixtures. Highly purified IgGs were obtained following separation of ascites protein in a pH 3-10 gradient (Fig 9). The procedure works equally well with cell culture supernatants where IgG makes up approximately 5% of the total protein (Ostrem et al. in preparation). It is not necessary to use urea or other denaturants during the focusing procedure. Consequently the immunoreactivity of the purified IgG, as determined by a solid phase antigen assay (Rhodes et al., 1989), is good (Rhodes, B.A., Ostrem J.A., unpublished observations). Furthermore, IgG is recovered in a neutral buffer, in contrast to affinity purification techniques that typically use low pH to temporarily disrupt antigen-antibody interactions.

One impediment to using isoelectric focusing as a preparative method has been the necessity of removing most salts from the protein solution before focusing. This is usually done by either dialyzing against a low ionic strength buffer, or by passing the protein solution over a gel filtration column equilibrated in a low ionic strength buffer. Both methods are time consuming, will increase the volume of the protein solution, and in some cases lead to unacceptable losses of protein or enzyme activity.

Salts in a protein buffer have two undesirable effects in an isoelectric focusing environment. First, conductive solutions generate heat during the focusing process. Long focusing times at low power levels, or an efficient heat exchanger are necessary to prevent heat denaturation or proteolytic activity in crude protein mixtures. Second, anions and cations present in a protein buffer accumulate in fractions next to the electrodes when exposed to an electric field. The ion selective membranes lining the focusing cell cavity prevent movement of these ions out of the ampholyte solution (Fig. 15, IEF mode).

Their charge is balanced by protons and hydroxide ions that migrate in from the electrode chambers. This results in regions with much higher pH near the cathode and much lower pH near the anode than expected.

Fortunately, there is a simple solution to the problem. If the polarity of the electric field and the electrolyte buffers are reversed (Fig. 15, ED mode) the instrument can be run in the electrodialysis configuration. This removes salt in the protein solution at the same time that focusing is taking place. One drawback to this method is that the most basic and the most acidic ampholytes in commercial carrier ampholytes also migrate into the electrode chambers. This limits the utility of the ED mode to proteins with pIs between pH 3 and pH 8 when carrier ampholytes are used to establish the pH gradient. An advantage of the method is that very linear and shallow gradients can be set up across all thirty fractions when the instrument is operated in the ED mode.

Commercial carrier ampholytes are convenient for setting up pH gradients with good buffering capacity over a wide range of pH values, or within specific and narrowly defined pH ranges. However, mixtures of amino acids, zwitterionic buffers, and peptides have also been used to generate pH gradients in gels and free-flow electrophoresis devices. One example of a simple three buffer system, arginine, cycloserine, and p-aminobenzoic acid, was used to separate two proteins injected into the RF3™ as a mixture (Fig. 5). Mathematical modeling shows promise in predicting the ability of various combinations of amino acids and peptides to generate pH gradients with useful separation characteristics (Bier et al., 1981).

References

Bier, M.; Egen, N.B.; Allgyer, G.E.; Twitty, G.E.; Mosher, R.A. In *Peptides: Structure and Biological Function*; Gross, E.; Meienhofer, J. Ed; Pierce Chemical Co., Rockford, IL, **1979**, pp 35-48.

Bier, M.; Mosher, R.A.; Palusinski, O.A. *J. Chromatog.* **1981**, *211*, 313-335.

Bier, M. In *Separation, Recovery, and Purification in Biotechnology*; Asenjo, J.A.; Hong, J. Ed; ACS Symposium Series 314, American Chemical Society, Washington, D.C., **1986**, pp 185-192.

Bier, M.; Twitty, G.E.; Sloan, J.E. *J. Chromatog.*, **1989**, *470*, 369-376.

Olesun, A.E.; Sasakuma, M. *Arch. Biochem. Biophys.*, **1980**, *204*, 361-370

Ostrem, J.A.; Van Oospree, T.R.; Marquez, R.; Barstow, L. *Electrophoresis*, **1990**, *11*, 953-957.

Rhodes, B.A.; Buckelew, J.M.; Pant, K.D.; Hinkle, G.H. *Biotechniques*, **1989**, *8*, 70-75.

Vogt, V.M. *Methods Enzymol.*, **1980**, *65*, 248-255.

Index

Index

414

Dextran–polyethylene glycol aqueous two-phase
systems—*Continued*
  size dependency, 340–341
  supercoiled plasmid, purification, 341
Diode lasers, surface-enhanced Raman spectroscopy,
  86–87,88*f*
Directed bioprocessing, description, 188,189*f*
Direct gene expression in preparative cell-free
  systems
  advantages, 41
  kinetics of expression
    chloramphenicol acetyltransferase encoding
      gene, 41*f*
    dihydrofolate reductase encoding gene, 39,40*f*,41
    β-lactamase and dihydrofolate reductase encoding
      genes, 38,39*f*
Downstream processing of biotechnology-derived
  proteins, role of functionalized membrane
  separation technology, 321–329
Drug(s)
  design procedure, 12–13
  targets, 11–12
  X-ray crystallographic design, 12*f*
Drug measurements, remote Raman spectroscopy,
  85,86–87*f*
Dynamic behavior of secondary metabolites,
  regulation via oligosaccharide elicitor
  addition, 188

E

Economic separation coefficient, calculation,
  371,372*t*
Efficiency, calculation, 371,372*t*
Electrokinetic separation of biomacromolecules,
  capabilities, 7
Elicitors, enhancement of secondary metabolite
  production, 188–197
Enantioselective enzymatic synthesis of
  prostaglandin synthons in multiphase
  reaction media
  amount of water, effect on reaction performance
    in organic media, 56–57
  chiral economic production of synthons, proposed
    procedure, 59,61*f*
  enantioselectivity
    enzymes used in transesterification of
      ketal–alcohol, 57,58*f*
    synthon hydrolysis vs. temperature, 54,55*f*
  experimental materials, 51
  future, 61
  hydrolysis of synthon, reaction scheme, 54,55*f*
  lipase reactions in organic media, possible, 55,56*f*
  liquid membrane emulsion systems, 51,52–54*f*
  organic solvent(s), effect on pig liver esterase
    activity, 54,55*f*
  organic solvent systems, 55,56*f*,57
  reaction under standard conditions, comparison
    with addition of molecular sieve, 59,60*f*
  remaining enzyme activity vs. reaction time, 59,60*f*
  synthons, structures, 51*f*
  time course of conversion, comparison with
    enantiomeric excess, 57,58*f*
  time course of enantiomeric excess vs. acyl donor,
    58,59*f*
  transesterification, reaction scheme, 57*f*

Enantioselective synthesis of chiral chemical
  substances
  importance, 50
  types of multiphase reaction media, 50
Engineering, interface with life science, 1
Enzymatic cell lysis, modeling, 363–367
Enzymatically catalyzed reactions, examples, 10
Enzyme(s), alteration for enhanced catalytic
  function, 10–11
Enzyme catalysts, design and production, 63
Enzyme design, requirements, 63–64
Erythropoietin, effect of oligosaccharides on specific
  activity, 223
Ethanol production during fermentation
  analytical methods, 91
  monitoring and control, 91–97
Expert system for selection and synthesis of protein
  purification process
  chromatography, use for high-resolution
    purification, 369
  cost factor, calculation, 371,372*t*
  developmental problems, 367
  economic separation coefficient, calculation,
    371,372*t*
  efficiency, calculation, 370–371,372*t*
  evolutionary systems, 378
  harvesting equipment selection, 368
  high-resolution purification operations
    first step, 369–370
    rationale for selection, 373,374*f*
  hybrid systems, 373,378
  parameters, 368
  protein bands in lysates and cell cultures,
    properties, 373,375–377*t*
  prototype system
    development, 367–370
    expansion, 370,371–372*t*
    implementation and testing, 373–377
  rules of knowledge, 368–370
  separation coefficient, calculation, 370–371,372*t*
  subprocesses, structuring of knowledge, 370
Expression of secreted proteins, by *Pichia pastoris*,
  163,164*t*
Extraction of lipases in aqueous two-phase systems,
  See Lipase extraction in aqueous two-phase systems

F

Fermentation
  ability to control, 90
  noninvasive spectroscopic monitoring, 90–97
  phases, 106
Fermentation broths, typical components, 304*t*
Fiber-optic biosensors incorporating sustained
  release of reagents
  general design, 116–117
  instrumentation, 118
  mathematical modeling, 120–121,122–123*f*
  open tubular heterogeneous enzyme reactor
    biosensor, 117*f*
  peroxide sensor development, 118,119*f*
  pH shift sensor, 118,119*t*,120*f*
  sustained release of intermediates, 122,124*f*
Fiber-optic Raman measurements
  difficulties, 75–76

419

Micromixing—*Continued*
  fermentation
    biological response to microscales, 279,280*f*
    carbon conversion–dilution rate–microscale
      experiments, 281*f*,282
    diazo system, development, 280
    fermentor characterization, 280
    grid-induced turbulence, 282,283*f*
    injection port design, 282*t*
    laser doppler velocimetry measurements, 283
    problems in understanding, 277–278
    reactions occurring in fermentor, 279
    yeast flip-flop experiment, 280*f*,281
    yeast metabolism, Chaos model, 283
Micromixing index, definition, 279
Microscales in fermentation, biological response to
    microscales, 279,280*f*
Modeling of antibody synthesis and assembly
    regulation in murine hybridoma
  antibody secretion in extracellular medium,
    251–252
  batch spinner culture, typical growth curve,
    243,247*f*
  experimental materials and procedure, 243
  future extensions, 256
  heavy and light chains
    assembly, 251
    half-lives, 247,249,250*f*
    intracellular balances, 250–251
    intracellular levels of mRNAs, 247,248–249*f*
  immunoglobulin assembly, simulation, 252*f*
  immunoglobulin secretion, simulation,
    253,254–256*f*,257
  initial structured kinetic model, 250–252
  model simulations, 252–256
  total cellular RNA content vs. hybridoma growth
    rate, 247,248*f*
Modeling of enzymatic cell lysis
  enzyme recovery and breakdown products,
    363,364*f*,366
  mechanism, 363*f*
  performance evaluation, 366
  pH
    effect on activity, 366*f*
    effect on concentration of extracted proteins,
      366,367*f*
  process conditions, 363,365*f*
Modeling of immobilized microbial cells
  biofilm thickness, prediction, 260–261
  product inhibition
    effect on biomass-specific activity, 263,266*f*
    effect on specific growth-rate profiles,
      263,264*f*,265
    effect on steady-state viable cell layer, 265*f*
    model, 261,262–263*t*
    reactor design, 265,267*f*
  substrate inhibition
    effect on location of viable cell layers, 273,274*f*
    effect on net specific growth rate, 271,272*f*,273
    effect on viable cell layer, 270,271*f*
    model, 267,268–269*t*,270
  substrate utilization, reactor design, 273,275*f*
Molecular biology, research at CARB, 17
Molecular modeling research studies at CARB
  design of inhibitors and site-directed mutants of
    proteins, 28

Molecular modeling research studies at CARB—
    Continued
  protein folding
    homologous modeling, 27–28
    systematic conformational search, 27
Monoclonal antibodies
  applications, 11
  purification by recycling free flow focusing,
    403,404–406*f*
Monoclonal antibody purification using
    functionalized membrane separation technology
  mouse immunoglobulin G, 323*f*
  purity evaluations, 324,325*f*
  scaleup conditions, 323,324*t*
Multiphase reaction media
  enantioselective enzymatic synthesis of
    prostaglandin synthons, 51–61
  types, 50
Murine monoclonal antibody
  accumulation during continuous perfusion in total
    recycle reactor, 243,244*f*
  commercial production methods, 241,242*f*
  hybridoma viability vs. dilution rate, 241,243*f*
  synthesis and assembly regulation, modeling,
    243–256
  total cell recycle reactor method, 243,245–246*f*
Murine monoclonal antibody synthesis, chain
    assembly, and secretion, schematic diagram,
    241,242*f*

N

Neuronal sensing devices
  advantages and disadvantages, 128
  analyte detection in solution, 129
  description, 128
  examples, 128–130
  extracellular recording in microfabricated
    electrode, 128–129
  multielectrode array, 129
  neuron-based sensing system, development of
    model, 130
  sensitivity to specific amino acids, 129–130
  temperature sensitivity of neuronal response to
    neurotransmitters, 130–135
N-linked oligosaccharides
  site microheterogeneity, 210,211*f*
  synthesis, 199–207
NMR spectroscopy
  perfused cell sample, 138
  mass transfer limitations in cell device, effect on
    spectra, 139–148
  experimental limitation, 138
  function, 138
  suitable perfused cell devices, 138,139*t*
Noninvasive real-time high-sensitivity devices,
    description, 4
Noninvasive spectroscopic monitoring of bioprocess
  advantages, 97
  aerobic–anaerobic status, on-line monitoring,
    96–97*f*
  calibration constants, applicability to succeeding
    fermentations, 95,96*f*
  data analysis, 92–93
  defined medium for fermentation, 91*t*,92

421

Thermodynamics of aqueous mixtures of salts and
    polymers—*Continued*
  polymer–polymer systems, phase behavior,
    385,387,388*f*
  salt–polymer systems, phase behavior, 387,389*f*
Threonine deaminase
  feedback, effect on steady-state kinetics, 20*f*,21
  molecular mechanisms that regulate catalysis,
    20–21
Time scale
  definition, 294
  micromixing
    definition, 278
    measurement difficulties, 279
    measurement using laser doppler velocimetry, 283
Tissue-type plasminogen activator, effect of
    oligosaccharides on specific activity, 223
Titration calorimetry, research at CARB, 18
Total cell recycle reactor
  photograph, 243,246*f*
  schematic diagram, 243,245*f*
Translation, description and stages, 31
Troponin, protein engineering studies, 21
Tryptophan C-methyltransferase, induction, 171,172*f*
Turbulence, difficulties in theoretical
    description, 277

U

Ultrafiltration
  membrane fouling, 304–309
  separation applications, 304*t*

V

Velocity scale of micromixing, definition, 278
Volumetric oxygen transfer rate, calculation, 287

W

Whole cell biosensors
  advantages and disadvantages, 126
  bacterial biosensors, 127
  development, 126–127
  mammalian tissue based biosensors, 127–128
  neuronal devices, 128–130
  plant tissue based biosensors, 127
  temperature sensitivity of neuronal response to
    neurotransmitters, 130–135

X

X-ray crystallography
  drug design, 12*f*,13
  protein crystal growth, 13–15
  surfactants, use in protein crystallization, 14

Y

Yeast, fluorescence, 100
Yeast metabolism, Chaos model, 283

Production: Margaret J. Brown and C. Buzzell-Martin
Indexing: Deborah H. Steiner
Acquisition: Robin Giroux and Cheryl Shanks
Cover design: Amy Meyer Phifer

Printed and bound by Victor Graphics, Baltimore, MD

Bestsellers from ACS Books

The ACS Style Guide: A Manual for Authors and Editors
Edited by Janet S. Dodd
264 pp; clothbound, ISBN 0–8412–0917–0; paperback, ISBN 0–8412–0943–X

Chemical Activities and Chemical Activities: Teacher Edition
By Christie L. Borgford and Lee R. Summerlin
330 pp; spiralbound, ISBN 0–8412–1417–4; teacher ed. ISBN 0–8412–1416–6

Chemical Demonstrations: A Sourcebook for Teachers,
Volumes 1 and 2, Second Edition
Volume 1 by Lee R. Summerlin and James L. Ealy, Jr.;
Vol. 1, 198 pp; spiralbound, ISBN 0–8412–1481–6;
Volume 2 by Lee R. Summerlin, Christie L. Borgford, and Julie B. Ealy
Vol. 2, 234 pp; spiralbound, ISBN 0–8412–1535–9

Writing the Laboratory Notebook
By Howard M. Kanare
145 pp; clothbound, ISBN 0–8412–0906–5; paperback, ISBN 0–8412–0933–2

Developing a Chemical Hygiene Plan
By Jay A. Young, Warren K. Kingsley, and George H. Wahl, Jr.
paperback, ISBN 0–8412–1876–5

Introduction to Microwave Sample Preparation: Theory and Practice
Edited by H. M. Kingston and Lois B. Jassie
263 pp; clothbound, ISBN 0–8412–1450–6

Principles of Environmental Sampling
Edited by Lawrence H. Keith
ACS Professional Reference Book; 458 pp;
clothbound; ISBN 0–8412–1173–6; paperback, ISBN 0–8412–1437–9

Biotechnology and Materials Science: Chemistry for the Future
Edited by Mary L. Good (Jacqueline K. Barton, Associate Editor)
135 pp; clothbound, ISBN 0–8412–1472–7; paperback, ISBN 0–8412–1473–5

Personal Computers for Scientists: A Byte at a Time
By Glenn I. Ouchi
276 pp; clothbound, ISBN 0–8412–1000–4; paperback, ISBN 0–8412–1001–2

Polymers in Aqueous Media: Performance Through Association
Edited by J. Edward Glass
Advances in Chemistry Series 223; 575 pp;
clothbound, ISBN 0–8412–1548–0

For further information and a free catalog of ACS books, contact:
American Chemical Society
Distribution Office, Department 225
1155 16th Street, NW, Washington, DC 20036
Telephone 800–227–5558

Highlights from ACS Books

Good Laboratory Practices: An Agrochemical Perspective
Edited by Willa Y. Garner and Maureen S. Barge
ACS Symposium Series No. 369; 168 pp; clothbound, ISBN 0–8412–1480–8

Silent Spring Revisited
Edited by Gino J. Marco, Robert M. Hollingworth, and William Durham
214 pp; clothbound, ISBN 0–8412–0980–4; paperback, ISBN 0–8412–0981–2

Insecticides of Plant Origin
Edited by J. T. Arnason, B. J. R. Philogène, and Peter Morand
ACS Symposium Series No. 387; 214 pp; clothbound, ISBN 0–8412–1569–3

Chemistry and Crime: From Sherlock Holmes to Today's Courtroom
Edited by Samuel M. Gerber
135 pp; clothbound, ISBN 0–8412–0784–4; paperback, ISBN 0–8412–0785–2

Handbook of Chemical Property Estimation Methods
By Warren J. Lyman, William F. Reehl, and David H. Rosenblatt
960 pp; clothbound, ISBN 0–8412–1761–0

The Beilstein Online Database: Implementation, Content, and Retrieval
Edited by Stephen R. Heller
ACS Symposium Series No. 436; 168 pp; clothbound, ISBN 0–8412–1862–5

Materials for Nonlinear Optics: Chemical Perspectives
Edited by Seth R. Marder, John E. Sohn, and Galen D. Stucky
ACS Symposium Series No. 455; 750 pp; clothbound; ISBN 0–8412–1939–7

Polymer Characterization:
Physical Property, Spectroscopic, and Chromatographic Methods
Edited by Clara D. Craver and Theodore Provder
Advances in Chemistry No. 227; 512 pp; clothbound, ISBN 0–8412–1651–7

From Caveman to Chemist: Circumstances and Achievements
By Hugh W. Salzberg
300 pp; clothbound, ISBN 0–8412–1786–6; paperback, ISBN 0–8412–1787–4

The Green Flame: Surviving Government Secrecy
By Andrew Dequasie
300 pp; clothbound, ISBN 0–8412–1857–9

For further information and a free catalog of ACS books, contact:
American Chemical Society
Distribution Office, Department 225
1155 16th Street, NW, Washington, DC 20036
Telephone 800–227–5558